Pressure Vessel Design

Donatello Annaratone

Pressure Vessel Design

With 275 Figures and 4 Tables

Springer

Donatello Annaratone
Via Ceradini 14
20129 Milano
Italy
e-mail: donatello@annaratone.eu

Library of Congress Control Number: 2006936077

ISBN-10 3-540-49142-2 Springer Berlin Heidelberg New York
ISBN-13 978-3-540-49142-2 Springer Berlin Heidelberg New York

Springer is a part of Springer Science+Business Media.

springer.com

Typesetting by author and SPi using a Springer LaTeX macro package
Cover design: eStudio Calamar, Girona, Spain.

Printed on acid-free paper SPIN 11607205 62/3100/SPi 5 4 3 2 1 0

Preface

1 Industrial Sectors Interested in Pressure Vessels

Pressure vessels are probably the most widespread "machines" within the different industrial sectors. In fact, there is no factory without pressure vessels, steam boilers, tanks, autoclaves, collectors, heat exchangers, pipes, etc. More specifically, pressure vessels represent fundamental components in sectors of enormous industrial importance, such as the nuclear, oil, petrochemical, and chemical sectors. There are periodic international symposia on the problems related to the verification of pressure vessels.

For many years an ISO committee was dedicated to pressure vessels design. There is also a technical committee of the EU specifically assigned to this field. All the industrialized countries have a code relative to pressure vessels design. However, even when the code includes specific regulations to determine the thickness of the different components, typically not all issues facing the designer are discussed. Finally, it is worth noting that a few regulations cause some perplexity.

In Italy, a specific area of ISPESL regulations (VSR collection) is devoted to pressure vessels.

2 Current Know-How with Regard to Resistance Verification

A pressure vessel is not an easy machinery in terms of resistance verification. A layman can easily make the mistake of considering somewhat simple structural forms that are in fact quite difficult to analyze, especially if one would like to apply the most modern criteria of verification (elastoplasticity, self-limiting stresses, etc.).

Regardless of the enormous interest in the topic and numerous efforts, many problems have not been studied in-depth, and there is still no agreement among scholars and the institutions of the various countries that define regulations. In addition, economical reasons and technical progress constantly present new challenges in connection with new forms and solutions, the necessity to reduce thickness to a minimum, etc.

Finally, the growth of the nuclear sector has highlighted the necessity of an investigation beyond the simplistic analysis of stresses, and this also led to a systematic analysis of the impact of fatigue phenomena. These accomplishments notwithstanding, much still needs to be done if, from a practical standpoint, one wishes to move from general principles to operational guidelines of calculation criteria that are as simple as possible.

3 Current State of Technical Literature

With regard to Italy, when pressure vessels are treated, they are included in general textbooks about mechanical engineering. This leads to a somewhat generic and often outdated treatment of the subject with regard to modern verification criteria, and hence the outcome is of little practical interest. Outside of Italy, there are textbooks and various publications specifically on pressure vessels.

However, these publications have a number of shortcomings:

(a) Simply a guide to apply the code's rules correctly.
(b) Sound scholarly framework that often does not extend itself to the point of analyzing the practical cases, thus becoming of little use to the designer.
(c) Lack of interest in problems that may seem marginal but are in reality those causing many obstacles to the designer, specifically those that are not analyzed in detail and also happen not to be included in regulations and codes.
(d) Experimental emphasis that for the cases under study is of obvious help. However, because the number of cases is considerable, and a theoretical background is lacking, the designer is unable to use the available data by applying "similarity approaches."
(e) Lack of interest in verification methodology which is essential for sizing; the designer is faced with values for stresses that he or she does not know how to evaluate; the situation becomes even more complex when the verification methodology exists but does not correspond to the modern verification criteria.

4 General Characteristics of the Book

The book focuses on general problems as well as fundamental ones derived from the previous ones, and on problems that may be incorrectly considered of secondary importance but are in fact crucial in the design phase.

The basic approach is rigorously scientific with a complete theoretical development of the topics treated, but the analysis is always pushed so far as to offer concrete and precise calculation criteria that can be immediately applied to actual designs. This is accomplished through appropriate algorithms that lead to final equations or to characteristic parameters defined through mathematical equations. Given the complexity of many of these, representative graphs are shown.

In other cases experimental graphs are shown. Their limit of applicability is discussed, also by including a basic theoretical treatment to justify their specific behavior. The result of this is a textbook with a large number of equations and graphs, both fundamental for the actual design of pressure vessels.

The topics treated are grouped in ten chapters.

The first chapter describes how to achieve verification criteria, the second analyzes a few general problems, such as stresses of the membrane in revolution solids and edge effects. The third chapter deals with cylinders under pressure from the inside, while the fourth focuses on cylinders under pressure from the outside. The fifth chapter covers spheres, and the sixth is about all types of heads. Chapter seven discusses different components of particular shape as well as pipes, with special attention to flanges. The eighth chapter discusses the influence of holes, while the ninth is devoted to the influence of supports. Finally, chapter ten illustrates the fundamental criteria regarding fatigue analysis.

5 Original Contributions of the Author to the Solution of Various Problems

Besides the rather unique approach to the entire work, see Sect. 4 above, original contributions can be found in most chapters, thanks to the author's numerous publications on the topic and to studies performed ad hoc for this book.

Specifically, we would like to draw your attention to the following topics:

3.4 Allowable out of Roundness
3.5 Stiffened Cylinders
3.6 Partially Plastic Deformed Cylinders
3.7 Stresses due to Thickness Variation
4.2 and 4.3 Cylinders Under External Pressure (special emphasis on ovalization)
5.4 Partially Plastic Deformed Spheres
6.4 Flat Heads
7.4 Flanges
7.6 Expansion Compensators
8.3 Isolated Holes on Cylinders, Spheres, and Cones: Y and T Branches
8.4 Flat Head with Central Hole
8.5 Drilled Plates
9.2 Spherical Vessels Resting on a Parallel

Contents

Notation

A	Cross-sectional area
C	Form factor (heads)
D, d	Diameter
D_e, d_e	Outside diameter
D_i, d_i	Inside diameter
D_m, d_m	Average diameter
E	Young's modulus; modulus of elasticity
F	Force, load
f	Basic allowable stress
f_c	Allowable stress for piping
f_a	Allowable stress for fatigue
H	Height
I	Moment of inertia
$k_1 \ldots\ldots\ldots k_{10}$	Dimensionless factors
L, l	Length, width
M	Bending moment
N	Normal force
n	Number of waves (buckling); number of cycles
p	Pressure
p_c, p_{ce}, p_{cp}	Critical pressures
R, r	Radius
R_e, r_e	Outside radius
R_i, r_i	Inside radius
R_m, r_m	Average radius
s	Thickness
T	Shear force
t	Temperature
u	Ovalization
W	Section modulus
z	Weld joint efficiency; efficiency of ligaments
α	Thermal expansion coefficient
δ	Deviation from roundness
ε	Deformation (strain)

ε_{id}	Ideal deformation
λ	Thermal conductivity coefficient
μ	Poisson's ratio
ν	Safety factor
σ	Normal stress
$\sigma_I, \sigma_{II}, \sigma_{III}$	Principal stresses
σ_a	Longitudinal stress, axial stress
σ_m	Meridian stress
σ_r	Radial stress
σ_t	Hoop stress, circumferential stress
σ_{id}	Ideal stress
σ_s	Yield strength
σ_R	Rupture stress
τ	Shear stress
$\alpha, \vartheta, \varphi, \omega$	Angles

1

Preliminary Considerations

1.1 Mechanical Characteristics of Steel

If we exert a tensile load on a specimen made of mild carbon steel, and we transfer on the x-axis the values of the elongation per unit of length between the references (ε) (called strain) and on the y-axis the values of the stress (σ) that equals the load applied to the specimen divided by its original cross-sectional area, we obtain a diagram qualitatively similar to the one shown in Fig. 1.1.

We notice that there is proportionality between stress and strain in the first portion of the curve, i.e., the steel follows Hooke's law that constitutes the basis of classic calculation in the elastic field. In fact, the steel behaves in an elastic fashion, i.e., the deformations completely disappear after removal of the load, and the specimen returns to its original shape.

The angular coefficient E of the straight portion given by the relationship σ/ε is called modulus of elasticity, or Young's modulus. The point on the curve at the end of the linear section identifies a value of σ which is called proportional limit.

Steel behaves in an elastic fashion even beyond the proportional limit, as long as another characteristic point corresponding to stress called elastic limit is not exceeded. Note that the two points mentioned above are near, and the second one is not easy to determine. In practice, we typically equate the proportional limit to the elastic limit.

By increasing the load applied to the specimen, we reach a point on the curve corresponding to a stress σ, called upper yield strength, that represents the maximum value of σ taking place at the onset of the yielding phenomenon. In fact, after reaching the upper yield strength the load decreases, and we reach a relative minimum of the curve that identifies the stress called lower yield strength.

The yielding phenomenon is characterized by large deformations (when compared to those typical of the elasticity field) under practically constant load.

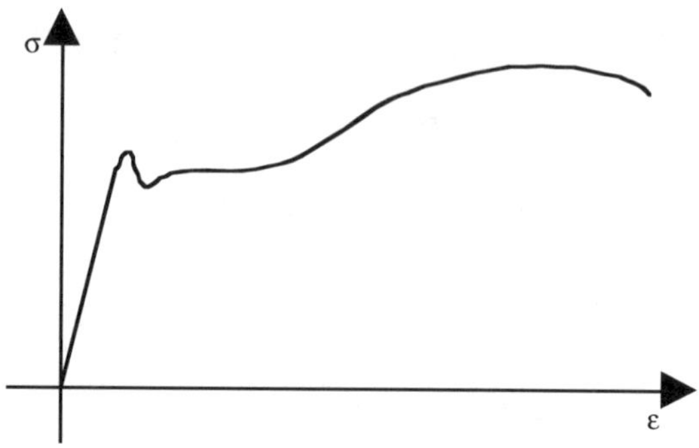

Fig. 1.1

This portion of the curve is then followed by a portion characterized by progressive increase in stress with large deformations. This is the well-known phenomenon of steel hardening, which persists until the stress reaches a maximum value called ultimate strength. After that σ decreases (again with regard to the original cross-sectional area of the specimen), and we reach rupture.

Conversely, if we consider the actual cross-sectional area of the specimen in the different stages, the highest value of σ is reached in correspondence with the rupture point. In fact, substantial elongations in correspondence with yielding and hardening areas happen together with a significant reduction of the cross-sectional area.

The lower value of the yield strength (simply known as "yield strength," σ_s) and the maximum value of σ that precedes the rupture are the most significant parameters of the steel's mechanical properties, and are therefore indicated in test certificates and represent the basis of resistance calculus.

The yield strength basically shows the condition under which the material starts yielding. At this point, the yielded fiber is not able to absorb growing stresses, and thus to contribute to the equilibrium of forces applied to the vessel. This is because we rule out the possibility that under safety conditions the deformations become so large that one is forced to consider the hardening phenomenon.

The fiber can be plastic deformed and, as we shall see, this has an important impact on the behavior of neighboring fibers, if we start from the assumption that they have not yet reached the yield strength. This leads to a different kind of calculation, somewhat different from the classic one based on the elasticity behavior of the entire component.

In view of the above considerations, one can replace the curve in Fig. 1.1 with that in Fig. 1.2, whereas in a first segment σ is proportional to ε (totally elastic behavior of material) followed by a segment parallel to the x-axis (perfectly plastic behavior of the material). Such simplification is most frequently used for resistance verification in the elastic–plastic field.

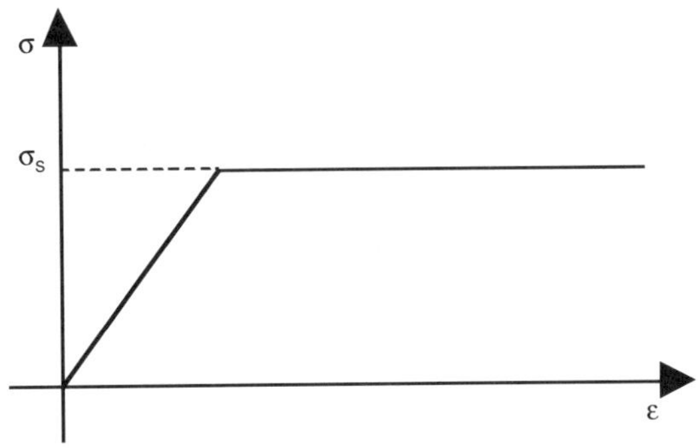

Fig. 1.2

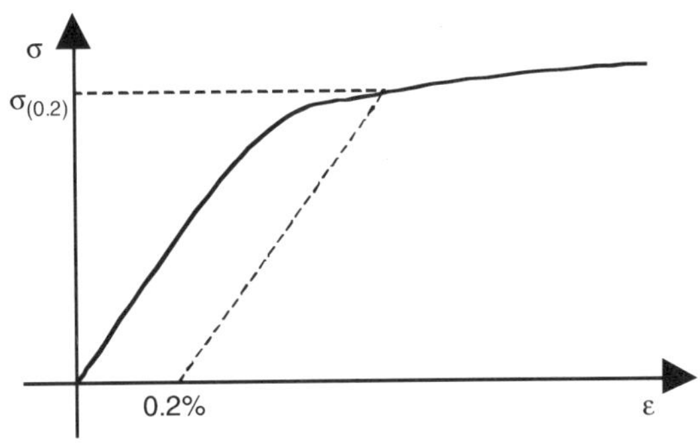

Fig. 1.3

The maximum value of σ in the tensile test is usually called rupture stress, although a better term would be unitary maximum load in the tensile test (σ_R). This is a characteristic parameter of the steel's behavior that cannot be ignored with regard to safety since it identifies its maximum level of resistance. It is therefore considered by adopting a relatively high factor of safety to ensure that the stresses present in the vessel are substantially distant from such value.

Steel does not always show a curve σ–ε similar to the one in Fig. 1.1; in the case of steel with a high content of carbon, for instance, the first segment of the curve has a shape similar to the one shown in Fig. 1.3. Moreover, this shape also characterizes steel used at high temperatures that exhibits a curve σ–ε, as in Fig. 1.1 at room temperature.

After the first linear segment the curve exhibits a substantial and progressive slope decrease, but the portion characterized by increasing deformations at basically constant stress is no longer present.

Since σ_s has not been found, we consider a conventional stress that substitutes for all practical purposes the classic yield strength with regard to calculus.

This is the stress that during the specimen's release that takes place according to Hooke's law causes a permanent deformation equal to 0.2% (Fig. 1.3). Therefore, it is indicated with the symbol $\sigma_{(0.2)}$.

Up to this point, we have discussed the steel's behavior at room temperature. It is, however, of the greatest importance to be aware of the influence of temperature on the mechanical characteristics of the material.

As we shall see, not only temperature but also time may have a strong influence, but right now we shall focus on the effects of temperature on the results of the classic tensile test.

The temperature also affects the values of resilience, of the elongation to rupture and of the area reduction. Limiting our focus to the values of $\sigma_{(0.2)}$ and σ_R, we notice that $\sigma_{(0.2)}$ decreases with the increase in temperature, while σ_R increases initially within a moderate range of temperatures and then decreases. In Fig. 1.4, we show the representative curves of carbon steel as a reference.

Moreover, the decrease of $\sigma_{(0.2)}$ has an important impact on the sizing of the vessel and, under certain conditions, on the selection of the steel to be used. In fact, the decrease of $\sigma_{(0.2)}$ can be more or less substantial for steel of different composition.

Knowing $\sigma_{(0.2)}$ and σ_R, however, is not always sufficient to identify the mechanical characteristics of steel under hot conditions, in order to calculate the allowable stress, as we shall see in Sect. 1.2.

When temperatures are typically below 300°C (570°F), the elongation of the specimen under tensile load does not increase over time, or it does so in a negligible fashion. When temperatures are higher than 300°C (570°F)

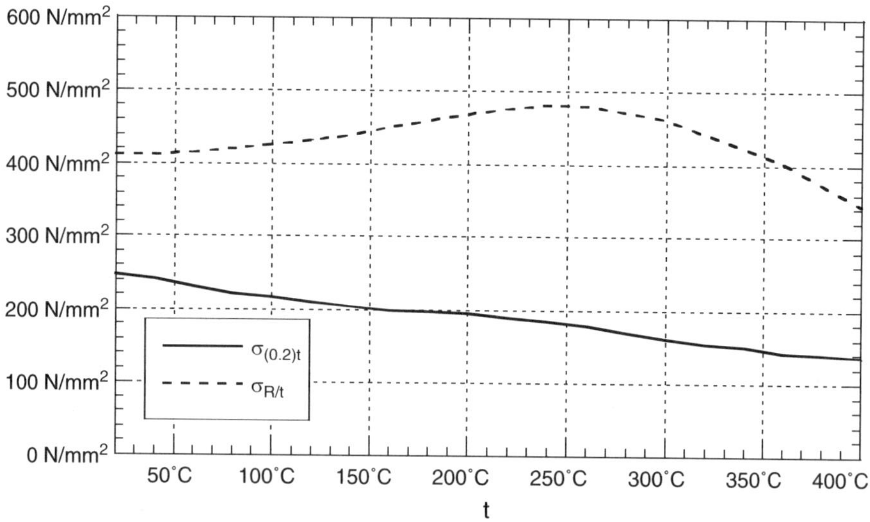

Fig. 1.4

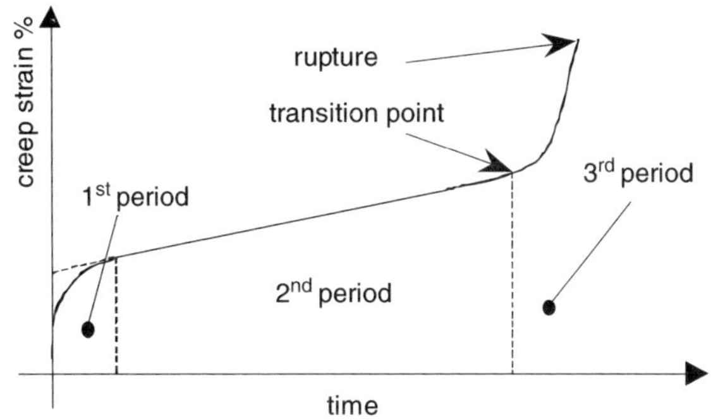

Fig. 1.5

the specimen is instead subject to an increased elongation over time; such elongation is also of variable entity as a function of temperature and stress applied to it.

Under certain temperature and load conditions the specimen can go into rupture over time. Such phenomenon, called creep, is clearly of great importance for the behavior of vessels over time for both safety and business reasons. If we examine the phenomenon in greater detail (see Fig. 1.5), we notice that the strain increases while the elongation's speed decreases over time until it reaches a minimum value.

This first portion of the curve identifies that which is called the first period. In the second period the elongation's speed remains practically constant; the representative point of the end of the second period is called point of transition. Finally, in the third period the elongation's speed and the values of the strain increase rapidly up to the rupture.

Amongst the three analyzed periods, the first and third are rather short, while the second takes most of the total time during which the specimen goes into rupture. As a reference point, please note that for a total time of about 100,000 h the first period represents at most a few hundred hours.

Finally, it is important to understand the great importance played by the point of transition that represents the beginning of the short period during which the specimen goes into rupture. The literature sometimes considers its reference time more important than the rupture time itself. In order to be able to take into account the creep in the sizing of components working under hot conditions, researchers and institutions initially developed different proposals that had in common the characteristic of forecasting tests of short duration, even though they differed with respect to the duration of the test and the evaluation of the results.

In Europe, the most popular proposal came from the German Metallurgic Union (DVM). Based on the name of this organization, the value of the stress derived from the test was called σ_{DVM} and was adopted for some time, and not only in Germany, to determine the allowable stress at high temperatures.

The stress σ_{DVM} is the one that between the 25th and the 35th hour causes an elongation's speed inside the specimen not higher than 0.001% h^{-1} and a permanent deformation after 45 h not greater than 0.2%.

The advantage of such tests is the short duration, but they were soon abandoned, mostly due to the extremely high variability of the values. In fact, the test occurs during the first period of the curve that may have a rather variable behavior from casting to casting of the same steel, even when the behavior of the curve during the second period is basically the same. The value of σ_{DVM} is a poor indicator of the performance of steel over long periods of time. Thus it became necessary to carry out tests over very long periods of time.

Nowadays, very many values of specimen made of steel of common usage that have reached the 100,000 h, and a relevant number of specimen that have been tested over even longer periods of time, are available. This allows us to determine with absolute certainty the behavior of steel of common usage up to 100,000 h of usage. Such is in fact the reference time that is typically taken into consideration for resistance verification purposes.

The value that is considered for resistance verification is the average of the rupture stress for a creep lasting 100,000 h. Such value is indicated with the symbol $\sigma_{R/100000/t}$ for a generic test temperature t. Note that this is an average value: these values are characterized by a certain amount of scattering. We admit that the variability may have a range around the average value considered for calculations equal to no more than 20% of the average value itself. If the minimum value is outside such range, we shall assume the minimum value multiplied by 1.25. If for a given steel and a given temperature we indicate the time on the x-axis and the values of the stress on the y-axis that cause the rupture of the specimen, which we call σ_R, in a doubly logarithmic diagram, we obtain curves that are qualitatively like the one shown in Fig. 1.6.

These curves look like broken lines with more or less evident knees. There are also instances – in most cases to be considered exceptional – where such knees are missing, and the broken line becomes a straight line. Therefore, the values of $\sigma_{R/100000/t}$ are derived from these curves for 100,000 h. As we said above, for steel of common usage there is no problem given the amount of values for the rupture stress per 100,000 h that are available. The problems arise when we deal with steel of new fabrication, or that has not yet undergone extensive tests; in such instances there is nothing else to do but to extrapolate on the basis of the values known for shorter periods of time. The extrapolation is possible with all the necessary cautionary tales, but it is sufficiently reliable only if it is not pushed too far. This is because of the presence of the knees, and also because even a small mistake in the determination of the values related to longer experimental time frames would result in substantial errors in the values obtained through extrapolation.

An acceptable extrapolation should not involve a timing ratio greater than 10; limiting such ratio to five leads to values that are sufficiently reliable.

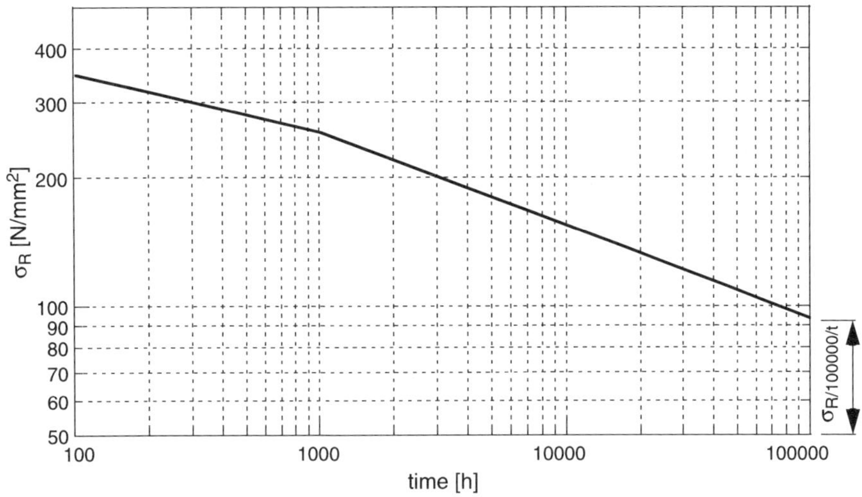

Fig. 1.6

In other words, to obtain reliable values for 100,000 h the longest test shall be at least of 20,000 h.

1.2 Allowable Stress

The following stresses are typically considered to determine the basic allowable stress of steel:

> σ_R = minimum value of the unitary maximum load during the tensile test (rupture stress) at room temperature.
> $\sigma_{(0.2)/t}$ = minimum value of the unitary load during the tensile test at temperature t with a permanent deformation equal to 0.2% of the initial length between references after removal of load.
> $\sigma_{R/100000/t}$ = average value of the unitary rupture stress for creep after 100,000 h at temperature t.

In the case of austenitic steel there is general agreement that instead of a permanent deformation of 0.2% we refer to a deformation of 1%. Note that the temperature t is the average wall temperature.

As discussed in Sect. 1.1, the meaning of these values is certainly clear to the reader; this not withstanding, further clarifications are due.

The rupture stress during the tensile test refers to room temperature. This may sound surprising since the resistance verification must be executed at design temperature t, to which the other two values above in fact refer. As pointed out in Sect. 1.1, the value of the rupture stress at moderate temperatures is greater than the one at room temperature. Within this temperature range, the adoption of this last value addresses basic safety criteria. The use of this value is in fact justified within the range of moderate temperatures: the goal here is to guarantee, through an adequate safety factor (as we shall see later on), that stresses acting upon the vessel do not cause a dangerous situation leading to rupture. Moreover, the value of σ_R is at times crucial, as far as allowable stress is concerned, when steel with high levels of yield strength are adopted. In this case if the design temperature is either room temperature or anyway moderate, the value of $\sigma_{(0.2)/t}$ is very close to σ_R. The determination of the allowable stress solely based on $\sigma_{(0.2)/t}$ may lead to a value that does not sufficiently protect against rupture.

Considering now the unitary load that causes a permanent deformation of 0.2% at release during the tensile test (we use the symbol $\sigma_{(0.2)/t}$ to remember that one should refer to the design temperature t), we pointed out in Sect. 1.1 that it practically replaces the yield strength when the steel does not exhibit the classic yielding phenomenon. If, on the other hand, this were the case, it would be easy to determine, due to the entity of the resulting deformations, that the value of $\sigma_{(0.2)}$ coincides with the lower value of the yield strength.

We shall use the symbol σ_s instead of $\sigma_{(0.2)/t}$ in the following chapters for simplicity purposes. The latter is in most cases crucial to determine the value of allowable stress. We will also use the term "yield strength", even though it is formally incorrect, to simplify the language. Finally, as far as the rupture stress per creep at 100,000 h is concerned, its importance is now evident, in view of what discussed in Sect. 1.1, if the design temperature is high. For such stress, given the dispersion of values present even for similar types of steel, one refers to the mean value of the range, generally assuming that the size of the range itself does not go beyond ±20%.

In order to obtain the allowable stress, the three characteristic stresses are associated to safety factors, the values of which lack a general consensus, and that have undergone numerous modifications over time. The general trend has been to reduce them, as the behavior of different kinds of steel became better understood, to require more stringent inspections and refine calculation methodologies. We therefore recommend the following criterion that will be applied throughout the book.

The basic allowable stress of the material that from now on we will call f is given by the smallest of the following values:

$$\frac{\sigma_R}{2.4}, \qquad \frac{\sigma_{(0.2)/t}}{1.5}, \text{ and } \quad \frac{\sigma_{R/100{,}000/t}}{1.5}.$$

From a conceptual point of view, the basic value is the one derived from $\sigma_{(0.2)/t}$. For this reason, we will always refer to the yield strength when we have to correlate the allowable stress with a value typical of the material's resistance. The other two values mentioned above occur in special instances, even though $\sigma_{R/100000/t}$ is found quite frequently.

We have already discussed elsewhere the reasons that suggest to take σ_R into consideration. $\sigma_{R/100000/t}$ is critical for f when its value is lower than the yield strength, since the safety factor has the same value of the one relative to $\sigma_{(0.2)/t}$. For instance, this happens for carbon steel at temperatures beyond 380–420°C (715–790°F), for low-alloy steel at temperatures higher than 470–500°C (880–930°F), and for austenitic steel at temperatures higher than 500°C (930°F).

Finally, note that we have defined the stress f as basic allowable stress of the material. This does not necessarily mean that it corresponds to the allowable stress during resistance verification of a specific piece in a specific position.

As we shall often return to this, in some cases it is acceptable that the ideal stress may reach the yield strength or even, in spite of being physically impossible, twice the yield strength (only from the point of view of calculation in the elastic field). We will discuss this issue in Sects. 1.4 and 1.5. The stress f, which in other cases actually corresponds to the maximum stress allowable for the piece, and at any rate to the maximum ideal stress of the membrane, represents a reference point, since it is present in all equations to compute the thickness of the various components. As a matter of fact, even when greater allowable values are assumed, they are correlated to the value of f.

1.3 Theories of Failure

This is a widely discussed topic in construction theory. In this book it is neither necessary nor relevant to examine all failure theories. We shall limit ourselves to consider those that directly relate to resistance verification of pressure vessels, and even for these we will highlight only those aspects that are required to understand what follows next.

Pressure vessels are characterized by the existence of stresses along three axis. First of all, due to pressure, there is a principal stress directed as the pressure itself and thus orthogonal to the wall of the vessel, while two additional principal stresses act on the plane orthogonal to the previous one.

In the case of cylindrical elements the first of such stresses is radial, the other two are directed, respectively, along the circumference and along the axis of the cylinder.

Similarly, in the case of spherical elements the first stress is radial, the other two are directed, respectively, along the meridian and the circumference orthogonal to the meridian, and they are obviously identical.

Different situations may occur with elements that are neither cylindrical nor spherical, e.g., in the case of a flat head the first of the three stresses mentioned above is orthogonal with respect to the head, and therefore along the same direction of the axis of the vessel, if the flat head is orthogonal to the latter. The other two have circumferential and radial direction.

In any case, we are faced with a state of stresses along three axis. This calls for a failure theory that allows one to correlate such state of stress with the resistance values of the material, derived from the tensile test that in turn is based on a single stress directed along the axis of the specimen.

The most generally accepted failure theories for ductile materials, such as steel used to build pressure vessels, are the well-known theory of maximum shear stress or Guest–Tresca, and the one known as distorsion energy theory or Huber–Hencky.

According to Guest–Tresca the level of danger is captured by the maximum shear stress, in other words all states having the same maximum shear stress are equivalent with respect to danger. The state of stress relative to the specimen being subjected to single tensile stress is represented in Mohr's plane by the only circle shown in Fig. 1.7. The maximum shear stress acting at 45° with respect to the only principal stress σ_{III} is equal to $\sigma_{III}/2$.

Therefore, if we associate a dangerous condition to the yielding of the material and we call σ_s the corresponding stress, the shear stress is given by

$$\tau_s = \frac{\sigma_s}{2}. \tag{1.1}$$

If the state of stress is along three axis, and we call σ_I, σ_{II}, and σ_{III} the three principal stresses, let us agree that the three increase in value from σ_I to σ_{III} (see Fig. 1.8).

The three maximum shear stresses on the three planes where they operate are hence given, respectively, by

$$\begin{aligned} \tau_{III,I} &= (\sigma_{III} - \sigma_I)/2; \\ \tau_{III,II} &= (\sigma_{III} - \sigma_{II})/2; \\ \tau_{II,I} &= (\sigma_{II} - \sigma_I)/2. \end{aligned} \tag{1.2}$$

With the above agreement the maximum value of the shear stress is given by $\tau_{III.I}$ and therefore the condition of danger is represented by the following

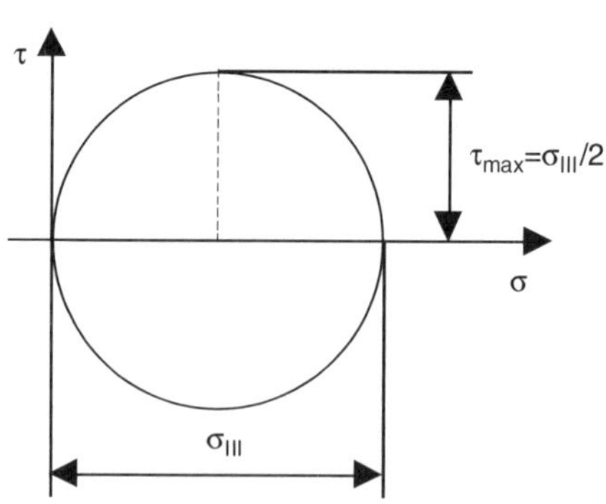

Fig. 1.7

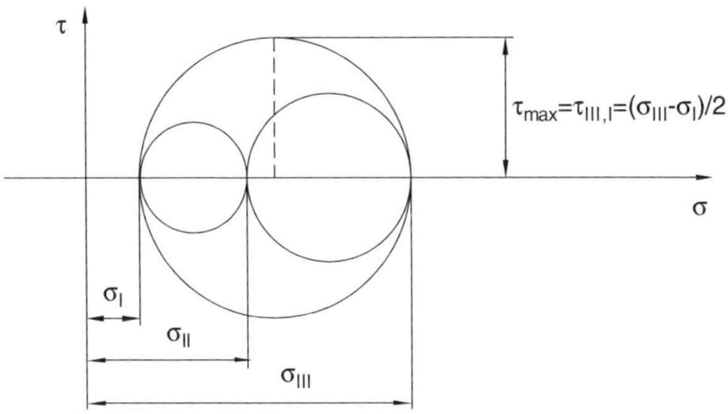

Fig. 1.8

expression derived from (1.1):

$$\tau_{III,I} = \frac{\sigma_{III} - \sigma_I}{2} = \tau_S = \frac{\sigma_S}{2}, \tag{1.3}$$

or

$$\sigma_{III} - \sigma_I = \sigma_S. \tag{1.4}$$

From a formal point of view (let us not forget that conceptually the stress at the basis of the theory is the shear one) a conventional stress (also called ideal stress or stress intensity) appears which is given by

$$\sigma_{id} = \sigma_{III} - \sigma_I. \tag{1.5}$$

Such stress, in the case of stress condition along more than one axis, takes over the same function that the only applied axial stress has inside the specimen, i.e., it can pinpoint the examined fiber's working condition with respect to danger.

In fact, when the ideal stress reaches the value of yield strength, in view of (1.4) and (1.5), according to the failure theory adopted here, we face a situation of danger. In practice, the ideal stress is therefore assumed to be the characteristic stress of the state of stress and to be limited to the allowable stress in order to obtain the sizing of the piece, as we shall see later on.

According to Huber and Hencky, the level of danger is captured by the distorsion energy, i.e., all conditions of stress that produce the same distorsion energy are equivalent to the condition of danger.

Defining again the three principal stresses as σ_I, σ_{II}, and σ_{III}, such energy is given by the following equation:

$$L = \frac{1}{6G}\left(\sigma_I^2 + \sigma_{II}^2 + \sigma_{III}^2 - \sigma_I\sigma_{II} - \sigma_{II}\sigma_{III} - \sigma_{III}\sigma_I\right). \tag{1.6}$$

In the specific case of the specimen since $\sigma_I = \sigma_{II} = 0$ we have

$$L = \frac{1}{6G}\sigma_{III}^2. \tag{1.7}$$

The condition of danger characterized by $\sigma_{III} = \sigma_s$ corresponds to a distorsion energy equal to

$$L_s = \frac{1}{6G}\sigma_s^2. \tag{1.8}$$

For a state of stresses along more than one axis the condition of danger is therefore given by

$$L = L_s = \frac{1}{6G}\sigma_s^2. \tag{1.9}$$

From (1.6) and (1.9) we obtain

$$\sqrt{\sigma_I^2 + \sigma_{II}^2 + \sigma_{III}^2 - \sigma_I\sigma_{II} - \sigma_{II}\sigma_{III} - \sigma_{III}\sigma_I} = \sigma_s. \tag{1.10}$$

Therefore, also in this case an ideal stress is determined

$$\sigma_{id} = \sqrt{\sigma_I^2 + \sigma_{II}^2 + \sigma_{III}^2 - \sigma_I\sigma_{II} - \sigma_{II}\sigma_{III} - \sigma_{III}\sigma_I}. \tag{1.11}$$

This ideal stress relates to the stress state and allows one to identify the working condition of the fiber under scrutiny. As for the previously discussed theory, when σ_{id} reaches the value of the yield strength a dangerous condition takes place, according to (1.10) and (1.11). Similarly, the sizing of the piece is obtained by making sure that σ_{id} does not go beyond the allowable stress.

According to von Mises, the condition of danger depends on the value of the following conventional stress:

$$\tau_{id} = \sqrt{\tau_{III,I}^2 + \tau_{III,II}^2 + \tau_{II,I}^2}. \tag{1.12}$$

It must not go above a threshold that depends on the average value of the principal stresses given by

$$\sigma_m = \frac{\sigma_I + \sigma_{II} + \sigma_{III}}{3}. \tag{1.13}$$

If we rule out the dependency of the value of danger of τ_{id} on σ_m, we realize that von Mises' theory formally corresponds to Huber–Hencky's, in the sense that the ideal stress σ_{id} is the same as that in (1.11).

In fact, on the basis of (1.2) we obtain from (1.12):

$$\tau_{id} = \frac{1}{\sqrt{2}}\sqrt{\sigma_I^2 + \sigma_{II}^2 + \sigma_{III}^2 - \sigma_I\sigma_{II} - \sigma_{II}\sigma_{III} - \sigma_{III}\sigma_I}. \tag{1.14}$$

In the case of the specimen

$$\tau_{id} = \frac{1}{\sqrt{2}}\sigma_{III}, \tag{1.15}$$

and the condition of danger is reached when $\sigma_{III} = \sigma_s$.

By replacing σ_{III} with σ_s in (1.15), and by comparing it with (1.14), we obtain (1.10). The ideal stress is represented by (1.11) in this case as well. Therefore, it is customary to talk about failure theory of Huber–Hencky–von Mises every time (1.11) is adopted, even though, as we have seen above, von Mises starts from assumptions that are completely different from a conceptual point of view.

Today this theory is the most generally accepted for resistance verification of pieces for which ductile materials are used, more than the Guest–Tresca theory. Note, though, that codes in the most important industrialized countries are based on the theory of Guest–Tresca.

The reason for this is because the theory of Guest–Tresca is more conservative than that of Huber–Hencky and easier to apply as well, as one can immediately realize by comparing the equations of σ_{id} in both cases.

We will generally refer to this theory of failure as well, without neglecting to refer to the theory of Huber–Hencky, however, every time it will be necessary or appropriate.

1.4 Plasticity Collaboration

In the sizing of pressure vessels the possibility of plastic collaboration of steel is widely exploited. This is a topic that, if dealt with in great depth, would result in a vast analysis that is in fact not justified considering what is of practical interest for resistance verification, a topic we will shortly introduce. Therefore, we will concentrate on reviewing the fundamental concepts by modeling the problems in a simple way, and by analyzing only those aspects of the phenomenon that find actual application in the analysis presented in the coming chapters.

The principle at the core of plastic collaboration consists of the possibility that less stressed fibers may contribute to the resistance of the piece by helping the most stressed ones. More precisely, the adoption of the criteria of plastic collaboration goes against the traditional concept of verification in the elastic field, which says that the condition of danger is reached when the most stressed fiber begins to show signs of yielding in the material. If the material is ductile it can withstand major deformations before rupturing. Therefore, the fact that the material yields in one area of the piece does not represent a condition of danger, if the nearby fibers are still far from yielding.

Let us consider a simplified stress–deformation curve, as is usually done in this cases, as in Fig. 1.9. The intuition is that the behavior of the steel is elastic–plastic, where the first section of the curve shows a perfectly elastic behavior, while the second section is parallel to the axis of deformations (i.e., a perfectly plastic behavior). In other words, we ignore the hardening, which actually has a favorable effect on resistance, and we assume that the material shows substantial yielding.

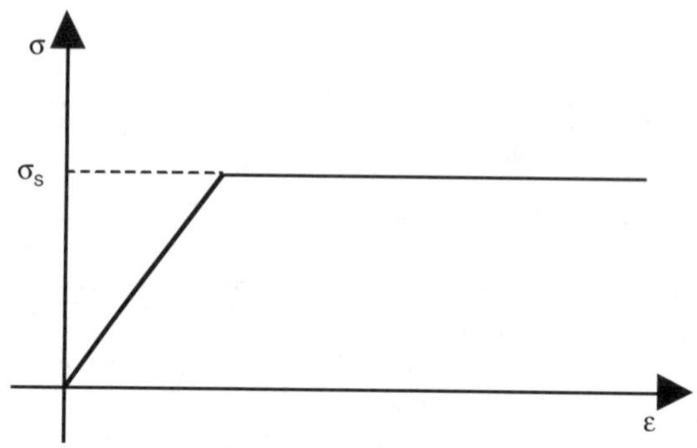

Fig. 1.9

Once yielding in a portion of the piece is reached, if we increase the external forces acting on the piece itself (in our case it is generally the pressure), the yielded fiber or fibers are unable to absorb another increase in stress. They undergo an increase in deformation, but the stress remains constant. At the same time, the nearby fibers that are still far from yielding are able to absorb increasing stresses. These are greater than those derived from calculation in the elastic field because of the greater deformation following the yielding of nearby fibers or, operating within an equilibrium framework, given the requirement to balance the increase in external forces for which the yielded fibers are no longer able to provide a contribution.

The condition of danger is reached when all fibers have exhausted the possibility to absorb an increase in stress; in other words, the condition of danger is represented by the plastic flow of the entire piece. At this point, while peaks of deformation are present given the constraint to the conditions of congruence, peaks of stress have disappeared since in every point the stress is equal to the yield strength. Following this simplified model, taking into account plastic collaboration corresponds in practice to ignoring the peaks of stress.

This issue is in fact more complex since there is vast case history, and every stress condition should be examined separately through a process known as stress analysis. This helps to identify the exact nature of the peak, in order to determine which criteria to apply to carry out verification. Such procedure can be found, e.g., in Sect. III of the ASME Code that deals with the verification of nuclear vessels, where the responsibility of the investigation of the stress condition is left to the designer. This can be done through traditional computational methodologies, if the theoretical modeling of the problem is possible, or through investigation criteria that are nowadays possible with the help of a computer and finite element analysis techniques.

The ASME code indicates at this point which criteria should be used to determine whether the values of stresses are compatible with the safety of the vessel, based on the nature of the stresses themselves, including general and local membrane stresses, those concerning bending, self-limiting ones, etc. We will focus on this topic in detail in Sect. 1.5 when we introduce the modern criteria of verification. We shall refer to them in the next chapters when we examine those cases that lend themselves to a theoretical analysis of the problem. Through the criteria illustrated in Sect. 1.5, we shall find confirmation to what stated in the beginning, i.e., the exploitation of plastic collaboration of the material according to various criteria that indeed depend on the different nature of peaks.

To understand the philosophy at the basis of the criteria in Sect. 1.5, it is therefore appropriate to try to model the most frequent situations. From this point of view a number of considerations can be made. The verification may involve a component for which an average stress not equal to zero or not localized is clearly identifiable. This is the case, for instance, of stresses in a cylinder subject to pressure from the inside without holes, or, if holes are present, in areas of the same but at such distance from the holes not to be influenced by them. In this case, as we shall see, the circumferential and the radial stress vary with the radius through the wall. Their difference represents the ideal stress according to Guest. This has a value that varies across the wall and shows a nonzero average value. With respect to the average value, the ideal stress shows both positive and negative peaks that are going to disappear if the cylinder becomes plastic, since the average stress, if ideally distributed over the entire thickness, is able to balance the acting pressure.

The methodology of plastic collaboration corresponds to neglecting these peaks by doing the calculations on the base of average values for the stresses. As we shall see, in the case at hand it is also possible to define a mathematical formulation of the distribution of stresses under total plastic conditions. This makes it possible to reason on such boundary situation by examining the resistance of the piece from a global viewpoint. Similar situations occur if there are drilling lines with holes that are so close to be considered non isolated, as we shall discuss in more detail later on. Even in this case the circumferential stress between holes, and in some instances the longitudinal stress as well, shows significant peaks in correspondence of the holes' edges. Adopting the criteria of plastic collaboration, these peaks are neglected by referring in the calculation to the average value of stresses that occur between holes.

A completely different situation occurs when stresses through the wall show a change in sign. In this case the average value may be even zero if there is pure bending. In reality this never takes place because of the simultaneous presence of stresses with a constant sign that overlap those caused by the bending. The average value is by the way very small with respect to the maximum values for stresses due to the clearly prevailing bending moment.

An example of this is a flat head or a vessel with a quadrangular section. It is obvious that even adopting plastic collaboration, one cannot here refer to the average value of stresses since it does not balance the external acting forces.

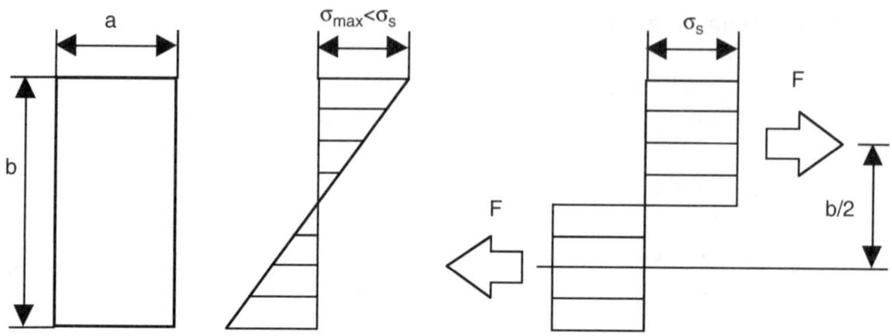

Fig. 1.10

Let us consider at this point how plastic collaboration occurs on a beam with a rectangular section under pure bending, as in Fig. 1.10.

In the elastic field stresses behave as shown on the left side of the figure. When increasing the bending moment up to the total plastic flow, the first fibers to yield are the external ones; subsequently, the yielding moves more and more towards the neutral axis, until the diagram becomes as shown on the right side of the figure.

At the limit of elastic behavior the moment is equal to

$$M_e = W\sigma_s = \frac{1}{6}ab^2\sigma_s, \tag{1.16}$$

where W is the section modulus. When the section is completely yielded, the corresponding moment is equal to

$$M_p = F\frac{b}{2} = a\frac{b}{2}\sigma_s\frac{b}{2} = \frac{1}{4}ab^2\sigma_s. \tag{1.17}$$

The coefficient of plastic collaboration is therefore given by the relationship between these two moments that represent the condition of danger, according to the calculation that takes plastic collaboration into account, and according to traditional computation practices in the elastic field, respectively. By indicating the coefficient of collaboration with ψ we have:

$$\psi = \frac{M_p}{M_e} = 1.5, \tag{1.18}$$

a well-known value in the literature. As far as stress peaks that occur in correspondence with isolated holes, the most appropriate way to approach the problem consists of examining the stress status around the edge of the hole in detail, and to apply the verification criteria discussed in Sect. 1.5.

If we are unable or do not want to perform such analysis, it is possible to adopt the criteria discussed in Chap. 8 that have been included with minor variations in the codes of the most industrialized countries for non-nuclear pressure vessels. As we shall see, even in this case plastic collaboration is factored in, by limiting the width of the area where the fibers collaborate with those subjected to most stress. In doing so, a limit is set to deformations in correspondence of peaks. In fact, an indiscriminate extension of the yielding to fibers far from reaching a peak, with regard to danger, would imply the onset of large deformations in correspondence with the peak itself, something clearly to be avoided.

It is also important to consider those peaks that are the consequence of the respect for congruence of deformations among pieces of different geometrical shape connected with each other. If they were ideally isolated, they would be characterized by different values of deformations in correspondence with the junction. Heads (whether flat, hemispherical or torospherical) connected to the ends of the cylinders represent a typical case.

In these cases, the stresses are, as we usually say, self-limiting, since they occur only out of necessity to respect congruence, and not to balance external forces. If we reach the yielding point and the subsequent deformations are relatively large, congruence is maintained without leading the material to rupture.

In fact, in these situations it would be more logical to carry out the analysis in terms of deformations. In practice this is done according to the laws of elasticity and peak stresses are obtained. We should also not forget the nature of such stresses, and we should not be surprised, if values for stress higher than the yield strength are introduced, which may seem absurd.

That is, the peak deformation takes on a value greater than that of the yield strength. The stress resulting from the calculation is greater than σ_s only from a formal point of view, as the product of deformation by the modulus of elasticity; in reality, it is obvious that the stress is equal to σ_s.

In these cases it is acceptable that the ideal stress computed according to the laws of elasticity may be even double that of the yield strength. Let us see now why this is the case. Let us look at Fig. 1.11 and assume that the deformation is greater than the yield point deformation in correspondence of the peak.

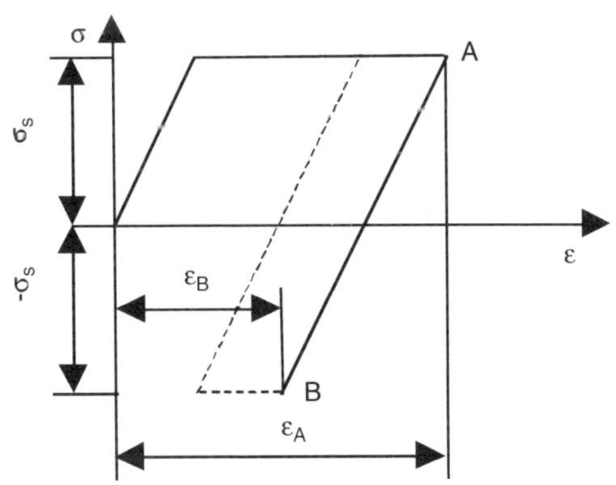

Fig. 1.11

The deformation is ε_A in correspondence of point A on the typical curve for perfectly elastic–plastic material. At release the material behaves elastically, and we also assume that the residual deformation is ε_B in correspondence of point B, and a residual stress equal to the negative yield strength.

If we next load the piece, we return to point A, and every subsequent cycle causes the stress to vary between $-\sigma_s$ and σ_s. In other words, except for the first cycle, for all the following ones the piece's behavior is elastic between point A and B, without further deformations since ε_A constitutes the deformation that takes congruence into account.

In order for this condition to occur, it is necessary and sufficient that the stress, computed according to the laws of elasticity, be double than the yield strength. Of course, if it happens to be lower the behavior of the material is completely analogous with the only difference that during release the negative yield strength is not reached. If, conversely, the stress should be more than double the yield strength, we would have the cycle that appears dashed in the figure. The material would therefore have an elastic–plastic behavior in those cycles following the first one, with the danger that incremental plastic deformations occur that the adopted model does not seem to justify, but that may actually happen in reality.

The calculation criterion introduced here is not limited to the junctions of pieces of different geometrical shape, but can be extended in general to all those situations where stresses originate from noncongruent deformations. Therefore, even the stresses due to thermal flux fall into this category, i.e., those stresses caused by variable thermal expansion through the wall of the piece. To conclude, it is noteworthy that the exploitation of plastic collaboration and the deriving lack of interest in peaks is possible and appropriate not only in relation with the type of steel used, but due to the absence of fatigue phenomena.

When an investigation about fatigue is required, the verification criteria are obviously different, and this is discussed in Sect. 1.5. Moreover, it is important to keep in mind that the verification criteria and the deriving equations for sizing discussed in the following chapters refer to work conditions that do not imply significant fatigue phenomena, due to the presence of a limited number of cycles.

If this were not the case, the validity of the equations to calculate the stresses that take peaks into account notwithstanding (thus ruling out the equations that consider the components of the vessel as membranes), the verification criteria must follow what discussed in Sect. 1.5 and in more detail in Chap. 10.

1.5 Verification Criteria

According to modern verification criteria, stresses can be divided into three categories: primary, secondary, and peak stresses. Primary stresses can then be divided into general membrane stresses, local membrane stresses, and primary bending stresses. In summary, there are the following types of stresses: general membrane, local membrane, primary bending, secondary, and peak.

1.5.1 General Membrane Stresses (σ_m)

They correspond to stresses derived from calculation when one considers the element under test as a membrane. More generally, they correspond to the average value of the stresses through the thickness of the vessel. In contrast to local membrane stresses, that we will discuss shortly, the fundamental characteristic of these stresses is that a potential yielding of the material does not cause a redistribution of the stresses, since the same stress is present in all the surrounding fibers.

A typical example of general membrane stresses is represented by the average values of the stresses acting in a cylinder without holes (or in an area that is not influenced by holes or by the junction with the heads). The same is true for the average values of the stresses acting on a sphere, or in the central area of a hemispheric or torospherical head.

1.5.2 Local Membrane Stresses (σ_{ml})

Here as well we are dealing with the average values of the stresses through the thickness in the analyzed section. In contrast with the previous ones, they involve a limited area of the component, and this means that the surrounding fibers are subject to membrane stress of lower value. A potential yielding of the material happens together with a redistribution of the stresses to the surrounding fibers that are still able to contribute to the local resistance of material, since they are not yielded.

Typical examples of stresses of this kind are the membrane stresses produced in the cylinder and in the dished heads in correspondence with their junction, or the membrane stresses that occur in the cylinder (or in the sphere), and in the nozzle welded on the same in correspondence of a hole.

Once again, these are the average values of the stresses. Stress peaks both in the junction cylinder-heads and in correspondence of the nozzles are generated that do not fall into this kind of stress category.

1.5.3 Primary Bending Stresses (σ_f)

These stresses belong to the category of primary stresses, such as the ones mentioned above, but they are characterized by the fact that their value is proportional to the distance of the fiber from the neutral axis of the section. As the previous ones, they derive from the balance conditions between internal stresses and external forces acting upon the vessel (pressure or mechanical loads). A typical example is represented by the stresses at the center of a flat head. The stresses produced by bending moments exerted on a vessel with a quadrangular section fall into this category, as well.

At this point we must discuss a very important concept in more detail. A bending moment may also generate a distribution of stresses that do not vary linearly across the thickness. An example can be seen at the corners of a vessel with a quadrangular section. Here, the primary bending stresses are those that balance the acting moment with linear variation through the thickness. With respect to these, the stresses that are actually present in the section, display positive or negative differences characterized by a resultant and a moment that are zero. These differential stresses do not balance the forces applied to the vessel, but only guarantee the congruence of the deformations. Therefore, they fall into the category of secondary stresses. The analogy with membrane stresses is evident: the latter ones represent the average value of the stresses produced by a normal load. The primary bending stresses represent the average behavior of the stresses caused by a bending moment.

The differential stresses versus the membrane ones (in the presence of a normal load), or with respect to the primary bending stresses (in the presence of a bending moment), are secondary stresses.

1.5.4 Secondary Stresses (σ_{sec})

Their fundamental characteristic is not to be involved in balancing the forces applied to the vessel, and to be for this reason self-limiting. Their only purpose is to guarantee the congruence of the deformations and, therefore, once the required deformations are produced (even though this happens through the yielding of the material) they do neither cause further deformations nor do they force the intervention of the surrounding fibers, as is the case instead for the local membrane stresses.

The stresses in correspondence of the junction between cylinder and heads belong to this category (not the membrane ones because in that case they are local membrane stresses); the stresses still not related to membranes in the cylinder or in the sphere and in the welded nipple in correspondence of a hole belong to this category, as well. In this last case local peaks due to the presence of sharp edges are excluded, as they belong to the next category.

The stresses due to thermal flux are secondary, as well. In fact, they are also self-limiting, since they are produced solely to reestablish the congruence of the deformations that differ in the various fibers because of the temperature gradient. Even though they originate elsewhere, some differences in stress levels also belong to this category (with respect to the average value that represents the general membrane stress); they occur in a cylinder or a sphere, especially in the case of great thickness, along the radius. These differences in stress do not contribute to balancing the pressure (as the balance is guaranteed by the general membrane stress), but only to ensure the congruence of the deformations.

1.5.5 Peak Stresses (σ_{pic})

Generally speaking, a peak stress is any type of stress that will show a maximum in the section under test. From this point of view peak stresses are the local membrane stresses (as they are locally greater than the general membrane stresses) and the secondary stresses. We will follow this general concept in the following chapters, as well.

Going back to the division of stresses in three categories (primary, secondary, and peak), peak stresses do have a different characterization. This category consists only of those stresses, present in a limited area of the component and in a limited area of the section under test, that do not produce significant deformations, and that represent a potential liability only insofar as the possibility of rupture due to fragility or fatigue.

We are therefore talking about peaks induced by special geometrical characteristics of the coupled elements that can often be eliminated or at least reduced through an appropriate management of the couplings. Moreover, the stresses due to a rapid variation in the temperature of the fluid inside the vessel belong to this category, as well.

We now introduce an easily applicable criterion that allows us to consider membrane stresses focused on an area of the vessel as local, i.e., as stresses σ_{ml}. The membrane stress can be considered local if the area in which its value is greater than 10% of the basic allowable stress f does not extend beyond $0.5\sqrt{rs}$, r being the average radius, and s the thickness (as we shall see the local membrane stress can be equal to $1.5f$). Moreover, this area must have a distance from another area where the membrane stress is greater than f, at least equal to $2.5\sqrt{rs}$.

Once the specific category for the stresses has been identified, it is necessary to establish for each category or group of categories which value is allowable for the stress (in the sense that, e.g., a primary bending stress or a secondary stress may overlap with a general membrane stress). It is important to notice that the stress to be compared with the allowable stress is at any rate the ideal stress or the comparison stress according to the Guest–Tresca failure theory, as discussed in Sect. 1.3.

The allowable stresses for the various categories are shown in Fig. 1.12. Here we identify with f the basic allowable stress of the material, as discussed in Sect. 1.2, and recall the analysis in Sect. 1.3 about the criteria to be followed to take the plastic collaboration of the material into account.

As one can see, the allowable stress is obviously equal to the basic allowable stress of the material for the general membrane stress. We can assume an allowable stress equal to $1.5f$ for local membrane stress, therefore generally equal to the yield strength, as seen in Sect. 1.2. The allowable stress is equal to $1.5f$ in the case of primary bending stresses; such criterion is equivalent to introducing the coefficient of plastic collaboration discussed in Sect. 1.4. If the bending stress is coupled with a general membrane stress or a local membrane stress, the resulting maximum ideal stress must not he higher than $1.5f$.

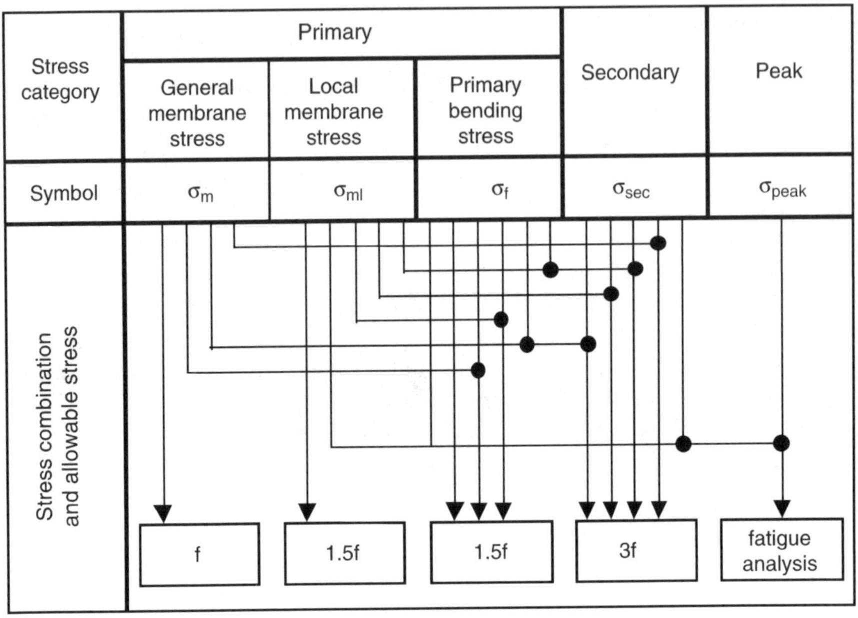

Fig. 1.12

If there are secondary stresses besides primary ones (general, local, and bending stresses), the resulting maximum ideal stress must not be greater than $3f$, a value which is generally equal to twice the yield strength, as we have seen in Sect. 1.2. The peak stresses can be neglected if we rule out significant fatigue phenomena. If this is not possible, one needs to perform a specific analysis of stresses under fatigue, and to take into account primary, secondary, and peak stresses. We will discuss this topic in Chap. 10.

2

General Calculation Criteria

2.1 Membrane Stresses in Revolution Shells

The knowledge of membrane stresses in revolution shells is particularly interesting for the analysis of stress conditions in pressure vessels. These membrane stresses correspond to stresses in cylinders, spheres, bended pipes, spherical heads and cones, provided they are thin. Moreover, they represent the average values of the stresses (called in fact membrane stresses) in vessels of great thickness. They are quite important as far as sizing is concerned, as we have already discussed in Sect. 1.5. Let us consider Fig. 2.1 that shows a generic revolution shell, where R_1 is the radius of curvature according to the meridian and R_2 is the radius of curvature on the plane orthogonal to the meridian.

Let σ_t and σ_m be the stresses along the parallel and the meridian, respectively, and let s be the thickness. The forces F_t are on sides A–B and C–D, and are as follows:

$$F_t = \sigma_t s dl_1. \tag{2.1}$$

Their resultant orthogonal to the element is

$$F_1 = 2\sigma_t s dl_1 \frac{d\vartheta_2}{2}. \tag{2.2}$$

The forces F_m on the sides $A - D$ and $B - C$ are

$$F_m = \sigma_m s dl_2. \tag{2.3}$$

The resultant orthogonal to the element is

$$F_2 = 2\sigma_m s dl_2 \frac{d\vartheta_1}{2}. \tag{2.4}$$

Pressure p produces a force that follows the same direction of F_1 and F_2. This force is

$$F_3 = p dl_1 dl_2. \tag{2.5}$$

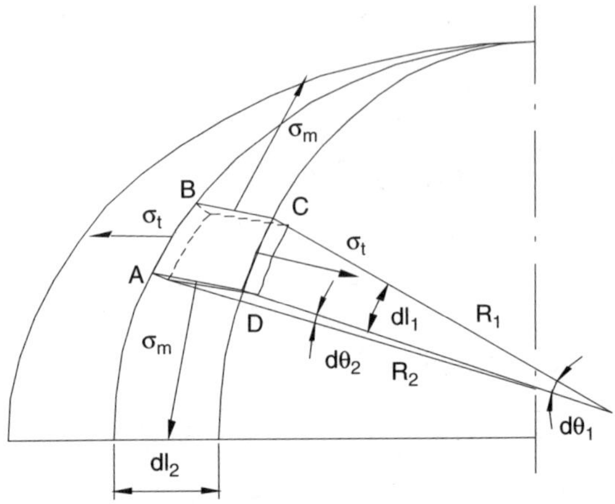

Fig. 2.1

For balance in the direction orthogonal to the element we have

$$F_3 = F_1 + F_2, \tag{2.6}$$

and therefore

$$\sigma_t s dl_1 d\vartheta_2 + \sigma_m s dl_2 d\vartheta_1 = p dl_1 dl_2. \tag{2.7}$$

On the other hand,

$$\begin{aligned} dl_1 &= R_1 d\vartheta_1 \\ dl_2 &= R_2 d\vartheta_2 \end{aligned} \tag{2.8}$$

Based on (2.7) we have

$$\sigma_t \frac{s}{R_2} dl_1 dl_2 + \sigma_m \frac{s}{R_1} dl_1 dl_2 = p dl_1 dl_2, \tag{2.9}$$

and finally

$$\frac{\sigma_m}{R_1} + \frac{\sigma_t}{R_2} = \frac{p}{s}. \tag{2.10}$$

Equation (2.10) is completely general, regardless of whether the axis crosses the membrane or not; a typical case of an axis not crossing the membrane occurs in a torus, or more generally, in a bent pipe. We will discuss this in Sect. 7.2.

If the axis crosses the membrane, two generally valid equations can be derived from (2.10) that allow us to calculate the meridian and circumferential stresses directly. As shown in Fig. 2.2, the thrust due to the pressure on the portion of the membrane located above the $A - B$ plane is

$$F = \pi R^2 p. \tag{2.11}$$

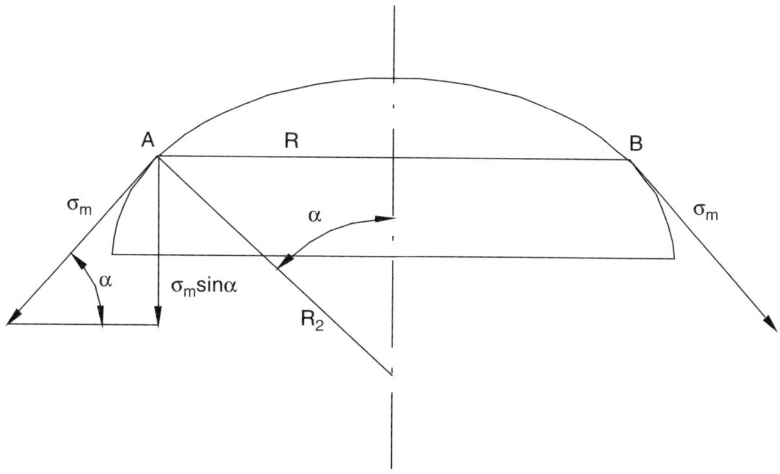

Fig. 2.2

The thrust is balanced by the vertical components of the stresses σ_m, that is

$$\pi R^2 p = 2\pi R s \sigma_m \sin\alpha. \tag{2.12}$$

We have

$$R = R_2 \sin\alpha. \tag{2.13}$$

Therefore,

$$\sigma_m = \frac{pR}{2s\sin\alpha} = \frac{pR_2}{2s}. \tag{2.14}$$

According to (2.10) we have

$$\sigma_t = \frac{pR_2}{s}\left(1 - \frac{R_2}{2R_1}\right). \tag{2.15}$$

Equations (2.14) and (2.15) allow one to calculate both the meridian and circumferential stresses of a revolution membrane, the axis of which crosses the membrane itself.

For instance, in the case of a cylinder we have $R_2 = D/2$ and $R_1 = \infty$, where D is the diameter of the cylinder. The meridian stress in the cylinder is more appropriately indicated as longitudinal or axial stress σ_a, and therefore

$$\sigma_a = \sigma_m = \frac{pD}{4s} \quad \text{and} \quad \sigma_t = \frac{pD}{2s}. \tag{2.16}$$

One may notice that the axis of the cylinder does not actually cross it. Still, the cylinder is enclosed by the heads that release the thrust F (see (2.11)) on the cylinder itself. Therefore, the conditions leading to (2.14) and (2.15) still hold.

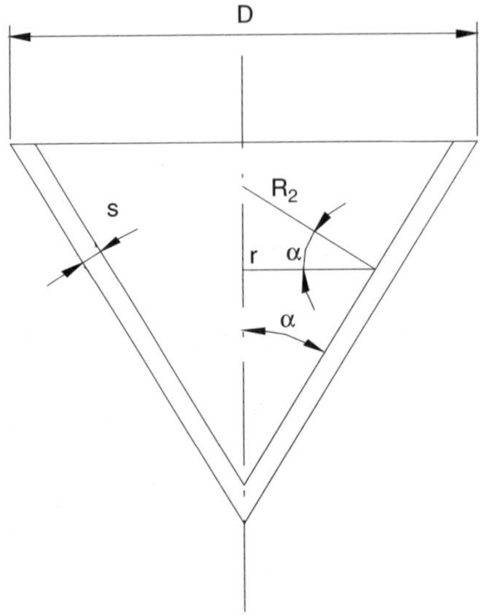

Fig. 2.3

As far as the sphere is concerned, we have $R_1 = R_2 = D/2$, and

$$\sigma_m = \sigma_t = \frac{pD}{4s}. \tag{2.17}$$

For the cone (see Fig. 2.3) we have $R_1 = \infty$ and $R_2 = r/\cos\alpha$, where r is the radius of the generic parallel and α the half angle at the tip of the cone. Therefore,

$$\sigma_m = \frac{pr}{2s\cos\alpha}, \text{ and } \sigma_t = \frac{pr}{s\cos\alpha}. \tag{2.18}$$

Indicating the diameter at the base of the cone with D, the maximum values of the meridian and the circumferential stresses are

$$\sigma_m = \frac{pD}{4s\cos\alpha}, \text{ and } \sigma_t = \frac{pD}{2s\cos\alpha}. \tag{2.19}$$

Comparing these equations with those of the cylinder (2.16), we notice the presence of α, the characteristic angle of the cone; when $\alpha = 0$ we obtain the equations of the cylinder again.

2.2 Edge Effects in Cylinders and Semispheres

The analysis of edge effects in cylinders and semispheres, i.e., the state of stress caused by forces and moments applied in a radial fashion to the edge of a cylinder or a semisphere, is fundamental for the study of pressure vessels. The laws that regulate the distribution of stresses and deformations in the cylinder due to the aforementioned forces and moments, determine the characteristics of the constraint at the junction between cylinder and heads. Moreover, as we shall see, these laws make it possible to determine the extent of the collaborating area (insofar as plastic collaboration is concerned) around a hole and along a nozzle welded to the cylinder in correspondence of a hole.

Figure 2.4 shows a cylinder where evenly distributed radial forces and moments are applied to its edge ($x = 0$); moreover, let us assume that the cylinder have ideally infinite length. Finally, let r be the average radius.

Let us isolate the beam of unitary width from the cylinder, as shown in Fig. 2.5, and let us indicate F_0 and M_0 the force and the bending moment on the edge of the beam (F_0 and M_0 are the force and the moment per unitary length of the circumference).

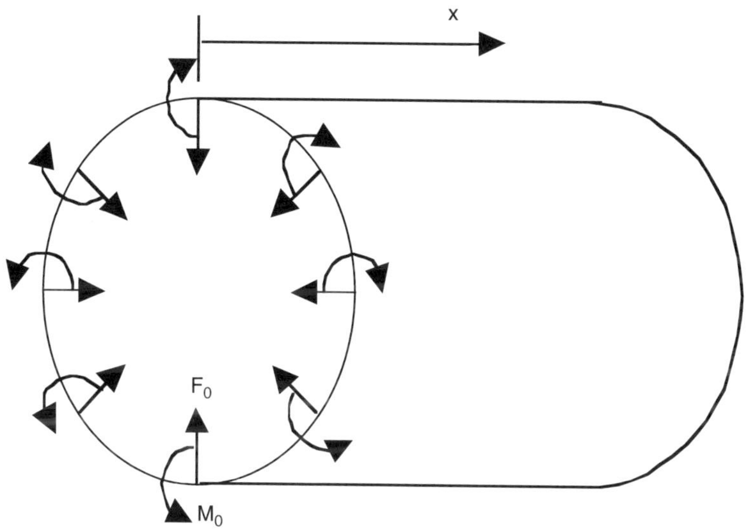

Fig. 2.4

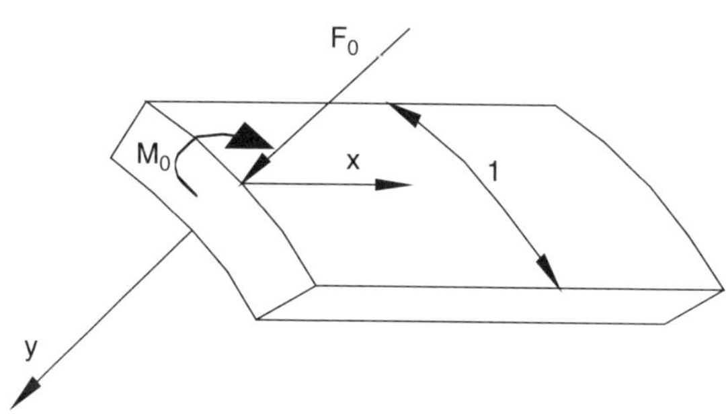

Fig. 2.5

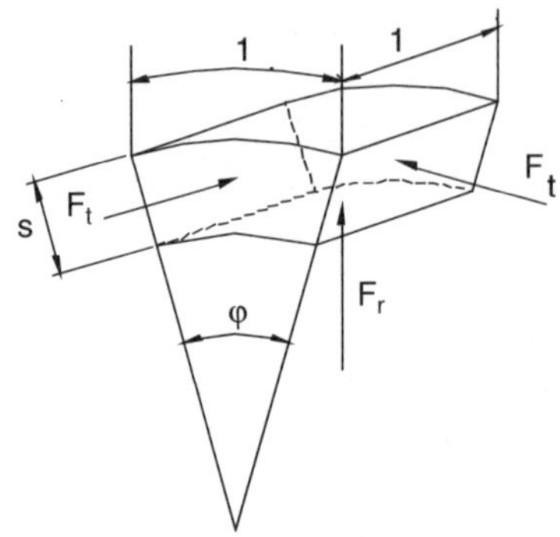

Fig. 2.6

F_0 and M_0 bend the beam; the value of the deflection y is a function of x. Due to this deflection the radius undergoes a variation in length Δr equal to y, accompanied by a circumferential deformation ε_t equal to

$$\varepsilon_t = \frac{\Delta r}{r} = \frac{y}{r}. \tag{2.20}$$

A stress σ_t corresponds to this deformation

$$\sigma_t = E\varepsilon_t = \frac{Ey}{r}. \tag{2.21}$$

Let us consider the element of beam of unitary length (see Fig. 2.6). Due to σ_t the forces F_t are on the radial sides, i.e.,

$$F_t = \sigma_t s = \frac{Es}{r}y. \tag{2.22}$$

A radial force F_r results from the presence of those forces that form the angle φ between them

$$F_r = 2F_t\frac{\varphi}{2} = F_t\varphi = \frac{Es}{r^2}y. \tag{2.23}$$

As one can see, the radial force F_r is proportional to y; the cylinder behaves similarly to a beam lying on an elastic foundation. This is a well-known topic in the literature.

Let us consider the element of length dx shown in Fig. 2.7.

The shear forces T and $T+dT$, as well as the moments M_x and M_x+dM_x, are on the sides perpendicular to the x-axis. In addition, the radial force dF_r is exerted on the element

$$dF_r = F_r dx = \frac{Es}{r^2}y dx. \tag{2.24}$$

Finally, the moments M_z act upon the radial sides. They are derived, as we shall see, from congruence requirements.

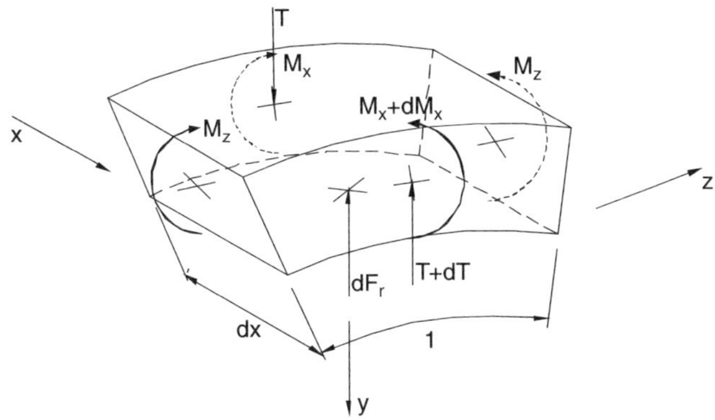

Fig. 2.7

As far as the balance along the y-axis, we have

$$T - \frac{Es}{r^2} y dx - (T + dT) = 0. \tag{2.25}$$

Therefore,

$$\frac{dT}{dx} = -\frac{Es}{r^2} y. \tag{2.26}$$

As far as the balance referred to rotation, we determine that

$$T dx - \frac{Es}{r^2} y \frac{dx^2}{2} + dM_x = 0. \tag{2.27}$$

By neglecting the term of higher order we have

$$T = -\frac{dM_x}{dx}, \tag{2.28}$$

and from (2.26)

$$-\frac{dT}{dx} = \frac{d^2 M_x}{dx^2} = \frac{Es}{r^2} y. \tag{2.29}$$

The maximum stress σ_x produced by M_x is

$$\sigma_x = \frac{M_x s}{2I}, \tag{2.30}$$

where I is the moment of inertia.

The maximum stress σ_z produced by M_z is

$$\sigma_z = \frac{M_z s}{2I}. \tag{2.31}$$

If M_z as well as M_x refer to the unit of length, the two moments of inertia occurring in the previous equations are identical and equal to $s^3/12$.

The deformations ε_x and ε_z are given by

$$\begin{aligned}\varepsilon_x &= \frac{1}{E}\left(\sigma_x - \mu\sigma_z\right)\\ \varepsilon_z &= \frac{1}{E}\left(\sigma_z - \mu\sigma_x\right).\end{aligned} \tag{2.32}$$

That is, ε_z must be zero since the radial sides must stay the same (radial); therefore, recalling (2.30) and (2.31) we have

$$\frac{s}{2EI}\left(M_z - \mu M_x\right) = 0, \tag{2.33}$$

from which

$$M_z = \mu M_x. \tag{2.34}$$

Moreover,

$$\varepsilon_x = \frac{M_x s}{2EI}\left(1-\mu^2\right). \tag{2.35}$$

Given that

$$\frac{d^2y}{dx^2} = -\frac{\varepsilon_x}{s/2}, \tag{2.36}$$

based on (2.35)

$$\frac{d^2y}{dx^2} = -\frac{M_x}{EI}\left(1-\mu^2\right). \tag{2.37}$$

From (2.29) and twice deriving (2.37) we have

$$\frac{d^4y}{dx^4} = -\frac{s\left(1-\mu^2\right)}{r^2 I}y. \tag{2.38}$$

With reference to the value of the moment of inertia I of the beam of unitary width, we finally have

$$\frac{d^4y}{dx^4} = -\frac{12\left(1-\mu^2\right)}{r^2 s^2}y. \tag{2.39}$$

Be the factor β given by

$$\beta = \frac{\sqrt[4]{3\left(1-\mu^2\right)}}{\sqrt{rs}}, \tag{2.40}$$

and with $\mu = 0.3$ for the steel

$$\beta = \frac{1.285}{\sqrt{rs}} = \frac{1.817}{\sqrt{Ds}}. \tag{2.41}$$

Based on (2.41), (2.39) becomes

$$\frac{d^4y}{dx^4} = -4\beta^4 y. \tag{2.42}$$

The general integral of (2.42) is

$$y = e^{-\beta x}\left(C_1 \sin\beta x + C_2 \cos\beta x\right) + e^{\beta x}\left(C_3 \sin\beta x + C_4 \cos\beta x\right), \tag{2.43}$$

with C_1, C_2, C_3 and C_4 as integration constants.

On the other hand, for $x = \infty$, $y = 0$ follows, and therefore

$$C_3 = C_4 = 0. \tag{2.44}$$

Equation (2.43) then becomes

$$y = e^{-\beta x} \left(C_1 \sin \beta x + C_2 \cos \beta x \right). \tag{2.45}$$

Equation (2.45) allows one to calculate the behavior of both the shear force and the bending moment along the axis of the cylinder, once the values of C_1 and C_2 have been determined according to the boundary conditions.

In fact, based on (2.37) and (2.38), we have

$$M_x = -\frac{Es^3}{12\left(1-\mu^2\right)} \frac{d^2 y}{dx^2} \tag{2.46}$$

and

$$T = \frac{Es^3}{12\left(1-\mu^2\right)} \frac{d^3 y}{dx^3}. \tag{2.47}$$

Based on (2.45) we have

$$\frac{d^2 y}{dx^2} = 2\beta^2 e^{-\beta x} \left(-C_1 \cos \beta x + C_2 \sin \beta x \right) \tag{2.48}$$

and

$$\frac{d^3 y}{dx^3} = 2\beta^3 e^{-\beta x} \left[\left(C_1 - C_2 \right) \sin \beta x + \left(C_1 + C_2 \right) \cos \beta x \right]. \tag{2.49}$$

Let us consider now the case $M_0 = 0$; it follows that the second derivative of y must be zero for $x = 0$; from (2.48), we determine that $C_1 = 0$; furthermore, based on (2.49)

$$\frac{d^3 y}{dx^3} = 2\beta^3 C_2. \tag{2.50}$$

Consequently, given that, for $x = 0$, $T = F_0$, we have from (2.49)

$$F_0 = \frac{Es^3}{6\left(1-\mu^2\right)} \beta^3 C_2 \tag{2.51}$$

and

$$C_2 = \frac{6\left(1-\mu^2\right)}{Es^3 \beta^3} F_0. \tag{2.52}$$

From (2.46) and (2.48), we have

$$M_x = -\frac{F_0}{\beta} e^{-\beta x} \sin \beta x. \tag{2.53}$$

The maximum value of M_x is obtained by deriving (2.53) and by making the derivative equal to zero

$$F_0 e^{-\beta x} \left(\sin \beta x - \cos \beta x \right) = 0. \tag{2.54}$$

Thus

$$\beta x = \frac{\pi}{4}, \tag{2.55}$$

and from (2.41)

$$x = \frac{\pi\sqrt{rs}}{4 \times 1.285} = 0.611\sqrt{rs} = 0.432\sqrt{Ds}.$$

The maximum absolute value of M_x is

$$\left|M_{x(\max)}\right| = \frac{F_0\sqrt{Ds}}{1.817}e^{-\pi/4}\frac{1}{\sqrt{2}} = 0.1773F_0\sqrt{Ds}. \tag{2.56}$$

If $F_0 = 0$, the third derivative of y must be zero for $x = 0$, and therefore

$$C_1 + C_2 = 0. \tag{2.57}$$

It follows that the moment M_x is:

$$M_x = \frac{Es^3C_1}{6\left(1-\mu^2\right)}\beta^2 e^{-\beta x}\left(\sin\beta x + \cos\beta x\right). \tag{2.58}$$

For $x = 0$ we have $M_x = M_0$, consequently

$$C_1 = \frac{6M_0\left(1-\mu^2\right)}{Es^3\beta^2}, \tag{2.59}$$

and

$$M_x = M_0 e^{-\beta x}\left(\sin\beta x + \cos\beta x\right). \tag{2.60}$$

We notice that

$$\beta x = 1.285\frac{x}{\sqrt{rs}} = 1.817\frac{x}{\sqrt{Ds}}. \tag{2.61}$$

Furthermore, (2.53) may also be written as

$$\frac{M_x}{-F_0x} = \frac{e^{-\beta x}\sin\beta x}{\beta x}. \tag{2.62}$$

The product $(-F_0x)$ represents the moment at distance x from the edge, if the elementary beam discussed earlier were not laterally constrained to the other elementary beams that make up the cylinder.

Similarly, (2.60) may be written as

$$\frac{M_x}{M_0} = e^{-\beta x}\left(\sin\beta x + \cos\beta x\right). \tag{2.63}$$

As one can see, both (2.62) and (2.63) are only a function of $x/\sqrt{Ds}$; the representative curves are shown in Fig. 2.8.

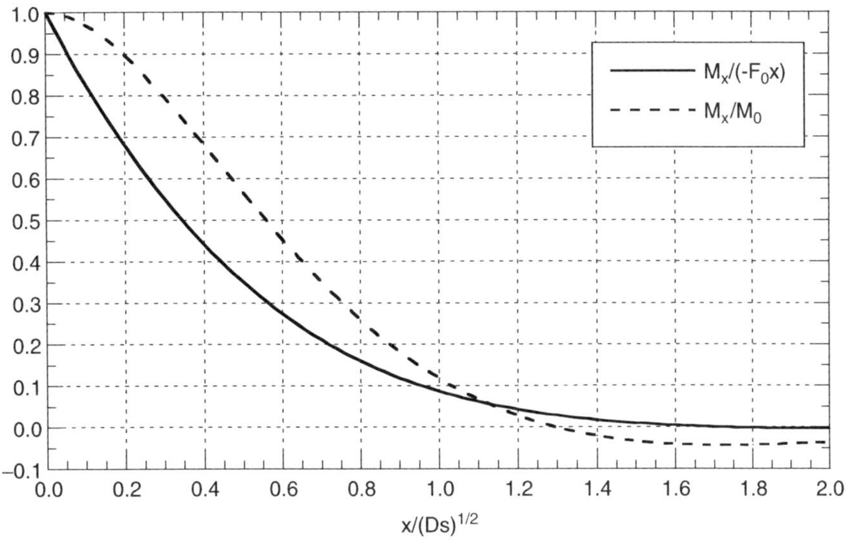

Fig. 2.8

First of all, they show the peculiar characteristic of the phenomenon in a significant way, i.e., the rapid damping of the bending moments, and therefore of the stresses when moving from the edge of the cylinder inward.

Moreover, the curves lead to the following considerations. As far as the damping of the edge effect, the parameter $\sqrt{Ds}$ plays a crucial role; at a distance from the edge equal to $\sqrt{Ds}$, the moments present in the cylinder are already reduced to about 1/10 compared to those without the damping; at a distance equal to $1.5\sqrt{Ds}$ the edge effect becomes almost insignificant. Note that if the thickness is equal, e.g., to $0.02D$, the parameter in question is equal to about $0.14D$; at a distance from the edge equal to $0.20D$ the edge effect has practically vanished.

This has important implications as far as the junctions between elements of different geometrical shape (e.g., cylinder and flat head or dished head) and plastic collaboration. As we shall see later, in correspondence of the junction mentioned above there are internal forces and moments induced by the need to restore the congruence of deformations. In fact, the deformations have different values for the two elements under pressure when considered ideally independent from one another.

These internal forces and moments at the edges of both elements cause stress peaks of relatively reduced extension due to the damping discussed above.

As far as plastic collaboration is concerned, the conclusions we reach as a result of the study of edge effects have highlighted the need to set limits to the extent of the collaborating area. In fact, it is intuitively clear that if deformations undergo rapid damping when moving away from peaks, the collaboration of too distant fibers could only occur at the expense of undesirably high deformations in correspondence of the peak. In addition, in agreement with our conclusions the extent of the collaborating area is correlated to the parameter $\sqrt{Ds}$.

Finally, it may be useful to know the equations of the y-deflection and of the rotation in correspondence of the edge, as well.

Based on (2.45) we determine that, for $x = 0$,

$$y_0 = C_2. \tag{2.64}$$

By deriving (2.45), we also determine that, for $x = 0$,

$$\vartheta_0 = \beta\,(C_1 - C_2)\,. \tag{2.65}$$

Recalling (2.41), (2.52), (2.57), and (2.59), we obtain the following equations:

$$y_0 = 0.909\frac{F_0}{E}\left(\frac{D}{s}\right)^{3/2} - 1.652\frac{M_0}{Es}\frac{D}{s} \tag{2.66}$$

and

$$\vartheta_0 = -1.652\frac{F_0}{Es}\frac{D}{s} + 6.007\frac{M_0}{Es^2}\sqrt{\frac{D}{s}}. \tag{2.67}$$

Up to this point we focused on the edge effects of the cylinder. As far as the semisphere is concerned, we look at Fig. 2.9 where we have identified a generic parallel through the angle φ that the tangent to the sphere forms with the equatorial plane.

By indicating with y the radial deflection generated by the forces F_0 and by the moments M_0, the variation of radius of the parallel is equal to $y \sin\varphi$.

If we indicate the radius of the parallel with r_p, in a similar way to (2.20) we have

$$\varepsilon_t = \frac{y}{r_p}\sin\varphi, \tag{2.68}$$

and, as seen for (2.21)

$$\sigma_t = E\varepsilon_t = \frac{Ey}{r_p}\sin\varphi. \tag{2.69}$$

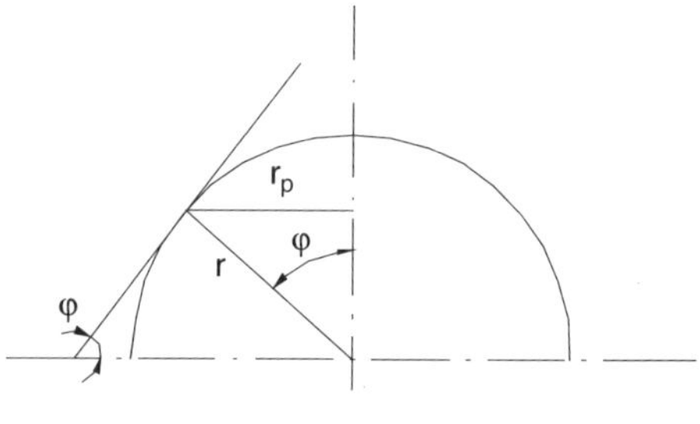

Fig. 2.9

On the other hand, if r is the radius of the sphere,

$$r_p = r \sin \varphi, \tag{2.70}$$

and, based on (2.69)

$$\sigma_t = \frac{Ey}{r}, \tag{2.71}$$

which coincides with (2.21).

Consequently, the forces F_t and the radial forces F_r do not depend on the position of the parallel and coincide with those of the cylinder of equal radius. The discussion about the cylinder is therefore equally valid for the semisphere. One could argue that in the case of the cylinder we have assumed an infinite length which, of course, does not hold in the case of a semisphere.

We should also keep in mind that the edge effects rapidly vanish: for this reason the discussion about the cylinder is true for the semisphere as well, provided the thickness of the latter is not great.

Let us clarify this concept a bit further.

Based on (2.62) we determine that the maximum and minimum values of the right-hand side occur for

$$\tan \beta x = \frac{\beta x}{\beta x + 1}. \tag{2.72}$$

The first two values that satisfy (2.72) are $\beta x \cong 3.81$ e $\beta x \cong 7$; for these values of βx the right-hand side in (2.62) is equal to -0.0036 and 0.000086, respectively.

Based on (2.63) the maximum and minimum values of the right-hand side occur for $\sin \beta x = 0$. The first two values to satisfy such condition (except $\beta x = 0$) are $\beta x = \pi$ and $\beta x = 2\pi$. For such values of βx the right-hand side of (2.63) is equal to -0.043 and 0.0019.

At a distance from the edge equal to $3\sqrt{Ds}$ the phenomenon may be considered exhausted. In fact, we have $\beta x = 1.817 \times 3 = 5.45$ at such a distance (see (2.61)).

In the worst case, i.e., in the presence of bending moments at the edge only, the right-hand side of (2.63) is equal to -0.0003, for $\beta x = 5.45$. In other words, the moment active in the section is equal in absolute value to 0.03% of the moment applied to the edge. On the other hand, we saw that even when $\beta x > 5.45$, the maximum acting moment in the cylinder or in the semisphere is equal to 0.19% of the moment applied to the edge. Therefore, we can practically consider it as equal to zero.

A quarter of the circumference must be equal to or greater than $3\sqrt{Ds}$ to ensure that the edge effects can be assimilated to those in the cylinder, therefore

$$\frac{\pi D}{4} \geq 3\sqrt{Ds}. \tag{2.73}$$

From (2.73) we have

$$\frac{s}{D} \leq \left(\frac{\pi}{12}\right)^2 = 0.068. \tag{2.74}$$

The resulting thickness is nothing but modest. The analysis for cylinders is therefore applicable to the great majority of semispheres (basically to hemispherical heads connected to the cylinders) with the only exception of very thick heads.

2.3 Stress Concentration Around Holes

Let us consider a hole with a radius r in a plate of ideally infinite size undergoing a uniform stress σ along the y-axis (see Fig. 2.10).

The generic element at distance ρ from the center of the hole that is located at an angle ϑ with the y-axis, undergoes the following stresses:

$$\begin{aligned}
\sigma_t &= \frac{\sigma}{2}\left(1+\frac{r^2}{\rho^2}\right) - \frac{\sigma}{2}\left(1+\frac{3r^4}{\rho^4}\right)\cos 2\vartheta \\
\sigma_\rho &= \frac{\sigma}{2}\left(1-\frac{r^2}{\rho^2}\right) + \frac{\sigma}{2}\left(1-\frac{4r^2}{\rho^2}+\frac{3r^4}{\rho^4}\right)\cos 2\vartheta \\
\tau &= -\frac{\sigma}{2}\left(1+\frac{2r^2}{\rho^2}-\frac{3r^4}{\rho^4}\right)\sin 2\vartheta
\end{aligned} \tag{2.75}$$

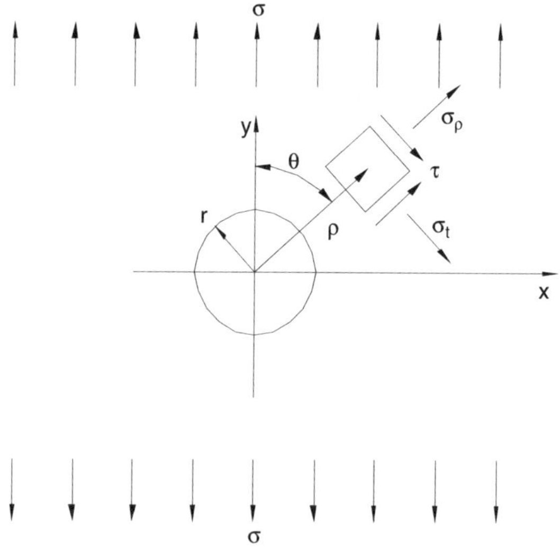

Fig. 2.10

Specifically for $\vartheta = \pi/2$ and hence along the x-axis

$$\begin{aligned} \sigma_{tx} &= \frac{\sigma}{2}\left(2 + \frac{r^2}{\rho^2} + \frac{3r^4}{\rho^4}\right) \\ \sigma_{\rho x} &= \frac{3\sigma}{2}\left(\frac{r^2}{\rho^2} - \frac{r^4}{\rho^4}\right) \\ \tau_x &= 0 \end{aligned} \tag{2.76}$$

whereas for $\vartheta = 0$ and hence along the y-axis

$$\begin{aligned} \sigma_{ty} &= \frac{\sigma}{2}\left(\frac{r^2}{\rho^2} - \frac{3r^4}{\rho^4}\right) \\ \sigma_{\rho y} &= \frac{\sigma}{2}\left(2 - \frac{5r^2}{\rho^2} + \frac{3r^4}{\rho^4}\right) \\ \tau_y &= 0 \end{aligned} \tag{2.77}$$

Figure 2.11 a,b shows σ_{tx}, $\sigma_{\rho x}$, σ_{ty}, and $\sigma_{\rho y}$.

For $\rho = \infty$ we have $\sigma_{tx} = 0$ and $\sigma_{\rho x} = 0$, and also $\sigma_{ty} = 0$ as well as $\sigma_{\rho y} = \sigma$. We determine that for $\rho = r$, therefore at the edge of the hole

$$\begin{aligned} \sigma_{tx} &= 3\sigma \\ \sigma_{ty} &= -\sigma \end{aligned} \tag{2.78}$$

In the direction of the σ active on the plate, and in correspondence of the edge of the hole, we have a stress peak equal to three times the stress in areas that are not influenced by the hole. The peak rapidly decreases. We observe that $\sigma_{tx} = 1.22\sigma$ for $\rho = 2r$, therefore at a distance from the edge equal to the radius of the hole. σ_{tx} is therefore greater than the basic stress in the plate by only 22%.

If two identical stresses σ are active in the plate along both the x- and y-axis (see Fig. 2.12) we can easily determine that we have

$$\begin{aligned} \sigma_{tx} &= \sigma_{ty} = \sigma\left(1 + \frac{r^2}{\rho^2}\right) \\ \sigma_{\rho x} &= \sigma_{\rho y} = \sigma\left(1 - \frac{r^2}{\rho^2}\right) \\ \tau_x &= \tau_y = 0 \end{aligned} \tag{2.79}$$

along both axes.

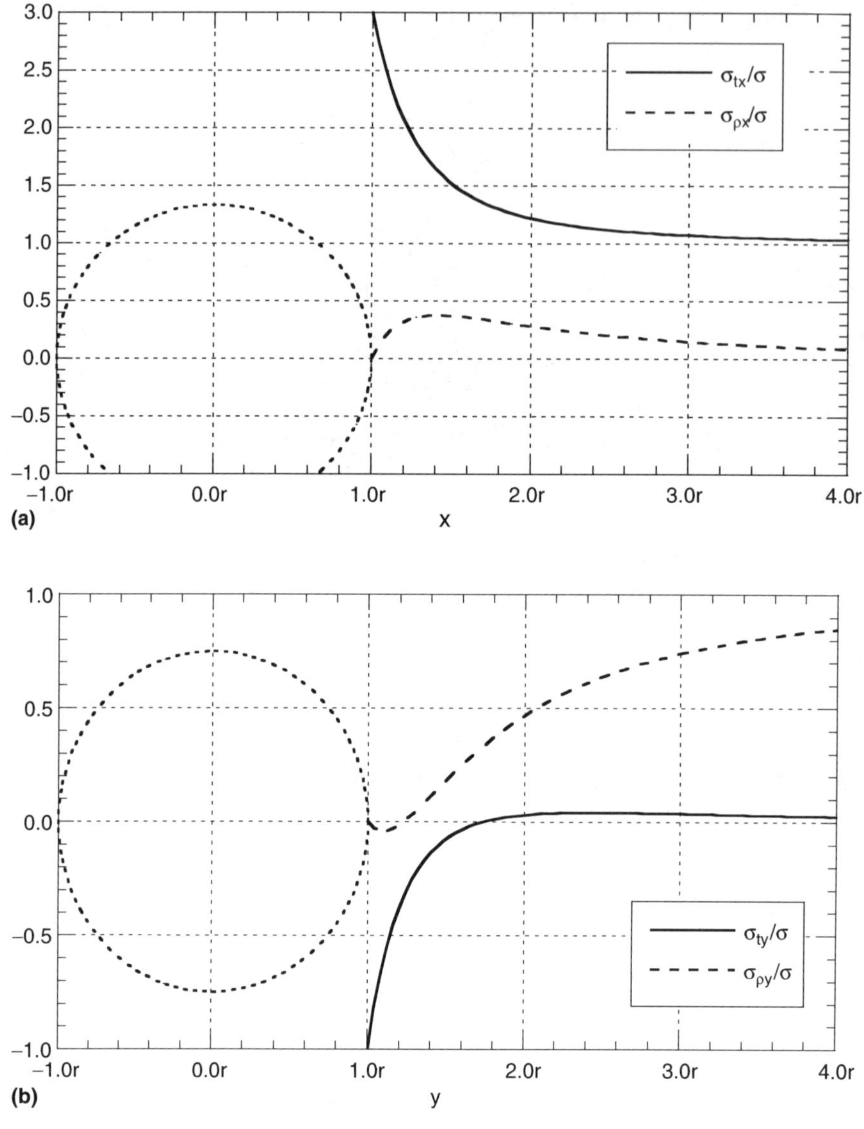

Fig. 2.11

Figure 2.13 shows σ_{tx} and $\sigma_{\rho x}$.
In correspondence of the edge of the hole we have

$$\sigma_{tx} = \sigma_{ty} = 2\sigma. \tag{2.80}$$

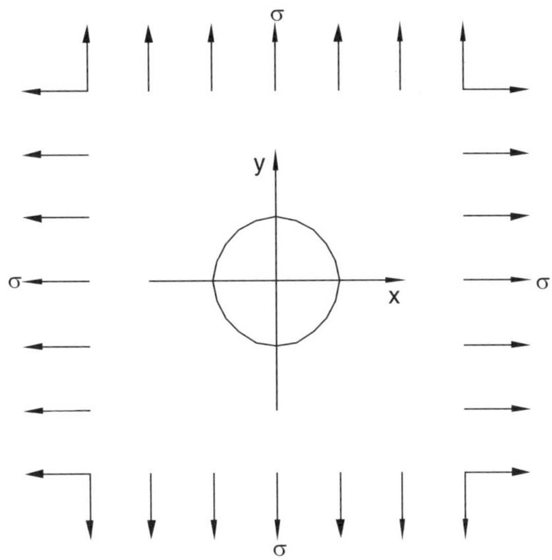

Fig. 2.12

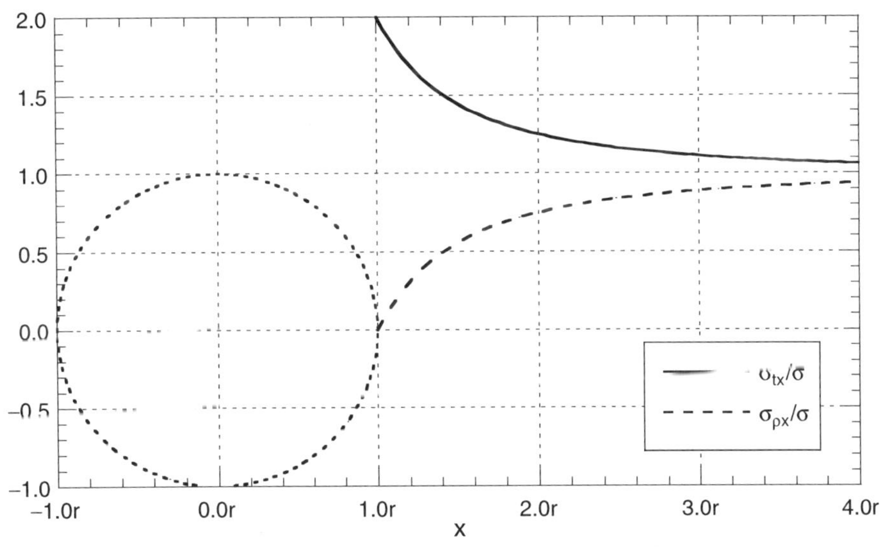

Fig. 2.13

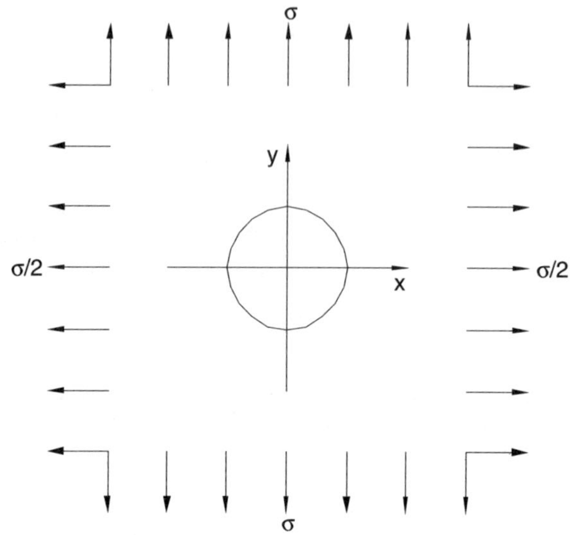

Fig. 2.14

Finally, if two stresses are on the plate, one equal to σ along the y-axis and the other equal to $\sigma/2$ along the x-axis (see Fig. 2.14), we have

$$\begin{aligned}\sigma_{tx} &= \frac{\sigma}{4}\left(4+\frac{3r^2}{\rho^2}+\frac{3r^4}{\rho^4}\right)\\ \sigma_{\rho x} &= \frac{\sigma}{4}\left(2+\frac{r^2}{\rho^2}-\frac{3r^4}{\rho^4}\right)\\ \tau_x &= 0.\end{aligned} \tag{2.81}$$

Moreover,

$$\begin{aligned}\sigma_{ty} &= \frac{\sigma}{4}\left(2+\frac{3r^2}{\rho^2}-\frac{3r^4}{\rho^4}\right)\\ \sigma_{\rho y} &= \frac{\sigma}{4}\left(4-\frac{7r^2}{\rho^2}+\frac{3r^4}{\rho^4}\right)\\ \tau_y &= 0.\end{aligned} \tag{2.82}$$

The pattern of behavior of such stresses is shown in Fig. 2.15 a,b.

Specifically, in correspondence of the edge of the hole we have:

$$\begin{aligned}\sigma_{tx} &= 2.5\sigma\\ \sigma_{ty} &= 0.5\sigma\end{aligned} \tag{2.83}$$

Figure 2.12 indicatively represents the situation in correspondence of a hole on a sphere, on a spherical head, or at the center of a torospherical head under

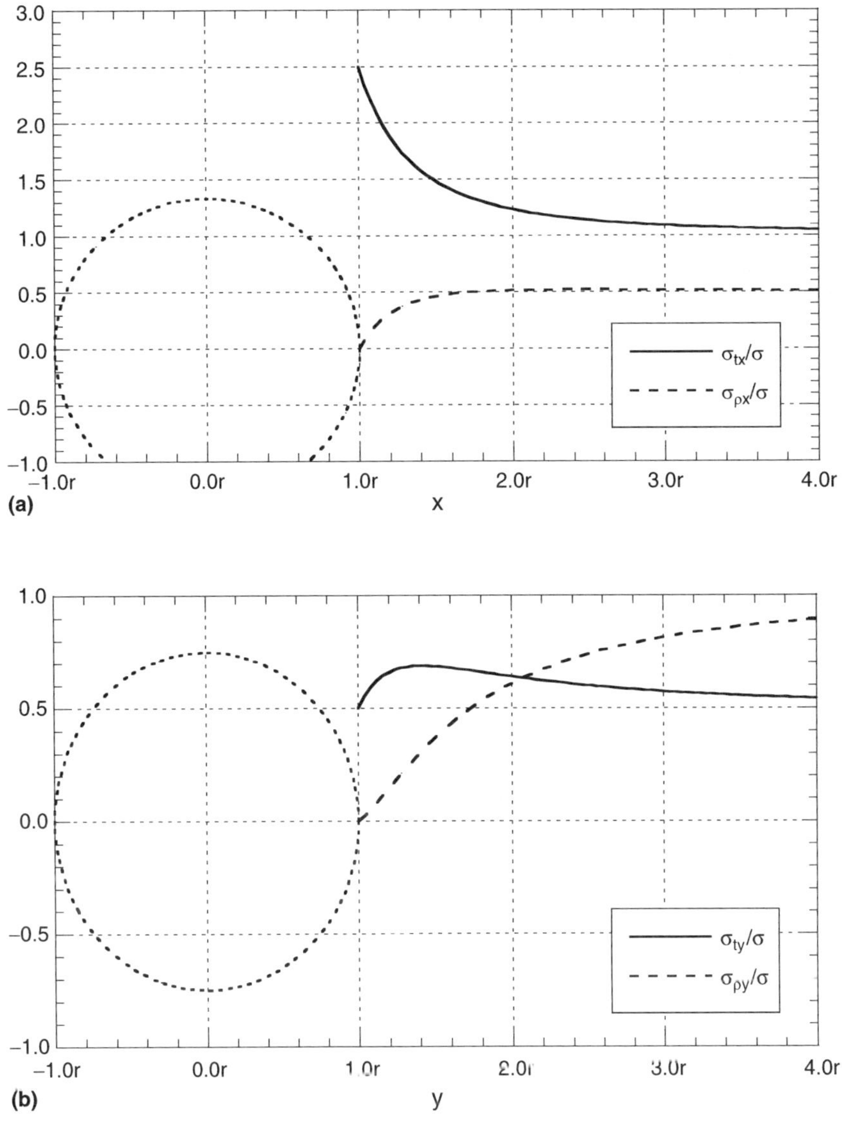

Fig. 2.15

pressure. In fact, the meridian and circumferential stresses are equal in the sphere, as we discussed in Sect. 2.1.

A stress peak equal to twice the circumferential or meridian stress in the sphere would be generated in correspondence of the hole.

Similarly, Fig. 2.14 may describe the situation in correspondence of a hole in a cylinder, with the y-axis along the circumference and the x-axis along the axis of the cylinder. In fact, as shown in Sect. 2.1, the longitudinal stress in the cylinder is equal to half the hoop stress. Therefore, a stress peak twice and a half the hoop stress in the cylinder would occur in correspondence of the hole.

From this point of view, there would be no difference between a hole with a small or large diameter compared to the diameter of the cylinder or the sphere, with respect to the amount of increase in membrane stress in correspondence of the hole.

This peak stress is a local membrane stress; according to the criteria shown in Sect. 1.5, it is necessary to limit its value to $1.5f$ that corresponds to the yield strength; therefore, we need to locally increase the thickness of the cylinder or the sphere, regardless of the diameter of the hole. Specifically, we observe that in the case of cylinder it would be necessary to increase the thickness and make it equal to 5/3 of the required thickness in the absence of holes. The hoop stress of reference σ would thus be equal to 3/5 of the basic allowable stress of the material. The value of the peak stress would therefore decrease to $2.5\frac{3}{5}f = 1.5f$, as required.

The size of the reinforced area may be quite small. We note that for $\rho = 1.5r$ the value of the peak is already lower than 1.5σ. Therefore, the reinforcement may be limited to a distance from the edge of the hole equal to $0.5r$, this notwithstanding the influence that local variations in thickness may exert on deformations and on the stress behavior. We would like to point out that by increasing the thickness of the cylinder by 66% for a width equal to $0.5r$, as required, the missing area due to the presence of the hole would not be compensated.

In fact, the reinforcement area is equal to $0.5r \times 0.66s = 0.33rs$ (where s is the thickness of the wall), while the corresponding area of half the hole is equal to rs. When we consider a sphere, the thickness should be locally equal to 4/3 of the cylinder's thickness, and the extent of the reinforcement should be equal to $0.5r$, as is the case for the cylinder. Its reinforcement area would be equal to $0.5r \times 0.33s = 0.165rs$, i.e., quite smaller than the corresponding area of half the hole. In the case of a hole on a cylinder, a sphere, or at the center of a torospherical head, we cannot disregard the curvature of the plate.

In fact, the peak at the hole's edge corresponds to a local increase in deformations that results in a local increase in radius greater than in the remaining component. The fibers around the hole go against such greater radial deflection in a similar way to what occurs in the case of edge effects discussed in Sect. 2.2. Moreover, if a nozzle is welded in correspondence of a hole (as is generally the case), the axial rigidity of the nipple counteracts the local increase in radius.

Experimental tests have shown that the area of the collaborating plate around the hole does not depend on its diameter, as the previous discussion about the flat plate would lead one to believe, but on the parameter $\sqrt{Ds}$, D being the diameter of the cylinder or the sphere, and s being its thickness. This bears similarity with the findings regarding edge effects.

In addition, the diameter of the hole strongly influences the entity of the peaks. In the case of a small hole, the phenomenon is localized to a limited area. The increase in radius (that goes with a local increase in hoop membrane stress) is counteracted by the surrounding areas, that maintain their rigidity with minor changes versus the component without holes. If the hole is large, the cylinder or the sphere turn out to be weaker, and at the edge of the hole, variations in radius that cause the stress peak are more likely.

Tests confirm this, and as we shall discuss in Chap. 8, we understand why the widely accepted calculation criterion consists of compensating for the area of the hole by correlating the thickness of the reinforcement to the diameter of the hole. It is also necessary to correlate the actual collaborating area of the reinforcement to the parameter $\sqrt{Ds}$, regardless of the diameter of the hole.

Finally, let us not forget that in pressure vessels the thrust corresponding to the hole itself is released on the edge of the hole. This generates peak primary stresses, partly of membrane type, partly of bending type. They relate to the parameter $\sqrt{Ds}$, as the edge effects discussed in Sect. 2.2.

There are also peaks of secondary stresses (see Sect. 1.5) on the outside or the inside fiber of substantial entity. The sum of primary and secondary stresses must not be greater than $3f$ (twice the yield strength). If we have real peak stresses, according to the classification in Sect. 1.5, this is of interest for a fatigue analysis only.

Finally, it may be of interest to examine the stress state that takes place at the edges of an elliptical opening in a flat plate. All due reservations are valid in this case with reference to the conclusions drawn about elliptical openings on a cylinder, a sphere or a dished head. Therefore, they only have some indicative value and may be of use mostly to highlight the qualitative differences of the stress state at the edges of an elliptical opening and at the edges of a circular hole.

Let us consider Fig. 2.16, where we have assumed that the stress σ in the plate acts parallel to the minor axis of the ellipsis (Case A), and that it acts parallel to the major axis (Case B).

By indicating with a and b the major and minor axis, respectively, and with y the axis along which the stress σ is present, we have, at the edge of the opening:

Case A

$$\begin{aligned} \sigma_{tx} &= \sigma\left(1 + 2\frac{a}{b}\right) \\ \sigma_{ty} &= -\sigma \end{aligned} \tag{2.84}$$

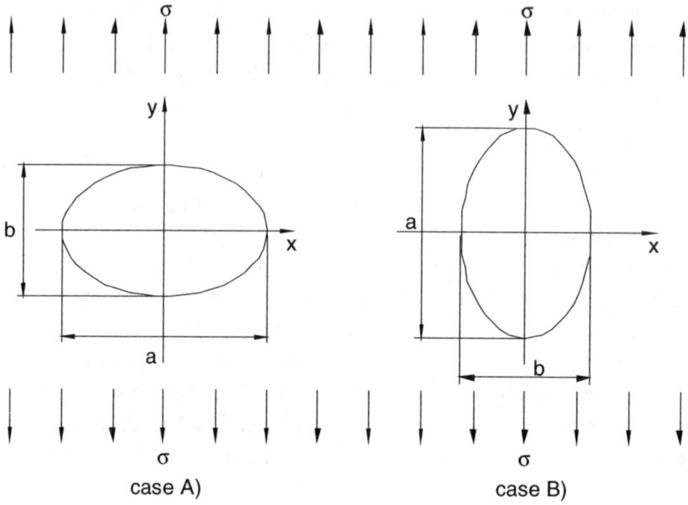

Fig. 2.16

Case B

$$\sigma_{tx} = \sigma\left(1 + 2\frac{b}{a}\right)$$
$$\sigma_{ty} = -\sigma \tag{2.85}$$

We now assume two equal stresses acting on the plate along x and y (Fig. 2.17). From (2.84) and (2.85) we have

Case A

$$\sigma_{tx} = 2\sigma\frac{a}{b}$$
$$\sigma_{ty} = 2\sigma\frac{b}{a} \tag{2.86}$$

Case B

$$\sigma_{tx} = 2\sigma\frac{b}{a}$$
$$\sigma_{ty} = 2\sigma\frac{a}{b} \tag{2.87}$$

If $a = b$ we refer to (2.80) about a round hole.

Let us consider Fig. 2.18. Stress σ is along y, while stress $\sigma/2$ is along x.

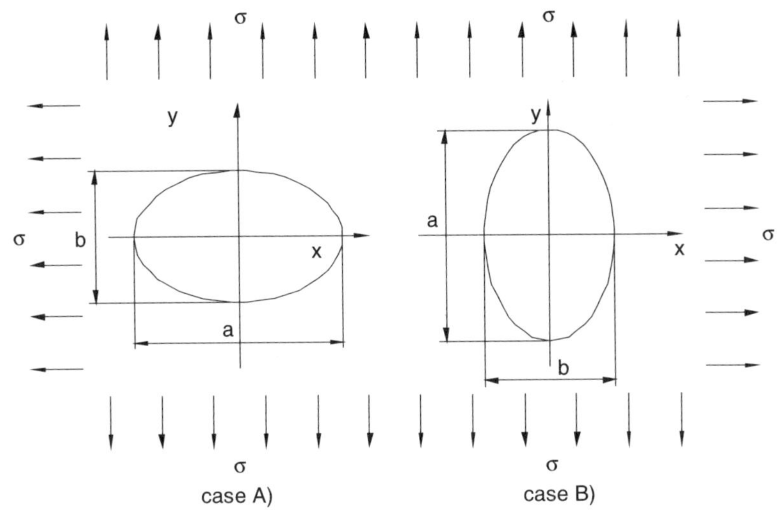

Fig. 2.17

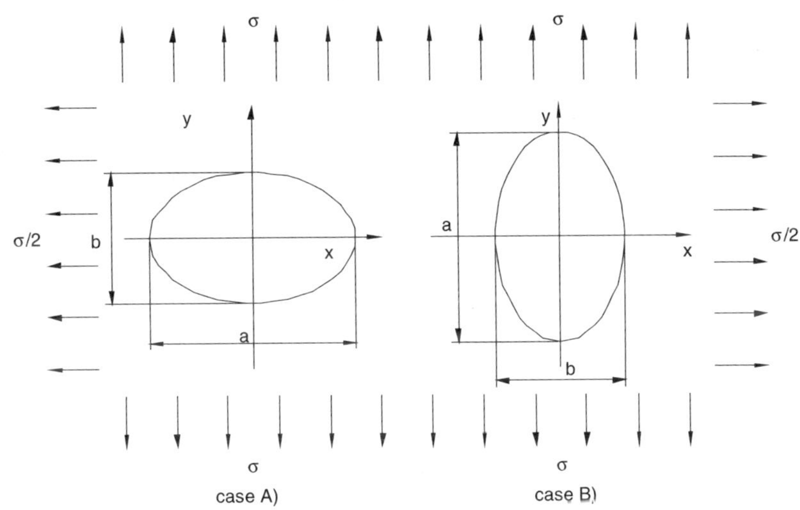

Fig. 2.18

Again from (2.84) and (2.85) we have:

Case A

$$\begin{aligned}\sigma_{tx} &= \sigma\left(1+2\frac{a}{b}\right)-\frac{\sigma}{2}=\sigma\left(\frac{1}{2}+2\frac{a}{b}\right)\\ \sigma_{ty} &= -\sigma+\frac{\sigma}{2}\left(1+2\frac{b}{a}\right)=\sigma\left(\frac{b}{a}-\frac{1}{2}\right)\end{aligned}\tag{2.88}$$

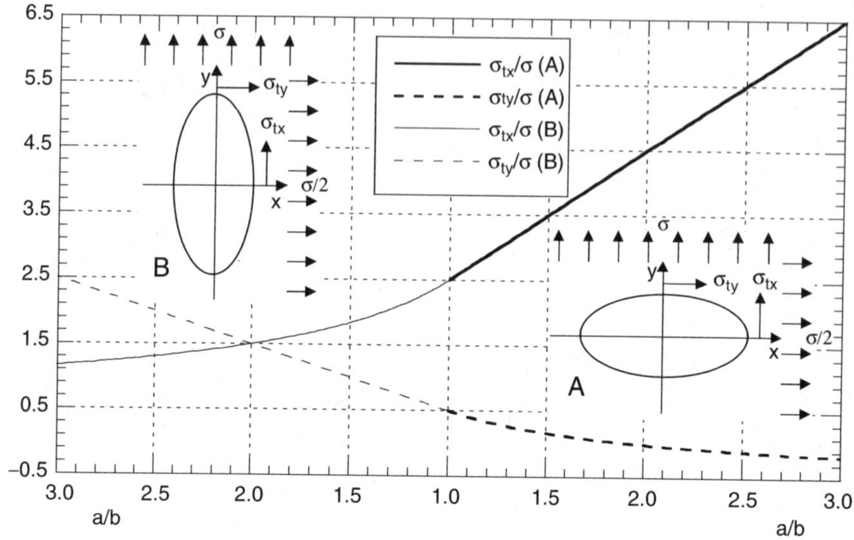

Fig. 2.19

Case B

$$\sigma_{tx} = \sigma\left(1 + 2\frac{b}{a}\right) - \frac{\sigma}{2} = \sigma\left(\frac{1}{2} + 2\frac{b}{a}\right)$$
$$\sigma_{ty} = -\sigma + \frac{\sigma}{2}\left(1 + 2\frac{a}{b}\right) = \sigma\left(\frac{a}{b} - \frac{1}{2}\right) \qquad (2.89)$$

In both cases, if $a = b$ we return to (2.83) about the round hole. Figure 2.19 shows the stresses in (2.88) and (2.89).

3

Cylinders Under Internal Pressure

3.1 General Design Criteria

Three principal stresses are generated by internal pressure: a hoop stress σ_t, a radial stress σ_r, and a longitudinal stress σ_a. The latter is due to the thrust of pressure on the heads of the cylinder. The value of the stresses σ_t and σ_r is not constant through the cylinder wall, whereas σ_a is in fact constant.

In the design phase it is therefore necessary to consider the stresses of the triaxial state and to derive the ideal stress via one of the theories of failure described in Sect. 1.3. Assuming that the ideal stress is equal to the basic allowable stress, we obtain an equation to compute the minimal required thickness. Before discussing the problem based on the above considerations, it may be useful to recall Mariotte's well-known method.

Let us consider the semicylinder of unitary length shown in Fig. 3.1.

The pressure resultant along x is

$$F = pD_i, \tag{3.1}$$

whereas it is obviously zero along y.

We must apply two equal forces equal to $F/2$ at the ends of the semicylinder to balance this thrust; if we assume that the hoop stress in the cylinder is constant through the thickness, we have:

$$\sigma_t = \frac{F/2}{s} = \frac{pD_i}{2s}, \tag{3.2}$$

where s is the thickness.

If the hoop stress is equal to the basic allowable stress f we obtain

$$s = \frac{pD_i}{2f}. \tag{3.3}$$

Equation (3.3) is Mariotte's formula, and it does not take into account the variation of σ_t through the thickness, as well as the presence of the other two

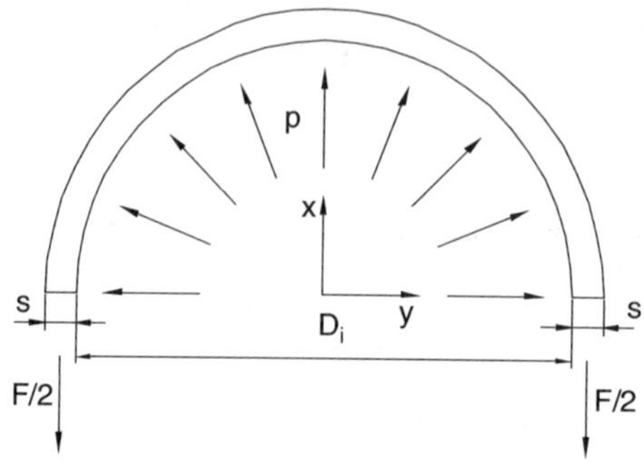

Fig. 3.1

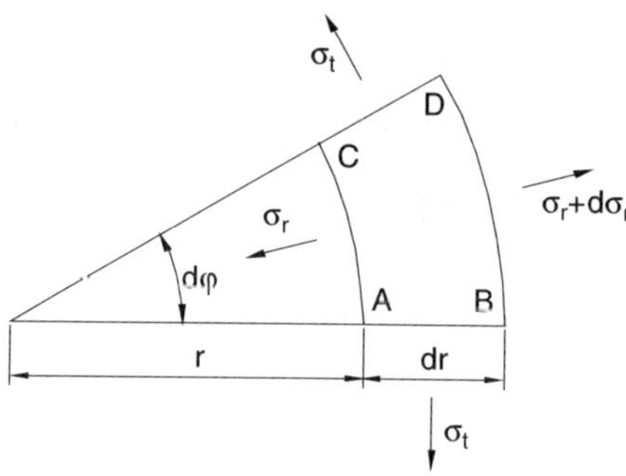

Fig. 3.2

principal stresses σ_r and σ_a; therefore, it cannot be used for the sizing of the cylinder.

Still, the simple procedure leading to (3.2) is nonetheless important. In fact, the value of σ_t obtained through this equation represents the hoop stress in a membrane; if the thickness is reduced to a point that it is possible to consider the cylinder as a membrane, σ_t is no longer a function of the radius (because the radius has in practice a unique value) and therefore the value of the hoop stress is constant. As we shall see, this value of the hoop stress is important because it corresponds to the average value of the hoop stress and does not depend on the thickness.

Let us now examine the generic element shown in Fig. 3.2. Its position is determined by the radius r, while its dimensions by dr and by the angle $d\varphi$. Let us assume that it has unitary dimension in the direction orthogonal to the figure.

The stress σ_t, constant along the circumference, is exercised on sides A–B and C–D; the equilibrium in this direction is therefore assured in any case. As far as the equilibrium in the radial direction, we note the presence of the force F_{AC} on the A–C side. We have

$$F_{AC} = \sigma_r r d\varphi. \tag{3.4}$$

Similarly, on the B–D side we have

$$F_{BD} = -(\sigma_r + d\sigma_r)(r + dr)d\varphi. \tag{3.5}$$

Neglecting the terms of higher order we have

$$F_{BD} = -\sigma_r r d\varphi - \sigma_r dr d\varphi - r d\sigma_r d\varphi. \tag{3.6}$$

Finally, on the sides A–B and C–D we have

$$F_{AB} = F_{CD} = \sigma_t dr. \tag{3.7}$$

The resultant based on such two forces in the radial direction is

$$F_{ABr} = 2F_{AB}\frac{d\varphi}{2} = \sigma_t dr d\varphi. \tag{3.8}$$

The equation of equilibrium with reference to the radius must be

$$F_{BD} + F_{AC} + F_{ABr} = 0. \tag{3.9}$$

According to the previous equations, with $d\varphi$ having a nonzero value, we obtain

$$-\sigma_r dr - r d\sigma_r + \sigma_t dr = 0. \tag{3.10}$$

Finally,

$$\sigma_t - \sigma_r - r\frac{d\sigma_r}{dr} = 0. \tag{3.11}$$

Equation (3.11) is the equilibrium equation of the cylinder. As far as the congruence of deformations, let us consider the circular ring of thickness dr shown in Fig. 3.3. Because of the circumferential deformation ε_t, the radius of circle α has an elongation Δr_α equal to

$$\Delta r_\alpha = \varepsilon_t r. \tag{3.12}$$

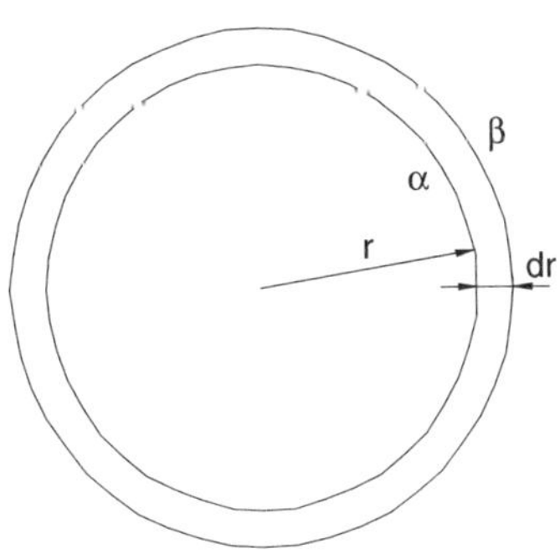

Fig. 3.3

The radius of circle β in turn has an elongation

$$\Delta r_\beta = (\varepsilon_t + d\varepsilon_t)(r + dr). \tag{3.13}$$

To impose congruence, the difference between these two elongations must correspond to the thickness' increment of the ring, i.e.,

$$\Delta s = \varepsilon_r dr. \tag{3.14}$$

In other words, it must be

$$\Delta r_\beta - \Delta r_\alpha = \Delta s. \tag{3.15}$$

From (3.12)–(3.14) we have

$$\varepsilon_r - \varepsilon_t - r\frac{d\varepsilon_t}{dr} = 0. \tag{3.16}$$

Equation (3.16) is the equation of congruence of the cylinder.

Let us remember that

$$\begin{aligned}
\varepsilon_t &= \frac{1}{E}\left[\sigma_t - \mu(\sigma_r + \sigma_a)\right] \\
\varepsilon_r &= \frac{1}{E}\left[\sigma_r - \mu(\sigma_t + \sigma_a)\right] \\
\varepsilon_a &= \frac{1}{E}\left[\sigma_a - \mu(\sigma_t + \sigma_r)\right]
\end{aligned} \tag{3.17}$$

ε_a being the longitudinal deformation, E the normal modulus of elasticity and μ Poisson's ratio. Applying (3.17), (3.16) becomes

$$\sigma_r - \sigma_t - r\frac{d\sigma_t}{dr} - \mu\left(\sigma_t - \sigma_r - r\frac{d\sigma_r}{dr}\right) + \mu r\frac{d\sigma_a}{dr} = 0. \tag{3.18}$$

This can also be written as

$$-(1+\mu)\left(\sigma_t - \sigma_r - r\frac{d\sigma_r}{dr}\right) - r\frac{d\sigma_r}{dr} - r\frac{d\sigma_t}{dr} + \mu r\frac{d\sigma_a}{dr} = 0. \tag{3.19}$$

Recalling (3.11), (3.19) becomes (due to the nonzero value of r)

$$\frac{d\sigma_r}{dr} + \frac{d\sigma_t}{dr} = \mu\frac{d\sigma_a}{dr}. \tag{3.20}$$

If we now derive the third equation in (3.17) we have

$$\frac{d\varepsilon_a}{dr} = \frac{1}{E}\left[\frac{d\sigma_a}{dr} - \mu\left(\frac{d\sigma_t}{dr} + \frac{d\sigma_r}{dr}\right)\right]; \tag{3.21}$$

and then (from (3.20))

$$\frac{d\varepsilon_a}{dr} = \frac{1-\mu^2}{E}\frac{d\sigma_a}{dr}. \tag{3.22}$$

For the sections of the cylinder to remain flat, ε_a must be constant and therefore the derivative of σ_a is zero, according to (3.22); the longitudinal stress is constant, as previously predicted; moreover, (3.20) simply becomes

$$\frac{d\sigma_r}{dr} + \frac{d\sigma_t}{dr} = 0. \tag{3.23}$$

If we now take the derivative of (3.11) we obtain

$$\frac{d\sigma_t}{dr} - \frac{d\sigma_r}{dr} - \frac{d\sigma_r}{dr} - r\frac{d^2\sigma_r}{dr^2} = 0. \tag{3.1.23 bis}$$

Finally, recalling (3.23),

$$\frac{d^2\sigma_r}{dr^2} = -\frac{3}{r}\frac{d\sigma_r}{dr}. \tag{3.24}$$

Assuming that

$$y = \frac{d\sigma_r}{dr}, \tag{3.25}$$

from (3.24) we obtain

$$y' = -\frac{3}{r}y. \tag{3.26}$$

This is a differential equation with separable variables; the general integral is

$$\int \frac{dy}{y} = -\int \frac{3}{r}dr + Z. \tag{3.27}$$

In other words,

$$\log_e y = -3\log_e r + Z = \log_e \frac{K}{r^3}, \tag{3.28}$$

with Z and K being constant. From (3.25) we have

$$\frac{d\sigma_r}{dr} = \frac{K}{r^3}. \tag{3.29}$$

Finally, by integrating we obtain

$$\sigma_r = A - \frac{B}{r^2}, \tag{3.30}$$

with A and B being constant. From (3.11) and (3.30) we also obtain

$$\sigma_t = A + \frac{B}{r^2}. \tag{3.31}$$

Note that if $r = r_e$, and r_e the external radius, we must have $\sigma_r = 0$, and therefore:

$$A = \frac{B}{{r_e}^2}; \tag{3.32}$$

then

$$\sigma_r = B\left(\frac{1}{{r_e}^2} - \frac{1}{r^2}\right). \tag{3.33}$$

On the other hand, if $r = r_i$, with r_i the inside radius, the radial compressive stress is equal to the pressure p, and therefore

$$-p = B\left(\frac{1}{{r_e}^2} - \frac{1}{{r_i}^2}\right), \tag{3.34}$$

and

$$B = \frac{p{r_e}^2}{\frac{{r_e}^2}{{r_i}^2} - 1}. \tag{3.35}$$

If we assume that

$$a = \frac{r_e}{r_i}, \tag{3.36}$$

then

$$\sigma_r = \frac{p}{a^2 - 1}\left(1 - \frac{{r_e}^2}{r^2}\right). \tag{3.37}$$

Similarly, from (3.31) one can easily determine that

$$\sigma_t = \frac{p}{a^2 - 1}\left(1 + \frac{{r_e}^2}{r^2}\right). \tag{3.38}$$

As far as σ_a which we know to be constant, and based on the thrust on the heads, one can easily determine its value

$$\sigma_a = \frac{\pi p {r_i}^2}{\pi({r_e}^2 - {r_i}^2)} = \frac{p}{a^2 - 1}. \tag{3.39}$$

To summarize, the three principal stresses are as follows:

$$\begin{aligned} \sigma_t &= \frac{p}{a^2 - 1}\left(1 + \frac{{r_e}^2}{r^2}\right) \\ \sigma_r &= \frac{p}{a^2 - 1}\left(1 - \frac{{r_e}^2}{r^2}\right) \\ \sigma_a &= \frac{p}{a^2 - 1} \end{aligned} \tag{3.40}$$

The equations shown in (3.40) are known as Lamè's equations.

Figures 3.4 and 3.5 illustrate the diagrams of the three principal stresses and deformations (computed from (3.17)), respectively; it is evident that σ_t reaches its maximum value in correspondence to the internal fiber. The same applies to the absolute value of σ_r, while the longitudinal stress has a value in between the two.

Therefore, the maximum value of the ideal stress, according to Guest, is to be found in correspondence to the internal fiber, and must be computed from σ_t and σ_r. From (3.40) we have

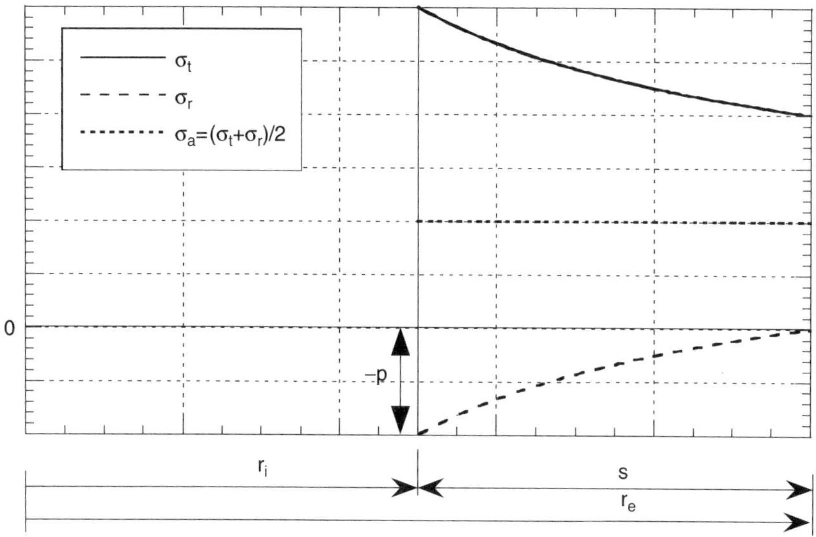

Fig. 3.4

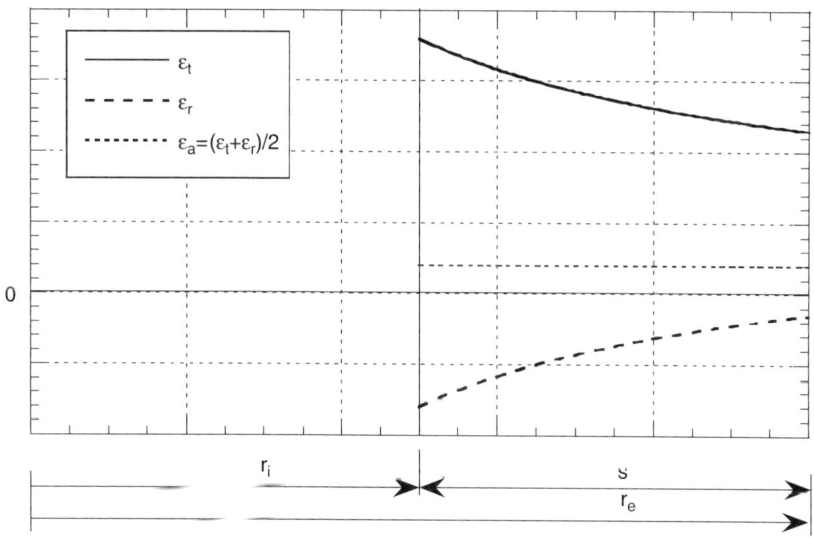

Fig. 3.5

$$\sigma_{ti} = p\frac{a^2+1}{a^2-1}, \text{ and } \sigma_{ri} = -p. \tag{3.41}$$

Therefore

$$\sigma_{id(i)} = \sigma_{ti} - \sigma_{ri} = \frac{2a^2}{a^2-1}. \tag{3.42}$$

If we were to design the cylinder according to Guest's theory of failure and traditional resistance criteria with regard to elasticity, we would set

$$p\frac{2a^2}{a^2-1}=f, \tag{3.43}$$

where f is the basic allowable stress.

From (3.43) we have

$$a=\sqrt{\frac{1}{1-2\frac{p}{f}}}. \tag{3.44}$$

Finally, with D_e being the outside diameter of the cylinder and s the thickness, we have

$$s=\frac{D_e}{2}\left(1-\sqrt{1-2\frac{p}{f}}\right). \tag{3.45}$$

Interestingly, if we use Huber–Hencky's theory of failure

$$s=\frac{D_e}{2}\left(1-\sqrt{1-\sqrt{3}\frac{p}{f}}\right). \tag{3.46}$$

Neither (3.45) nor (3.46) are used to size the cylinder.

As shown in Sect. 1.4, there is general consensus that one should not look at the peak stress located in the internal fiber during the design, but instead take into account the possibility that other less stressed fibers may cooperate with the latter.

According to this criterion, the calculation can be carried out in different ways; let us examine the first one, which accounts for plastic collaboration in a streamlined and global way.

If one downplays the danger determined by the maximum stress value, the danger is no longer linked to the yielding of the internal fiber but rather to the whole plastic flow of the cylinder, which takes place when the ideal stress is equal to the yield strength everywhere. In other words, following Guest's theory one must have everywhere:

$$\sigma_{id}=\sigma_t-\sigma_r=\sigma_s, \tag{3.47}$$

where σ_s is the yield strength. Therefore, the global plastic flow from (3.11) can be represented as follows:

$$\sigma_s-r\frac{d\sigma_r}{dr}=0. \tag{3.48}$$

Integrating the above equation results in

$$\sigma_r=\sigma_s\log_e r+A, \tag{3.49}$$

where A is constant.

Recalling that if $r = r_e$ then $\sigma_r = 0$, we have

$$A = -\sigma_s \log_e r_e, \tag{3.50}$$

and

$$\sigma_r = \sigma_s \log_e \frac{r}{r_e}. \tag{3.51}$$

On the other hand, for $r = r_i$ we have $\sigma_r = -p$, and therefore

$$p = \sigma_s \log_e \frac{r_e}{r_i} = \sigma_s \log_e a. \tag{3.52}$$

Finally,

$$a = \frac{r_e}{r_i} = e^{p/\sigma_s}. \tag{3.53}$$

Equation (3.53) means that given a certain pressure p and a material having a yield strength equal to σ_s, if the ratio between outside and inside radii is equal to the value given by (3.53), the cylinder is completely plasticized.

Of course, this approaches a danger condition and therefore a safety factor must be applied. The safety condition can be obtained by replacing σ_s with the allowable stress f. We then have

$$a = \frac{r_e}{r_i} = e^{p/f}. \tag{3.54}$$

Equation (3.54) lets us safely size the cylinder while taking into account the plastic collaboration. From (3.54)

$$s = \frac{e^{p/f} - 1}{2e^{p/f}} D_e, \tag{3.55}$$

which lets us compute the minimum required thickness.

If the cylinder is fully plasticized the radial stress is as in (3.51); σ_t can instead be computed using (3.47) that is

$$\sigma_t = \sigma_s \left(1 - \log_e \frac{r_e}{r}\right). \tag{3.56}$$

To determine instead the behavior of σ_a, which is no longer constant, we will make the following considerations.

Any stress state can be considered as the sum of a hydrostatic stress state and a stress deviator; the former is characterized by three equal principal stresses, as a function of the principal stresses in the state under investigation

$$\sigma'_I = \sigma'_{II} = \sigma'_{III} = \frac{\sigma_I + \sigma_{II} + \sigma_{III}}{3} = \sigma_m. \tag{3.57}$$

These stresses produce all the cubic dilations of the stress state $\sigma_I, \sigma_{II}, \sigma_{III}$, i.e., as can easily be ascertained

$$\varepsilon'_I + \varepsilon'_{II} + \varepsilon'_{III} = \varepsilon_I + \varepsilon_{II} + \varepsilon_{III}. \tag{3.58}$$

The stress deviator can be expressed by the following stresses:

$$\begin{aligned} \sigma_I'' &= \sigma_I - \frac{\sigma_I + \sigma_{II} + \sigma_{III}}{3} = \frac{2}{3}\sigma_I - \frac{\sigma_{II}}{3} - \frac{\sigma_{III}}{3} \\ \sigma_{II}'' &= \sigma_{II} - \frac{\sigma_I + \sigma_{II} + \sigma_{III}}{3} = \frac{2}{3}\sigma_{II} - \frac{\sigma_I}{3} - \frac{\sigma_{III}}{3} \\ \sigma_{III}'' &= \sigma_{III} - \frac{\sigma_I + \sigma_{II} + \sigma_{III}}{3} = \frac{2}{3}\sigma_{III} - \frac{\sigma_I}{3} - \frac{\sigma_{II}}{3} \end{aligned} \tag{3.59}$$

The stress deviator produces a deformation with a zero cubic dilation. When the material yields the deformations are characterized by a zero cubic deformation, i.e., they are proportional to the stress deviator. Examining the fully plasticized cylinder we can therefore write that

$$\begin{aligned} \varepsilon_t &= f(r)\left(\sigma_t - \frac{\sigma_r - \sigma_a}{2}\right) \\ \varepsilon_r &= f(r)\left(\sigma_r - \frac{\sigma_t - \sigma_a}{2}\right) \\ \varepsilon_a &= f(r)\left(\sigma_a - \frac{\sigma_t - \sigma_r}{2}\right) \end{aligned} \tag{3.60}$$

with $f(r)$ a function of the radius.

On the other hand, because the cubic dilation is zero, we have

$$\varepsilon_t + \varepsilon_r + \varepsilon_a = 0. \tag{3.61}$$

In order to keep the cylinder sections flat

$$\varepsilon_a = -2B, \tag{3.62}$$

where B is constant.

According to (3.61), (3.62), and recalling (3.6) (congruence equation)

$$r\frac{d\varepsilon_t}{dr} + 2\varepsilon_t - 2B = 0. \tag{3.63}$$

Integrating the above equation we obtain

$$\varepsilon_t = B + \frac{C}{r^2}, \tag{3.64}$$

where C is constant.

From (3.64), (3.61), and (3.62) we also have

$$\varepsilon_r = B - \frac{C}{r^2}. \tag{3.65}$$

From (3.64) and (3.65), and recalling (3.60) and (3.47) we finally have

$$\varepsilon_t - \varepsilon_r = \frac{3}{2}f(r)\sigma_s = 2\frac{C}{r^2}, \tag{3.66}$$

and

$$f(r) = \frac{4}{3}\frac{C}{\sigma_s r^2}. \tag{3.67}$$

Recalling the third equation of (3.60) together with (3.49), (3.56), and (3.62), we obtain

$$\varepsilon_a = \frac{4C}{3\sigma_s r^2}\left[\sigma_a - \sigma_s\left(\frac{1}{2} - \log_e \frac{r_e}{r_i}\right)\right] = -2B, \tag{3.68}$$

hence

$$\sigma_a = -\frac{3B\sigma_s r^2}{2C} + \sigma_s\left(\frac{1}{2} - \log_e \frac{r_e}{r}\right). \tag{3.69}$$

To achieve equilibrium in the longitudinal direction

$$2\pi\int_{r_i}^{r_e} \sigma_a r dr = \pi r_i^2 p, \tag{3.70}$$

and after integrating

$$-\frac{3\pi B\sigma_s}{4C}(r_e^4 - r_i^4) + \pi r_i^2 \sigma_s \log_e \frac{r_e}{r_i} = \pi r_i^2 p. \tag{3.71}$$

On the other hand, according to (3.53)

$$\pi r_i^2 p = \pi r_i^2 \sigma_s \log_e \frac{r_e}{r_i}. \tag{3.72}$$

We can conclude that $B = 0$ (see (3.71) and (3.72)), and based on (3.69) we have

$$\sigma_a = \sigma_s\left(\frac{1}{2} - \log_e \frac{r_e}{r}\right). \tag{3.73}$$

Because $B = 0$ and using (3.64) and (3.65), we determine the following system of stresses and deformations in a fully plasticized cylinder:

$$\begin{aligned}
\sigma_t &= \sigma_s\left(1 - \log_e \frac{r_e}{r}\right)\\
\sigma_r &= -\sigma_s \log_e \frac{r_e}{r}\\
\sigma_a &= \sigma_s\left(\frac{1}{2} - \log_e \frac{r_e}{r}\right)\\
\varepsilon_t &= \frac{C}{r^2}\\
\varepsilon_r &= -\frac{C}{r^2}\\
\varepsilon_a &= 0
\end{aligned} \tag{3.74}$$

Figures 3.6 and 3.7 show the stresses and deformations as in (3.74). Note that the behavior of σ_t and σ_a is completely different in the plastic condition compared to the elastic condition (Fig. 3.4); conversely, the behavior of

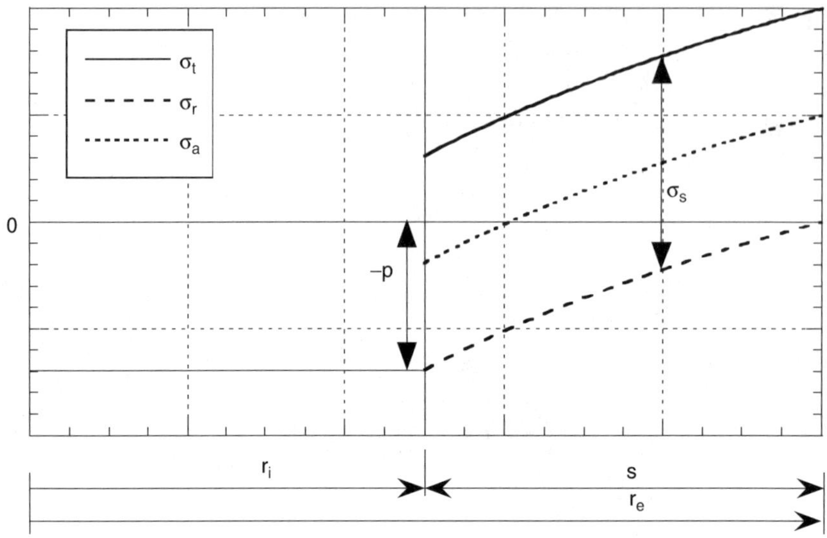

Fig. 3.6

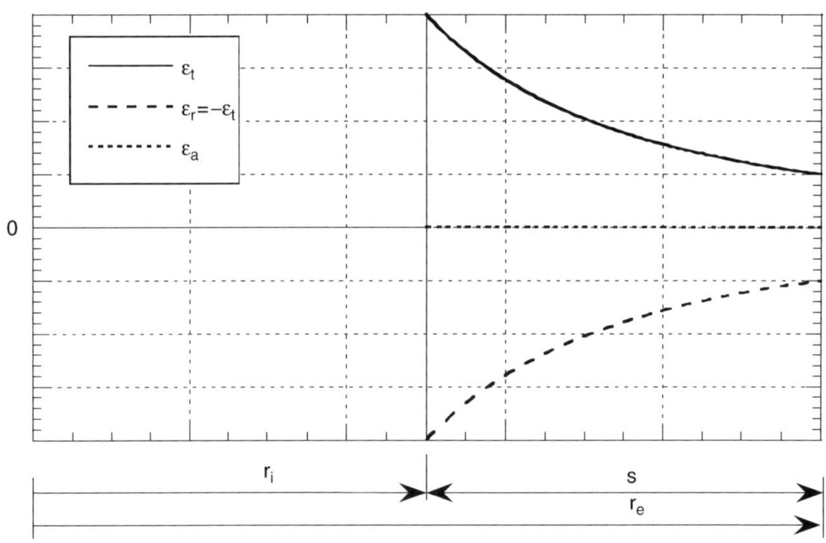

Fig. 3.7

the deformations is "qualitatively" similar to the one in the elastic condition (Fig. 3.5).

We mentioned earlier that (3.53) may also be obtained in a different way. Recalling (3.51) and (3.52), the second equation of (3.74) can be rewritten as

$$\sigma_r = \frac{p}{\log_e a} \log_e \frac{r}{r_e}. \tag{3.75}$$

Its average value is

$$\sigma_{rm} = \frac{p}{\log_e a(r_e - r_i)} \int_{r_i}^{r_e} \log_e \frac{r}{r_e} dr. \tag{3.76}$$

Resolving the integral we have

$$\sigma_{rm} = p\left(\frac{1}{a-1} - \frac{1}{\log_e a}\right). \tag{3.77}$$

The average value of σ_t is simply the membrane stress σ mentioned at the beginning; thus, recalling (3.2) we have

$$\sigma_{tm} = \frac{pD_i}{2s} = \frac{pr_i}{r_e - r_i} = \frac{p}{a-1}. \tag{3.78}$$

According to Guest, considering the ideal stress relative to the average values we have

$$\sigma_{id(m)} = \sigma_{tm} - \sigma_{rm} = \frac{p}{\log_e a}. \tag{3.79}$$

Assuming that $\sigma_{id(m)} = f$, we go back to (3.53) from (3.79).

We can obtain the equation that includes plastic collaboration by computing the ideal stress based on the average values of σ_t and σ_r instead of considering peak values, as we would do according to the traditional verification in the elastic field. As mentioned earlier in Sect. 1.4, this type of computation has general validity, once we can find those membrane stresses that determine the resistance of the vessel.

Finally, note that by adopting the Huber–Hencky theory of failure, we would reach the following sizing equation (instead of (3.55))

$$s = \frac{e^{\frac{\sqrt{3}}{2}\frac{p}{f}} - 1}{2e^{\frac{\sqrt{3}}{2}\frac{p}{f}}} D_e. \tag{3.80}$$

Equation (3.55) can be replaced by

$$s = \frac{pD_e}{2f + \left(1 + 0.15\frac{p}{f}\right)p}. \tag{3.81}$$

Equations (3.55) and (3.81) practically coincide; adopting (3.81) instead of (3.55) causes overestimation of the thickness equal to 0.03% for $p/f = 0.2$; it equals 0.10% for $p/f = 0.4$, and it equals 0.21% for $p/f = 0.6$; note that, for example, with $f = 50\,\mathrm{N\,mm^{-2}}$ and $p/f = 0.6$, the pressure p is equal to $30\,\mathrm{N\,mm^{-2}}(= 300\,\mathrm{bar})$.

If we assume as average value of radial stress σ_r the average between the extreme values (for $r = r_i$ and $r = r_e$) instead of the value obtained from (3.7), we have

$$\sigma_{rm} = -\frac{p}{2}. \tag{3.82}$$

Recalling (3.78) the ideal average stress that represents the general membrane stress according to Sect. 1.5, will be given by

$$\sigma_{id(m)} = \frac{p}{a-1} + \frac{p}{2} = \frac{p}{2}\frac{a+1}{a-1}. \tag{3.83}$$

With $\sigma_{id(m)} = f$ and referring to (3.83), we obtain

$$\frac{a+1}{a-1} = \frac{\dfrac{r_e}{r_i}+1}{\dfrac{r_e}{r_i}-1} = \frac{r_e + r_i}{r_e - r_i} = \frac{2f}{p}. \tag{3.84}$$

From (3.84) we obtain the following equation, where D_m is the average diameter:

$$\frac{D_m}{s} = \frac{2f}{p}, \tag{3.85}$$

and

$$s = \frac{pD_m}{2f}. \tag{3.86}$$

Equation (3.86) is the so-called average diameter equation; let us compare it with Mariotte's (3.3); with regard to the latter, the average diameter substitutes the inside one.

Equation (3.86) can also be rewritten as follows: from (3.85), and considering the outside diameter

$$p(D_e - s) = 2fs, \tag{3.87}$$

and hence

$$s = \frac{pD_e}{2f + p}. \tag{3.88}$$

Equation (3.88) is used in many national codes and also in the ISO Code; a comparison with (3.81) shows that they differ in the term between parentheses absent in (3.88).

Equation (3.88) is more conservative than (3.81), and therefore definitely acceptable; it is easier to use and results in walls only slightly thicker than strictly required. Figure 3.8 summarizes the various curves from the different equations that have been obtained thus far. Until now we have implicitly assumed that the cylinder does not have holes (or, as we shall see later on, that the holes are completely compensated for) and that welding does not cause a reduction of the cylinder's resistance in the seams.

These situations are practically possible, yet impossible to generalize. Even though we shall examine the influence of holes in Chap. 8, we point out that welded joints can generally be less resistant compared to the basic metal that constitutes the cylinder. We therefore introduce a weld joint efficiency defined as z; z identifies the allowable stress at the weld joint as a function of the allowable stress of the basic metal through the relationship

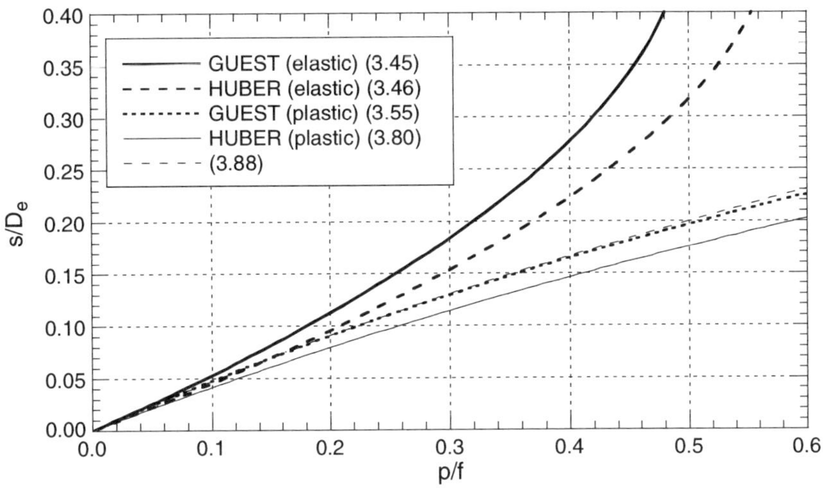

Fig. 3.8

$$f' = fz, \tag{3.89}$$

where f' is the allowable stress in the weld joint, and z is always less than or equal to one. Therefore, (3.88) can be written in more general terms as

$$s = \frac{pD_e}{2fz + p}. \tag{3.90}$$

3.2 Thick Cylinders

The sizing of the cylinder carried out with (3.88) (or (3.90)) is based, as discussed in Sect. 3.1, on a methodology that utilizes average values of stresses, while neglecting the peaks occurring in correspondence to the internal fiber.

The higher the ratio between thickness and outside diameter or, similarly, the ratio between the outside and the inside radius, the higher the value of such peaks. To appreciate this, one has to consider the ratio between the ideal maximum stress, as indicated in (3.42), and the ideal average (approximated) stress given by (3.83).

Such ratio, that we indicate as γ, is

$$\gamma = \frac{p\dfrac{2a^2}{a^2-1}}{\dfrac{p}{2}\dfrac{a+1}{a-1}} = 4\frac{a^2}{(a+1)^2}, \tag{3.91}$$

where a is the ratio between outside and inside radius of the cylinder. For instance, for $a = 1.2$ we have $\gamma = 1.19$; for $a = 1.5$ we have $\gamma = 1.44$; for $a = 2$ we have $\gamma = 1.78$.

If the value of γ is equal to or greater than the safety factor adopted to obtain the basic allowable stress f, the yield strength or the unitary load that causes creep rupture after 100,000 h is clearly reached or surpassed, depending on whether for the value f one or the other of the two stresses mentioned above be determinant.

In other words, in the case of very thick cylinders (3.88) may lead to partial plastic flow in correspondence to the internal fiber. The phenomenon may not be of concern if the working conditions of the cylinder and its geometric characteristics are such to rule out further stress peaks (absence of holes, absent or reduced stresses due to thermal flux, absence of external forces, and so on).

In Sect. 3.6 we will return to this topic and show how it is possible to adopt (3.88) even in the case of extremely high values for the ratio between outside and inside radius if the cylinder has no holes, and if the stresses due to thermal flux are absent or smaller than certain values that depend on both the ratio mentioned above and the direction of the flux.

Generally, in the case of particularly thick cylinders it is necessary, or at least advisable, to perform a direct analysis of the stresses through a finite element analysis programs or through experimental research. Lacking such investigations, it is advisable to limit the maximum value of the ideal stress due to pressure in correspondence to the internal fiber below the value σ_s, or $\sigma_{R/100000/t}$. Considering that the safety factor is typically assumed to be equal to 1.5, recalling (3.42) we must write that at the limit

$$p\frac{2a^2}{a^2-1} = 1.5f, \tag{3.92}$$

where $1.5f$ is equal to the yield strength σ_s or the creep rupture stress $\sigma_{R/100000/t}$.

From (3.92) we derive

$$\frac{1}{a} = \sqrt{\frac{f-1.33p}{f}}. \tag{3.93}$$

Recalling the meaning of a, and through steps that are not shown here, we obtain the following equation (where s is the thickness):

$$s = \left(1 - \sqrt{\frac{f-1.33p}{f}}\right)\frac{D_e}{2}. \tag{3.94}$$

The latter that may be adopted, according to elementary criteria, for sizing of very thick cylinders, is valid only insofar the resulting value of the thickness is greater than the one obtained through (3.88).

If we consider the weld joint efficiency lower than one, (3.94) then becomes

$$s = \left(1 - \sqrt{\frac{fz-1.33p}{fz}}\right)\frac{D_e}{2}. \tag{3.95}$$

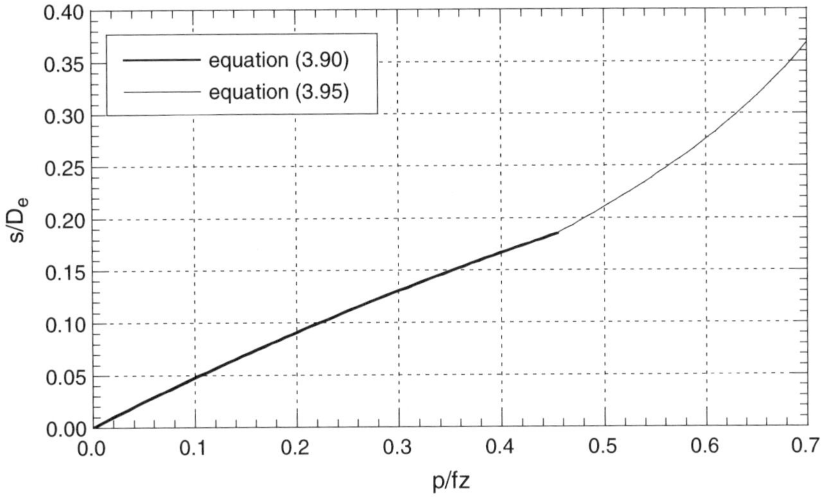

Fig. 3.9

Figure 3.9 shows the two curves representative of both (3.90) and (3.95) in order to highlight the respective range of validity.

When we examine the two equations above, we observe that (3.95) has operational validity when one of the following conditions holds:

$$p/fz > 0.45$$
$$s/D_e > 0.1825$$
$$a > 1.58$$

3.3 Thermal Stresses

If the wall of the cylinder is subjected to a centrifugal or centripetal thermal flux, stresses are created that re-establish the congruence of deformations; the congruence is, however, perturbed by thermal dilations that vary throughout the wall. In fact, due to the heat transmission, the temperature of the metal is a function of the radius.

The law of variations in temperature can be found via the following considerations.

Let us examine the cylinder of unitary length shown in Fig. 3.10.

The generic surface S corresponding to radius r is

$$S = 2\pi r. \tag{3.96}$$

The heat q traversing this surface is

$$q = \lambda S \frac{dt}{dr}, \tag{3.97}$$

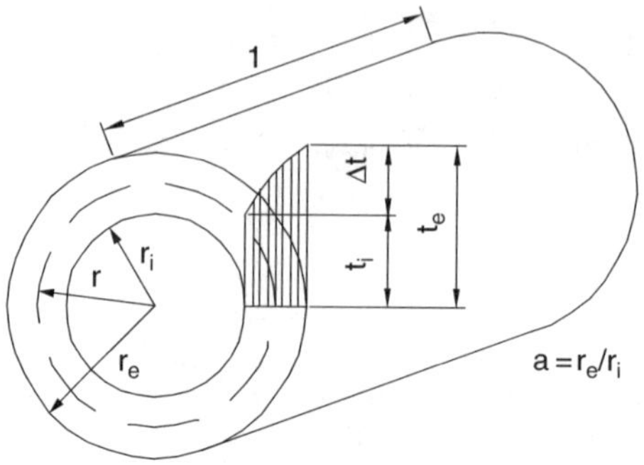

Fig. 3.10

where dt/dr is the temperature gradient and λ is the thermal conductivity coefficient of the material. Recalling (3.96) we have

$$dt = \frac{q}{2\pi\lambda}\frac{dr}{r}. \tag{3.98}$$

Therefore, by integrating we obtain

$$t = \frac{q}{2\pi\lambda}\log_e r + C, \tag{3.99}$$

C being a constant. By indicating with t_e and t_i the temperatures corresponding to r_e and r_i, respectively, and with Δt their difference, we have

$$\Delta t = t_e - t_i = \frac{q}{2\pi\lambda}\log_e \frac{r_e}{r_i}. \tag{3.100}$$

As usual, a will be the ratio between outside and inside radius; from (3.100) and (3.98) we obtain

$$\frac{dt}{dr} = \frac{\Delta t}{r\log_e a}. \tag{3.101}$$

The expression of the gradient in (3.101) will be useful below. To compute the stresses on the cylinder due to the effect of Δt, let us recall the equations of equilibrium and congruence shown earlier ((3.11) and (3.16)), that we rewrite as

$$\sigma_t - \sigma_r - r\frac{d\sigma_r}{dr} = 0 \tag{3.102}$$

and

$$\varepsilon_r - \varepsilon_t - r\frac{d\varepsilon_t}{dr} = 0. \tag{3.103}$$

Instead of the equations in (3.17), we now have

$$\begin{aligned} \varepsilon_t &= \frac{1}{E}\left[\sigma_t - \mu\left(\sigma_r + \sigma_a\right)\right] + \alpha t \\ \varepsilon_r &= \frac{1}{E}\left[\sigma_r - \mu\left(\sigma_t + \sigma_a\right)\right] + \alpha t \\ \varepsilon_a &= \frac{1}{E}\left[\sigma_a - \mu\left(\sigma_t + \sigma_r\right)\right] + \alpha t \end{aligned} \tag{3.104}$$

where α is the thermal expansion coefficient of the material. Based on (3.103),

$$-(1+\mu)\left(\sigma_t-\sigma_r-r\frac{d\sigma_r}{dr}\right)-r\frac{d\sigma_t}{dr}+\mu r\frac{d\sigma_a}{dr}-r\frac{d\sigma_r}{dr}-Er\alpha\frac{dt}{dr}=0. \quad (3.105)$$

By recalling (3.102) the latter now becomes

$$\frac{d\left(\sigma_t+\sigma_r\right)}{dr}=\mu\frac{d\sigma_a}{dr}-E\alpha\frac{dt}{dr}. \quad (3.106)$$

We derive the third equation of (3.104), that is

$$\frac{d\varepsilon_a}{dr}=\frac{1}{E}\left(\frac{d\sigma_a}{dr}-\mu\frac{d\left(\sigma_t+\sigma_r\right)}{dr}\right)+\alpha\frac{dt}{dr}. \quad (3.107)$$

With the constraint that the sections of the cylinder orthogonal to the axis remain flat $d\varepsilon_a/dr=0$; therefore, we obtain from (3.107)

$$\frac{d\sigma_a}{dr}=\mu\frac{d\left(\sigma_t+\sigma_r\right)}{dr}-E\alpha\frac{dt}{dr}. \quad (3.108)$$

With reference to (3.106) we then have

$$\frac{d\left(\sigma_t+\sigma_r\right)}{dr}=\mu^2\frac{d\left(\sigma_t+\sigma_r\right)}{dr}-E\alpha\mu\frac{dt}{dr}-E\alpha\frac{dt}{dr}. \quad (3.109)$$

Finally,

$$\frac{d\left(\sigma_t+\sigma_r\right)}{dr}=-\frac{E\alpha}{1-\mu}\frac{dt}{dr}. \quad (3.110)$$

From (3.101) and with

$$K=\frac{E\alpha\Delta t}{2\left(1-\mu\right)}, \quad (3.111)$$

(3.110) becomes

$$\frac{d\left(\sigma_t+\sigma_r\right)}{dr}=-\frac{2K}{r\log_e a}; \quad (3.112)$$

and the integral holds

$$\sigma_t+\sigma_r=\frac{2K}{\log_e a}\log_e r+2A, \quad (3.113)$$

where A is constant. Moreover,

$$\sigma_t=-\frac{2K}{\log_e a}\log_e r+2A-\sigma_r. \quad (3.114)$$

Based on (3.102) we have

$$2\sigma_r-r\frac{d\sigma_r}{dr}+\frac{2K}{\log_e a}\log_e r-2A=0. \quad (3.115)$$

The general integral in (3.115) is

$$\sigma_r = A - \frac{B}{r^2} + \frac{K}{2\log_e a}(1 - 2\log_e r), \tag{3.116}$$

where B is constant. For $r = r_i$ and for $r = r_e$, we must have $\sigma_r = 0$, and therefore

$$\begin{aligned} A - \frac{B}{r_e^2} + \frac{K}{2\log_e a}(1 - 2\log_e r_e) = 0 \\ A - \frac{B}{r_i^2} + \frac{K}{2\log_e a}(1 - 2\log_e r_i) = 0 \end{aligned} \tag{3.117}$$

By subtracting the second equation of (3.117) from the first one, we have

$$B\left(\frac{1}{r_e^2} - \frac{1}{r_i^2}\right) + K = 0; \tag{3.118}$$

that is

$$B = -\frac{K}{\dfrac{1}{r_e^2} - \dfrac{1}{r_i^2}} = \frac{K r_e^2}{a^2 - 1}. \tag{3.119}$$

Furthermore,

$$A = K\frac{a^2}{a^2 - 1} - \frac{K}{2\log_e a}(1 - 2\log_e r_i). \tag{3.120}$$

Finally, from (3.116),

$$\sigma_r = K\left[\frac{a^2 - \left(\dfrac{r_e}{r}\right)^2}{a^2 - 1} - \frac{\log_e \dfrac{r}{r_i}}{\log_e a}\right]. \tag{3.121}$$

Based on (3.114)

$$\sigma_t = K\left[\frac{2a^2}{a^2 - 1} - \frac{1 - 2\log_e r_i}{\log_e a} - \frac{2\log_e r}{\log_e a} - \frac{a^2 - \left(\dfrac{r_e}{r}\right)^2}{a^2 - 1} + \frac{\log_e \frac{r}{r_i}}{\log_e a}\right], \tag{3.122}$$

from which through several steps we obtain

$$\sigma_t = K\left[\frac{a^2 + \left(\dfrac{r_e}{r}\right)^2}{a^2 - 1} - \frac{1 + \log_e \frac{r}{r_i}}{\log_e a}\right]. \tag{3.123}$$

As far as σ_a, let us recall (3.108) and (3.110) that lead to

$$\frac{d\sigma_a}{dr} = -\frac{1}{1-\mu} E\alpha \frac{dt}{dr} = -\frac{2K}{2\log_e a}. \tag{3.124}$$

By integrating,

$$\sigma_a = -\frac{2K}{\log_e a} \log_e r + KC, \tag{3.125}$$

where C is constant. On the other hand, the resultant of σ_a must be zero given the absence of external forces. Therefore,

$$2\pi \int_{r_i}^{r_e} \sigma_a r dr = 2\pi K \int_{r_i}^{r_e} \left(Cr - \frac{2r \log_e r}{\log_e a} \right) dr = 0. \tag{3.126}$$

The integral has the following equation:

$$\int_{r_i}^{r_e} \left(Cr - \frac{2r \log_e r}{\log_e a} \right) dr = \left| \frac{Cr^2}{2} - \frac{r^2 \log_e r - \frac{1}{2} r^2}{\log_e a} \right|_{r_i}^{r_e}; \tag{3.127}$$

thus

$$\frac{C}{2} \left(r_e^2 - r_i^2 \right) = \frac{r_e^2 \log_e r_e - r_i^2 \log_e r_i - \frac{r_e^2}{2} + \frac{r_i^2}{2}}{\log_e a}; \tag{3.128}$$

then

$$C = \frac{2r_e^2 \log_e r_e - 2r_i^2 \log_e r_i - r_e^2 + r_i^2}{\left(r_e^2 - r_i^2 \right) \log_e a}. \tag{3.129}$$

Equation (3.129) may also be written as follows:

$$C = \frac{2a^2 \log_e a + \left(a^2 - 1 \right) \left(2 \log_e r_i - 1 \right)}{\left(a^2 - 1 \right) \log_e a}. \tag{3.130}$$

Finally,

$$C = \frac{2a^2}{a^2 - 1} + \frac{2 \log_e r_i - 1}{\log_e a}. \tag{3.131}$$

Therefore, based on (3.125)

$$\sigma_a = K \left(\frac{2a^2}{a^2 - 1} - \frac{2 \log_e \frac{r}{r_i} + 1}{\log_e a} \right), \tag{3.132}$$

and it is easy to verify that

$$\sigma_a = \sigma_t + \sigma_r. \tag{3.133}$$

By recalling the meaning of K, we finally obtain the following equations for σ_t, σ_a and σ_r:

$$\sigma_t = \frac{E\alpha\Delta t}{2(1-\mu)}\left[\frac{a^2+\left(\frac{r_e}{r}\right)^2}{a^2-1} - \frac{1+\log_e\frac{r}{r_i}}{\log_e a}\right], \tag{3.134}$$

$$\sigma_a = \frac{E\alpha\Delta t}{2(1-\mu)}\left[\frac{2a^2}{a^2-1} - \frac{1+2\log_e\dfrac{r}{r_i}}{\log_e a}\right], \tag{3.135}$$

and

$$\sigma_r = \frac{E\alpha\Delta t}{2(1-\mu)}\left[\frac{a^2-\left(\frac{r_e}{r}\right)^2}{a^2-1} - \frac{\log_e\frac{r}{r_i}}{\log_e a}\right]. \tag{3.136}$$

Specifically, in correspondence to the internal fiber ($r = r_i$)

$$\begin{aligned}\sigma_{ti} = \sigma_{ai} &= \frac{E\alpha\Delta t}{2(1-\mu)}\left(\frac{2a^2}{a^2-1} - \frac{1}{\log_e a}\right). \\ \sigma_{ri} &= 0\end{aligned} \tag{3.137}$$

In correspondence to the external fiber ($r = r_e$)

$$\begin{aligned}\sigma_{te} = \sigma_{ae} &= \frac{E\alpha\Delta t}{2(1-\mu)}\left(\frac{2}{a^2-1} - \frac{1}{\log_e a}\right). \\ \sigma_{re} &= 0\end{aligned} \tag{3.138}$$

Figure 3.11 shows the three stresses assuming that Δt is positive. Based on (3.100), this means that $t_e > t_i$, and that the thermal flux is centripetal. Of course, the sign of the stresses changes when the thermal flux is centrifugal.

We note that the maximum positive values of σ_t and σ_a occur in correspondence to the internal fiber if the flux is centripetal, and in correspondence to the external fiber if the flux is centrifugal. The simultaneous presence of internal pressure and thermal flux produces the maximum values for σ_t in correspondence to the internal fiber in the case of centripetal flux, and of the external fiber in the case of centrifugal flux, respectively. From this point of view, the centripetal thermal flux is more dangerous, because for the same Δt the σ_{ti} caused by both pressure and centripetal flux is greater than the σ_{te} caused by both the pressure and the centrifugal flux.

In fact, recalling Sect. 3.1 in the case of centripetal flux, we have a total of

$$\sigma_{ti} = p\frac{a^2+1}{a^2-1} + \frac{E\alpha\Delta t}{2(1-\mu)}\left(\frac{2a^2}{a^2-1} - \frac{1}{\log_e a}\right), \tag{3.139}$$

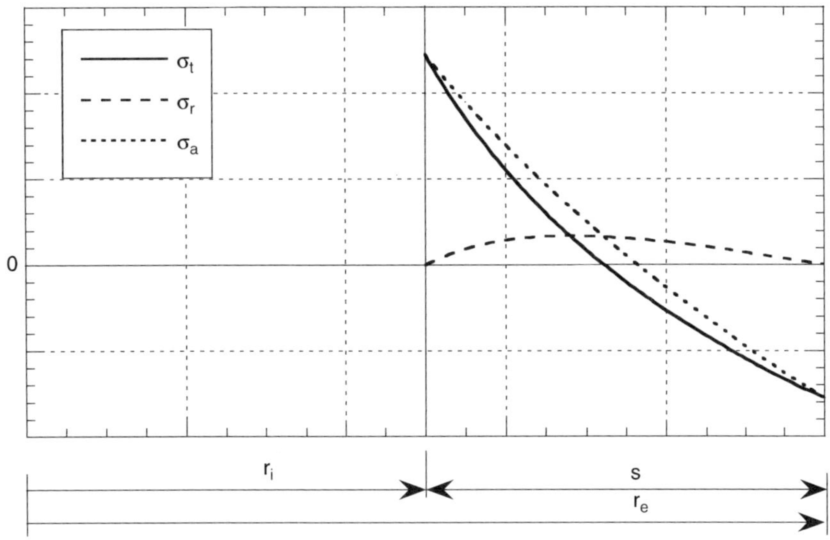

Fig. 3.11

whereas in the case of centrifugal flux we have a total of

$$\sigma_{te} = p\frac{2}{a^2 - 1} + \frac{E\alpha\Delta t}{2\,(1-\mu)}\left(\frac{2}{a^2-1} - \frac{1}{\log_e a}\right), \tag{3.140}$$

and Δt is negative in this case. If

$$\gamma = \pm\frac{E\alpha\Delta t}{2p\,(1-\mu)} \tag{3.141}$$

with the positive sign for the centripetal flux and the negative sign for the centrifugal sign, both (3.139) and (3.140) can be written as follows:

$$\frac{\sigma_{ti}}{p} = \frac{a^2+1}{a^2-1} + \gamma\left(\frac{2a^2}{a^2-1} - \frac{1}{\log_e a}\right), \tag{3.142}$$

and

$$\frac{\sigma_{te}}{p} = \frac{2}{a^2-1} - \gamma\left(\frac{2}{a^2-1} - \frac{1}{\log_e a}\right). \tag{3.143}$$

The values of σ_{ti}/p and σ_{te}/p derived from the equations above are graphically depicted in Figs. 3.12 and 3.13.

The values for σ_{ti} and σ_{te} for the two types of flux show the greatest differences for large values of a (i.e., the ratio between outside and inside radius), as well as for γ. Some portions of the curves have been dashed in Fig. 3.13, since the stress σ_{te} does not show the highest absolute value under certain conditions, which means that it is not significant as far as the identification of the

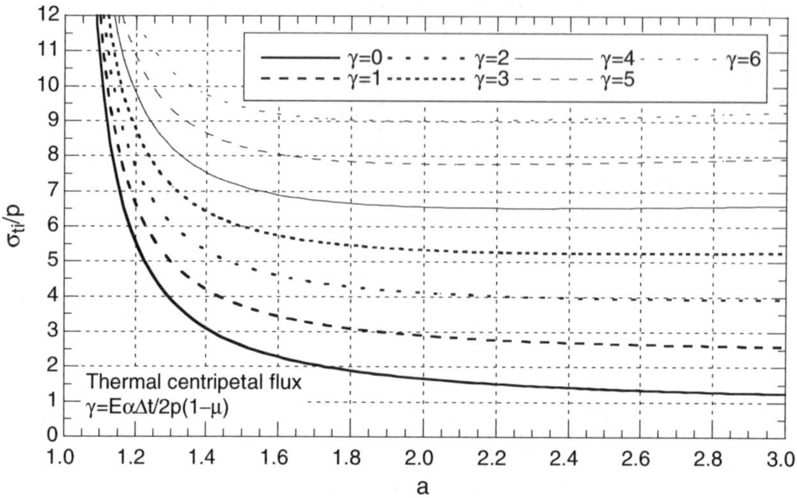

Fig. 3.12

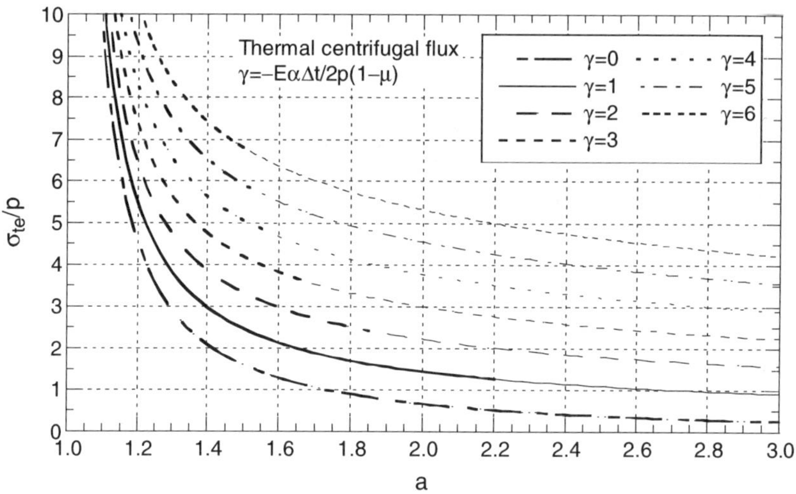

Fig. 3.13

most dangerous stress. Indeed, it may be the case that the longitudinal stress with negative sign is greater in absolute value than σ_{te} in correspondence to the internal fiber.

Let us examine the conditions required for this to occur.

Taking into account the simultaneous presence of pressure and centrifugal flux, and referring to what we discussed in Sect. 3.1, the longitudinal stress in correspondence to the internal fiber is given by

$$\frac{\sigma_{ai}}{p} = \frac{1}{a^2-1} - \gamma\left(\frac{2a^2}{a^2-1} - \frac{1}{\log_e a}\right). \tag{3.144}$$

We are interested in analyzing the cases where σ_{ai} is negative, therefore we consider the absolute value of σ_{ai} to be

$$\left|\frac{\sigma_{ai}}{p}\right| = -\frac{1}{a^2-1} + \gamma\left(\frac{2a^2}{a^2-1} - \frac{1}{\log_e a}\right). \tag{3.145}$$

Recalling (3.143) we set the absolute value of σ_{ai} to be greater than σ_{te}. Therefore, we have

$$-\frac{1}{a^2-1} + \gamma\left(\frac{2a^2}{a^2-1} - \frac{1}{\log_e a}\right) > \frac{2}{a^2-1} - \gamma\left(\frac{2}{a^2-1} - \frac{1}{\log_e a}\right), \tag{3.146}$$

and then

$$\gamma > \frac{3/2}{a^2+1-\dfrac{a^2-1}{\log_e a}}. \tag{3.147}$$

Figure 3.14 shows the right-hand side of (3.147); if γ is greater than the values obtained from the curve, it is necessary to compute σ_{ai} as well, in order to correctly identify the worst working conditions of the cylinder. Values even greater than σ_{ai} occur if pressure is missing. In this case we have to write

$$\gamma\left(\frac{2a^2}{a^2-1} - \frac{1}{\ln a}\right) > \frac{2}{a^2-1} - \gamma\left(\frac{2}{a^2-1} - \frac{1}{\ln a}\right), \tag{3.148}$$

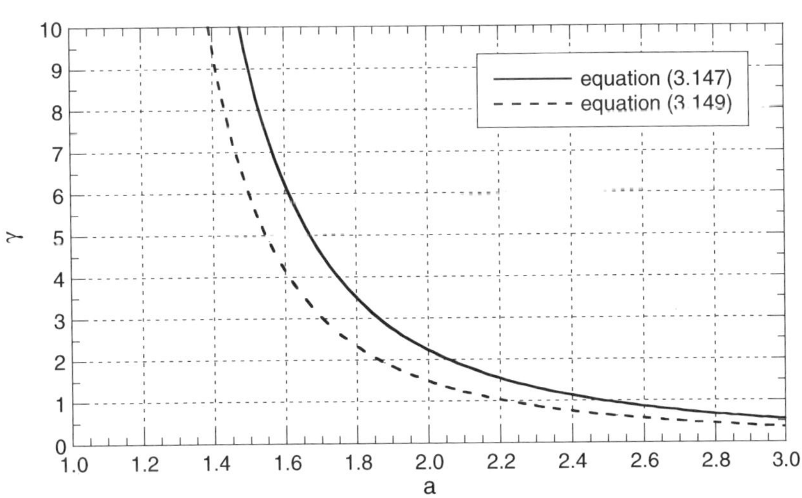

Fig. 3.14

and therefore

$$\gamma > \frac{1}{a^2 + 1 - \dfrac{a^2 - 1}{\ln a}}. \tag{3.149}$$

Figure 3.14 also shows the right-hand side of (3.149).

Finally, Fig. 3.15 shows σ_{ai}/p based on (3.144).

Looking at the curves in Fig. 3.15 and comparing them with the ones in Fig. 3.13, it is immediate to notice that longitudinal stresses may be more dangerous than hoop ones.

When there is no pressure the stress σ_{ai} is even greater and may be computed according to (3.137), with a negative Δt.

Up to this point we assumed, as it is customary under steady-state conditions, that the distribution of temperature is represented by (3.99), and that consequently it is possible to introduce in the computation the value for Δt given by (3.100) leading to equations shown above.

If we want to generalize the analysis of the phenomenon, especially in view of some considerations we will make about transient thermal conditions, we observe that, based on (3.108) and (3.110),

$$\frac{d\sigma_a}{dr} = -\frac{E\alpha}{1-\mu}\frac{dt}{dr}, \tag{3.150}$$

and integrating

$$\sigma_a - \sigma_{ai} = -\frac{E\alpha}{1-\mu}(t - t_i), \tag{3.151}$$

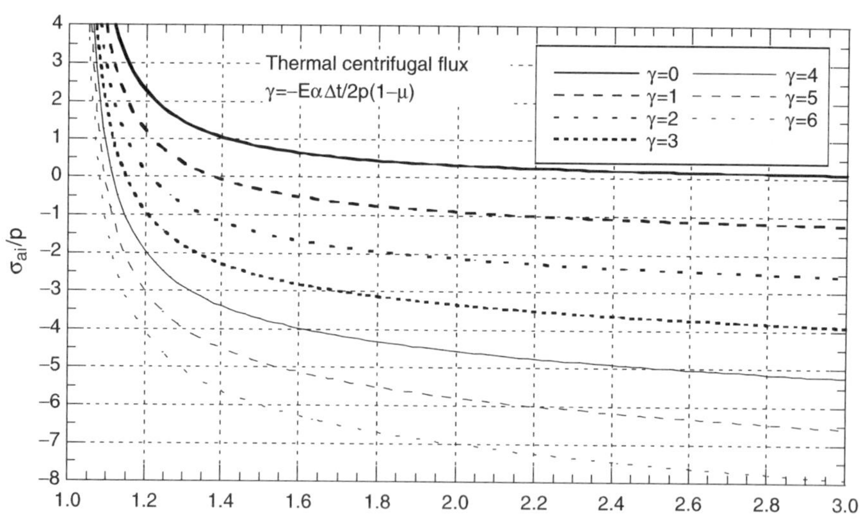

Fig. 3.15

where t_i is the temperature that corresponds to r_i. The resultant of the longitudinal stresses σ_a must be zero. That is,

$$2\pi \int_{r_i}^{r_e} \sigma_a r dr = 0. \tag{3.152}$$

Thus

$$\sigma_{ai} \int_{r_i}^{r_e} r dr - \frac{E\alpha}{1-\mu} \int_{r_i}^{r_e} t r dr + \frac{E\alpha t_i}{1-\mu} \int_{r_i}^{r_e} r dr = 0. \tag{3.153}$$

By solving the integrals that do not include the temperature t we obtain

$$\sigma_{ai} = \frac{2E\alpha}{(1-\mu)\left(r_e^2 - r_i^2\right)} \int_{r_i}^{r_e} t r dr - \frac{E\alpha}{1-\mu} t_i. \tag{3.154}$$

From (3.154), (3.151) can therefore be written as

$$\sigma_a = \frac{E\alpha}{1-\mu} \left(\frac{2}{r_e^2 - r_i^2} \int_{r_i}^{r_e} t r dr - t \right). \tag{3.155}$$

The comparison between (3.150) and (3.110) shows that

$$\sigma_t + \sigma_r = \sigma_a. \tag{3.156}$$

Therefore,

$$\sigma_t = \frac{E\alpha}{1-\mu} \left(\frac{2}{r_e^2 - r_i^2} \int_{r_i}^{r_e} t r dr - t \right) - \sigma_r, \tag{3.157}$$

and based on (3.102)

$$\frac{E\alpha}{1-\mu} \left(\frac{2}{r_e^2 - r_i^2} \int_{r_i}^{r_e} t r dr - t \right) - 2\sigma_r - r\frac{d\sigma_r}{dr} = 0. \tag{3.158}$$

The general integral of (3.158) is

$$\sigma_r = \frac{E\alpha}{1-\mu} \left[\left(\frac{1}{r_e^2 - r_i^2} - \frac{A}{r^2} \right) \int_{r_i}^{r_e} t r dr - \frac{1}{r^2} \int_0^r t r dr \right], \tag{3.159}$$

where A is a constant of integration.

In fact, we observe that

$$r\frac{d\sigma_r}{dr} = \frac{E\alpha}{1-\mu} \left(\frac{2A}{r^2} \int_{r_i}^{r_e} t r dr + \frac{2}{r^2} \int_0^r t r dr - t \right); \tag{3.160}$$

and we determine that (3.158) has been satisfied. As far as the constant of integration we observe that for $r = r_i$ we must have $\sigma_r = 0$; therefore

$$\left(\frac{1}{r_e^2 - r_i^2} - \frac{A}{r_i^2} \right) \int_{r_i}^{r_e} t r dr - \frac{1}{r_i^2} \int_0^{r_i} t r dr = 0. \tag{3.161}$$

Then

$$A = \frac{r_i^2}{r_e^2 - r_i^2} - \frac{\int_0^{r_i} t r dr}{\int_{r_i}^{r_e} t r dr}. \tag{3.162}$$

By analogy, for $r = r_e$ it is required that $\sigma_r = 0$; it follows that:

$$\left(\frac{1}{r_e^2 - r_i^2} - \frac{A}{r_e^2}\right)\int_{r_i}^{r_e} trdr - \frac{1}{r_e^2}\int_0^{r_e} trdr = 0, \tag{3.163}$$

and

$$A = \frac{r_e^2}{r_e^2 - r_i^2} - \frac{\int_0^{r_e} trdr}{\int_{r_i}^{r_e} trdr}. \tag{3.164}$$

We observe that the integral of the product tr for $0 < r < r_i$ bears no physical meaning; conventionally assuming $t = 0$ for this interval of r

$$\int_0^{r_i} trdr = 0, \tag{3.165}$$

and

$$\int_0^{r_e} trdr = \int_{r_i}^{r_e} trdr. \tag{3.166}$$

We obtain the same expression for A from (3.162) as well as (3.164)

$$A = \frac{r_i^2}{r_e^2 - r_i^2}. \tag{3.167}$$

Equation (3.159) can thus be written as follows by taking into account (3.165)

$$\sigma_r = \frac{E\alpha}{1-\mu}\left(\frac{1-(r_i/r)^2}{r_e^2 - r_i^2}\int_{r_i}^{r_e} trdr - \frac{1}{r^2}\int_{r_i}^{r} trdr\right). \tag{3.168}$$

By recalling (3.156) we obtain the following general equations of the three stresses σ_t, σ_a and σ_r

$$\begin{aligned}
\sigma_t &= \frac{E\alpha}{1-\mu}\left(\frac{1-(r_i/r)^2}{r_e^2 - r_i^2}\int_{r_i}^{r_e} trdr + \frac{1}{r^2}\int_{r_i}^{r} trdr - t\right) \\
\sigma_a &= \frac{E\alpha}{1-\mu}\left(\frac{2}{r_e^2 - r_i^2}\int_{r_i}^{r_e} trdr - t\right) \\
\sigma_r &= \frac{E\alpha}{1-\mu}\left(\frac{1-(r_i/r)^2}{r_e^2 - r_i^2}\int_{r_i}^{r_e} trdr - \frac{1}{r^2}\int_{r_i}^{r} trdr\right)
\end{aligned} \tag{3.169}$$

We note that

$$\frac{2}{r_e^2 - r_i^2}\int_{r_i}^{r_e} trdr = \frac{\int_{r_i}^{r_e} 2\pi trdr}{\pi\left(r_e^2 - r_i^2\right)}. \tag{3.170}$$

We realize that such term is none other than the average temperature t_m in the wall of the cylinder; consequently

$$\frac{1 \pm (r_i/r)^2}{r_e^2 - r_i^2}\int_{r_i}^{r_e} trdr = \frac{1 \pm (r_i/r)^2}{2} t_m. \tag{3.171}$$

Furthermore, we observe that, taking into account (3.165),

$$\frac{1}{r^2}\int_{r_i}^{r} trdr = \frac{\int_0^r 2\pi trdr}{2\pi r^2}. \tag{3.172}$$

The term represents half the average temperature in the ideal cylinder with radius r, under the constraint that for r $< r_i$ we must assume $t = 0$; we call such virtual average temperature t^*_{mr}. Then

$$\begin{aligned}
\sigma_t &= \frac{E\alpha}{1-\mu}\left(\frac{1-(r_i/r)^2}{2}t_m + \frac{1}{2}t^*_{mr} - t\right) \\
\sigma_a &= \frac{E\alpha}{1-\mu}\left(t_m - t\right) \\
\sigma_r &= \frac{E\alpha}{1-\mu}\left(\frac{1-(r_i/r)^2}{2}t_m - \frac{1}{2}t^*_{mr}\right)
\end{aligned} \tag{3.173}$$

If the thickness of the cylinder is small compared to the diameter

$$\frac{1+(r_i/r)^2}{2} \cong 2, \tag{3.174}$$

$$\frac{1-(r_i/r)^2}{2} \cong 0, \tag{3.175}$$

and

$$t^*_{mr} \cong 0. \tag{3.176}$$

Equations (3.173) may generally be simplified with sufficient approximation for practical use as

$$\sigma_t = \sigma_a = \frac{E\alpha}{1-\mu}\left(t_m - t\right), \tag{3.177}$$

and

$$\sigma_r = 0. \tag{3.178}$$

Let us look at Fig. 3.11. In fact, σ_r is basically negligible, and the values of σ_t do not differ much from σ_a ((3.177) is correct for the latter). Specifically, for $r = r_i$ and for $r = r_e$ (3.177) is valid for σ_t, as well.

Equations (3.177) and (3.178) differ from the ones in (3.173) only as far as the behavior of the stresses σ_t and σ_r inside the wall. Besides, these differences are modest, except in the case of thick cylinders.

Finally, we observe that during a transient thermal condition (let us assume a sudden temperature increase on the internal side of the cylinder) the temperature flow is qualitatively represented by the curves in Fig. 3.16, where the lower curves correspond to the sudden increase in temperature, while the upper curve corresponds to steady-state conditions. At the beginning

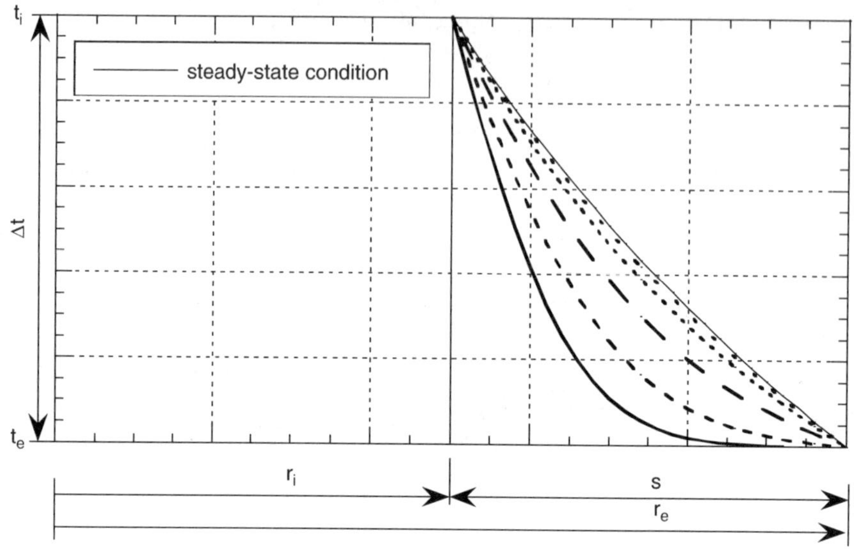

Fig. 3.16

of the temperature increase we can assume that the average temperature is approximately equal to the temperature on the external side. Therefore, based on (3.177)

$$\sigma_{ti} = \sigma_{ai} \cong \frac{E\alpha\Delta t}{1-\mu}, \tag{3.179}$$

and

$$\sigma_{te} = \sigma_{ae} \cong 0, \tag{3.180}$$

where Δt is the difference in temperature between the external and internal side.

Under steady-state conditions we have

$$\sigma_{ti} = \sigma_{ai} \cong \frac{E\alpha\Delta t}{2\,(1-\mu)}, \tag{3.181}$$

and

$$\sigma_{te} = \sigma_{ae} \cong -\frac{E\alpha\Delta t}{2\,(1-\mu)}. \tag{3.182}$$

Therefore, it is most important to heat gradually, in order to avoid the onset of high stresses during transients.

A peculiar characteristic of stresses due to the thermal flux consists of having a zero average value, since they are not generated to balance external forces, but only to re-establish the congruence of deformations instead. Based on the computation criterion that takes only the average values of the stresses into account and neglects peaks (see Sect. 3.1), the stresses generated by flux should not be taken into consideration, and such criterion is often, and rightfully, adopted when the values of Δt are modest, and the cylinders are thin. Remember, however, that the presence of flux sometimes leads to an increase in peaks that cannot be ignored.

The stresses generated by flux are self-limiting and belong to the special category of stresses for which a high ideal allowable stress may be adopted, according to the criteria discussed in Sect. 1.5. During this process, of course, one needs to consider stresses caused by pressure, structural discontinuity and potential external loads, too. A specific discussion of the problem, according to these criteria, is carried out in Sect. 3.6, under the hypothesis that, besides stresses generated by flux, there may only be stresses due to internal pressure in the cylinder, and that fatigue phenomena are ruled out.

In conclusion, let us keep in mind that such peaks, as well as the ones caused by pressure, are crucial whenever the working conditions of the vessel are such to require a fatigue analysis (see Chap. 10).

3.4 Allowable Out of Roundness

The profile of the cylinder differs from a perfectly circular one. This is due to the irregularity of the metal sheet, to its imperfect curvature, especially in correspondence to welded joints, to the welding shrinkage, and so on. The deviation of the profile from the theoretically circular one causes bending moments that add in terms of stress to the ones given by normal forces, thus generating a different hoop stress distribution, compared to the ones to be expected if the cylinder were perfectly circular. Therefore, it is necessary to determine up to which point these issues, particularly present in thin cylinders, are allowable with regard to resistance.

The characteristics of the irregularities vary widely, which explains why it is difficult to establish general evaluation criteria of their impact, especially if the criteria are to be practical and easy to apply. Generally, these irregularities are modelled assuming that the profile will be oval, hence setting the value of allowable ovalization.

Referring to Fig. 3.17 the ovalization u can be expressed as

$$u = \frac{D_{m(\max)} - D_{m(\min)}}{D_m} = 2\frac{D_{m(\max)} - D_{m(\min)}}{D_{m(\max)} + D_{m(\min)}}, \tag{3.183}$$

where D_m is the diameter of a mean fiber. Let δ be the deviation of the cylinder's profile

$$\delta = \delta_0 \cos 2\omega, \tag{3.184}$$

where δ_0 is the maximum deviation at point A. The bending moment has a similar expression, specifically

$$M = M_0 \cos 2\omega, \tag{3.185}$$

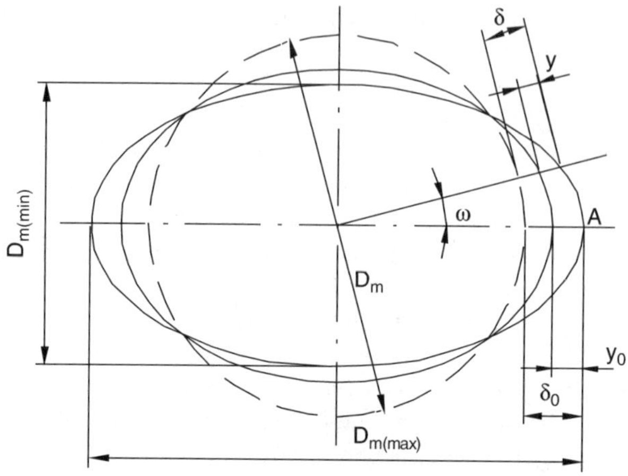

Fig. 3.17

where M_0 is the bending moment at point A. As a result of these moments the cylinder is deformed; the corresponding radial deflections y are related to the moments through the following equation:

$$M = \frac{EI}{r_m^2(1-\mu^2)}\left(\frac{d^2y}{d\omega^2} + y\right). \tag{3.186}$$

I is the moment of inertia and the presence of Poisson factor μ originates from the congruence of deformations, i.e., the cylinder sections must remain flat after the deformation.

Considering (3.185) implies that

$$\frac{d^2y}{d\omega^2} + y = -\frac{M_0 r_m^2}{EI}(1-\mu^2)\cos 2\omega, \tag{3.187}$$

and solving (3.187) for y we have

$$y = \frac{M_0 r_m^2}{3EI}(1-\mu^2)\cos 2\omega. \tag{3.188}$$

For $\omega = 0$ (at point A)

$$y_0 = \frac{M_0 r_m^2}{3EI}(1-\mu^2). \tag{3.189}$$

On the other hand, the internal pressure tends to make the profile become circular; in other words, the radial deflections y have the opposite sign of deviations δ. In addition, the normal force per unitary length of the cylinder is

$$N = pr_i \cong pr_m. \tag{3.190}$$

In case of minor deviations from circularity the normal force can still be assumed to be equal to the value indicated in (3.190). Consequently, the bending moment at point A is

$$M_0 = N(\delta_0 - y_0) = pr_m(\delta_0 - y_0). \tag{3.191}$$

Recalling (3.189)

$$M_0 = pr_m \left[\delta_0 - \frac{M_0 r_m^2}{3EI}(1-\mu^2) \right]. \tag{3.192}$$

Finally,

$$M_0 = \frac{pr_m\delta_0}{1 + \frac{pr_m^2}{3EI}(1-\mu^2)}. \tag{3.193}$$

According to (3.183)

$$u = \frac{2\delta_0}{r_m}; \tag{3.194}$$

therefore, given that $I = s^3/12$, where s is the thickness,

$$M_0 = \frac{1}{2}\frac{upr_m^2}{1 + 4\frac{p}{E}\left(\frac{r_m}{s}\right)^3(1-\mu^2)}. \tag{3.195}$$

The corresponding maximum stress with $W = s^2/6$ is

$$\sigma_f = \pm\frac{3up\left(\frac{r_m}{s}\right)^2}{1 + 4\frac{p}{E}\left(\frac{r_m}{s}\right)^3(1-\mu^2)}. \tag{3.196}$$

Let us now assume that the thickness is the minimum required value and that the verification in correspondence to weld joints with efficiency z is required. Recalling Sect. 3.1 the thickness s is

$$s = \frac{pr_m}{fz}, \tag{3.197}$$

where f is the basic allowable stress, and

$$p = \frac{s}{r_m}fz. \tag{3.198}$$

From (3.196) we obtain

$$\sigma_f = \pm\frac{3ufz\frac{r_m}{s}}{1 + 4\frac{fz}{E}\left(\frac{r_m}{s}\right)^2(1-\mu^2)}. \tag{3.199}$$

The maximum hoop stress caused by pressure is given, as shown in Sect. 3.1, by

$$\sigma_{ti} = p\frac{a^2+1}{a^2-1}; \tag{3.200}$$

as usual, a is the ratio between outside and inside radius. Through several steps the following equation is derived from (3.200)

$$\sigma_{ti} = p\left(\frac{r_m}{s} + \frac{1}{4}\frac{s}{r_m}\right), \tag{3.201}$$

and, recalling (3.198),

$$\sigma_{ti} = fz\left[1 + \frac{1}{4}\left(\frac{s}{r_m}\right)^2\right]. \tag{3.202}$$

The radial stress on the internal fiber is

$$\sigma_{ri} = -p = -\frac{s}{r_m}fz. \tag{3.203}$$

If we assume that the structure develops the maximum deviation from the theoretical circular profile in correspondence to a longitudinal weld (which is to be expected, considering what we have discussed about sheet curvature and welding shrinkage) the maximum ideal stress, according to Guest's theory, is

$$\sigma_{id} = \sigma_f + \sigma_{ti} - \sigma_{ri}. \tag{3.204}$$

Applying the previous concepts

$$\sigma_{id} = \frac{3ufz\frac{r_m}{s}}{1 + 4\frac{fz}{E}\left(\frac{r_m}{s}\right)^2(1-\mu^2)} + fz\left[1 + \frac{1}{4}\left(\frac{s}{r_m}\right)^2\right] + fz\frac{s}{r_m}. \tag{3.205}$$

If σ_{id} is greater than the yield strength, the deformations y are greater than the ones assumed under the hypothesis of elastic behavior of the material. In this case, the ovalized cylinder tends more towards a circular shape, and therefore the moment M is smaller than the one computed; this is a typical situation of self-limiting stresses, and therefore it would be reasonable to allow a very high value for the ideal stress computed from the elasticity laws, i.e., typically up to twice the yield strength (see Sect. 1.5).

Nevertheless, due to the potential presence of stress peaks caused by other reasons, it is best to conservatively limit the value of σ_{id} to the yield strength. The yield strength is equal to $1.5fz$, where z is the weld joint efficiency, and the basic allowable stress f is equal to the yield strength divided by 1.5. Consequently, based on (3.205),

$$\frac{3u\frac{r_m}{s}}{1 + 4\frac{fz}{E}\left(\frac{r_m}{s}\right)^2(1-\mu^2)} + 1 + \frac{1}{4}\left(\frac{s}{r_m}\right)^2 + \frac{s}{r_m} \leq 1.5. \tag{3.206}$$

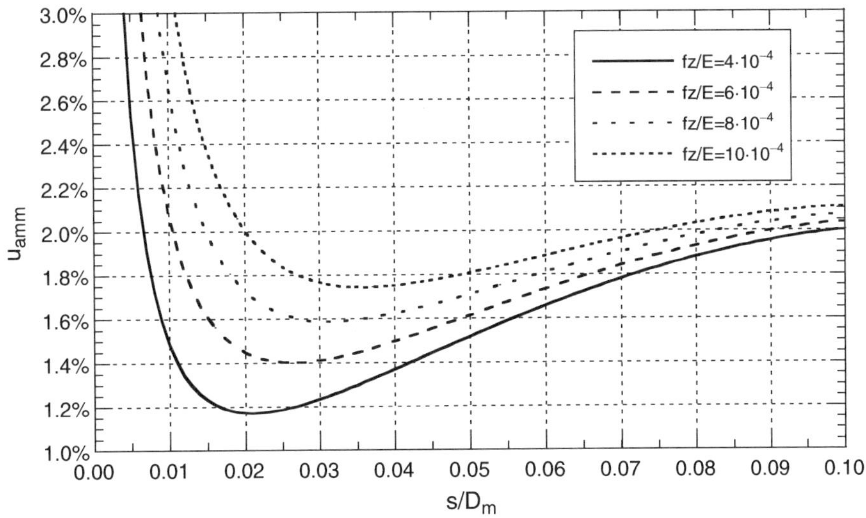

Fig. 3.18

The allowable ovalization u_{amm} may be written as

$$u_{amm} = \frac{\left[0.5 - \dfrac{s}{r_m} - \dfrac{1}{4}\left(\dfrac{s}{r_m}\right)^2\right]\left[1 + 4(1-\mu^2)\dfrac{fz}{E}\left(\dfrac{r_m}{s}\right)^2\right]}{3\dfrac{r_m}{s}}. \qquad (3.207)$$

Introducing D_m as the average diameter and considering u_{amm} as a percentage, with $\mu = 0.3$:

$$u_{amm} = \frac{2}{3}\left[0.5 - 2\frac{s}{D_m} - \left(\frac{s}{D_m}\right)^2\right]\left[\frac{s}{D_m} + 0.91\frac{fz}{E}\frac{D_m}{s}\right]100. \qquad (3.208)$$

Figure 3.18 shows the curves corresponding to (3.208).

3.5 Two-Wall, Multilayer, and Stiffened Cylinders

The cylinder can be built from two coaxial cylinders, thus the name two-wall cylinder, threading the outer cylinder into the inner one; this is achieved by heating the external cylinder or by cooling the internal cylinder. This construction procedure can be recommended, for instance, to limit the necessity to use an expensive material for the inner cylinder (because of the corrosive fluid present inside). In addition, assuming an interference between the cylinders, it is possible to adjust the behavior of hoop stresses compared to those in the monolithic cylinder.

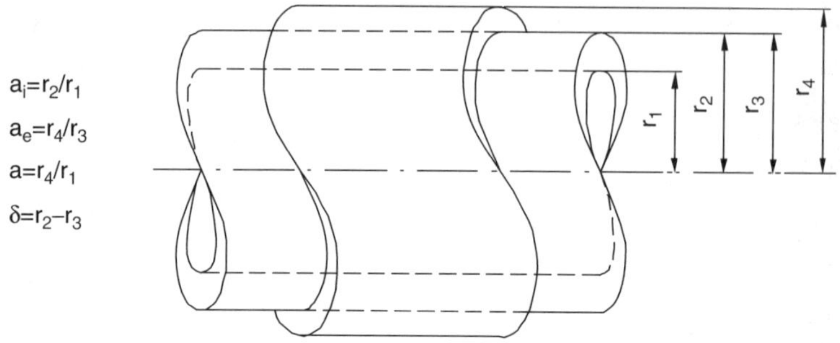

Fig. 3.19

In Sect. 3.1 we extensively explained the calculation of stresses in the cylinder under internal pressure. Specifically, with reference to Fig. 3.19, we indicate with r_1 and r_2 the inside and outside radius of the inner cylinder, respectively, and with r_3 and r_4 the inside and outside radius of the outer cylinder. Because of pressure p, in correspondence to r_2 in the inner cylinder we have

$$\begin{aligned} \sigma_{t2} &= \frac{2p}{{a_i}^2 - 1} \\ \sigma_{a2} &= \frac{p}{{a_i}^2 - 1} \\ \sigma_{r2} &= 0 \end{aligned} \tag{3.209}$$

where

$$a_i = \frac{r_2}{r_1}. \tag{3.210}$$

The circumferential deformation is therefore

$$\varepsilon_{t2} = \frac{p}{E}\left(\frac{2}{{a_i}^2 - 1} - \frac{\mu}{{a_i}^2 - 1}\right) = \frac{p}{E}\frac{2 - \mu}{{a_i}^2 - 1}. \tag{3.211}$$

Because of the deformation in the inner cylinder and the possible interference caused by their construction, a pressure p' is generated between the cylinders. No axial load is generated.

The stresses on the cylinder under external pressure will be examined in Sect. 4.1. Ruling out the axial load, because of p' in correspondence to r_2

$$\begin{aligned} \sigma_{t2} &= -p'\frac{{a_i}^2 + 1}{{a_i}^2 - 1} \\ \sigma_{r2} &= -p' \\ \sigma_{a2} &= 0 \end{aligned} \tag{3.212}$$

Therefore,

$$\varepsilon'_{t2} = -\frac{p'}{E}\left(\frac{{a_i}^2 + 1}{{a_i}^2 - 1} - \mu\right) = -\frac{p'}{E}\frac{{a_i}^2(1 - \mu) + 1 + \mu}{{a_i}^2 - 1}. \tag{3.213}$$

The outer cylinder is exposed to the following stresses in r_3 caused by pressure p':

$$\begin{aligned} \sigma_{t3} &= p' \frac{a_e{}^2 + 1}{a_e{}^2 - 1} \\ \sigma_{r3} &= -p' \\ \sigma_{a3} &= 0 \end{aligned} \tag{3.214}$$

where

$$a_e = \frac{r_4}{r_3}. \tag{3.215}$$

Therefore

$$\varepsilon_{t3} = \frac{p'}{E} \left(\frac{a_e{}^2 + 1}{a_e{}^2 - 1} + \mu \right) = \frac{p'}{E} \frac{a_e{}^2 (1 + \mu) + 1 - \mu}{a_e{}^2 - 1}. \tag{3.216}$$

If δ is the radial interference between the cylinders

$$\delta = r_2 - r_3, \tag{3.217}$$

it follows that:

$$\varepsilon_{t3} r_3 - (\varepsilon_{t2} + \varepsilon'_{t2}) r_2 = \delta. \tag{3.218}$$

The value of δ is very low compared to r_2 or r_3. Hence, it is possible to simplify the equation:

$$\varepsilon_{t3} - (\varepsilon_{t2} + \varepsilon'_{t2}) = \frac{\delta}{r_2}, \tag{3.219}$$

given that, according to (3.217)

$$\frac{\delta}{r_2} = 1 - \frac{r_3}{r_2}. \tag{3.220}$$

Let us introduce the dimensionless quantity α

$$\alpha = p'/p \tag{3.221}$$

According to (3.211), (3.213), (3.216), and (3.219)

$$p \frac{\alpha \left[a_e{}^2 (1 + \mu) + 1 - \mu \right]}{a_e{}^2 - 1} - p \frac{2 - \mu - \alpha \left[a_i{}^2 (1 - \mu) + 1 + \mu \right]}{a_i{}^2 - 1} = \frac{E\delta}{r_2}; \tag{3.222}$$

thus

$$\alpha \left(\frac{a_e{}^2 (1 + \mu) + 1 - \mu}{a_e{}^2 - 1} + \frac{a_i{}^2 (1 - \mu) + 1 + \mu}{a_i{}^2 - 1} \right) = \frac{E\delta}{p r_2} + \frac{2 - \mu}{a_i{}^2 - 1}. \tag{3.223}$$

Finally,

$$\alpha = \frac{\dfrac{E\delta}{pr_2} + \dfrac{2-\mu}{a_i{}^2 - 1}}{\dfrac{a_e{}^2\left(1+\mu\right)+1-\mu}{a_e{}^2-1} + \dfrac{a_i{}^2\left(1-\mu\right)+1+\mu}{a_i{}^2-1}}. \tag{3.224}$$

Specifically, in the case of steel with $\mu = 0.3$ (3.224) is reduced to

$$\alpha = \frac{\dfrac{E\delta}{pr_2} + \dfrac{1.7}{a_i{}^2 - 1}}{\dfrac{1.3a_e{}^2+0.7}{a_e{}^2-1} + \dfrac{0.7a_i{}^2+1.3}{a_i{}^2-1}}. \tag{3.225}$$

Let us introduce the following dimensionless quantities:

$$\rho = \frac{E\delta}{pr_2}, \tag{3.226}$$

$$a = \frac{r_4}{r_1}, \tag{3.227}$$

and

$$\varphi = \frac{r_2{}^2}{r_1 r_4}. \tag{3.228}$$

We observe that r_3 and r_2 almost coincide, therefore

$$a = \frac{r_4}{r_3}\frac{r_2}{r_1} = a_e a_i, \tag{3.229}$$

and

$$\varphi = a_i/a_e. \tag{3.230}$$

Equation (3.225) can be written as

$$\alpha = \frac{\rho + \dfrac{1.7}{a\varphi - 1}}{\dfrac{1.3a/\varphi+0.7}{a/\varphi-1} + \dfrac{0.7a\varphi+1.3}{a\varphi-1}}. \tag{3.231}$$

Consequently to these assumptions, we have for the inner cylinder

$$\begin{aligned} \sigma_{t2} &= \frac{p}{a\varphi - 1}\left[2 - \alpha\left(a\varphi + 1\right)\right], \\ \sigma_{a2} &= \frac{p}{a\varphi - 1}, \\ \sigma_{r2} &= -\alpha p. \end{aligned} \tag{3.232}$$

According to Sect. 4.1, we can conclude that in correspondence to the internal fiber

$$\begin{aligned} \sigma_{t1} &= \frac{p}{a\varphi - 1}\left[a\varphi + 1 - 2\alpha a\varphi\right], \\ \sigma_{a1} &= \frac{p}{a\varphi - 1}, \\ \sigma_{r1} &= -p. \end{aligned} \tag{3.233}$$

Similarly, for the outer cylinder

$$\begin{aligned} \sigma_{t3} &= \alpha p \frac{a/\phi + 1}{a/\phi - 1}, \\ \sigma_{a3} &= 0, \\ \sigma_{r3} &= -\alpha p. \end{aligned} \tag{3.234}$$

And in correspondence to the external fiber

$$\begin{aligned} \sigma_{t4} &= \alpha p \frac{2}{a/\varphi - 1}, \\ \sigma_{a4} &= 0, \\ \sigma_{r4} &= 0. \end{aligned} \tag{3.235}$$

Figures 3.20 and 3.21 show (3.231) for two values of φ. The curves allow us to compute α and through α the values of the different stresses in the cylinders. We have considered thus far that the value of ρ may also carry

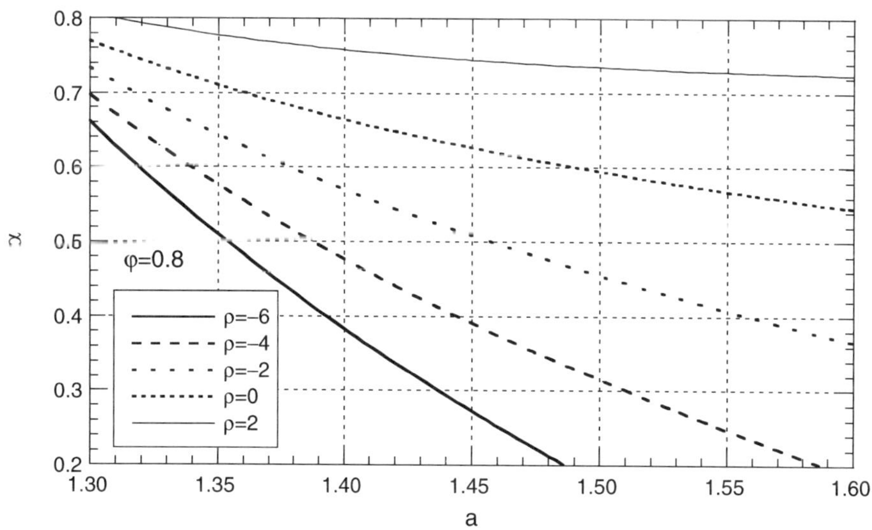

Fig. 3.20

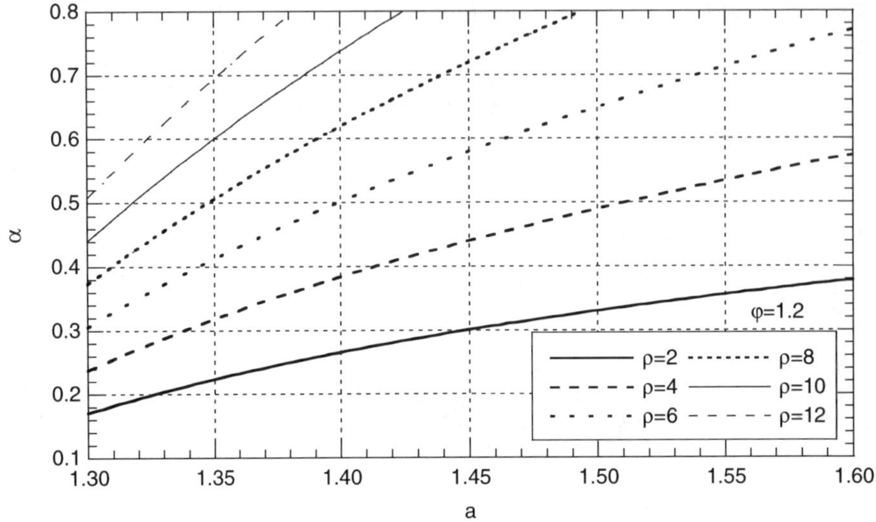

Fig. 3.21

a negative sign; in this case, not only is there no interference between the cylinders, but the radius r_3 is also greater than the radius r_2. To conclude, once the desired value of α is set (and therefore the value of pressure p'), we obtain ρ from the curves, and from (3.226) as follows:

$$\frac{\delta}{r_2} = \frac{p\rho}{E}. \tag{3.236}$$

Finally, from (3.220)

$$r_3 = \left(1 - \frac{p\rho}{E}\right) r_2. \tag{3.237}$$

Equation (3.237) lets us determine the inside radius of the outer cylinder for a specific pressure p' between the cylinders. The multilayer cylinders are made up of several coaxial cylinders and manufactured with special techniques. They are not conceptually different from the two-wall cylinders except for the fact that the computation is obviously more complex, since one shall consider the various pressures generated between the cylinders.

Apart from the more cumbersome computation, the same considerations made for the two-wall cylinders still apply and can be in fact extended to a number of cylinders greater than two. Stiffened cylinders are also conceptually similar to two-wall cylinders (Fig. 3.22). However, particular issues arise here due to the presence of cylinder portions that are not reinforced by stiffening rings.

This type of structure obviously aims at limiting the cylinder thickness through reinforcement rings of light weight and elevated resistance. To analyze the stress conditions of both the cylinder and the rings it is necessary, first

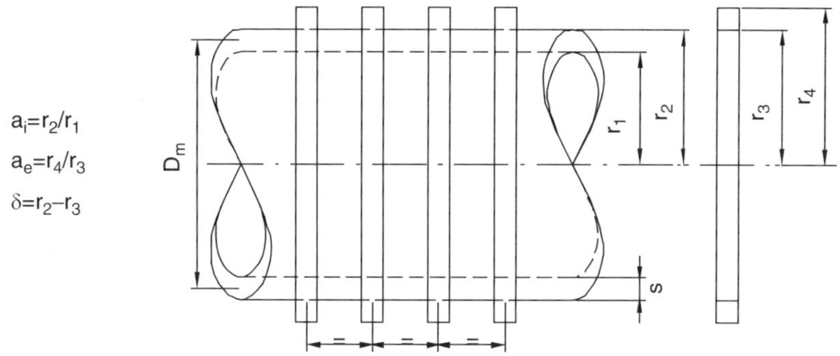

Fig. 3.22

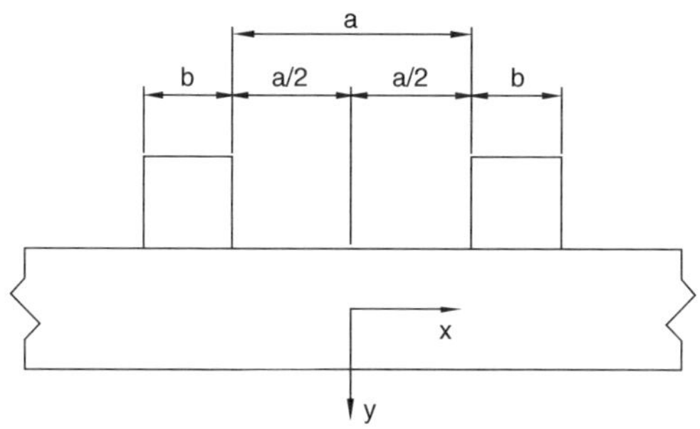

Fig. 3.23

of all, to examine the behavior of the cylinder in the areas between stiffening rings. In order to do so, we go back to Sect. 2.2 and (2.43).

Let us examine Fig. 3.23 where the origin is located in the middle of the nonreinforced area of the cylinder, and its width is a.

From (2.43)

$$y = e^{-\beta x}\left(C_1 \sin\beta x + C_2 \cos\beta x\right) + e^{\beta x}\left(C_3 \sin\beta x + C_4 \cos\beta x\right). \qquad (3.238)$$

First of all, when considering an ideally infinite series of rings at a constant pitch, if the x sign is reversed the radial deformations y must be equal for symmetry with respect to the origin.

We verify that

$$\begin{aligned} C_3 &= -\,C_1 \\ C_4 &= C_2 \end{aligned} \qquad (3.239)$$

Therefore, (3.238) now becomes

$$y = C_2 \cos\beta x\left(e^{\beta x} + e^{-\beta x}\right) - C_1 \sin\beta x\left(e^{\beta x} - e^{-\beta x}\right). \qquad (3.240)$$

We derive (3.240) obtaining

$$\frac{dy}{dx} = -\beta \left[\cos \beta x \left(C_1 - C_2\right) \left(e^{\beta x} - e^{-\beta x}\right) + \sin \beta x \left(C_1 + C_2\right) \left(e^{\beta x} + e^{-\beta x}\right)\right]. \tag{3.241}$$

If we reverse the sign of x, again for symmetry reasons, the value of the derivative should change sign, and in fact this happens to be the case. We can also assume that, for $x = a/2$ and therefore for $x = -a/2$ as well, the derivative is zero, i.e., no appreciable rotations are present in the cylinder in correspondence to the edges of the rings.

From (3.241) we obtain

$$\cos \frac{\beta a}{2} \left(C_1 - C_2\right) \left(e^{\beta a/2} - e^{-\beta a/2}\right) + \sin \frac{\beta a}{2} \left(C_1 + C_2\right) \left(e^{\beta a/2} - e^{-\beta a/2}\right) = 0. \tag{3.242}$$

Let us now remember the meaning of β. According to Sect. 2.2 we find that with regard to steel

$$\beta = \frac{1.817}{\sqrt{D_m s}}, \tag{3.243}$$

assuming as reference diameter the average diameter of the cylinder, and s being the thickness.

We now introduce the coefficient χ

$$\chi = \frac{a}{\sqrt{D_m s}}. \tag{3.244}$$

Therefore, by introducing the coefficient γ,

$$\gamma = \frac{\beta a}{2} = 0.909\chi. \tag{3.245}$$

Let us introduce the dimensionless quantity k_1:

$$k_1 = \frac{\sin \gamma \left(e^{\gamma} + e^{-\gamma}\right) - \cos \gamma \left(e^{\gamma} - e^{-\gamma}\right)}{\sin \gamma \left(e^{\gamma} + e^{-\gamma}\right) + \cos \gamma \left(e^{\gamma} - e^{-\gamma}\right)}. \tag{3.246}$$

From (3.242) and through various steps one determines that

$$C_1 = -k_1 C_2. \tag{3.247}$$

Figure 3.24 shows factor k_1 as a function of χ from (3.244).

Based on (3.247), (3.240) is written as

$$y = C_2 \left[\cos \beta x \left(e^{\beta x} + e^{-\beta x}\right) + k_1 \sin \beta x \left(e^{\beta x} - e^{-\beta x}\right)\right]; \tag{3.248}$$

for $x = 0$ the corresponding value of y which we call y_0 is

$$y_0 = 2C_2. \tag{3.249}$$

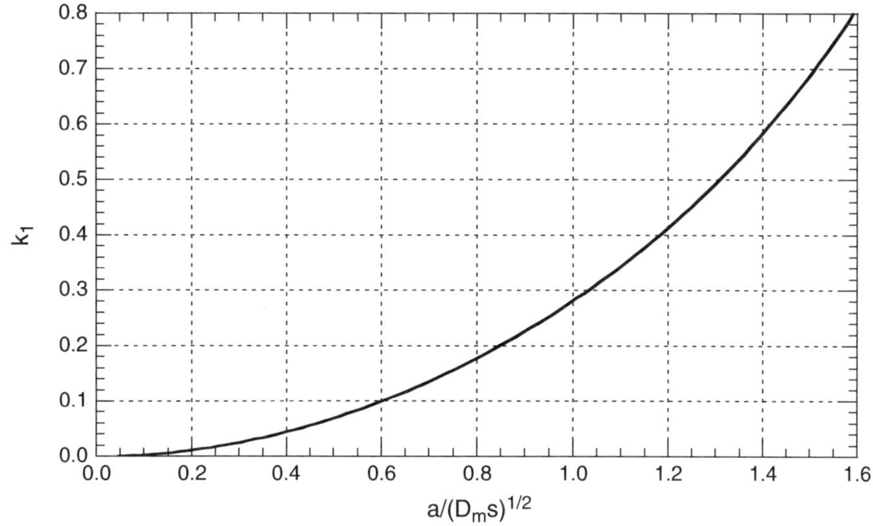

Fig. 3.24

Based on (3.245), when $x = a/2$, we obtain instead the following value of y, which we call y_1:

$$y_1 = C_2 \left[\cos\gamma \left(e^{\gamma} + e^{-\gamma}\right) + k_1 \sin\gamma \left(e^{\gamma} - e^{-\gamma}\right)\right]. \tag{3.250}$$

Let us now introduce the following dimensionless factor k_2:

$$k_2 = \frac{2}{\cos\gamma \left(e^{\gamma} + e^{-\gamma}\right) + k_1 \sin\gamma \left(e^{\gamma} - e^{-\gamma}\right)}. \tag{3.251}$$

Therefore,

$$y_1 = 2C_2/k_2. \tag{3.252}$$

Finally,

$$k_2 = y_0/y_1. \tag{3.253}$$

Figure 3.25 shows k_2.

The curve clearly indicates the influence of the ratio between length a and $\sqrt{D_m s}$ on the behavior of the metal sheet between the rings.

In fact, for $\chi = 0$, i.e., if we assume that the rings are located side by side (in this case we would have a two-wall cylinder) $k_2 = 1$ and $y_0 = y_1$, and in general the radial deformation of the cylinder is constant along the generatrix. As the rings are placed more apart, the ratio between y_0 and y_1 decreases as the distance between the rings increases. In other words, the central area of the not reinforced wall is less influenced by the presence of the rings.

In order to account for the phenomenon in a tractable way, let us integrate the (3.248) from $-a/2$ to $a/2$ and divide the result by y_1. This lets us identify a

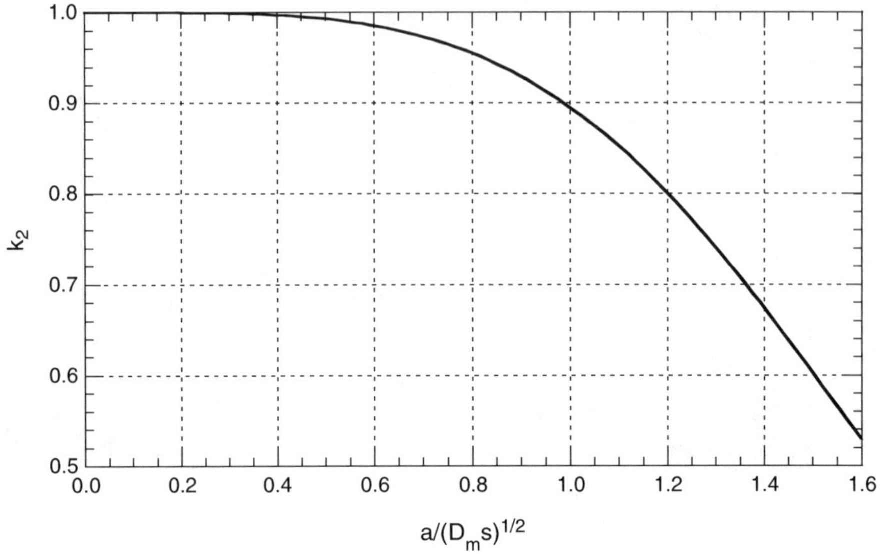

Fig. 3.25

virtual width L_{id} of the cylinder area without rings that is completely affected by the presence of the rings, as the area of the cylinder where the ring is present. In other words

$$L_{id} = \frac{1}{y_1} \int_{-a/2}^{a/2} y dx. \tag{3.254}$$

By executing the integration we obtain the following value for the ratio between the virtual width L_{id} and the actual width a:

$$\frac{L_{id}}{a} = \frac{k_2}{4\gamma} \left[(1 - k_1) \cos\gamma \left(e^{\gamma} - e^{-\gamma}\right) + (1 + k_1) \sin\gamma \left(e^{\gamma} + e^{-\gamma}\right)\right]. \tag{3.255}$$

Figure 3.26 shows this ratio as a function of χ. If the value of χ is known through a, D_m and s, the curve in the figure above helps us determine L_{id}. The deformations y we are discussing here generate some bending moments in the cylinder that we can identify through the second derivative of y. Considering (3.240) we obtain the following:

$$\frac{d^2y}{dx^2} = -2\beta^2 C_2 \left[\sin\beta x \left(e^{\beta x} - e^{-\beta x}\right) - k_1 \cos\beta x \left(e^{\beta x} + e^{-\beta x}\right)\right]; \tag{3.256}$$

for $x = 0$ we obtain

$$\left(\frac{d^2y}{dx^2}\right)_{x=0} = 4k_1\beta^2 C_2 = 2k_1k_2\beta^2 y_1. \tag{3.257}$$

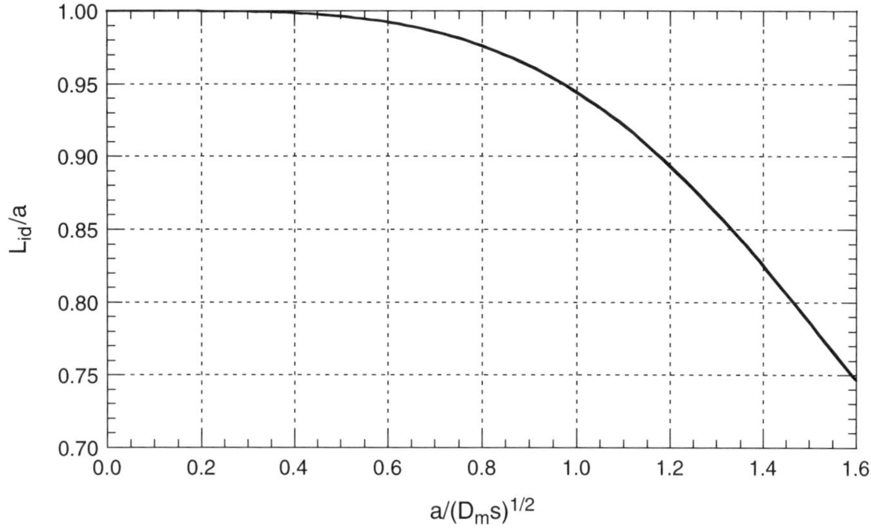

Fig. 3.26

Let us introduce k_3

$$k_3 = \frac{1}{2}\left[k_1 \cos\gamma\left(e^{\gamma} + e^{-\gamma}\right) - \sin\gamma\left(e^{\gamma} - e^{-\gamma}\right)\right]. \tag{3.258}$$

The second derivative for $x = a/2$ is given by

$$\left(\frac{d^2y}{dx^2}\right)_{x=a/2} = 2k_3k_2\beta^2 y_1. \tag{3.259}$$

The factor k_3 is shown in Fig. 3.27.
Finally, let us introduce factors k_4 and k_5, defined, respectively, as

$$k_4 = k_1k_2, \tag{3.260}$$

and

$$k_5 = k_3k_2. \tag{3.261}$$

They are shown in Fig. 3.28.

Based on Sect. 2.2 the generic bending moment M is

$$M = -\frac{EI}{1-\mu^2}\frac{d^2y}{dx^2}. \tag{3.262}$$

Let us examine the stresses due to the bending moment for $x = 0$. Let M_0 be this moment and σ_{fo} the corresponding maximum stress. Given that $I = s^3/12$

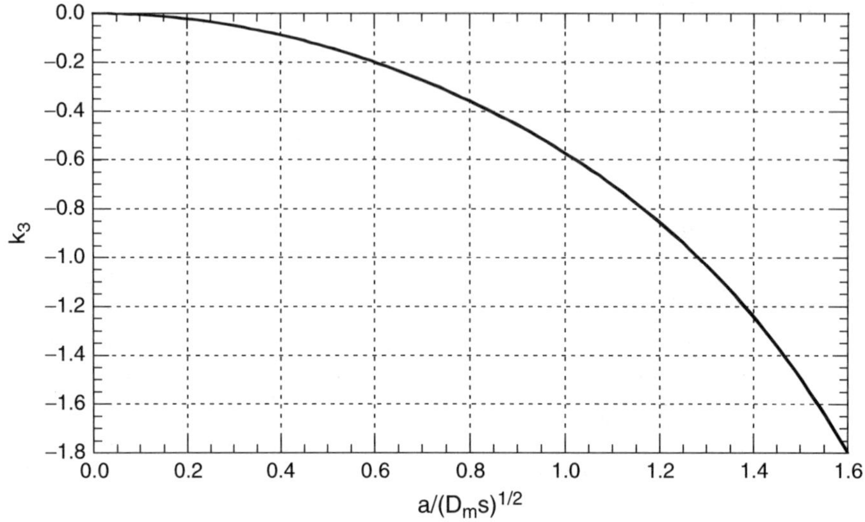

Fig. 3.27

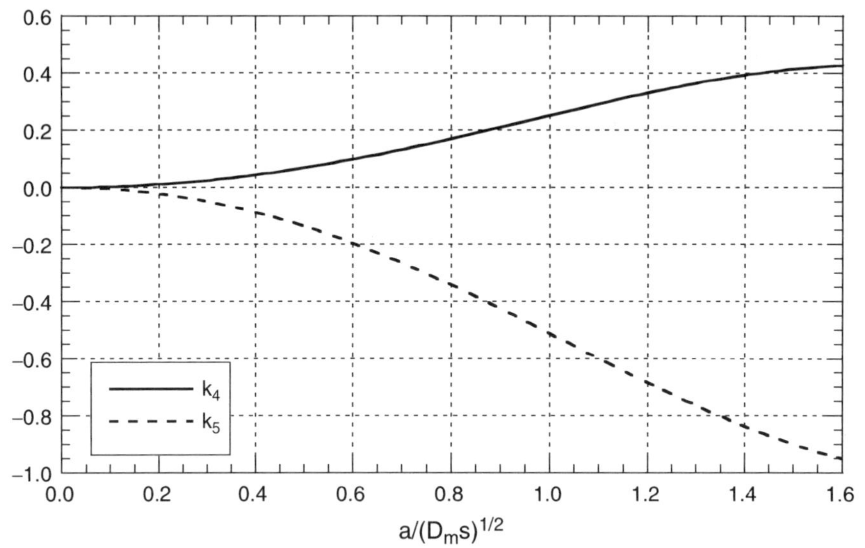

Fig. 3.28

$$M_0 = -\frac{Es^3}{6\,(1-\mu^2)} k_4 \beta^2 y_1. \tag{3.263}$$

Considering that $W = s^2/6$

$$\sigma_{f0} = \pm \frac{Es}{1-\mu^2} k_4 \beta^2 y_1. \tag{3.264}$$

Reconsidering (3.263) for $\mu = 0.3$:

$$\sigma_{f0} = \pm 3.63 E k_4 \frac{y_1}{D_m}. \tag{3.265}$$

Similarly, the stress σ_{fl} due to the bending moment at the edge of the ring is given by

$$\sigma_{fl} = \pm 3.63 E k_5 \frac{y_1}{D_m}. \tag{3.266}$$

In (3.265) and (3.266) the positive sign refers to the external fiber, and the negative sign to the internal fiber. At this point we have all the necessary elements to define the stress values in the cylinder and in the rings. The deformation of the cylinder ε_{t2} due to the pressure p is always computed from (3.211).

If we indicate with p' the pressure caused by the rings on the external fibers of the cylinder, we must consider that such pressure is present only in one area; the cylinder cannot deform, according to (3.213), due to the presence of the not ringed cylinder area which prevents this deformation.

It is reasonable to argue that the deformation, according to (3.213), shall be corrected through the ratio between the ring width b (see Fig. 3.24) and the sum of this width and the virtual width L_{id}.

In other words

$$\varepsilon'_{te} = -\frac{p'}{E} \frac{{a_i}^2 (1-\mu) + 1 + \mu}{{a_i}^2 - 1} \frac{b}{b + L_{id}}. \tag{3.267}$$

We can say with sufficient approximation for practical applications that the radial deformation y_1, is equal to

$$y_1 = -\varepsilon'_{te} r_2; \tag{3.268}$$

r_2 is the outside radius of the cylinder; therefore,

$$\frac{y_1}{D_m} = -\varepsilon'_{te} \frac{r_2}{D_m}. \tag{3.269}$$

Let us introduce the dimensionless factor η given by

$$\eta = \frac{L_{id}}{b}. \tag{3.270}$$

Reconsidering (3.221) and (3.267), for $\mu = 0.3$ (3.269) becomes

$$\frac{y_1}{D_m} = \frac{\alpha p}{E} \frac{0.7{a_i}^2 + 1.3}{{a_i}^2 - 1} \frac{a_i}{a_i + 1} \frac{1}{1 + \eta}. \tag{3.271}$$

If we now introduce k_6

$$k_6 = \frac{0.7{a_i}^2 + 1.3}{{a_i}^2 - 1} \frac{a_i}{a_i + 1} \frac{1}{1 + \eta}, \tag{3.272}$$

we obtain

$$\frac{y_1}{D_m} = k_6 \frac{\alpha p}{E}. \tag{3.273}$$

The knowledge of y_1/D_m allows us to compute the stresses σ_{f0} and σ_{fl} (from (3.265) and (3.266)).

The factor k_6 is shown in Fig. 3.29. Similarly to what we did in the case of a two-wall cylinder, the ratio between pressure p' and pressure p is determined by

$$\alpha = \frac{\rho + \dfrac{1.7}{{a_i}^2 - 1}}{\dfrac{1.3{a_e}^2 + 0.7}{{a_e}^2 - 1} + \dfrac{0.7{a_i}^2 + 1.3}{{a_i}^2 - 1} \dfrac{1}{1 + \eta}}. \tag{3.274}$$

Let us remember that ρ depends on the interference between rings and cylinder, as captured in (3.226); a_i is equal to the ratio between outside radius r_2 and the inside radius r_1 of the cylinder; a_e is the ratio between the outside radius r_4 and the inside radius r_3 of the rings; η is the ratio between L_{id} and the width b of the rings.

When comparing the stiffened cylinder with the two-wall cylinder we shall make the following additional considerations. First, the bending moments computed through (3.262) generate longitudinal stresses that in the more significant positions ($x = 0$ and $x = a/2$) are equal to σ_{f0} and σ_{fl}, according to (3.265) and (3.266). Second, it must also be said that for the deformation's congruence perpendicular moments equal to μM that generate hoop stresses correspond to these moments; with reference to the two positions above as

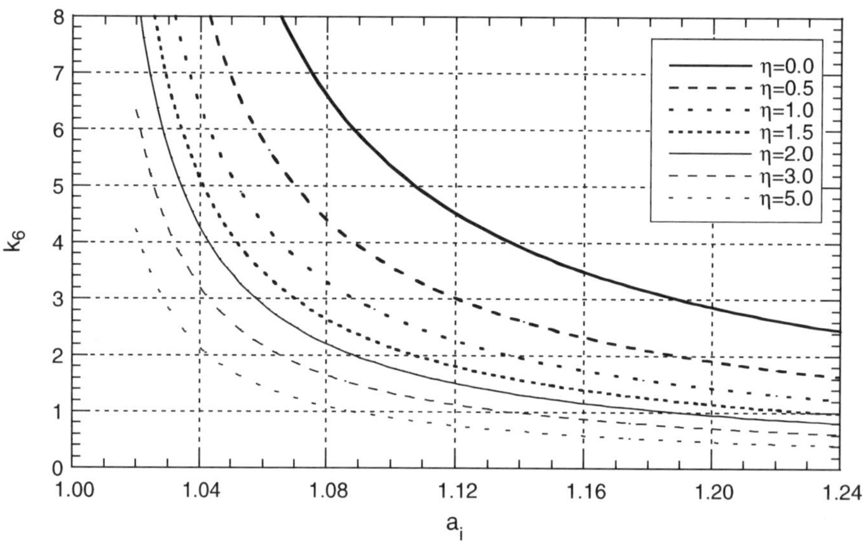

Fig. 3.29

well as $\mu = 0.3$, and following the same sign rule adopted for σ_{f0} and σ_{fl}, we obtain, respectively,

$$\sigma'_{f0} = \pm 1.09 E k_4 \frac{y_1}{D_m}, \tag{3.275}$$

and

$$\sigma'_{f1} = \pm 1.09 E k_5 \frac{y_1}{D_m}. \tag{3.276}$$

With regard to the average hoop stresses (membrane stresses) we can assume a virtual external pressure that takes the nonstiffened cylinder area into consideration. Specifically, we assume that in the area of the stiffened cylinder

$$p'_{id} = p' \frac{b}{b + L_{id}} = \frac{\alpha p}{1+\eta}. \tag{3.277}$$

In the middle of the nonstiffened area, remembering the meaning of k_2, we assume

$$p'_{id} = p' k_2 \frac{b}{b + L_{id}} = k_2 \frac{\alpha p}{1+\eta}. \tag{3.278}$$

The average stresses in the areas corresponding to the rings are therefore determined by

$$\begin{aligned} \sigma_{tm} &= \frac{p r_1}{s} - \frac{\alpha p r_2}{(1+\eta)\, s} = \frac{p}{a_i - 1}\left(1 - \frac{\alpha a_i}{1+\eta}\right), \\ \sigma_{am} &= \frac{p}{{a_i}^2 - 1}, \\ \sigma_{rm} &= -\frac{p}{2} - \frac{p'}{2} = -p\frac{1+\alpha}{2}. \end{aligned} \tag{3.279}$$

The average stresses in the middle of the nonstiffened area are determined by

$$\begin{aligned} \sigma_{tm} &= \frac{p}{{a_i}^2 - 1}\left(1 - k_2 \frac{\alpha a_i}{1+\eta}\right), \\ \sigma_{am} &= \frac{p}{{a_i}^2 - 1}, \\ \sigma_{rm} &= -\frac{p}{2}. \end{aligned} \tag{3.280}$$

Finally, in the rings the average stresses are determined by

$$\begin{aligned} \sigma_{tm} &= \frac{p' r_3}{r_4 - r_3} = \frac{\alpha p}{a_e - 1}, \\ \sigma_{am} &= 0, \\ \sigma_{rm} &= -\frac{p'}{2} = -\frac{\alpha p}{2}. \end{aligned} \tag{3.281}$$

With regard to local peaks, it is possible to operate in a similar manner, of course always under consideration of bending moments. As far as the cylinder, the situation is as follows.

In correspondence to the rings

$$\begin{aligned}
\sigma_{t1} &= \frac{p}{{a_i}^2-1}\left({a_i}^2+1-\frac{2\alpha {a_i}^2}{1+\eta}\right)-1.09k_5k_6\alpha p\\
\sigma_{a1} &= \frac{p}{{a_i}^2-1}-3.63k_5k_6\alpha p\\
\sigma_{r1} &= -p\\
\sigma_{t2} &= \frac{p}{{a_i}^2-1}\left(2-\frac{\alpha\left({a_i}^2+1\right)}{1+\eta}\right)-1.09k_5k_6\alpha p\\
\sigma_{a2} &= \frac{p}{{a_i}^2-1}+3.63k_5k_6\alpha p\\
\sigma_{r2} &= -\alpha p
\end{aligned} \quad . \tag{3.282}$$

In correspondence to the middle of the nonstiffened area

$$\begin{aligned}
\sigma_{t1} &= \frac{p}{{a_i}^2-1}\left({a_i}^2+1-\frac{2\alpha k_2 {a_i}^2}{1+\eta}\right)-1.09k_4k_6\alpha p\\
\sigma_{a1} &= \frac{p}{{a_i}^2-1}-3.63k_4k_6\alpha p\\
\sigma_{r1} &= -p\\
\sigma_{t2} &= \frac{p}{{a_i}^2-1}\left(2-\frac{\alpha k_2\left({a_i}^2+1\right)}{1+\eta}\right)+1.09k_4k_6\alpha p\\
\sigma_{a2} &= \frac{p}{{a_i}^2-1}+3.63k_4k_6\alpha p\\
\sigma_{r2} &= 0
\end{aligned} \tag{3.283}$$

As far as the rings

$$\begin{aligned}
\sigma_{t3} &= \alpha p\frac{{a_e}^2+1}{{a_e}^2-1}\\
\sigma_{a3} &= 0\\
\sigma_{r3} &= -\alpha p\\
\sigma_{t4} &= \alpha p\frac{2}{{a_e}^2-1}\\
\sigma_{a4} &= 0\\
\sigma_{r4} &= 0
\end{aligned} \tag{3.284}$$

With reference to the average stresses the maximum ideal stress σ_{id} (equal to the difference between σ_{tm} and σ_{rm}, or between σ_{am} and σ_{rm}) should not be greater than the basic allowable stress for the material, while for the local stresses it is allowable to reach the yield strength.

It would also be possible to allow for a local plastic flow of the material, but in this case the different calculation theories that assume the elasticity of the material would become invalid. Specifically, the value k_2 and the value of the virtual width L_{id} are not applicable anymore.

The stress status of the cylinder could still be considered acceptable, given that the nature of the peaks are mostly due to the bending moments, but a specific investigation that accounts for the effects of plasticity should be carried out. It is useful to illustrate a practical example (Fig. 3.30). The starting data are as follows:

$$p = 17.5\,\text{N}\,\text{mm}^{-2}(175\text{ bar})$$
$$E = 210,000\,\text{N}\,\text{mm}^{-2}$$
$$r_1 = 470\,\text{mm}$$
$$r_2 = 500\,\text{mm}$$
$$D_m = 970\,\text{mm}$$

The yield strength of the material of the cylinder is 240 N mm^{-2}, assuming an allowable stress of 160 N mm^{-2}. If the cylinder with an outside diameter equal to 1,000 mm were not stiffened, the minimum allowable thickness would be

$$s = \frac{17.5 \times 1000}{2 \times 160 + 17.5} = 51.5\,\text{mm}$$

The actual thickness of 30 mm is therefore clearly lower than the required one. We can introduce rings with a section equal to 60 × 60 mm and with a pitch of 210 mm, that is

$$r_4 = 560\,\text{mm}$$
$$a = 150\,\text{mm}$$
$$b = 60\,\text{mm}$$

Let the yield strength of the ring's material be 480 N mm^{-2} and the allowable stress be 320 N mm^{-2}.

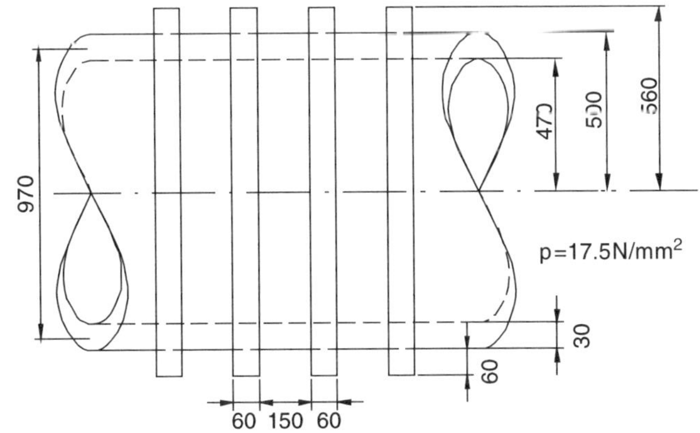

Fig. 3.30

In addition, we have

$$a_i = 500/470 = 1.064$$
$$a_e = 560/500 = 1.12$$
$$\sqrt{D_m s} = 970 \times 30 = 170.59\,\text{mm}$$
$$\chi = a/\sqrt{D_m s} = 150/170.59 = 0.8793$$

We assume an interference δ equal to 0.5 mm, that is

$$\rho = \frac{210000 \times 0.5}{17.5 \times 500} = 12$$

Based on Fig. 3.26 we obtain

$$L_{id}/a = 0.9656$$

Therefore

$$L_{id} = 0.9656 \times 150 = 144.84\,\text{mm}$$
$$\eta = 144.84/60 = 2.414$$

Moreover, from the various factors we obtain

$$k_2 = 0.9355$$
$$k_4 = 0.2013$$
$$k_5 = -0.4066$$
$$k_6 = 2.398$$

From (3.274)

$$\alpha = \frac{12 + 12.905}{9.1616 + 4.6521} = 1.803$$

The membrane stresses in the cylinder corresponding to the ring edge are as follows:

$$\sigma_{tm} = \frac{17.5}{0.064}\left(1 - \frac{1.803 \times 1.064}{3.414}\right) = 120.14\,\text{N}\,\text{mm}^{-2}$$
$$\sigma_{am} = \frac{17.5}{1.1317 - 1} = 132.84\,\text{N}\,\text{mm}^{-2}$$
$$\sigma_{rm} = -17.5\frac{2.803}{2} = -24.53\,\text{N}\,\text{mm}^{-2}$$
$$\sigma_{id(m)} = 132.84 + 24.53 = 157.37\,\text{N}\,\text{mm}^{-2} < 160$$

In correspondence to the middle of the nonstiffened area

$$\sigma_{tm} = \frac{17.5}{0.1317}\left(1 - \frac{0.9355 \times 1.803 \times 1.064}{3.414}\right) = 63.02\,\text{N}\,\text{mm}^{-2}$$
$$\sigma_{am} = 132.84\,\text{N}\,\text{mm}^{-2}$$
$$\sigma_{rm} = -17.5/2 = -8.75\,\text{N}\,\text{mm}^{-2}$$
$$\sigma_{id(m)} = 132.84 + 8.75 = 141.59\,\text{N}\,\text{mm}^{-2} < 160$$

Finally, in the rings

$$\sigma_{tm} = \frac{1.803 \times 17.5}{1.12 - 1} = 262.92\,\mathrm{N\,mm^{-2}}$$

$$\sigma_{am} = 0$$

$$\sigma_{rm} = -1.803 \times 17.5/2 = -15.78\,\mathrm{N\,mm^{-2}}$$

$$\sigma_{id(m)} = 262.92 + 15.78 = 278.70\,\mathrm{N\,mm^{-2}} < 320$$

Let us now examine the local stresses. In correspondence to the ring edge in the cylinder

$$\sigma_{t1} = \frac{17.5}{0.1317}\left(1.1317+1-\frac{2\times1.803\times1.1317}{3.414}\right)+1.09\times0.4066\times2.398$$
$$\times1.803\times17.5=124.39+33.54=157.93\mathrm{N mm^{-2}}$$

$$\sigma_{a1} = \frac{17.5}{0.1317}+3.63\times0.4066\times2.398\times1.803\times17.5=132.84+111.68$$
$$= 244.52\mathrm{N mm^{-2}}$$

$$\sigma_{r1} = -17.5\mathrm{N mm^{-2}}$$

$$\sigma_{t2} = \frac{17.5}{0.1317}\left(2-\frac{1.803\times2.1317}{3.414}\right)-33.54=116.14-33.54=82.60\mathrm{N mm^{-2}}$$

$$\sigma_{a2} = \frac{17.5}{0.1317}-111.68=21.16\mathrm{N mm^{-2}}$$

$$\sigma_{r2} = -1.803\times17.5=-31.56\mathrm{N mm^{-2}}$$

$$\sigma_{id} = 244.52+17.5=262.02\mathrm{N mm^{-2}}>240$$

In correspondence to the middle of the nonstiffened area

$$\sigma_{t1}=\frac{17.5}{0.1317}\left(1.1317+1-\frac{2\times1803\times0.9355\times1.1317}{3.414}\right)$$
$$-1.09\times0.2013\times2.398\times1.803\times17.5=134.63-16.60=118.03\mathrm{N mm^{-2}}$$

$$\sigma_{a1}=132.84-3.63\times0.2013\times2.398\times1.803\times17.5=132.84-55.30$$
$$=77.54\mathrm{N mm^{-2}}$$

$$\sigma_{r1}=-17.5\mathrm{N mm^{-2}}$$

$$\sigma_{t2}=\frac{17.5}{0.1317}\left(2-\frac{1.803\times0.9355\times2.1317}{3.414}\right)+16.60=125.78+16.60$$
$$=142.38\mathrm{N mm^{-2}}$$

$$\sigma_{a2}=132.84+55.30=188.14\mathrm{N mm^{-2}}$$

$$\sigma_{r2}=0$$

$$\sigma_{id}=188.14\mathrm{N mm^{-2}}$$

As far as the rings, computing the local stresses is not necessary because of the small value of a_e. As one can see, the ideal mean stresses are below the allowable stress of materials. As far as the peaks in the middle of the nonstiffened area of the cylinder, the ideal maximum stress is below the yield strength, whereas this is not true in correspondence to the edge of the rings. To reduce this stress it is necessary to move the rings closer and to proportionally reduce the width b. One could set, for instance, $a = 130\,\text{mm}$ and $b = 50\,\text{mm}$ and leave everything else unchanged.

3.6 Partially Plastic Deformed Cylinders

In Sect. 3.2 we discussed a sizing criterion for thick cylinders that does not include the plastic flow of internal fibers, limiting the maximum value of the ideal stress to the yield strength. We also specified that such limitation implicitly considered the potential presence of stress peaks in the cylinder beyond the one under investigation due to σ_t and σ_r generated by the pressure in the cylinder without holes.

In fact, under certain conditions, a partial plastic flow of the cylinder may be allowed provided, of course, that the average value of the ideal stress is not greater than the basic allowable stress of the material. In other words, we must determine if the fundamental equation for the sizing of the cylinder in Sect. 3.1 can be used when the ratio between outside and inside radius is very large and, if so, under what conditions.

First of all, the equation we are examining is

$$s = \frac{pD_e}{2f + p}, \tag{3.285}$$

and it can also be written as

$$\frac{a-1}{a+1} = \frac{p}{2f}. \tag{3.286}$$

Once again

$$a = \frac{r_e}{r_i}. \tag{3.287}$$

As we have seen in Sect. 3.1, the maximum value of the ideal stress in the internal fiber, according to Guest–Tresca, is given by

$$\sigma_{id(i)} = p\frac{2a^2}{a^2 - 1}. \tag{3.288}$$

From (3.286) we have

$$\sigma_{id(i)} = \frac{4a^2}{(a+1)^2} f, \tag{3.289}$$

where f is the basic allowable stress of the material. The nature of this peak is such that based on the criteria discussed in Sect. 1.5 we can allow its value to be up to twice the yield strength. Since the allowable stress f is typically equal to the yield strength divided by the safety factor 1.5, the value of $\sigma_{id(i)}$ may even be $3f$.

From (3.288) we obtain

$$\frac{4a^2}{(a+1)^2} f = 3f; \tag{3.290}$$

thus

$$a^2 - 6a - 3 = 0, \tag{3.291}$$

and, finally,

$$a = 3 + \sqrt{12} = 6.464.$$

As we can see, we obtain a very large allowable value for a, a value much greater than the value (1.58) above which in Sect. 3.2 we recommended the adoption of (3.94) instead of (3.285).

If the value of α is greater than 1.58, the cylinder shows sign of partial plastic flow, and residual stresses are generated at release that will be discussed later on. To this end, let us examine the behavior of the cylinder in the elastic–plastic field in view of Sect. 3.1. Based on (3.10) the radial stress σ_r in the area characterized by elastic behavior is given by

$$\sigma_r = A - \frac{B}{r^2}, \tag{3.292}$$

where A and B are constant. On the other hand, for $r = r_e$ we must have $\sigma_r = 0$, and therefore

$$A = \frac{B}{{r_e}^2}; \tag{3.293}$$

thus

$$\sigma_r = B\left(\frac{1}{{r_e}^2} - \frac{1}{r^2}\right). \tag{3.294}$$

In the area showing plastic flow we have (from (3.49) and with F being constant)

$$\sigma_r = \sigma_s \log_e r + F, \tag{3.295}$$

where σ_s is the yield strength. For $r = r_i$ we must have $\sigma_r = -p$, therefore

$$F = -p - \sigma_s \log_e r_i, \tag{3.296}$$

then

$$\sigma_r = \sigma_s \log_e \frac{r}{r_i} - p. \tag{3.297}$$

Let r_o be the radius corresponding to the circumference that separates the area of the cylinder with plastic flow from the elastic one.

Let us also introduce the following dimensionless factor:

$$a_0 = \frac{r_0}{r_i}; \tag{3.298}$$

for $r = r_o$ the two values of σ_r must coincide, that is

$$B\left(\frac{1}{{r_e}^2} - \frac{1}{{r_0}^2}\right) = \sigma_s \log_e \frac{r_0}{r_i} - p; \tag{3.299}$$

then

$$B = \frac{\sigma_s \log_e \dfrac{r_0}{r_i} - p}{\dfrac{1}{{r_e}^2} - \dfrac{1}{{r_0}^2}}. \tag{3.300}$$

From (3.294) the radial stress in the elastic area is given by

$$\sigma_r = \left(p - \sigma_s \log_e \frac{r_0}{r_i}\right) \frac{\dfrac{1}{{r_e}^2} - \dfrac{1}{r^2}}{\dfrac{1}{{r_0}^2} - \dfrac{1}{{r_e}^2}}, \tag{3.301}$$

or by

$$\sigma_r = (p - \sigma_s \log_e a_0) \frac{1 - \dfrac{{r_e}^2}{r^2}}{\dfrac{a^2}{a_0^2} - 1}. \tag{3.302}$$

From (3.31) and recalling both (3.293) and (3.300) we also obtain the following equation of σ_t in the elastic field

$$\sigma_t = (p - \sigma_s \log_e a_0) \frac{1 + \dfrac{{r_e}^2}{r^2}}{\dfrac{a^2}{a_0^2} - 1}. \tag{3.303}$$

On the other hand, in the plastic area we know that

$$\sigma_t - \sigma_r = \sigma_s, \tag{3.304}$$

where σ_s is the yield strength. From (3.297)

$$\sigma_t = \sigma_s \left(1 + \log_e \frac{r}{r_i}\right) - p. \tag{3.305}$$

For $r = r_o$ the two values of σ_t must coincide, therefore

$$(p - \sigma_s \log_e a_0) \frac{\dfrac{a^2}{a_0^2} + 1}{\dfrac{a^2}{a_0^2} - 1} = \sigma_s (1 + \log_e a_0) - p, \tag{3.306}$$

then

$$p\frac{2\frac{a^2}{a_0^2}}{\frac{a^2}{a_0^2}-1} = \sigma_s \left(\frac{2\frac{a^2}{a_0^2}}{\frac{a^2}{a_0^2}-1} \log_e a_0 + 1 \right). \tag{3.307}$$

Finally, recalling (3.286) and the fact that $\sigma_s = 1.5f$,

$$\frac{2\frac{a^2}{a_0^2}}{\frac{a^2}{a_0^2}-1} \left(2\frac{a-1}{a+1} - 1.5 \log_e a_0 \right) = 1.5. \tag{3.308}$$

Equation (3.308) allows us to obtain through trial-and-error the value of a_0 as a function of a. These values are shown in Fig. 3.31. We considered the values of a between 1.58 and 6.5 that correspond to the beginning of plastic flow of the cylinder, and the value of a that we obtain from (3.291), respectively.

Always taking into account (3.286), the equations of stresses σ_t, σ_r, and σ_{id} are as follows: Area with plastic flow ((3.297) and (3.305))

$$\begin{aligned} \sigma_t &= f\left(1.5 + 1.5\log_e \frac{r}{r_i} - 2\frac{a-1}{a+1}\right) \\ \sigma_r &= f\left(1.5\log_e \frac{r}{r_i} - 2\frac{a-1}{a+1}\right) \\ \sigma_{id} &= 1.5f = \sigma_s \end{aligned} \tag{3.309}$$

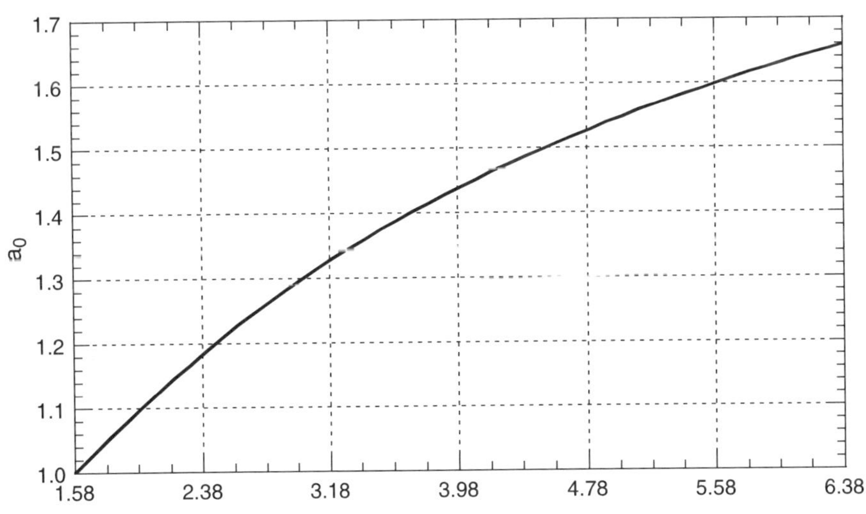

Fig. 3.31

Elastic field ((3.302) and (3.303))

$$\sigma_t = f\left(2\frac{a-1}{a+1} - 1.5\log_e a_0\right)\frac{1+\dfrac{{r_e}^2}{r^2}}{\dfrac{a^2}{a_0^2}-1}$$

$$\sigma_r = f\left(2\frac{a-1}{a+1} - 1.5\log_e a_0\right)\frac{1-\dfrac{{r_e}^2}{r^2}}{\dfrac{a^2}{a_0^2}-1} \tag{3.310}$$

$$\sigma_{id} = 2f\left(2\frac{a-1}{a+1} - 1.5\log_e a_0\right)\frac{\dfrac{{r_e}^2}{r^2}}{\dfrac{a^2}{a_0^2}-1}$$

Eliminating the pressure, the release occurs according to the laws of elasticity. Recalling (3.40) and (3.286) the corresponding stresses are given by

$$\sigma_t = -\frac{2f}{(a+1)^2}\left(1+\frac{{r_e}^2}{r^2}\right)$$

$$\sigma_r = -\frac{2f}{(a+1)^2}\left(1-\frac{{r_e}^2}{r^2}\right) \tag{3.311}$$

$$\sigma_{id} = -\frac{4f}{(a+1)^2}\frac{{r_e}^2}{r^2}$$

The residual stresses generated at release are the sum of the values obtained from (3.309), or (3.310), and those obtained from (3.311).

Figure 3.32 shows the stresses σ_t and σ_r relative to the pressurization of the cylinder and at release; moreover, it shows the residual stresses generated at release. We have assumed $a = 5$ with the corresponding $a_o = 1.55$, based on (3.308) and the curve in Fig. 3.31.

Figure 3.33 shows the ideal stresses σ_{id}. We determine that the maximum absolute value of the ideal stress corresponding to the release is smaller than $3f$ since $a < 6.464$. Consequently, the residual ideal stress is smaller than $1.5f$ as an absolute value, i.e., smaller than the yield strength. In terms of deformations, as we discussed in Sect. 3.1, the difference between ε_t and ε_r in the area showing plastic flow is proportional to the ideal deformation, as well as inversely proportional to the square of the radius.

We can therefore write

$$\varepsilon_{id(i)} = \varepsilon_{id(0)}\frac{{r_0}^2}{{r_i}^2}, \tag{3.312}$$

where $\varepsilon_{id(i)}$ and $\varepsilon_{id(0)}$ are the ideal deformations in correspondence to the internal fiber and of the radius r_0, respectively. Recalling (3.298) we have

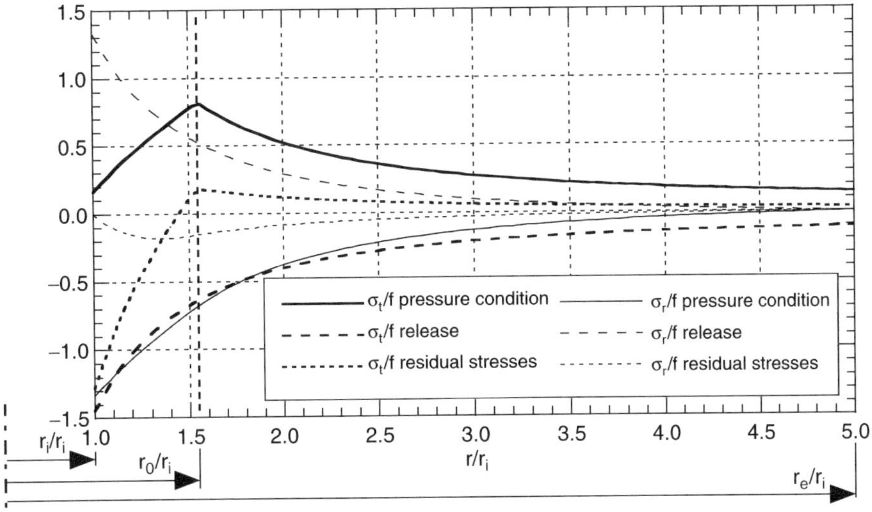

Fig. 3.32

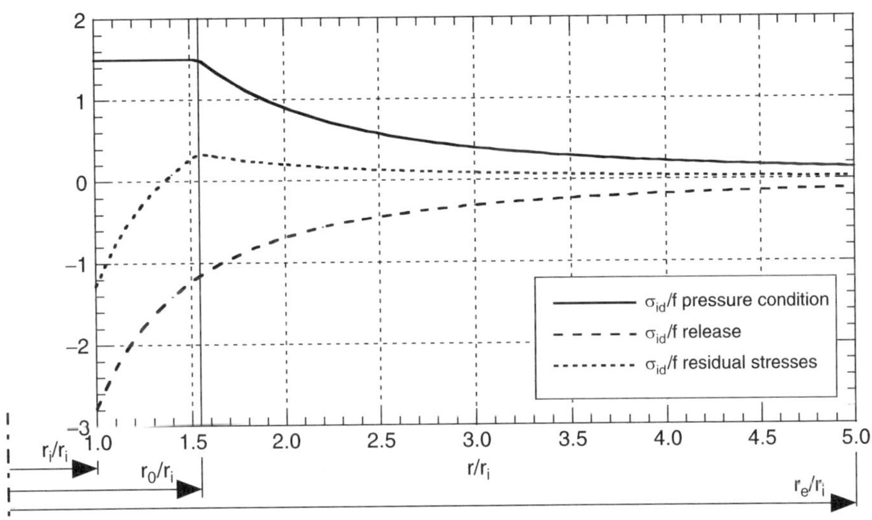

Fig. 3.33

$$\frac{\varepsilon_{id(i)}}{\varepsilon_{id(0)}} = a_0^2. \tag{3.313}$$

It is possible to compute the ratio between the deformations discussed above and in (3.313) from the curve in Fig. 3.31. This ratio is shown in Fig. 3.34.

This lets us highlight the influence of a on maximum deformations occurring in correspondence to the internal fiber.

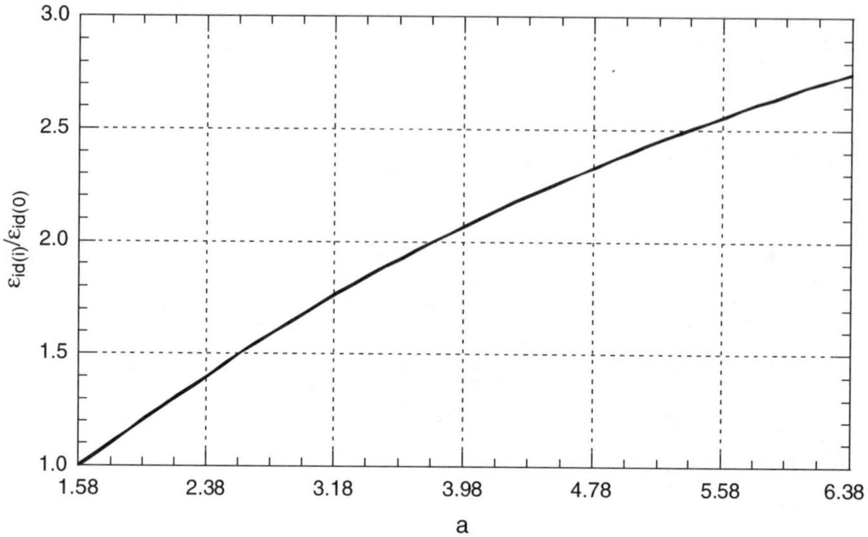

Fig. 3.34

We note that $\varepsilon_{id(o)}$ corresponds to the yield strength. It is better to limit the deformations in the area with plastic flow, and our recommendation is that their value be no greater than twice the deformation that corresponds to the yield strength. Consequently, in view of Fig. 3.34, the maximum allowable value for a we obtain is $a = 3.8$.

This value substitutes the limit of 6.464, previously obtained through (3.291) as a condition for the validity of (3.285). Note that applying this conservative measure results in $p = 1.17f$ (i.e., smaller than the yield strength), and that the residual ideal stress at release practically coincides with the allowable stress of the material. In other words, this is equivalent to assuming within the elastic field an allowable stress of $2.5f$ instead of $3f$, as shown in (3.289).

Let us now look at the influence of a thermal flux in relation to the allowable value of a. Recalling the discussion in Sect. 3.3, if the flux is centripetal the worst condition occurs in correspondence to the internal fiber.

The ideal stress is given by

$$\sigma_{id(i)} = p\frac{2a^2}{a^2-1} + \frac{E\alpha\Delta t}{2(1-\mu)}\left(\frac{2a^2}{a^2-1} - \frac{1}{\log_e a}\right). \tag{3.314}$$

Let us introduce the dimensionless factor β

$$\beta = \frac{E\alpha\Delta t}{2f(1-\mu)}. \tag{3.315}$$

Recalling (3.286) and assuming that $\sigma_{id(i)} = 3f$, from (3.314) we obtain

$$\frac{4a^2}{(a+1)^2} + \beta\left(\frac{2a^2}{a^2-1} - \frac{1}{\log_e a}\right) = 3, \tag{3.316}$$

then

$$\beta = \frac{3 - \dfrac{4a^2}{(a+1)^2}}{\dfrac{2a^2}{a^2-1} - \dfrac{1}{\log_e a}}. \tag{3.317}$$

Equation (3.317) allows us to compute the maximum allowable value of β, and therefore of Δt as a function of a. In the case of centrifugal flux, we shall first look at the value of the circumferential stress σ_{te} in correspondence to the external fiber. This circumferential stress is equal to the ideal stress since $\sigma_{re} = 0$ and is given by

$$\sigma_{id(e)} = \sigma_{te} = p\frac{2}{a^2-1} + \frac{E\alpha\Delta t}{2(1-\mu)}\left(\frac{2}{a^2-1} - \frac{1}{\log_e a}\right). \tag{3.318}$$

Let us remember that in this case Δt is negative, and the same holds true for β; following the same procedure as before we have

$$\frac{4}{(a+1)^2} + \beta\left(\frac{2}{a^2-1} - \frac{1}{\log_e a}\right) = 3, \tag{3.319}$$

then

$$\beta = \frac{3 - \dfrac{4}{(a+1)^2}}{\dfrac{2}{a^2-1} - \dfrac{1}{\log_e a}}. \tag{3.320}$$

Equation (3.320) allows us to compute the values of β for $\sigma_{te} = 3f$; values of β that are greater (smaller as an absolute value) than those computed from (3.320) are allowed.

As we discussed in Sect. 3.3, we cannot ignore the value that the longitudinal stress σ_a takes in correspondence to the internal fiber. Conservatively assuming that the pressure may not be present, such (negative) stress becomes identical to the circumferential stress, and we obtain

$$\sigma_{ai} = \sigma_{ti} = \frac{E\alpha\Delta t}{2(1-\mu)}\left(\frac{2a^2}{a^2-1} - \frac{1}{\log_e a}\right). \tag{3.321}$$

Let us now set $\sigma_{ai} = -3f$. Through a procedure similar to the one shown above, we obtain

$$\beta = -\frac{3}{\dfrac{2a^2}{a^2-1} - \dfrac{1}{\log_e a}}. \tag{3.322}$$

Keeping in mind the limitation of a in relation to pressure only, and referring back to $a = 3.8$, based on (3.317), (3.320), and (3.322), we identify a field

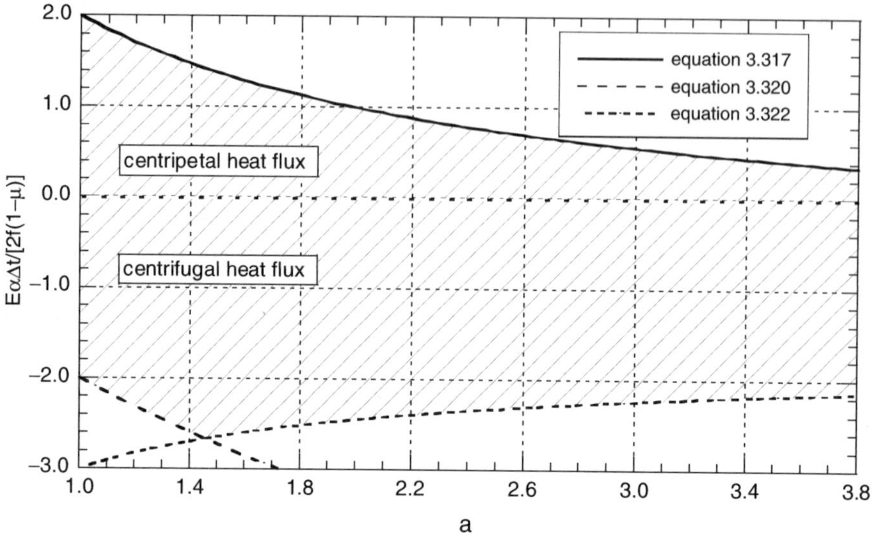

Fig. 3.35

where it is possible to apply (3.285) for the sizing of the cylinder. This field is shown in Fig. 3.35.

Let us remember that positive values of β correspond to a centripetal thermal flux, whereas negative values of β correspond to a centrifugal flux. Specifically, we determine that the centripetal flux may considerably reduce the allowable value of a (even below 1.58 discussed in Sect. 3.2). Moreover, we should keep in mind that small values of a generally correspond to small values of the thickness (unless the diameter is very large); they seldom lead to large Δt and β.

The criteria discussed in this section are applicable to the sizing of thick cylinders if we can disregard stress peaks other than the ones discussed. We will discuss stress conditions occurring in correspondence to the junction between cylinder and heads (most likely hemispherical ones).

As we have seen, the use of (3.285) is justified because self-stresses generated at release are exploited, thus the cylinder subsequently put under pressure (and temperature) works in the elastic field at the limit between stresses $-\sigma_s$ and σ_s. Of course, we should not forget that such situation occurs only in a limited area of the cylinder, and that in any case the average ideal stress is not greater than the allowable stress of the material. It is probably useful to remember that in the case of large values of a, the average ideal stress is in fact smaller than the allowable stress of the material (3.285), as discussed in Sect. 3.1.

Specifically, from (3.54) we should have

$$a = e^{p/f}; \tag{3.323}$$

instead, from (3.286) we have

$$a = \frac{1 + \dfrac{p}{2f}}{1 - \dfrac{p}{2f}}. \tag{3.324}$$

From (3.324) for $a = 3.8$ we have $p/f = 1.16$, while from (3.323) we have $p/f = 1.335$. This means that the average ideal stress is not equal to f but rather to $1.167f/1.335 = 0.874f$. The approach to tolerate large local stress peaks appears all the more justified, therefore.

Finally, we observe that if Δt does not refer to static conditions but to an abrupt increase in temperature on the internal or external side of the cylinder, the curve in Fig. 3.35 may still be used, but with a conservative value Δt twice the actual one, as long as the number of cycles does not impose a fatigue analysis: indeed, let us remember the end of Sect. 3.3, and specifically (3.179) and (3.181).

3.7 Stresses Due to Thickness Variation

Sheets of different thickness may be used during production of cylinders. A specific example is shown in Fig. 3.36. The cylinder is made out of two semicylinders of different thickness. In general, this happens when there are holes concentrated in an area which is only half the circumference of the cylinder. Therefore, it may be appropriate to design a thicker piece because of the holes instead of increasing the thickness of the entire cylinder. At the junction of the two semicylinders that we assume to be disconnected, the variations in radius are different due to the different rigidity of the two elements. Forces F_o and moments M_o are generated to restore congruence. Let us now compute both. The circumferential deformation ε_t in a cylinder seen as a membrane is given by

$$\varepsilon_t = \frac{1}{E}\left(\sigma_t - \mu\sigma_a\right). \tag{3.325}$$

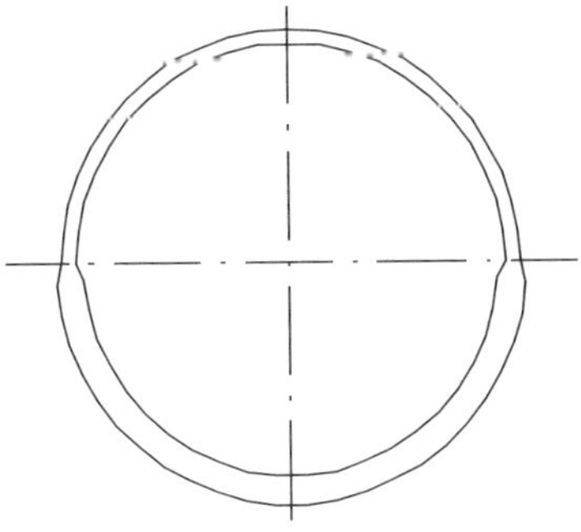

Fig. 3.36

From (2.16) we have

$$\varepsilon_t = \frac{pD}{4Es}(2-\mu), \tag{3.326}$$

where D is the average diameter. The variation in radius Δr is therefore

$$\Delta r = \varepsilon_t \frac{D}{2} = \frac{pD^2}{8Es}(2-\mu). \tag{3.327}$$

Let us indicate with subscript 1 all parameters for the semicylinder of thickness s_1, and with subscript 2 all those for the semicylinder of thickness s_2 (see Fig. 3.37). Then

$$\begin{aligned} \Delta r_1 &= \frac{pD^2}{8Es_1}(2-\mu) \\ \Delta r_2 &= \frac{pD^2}{8Es_2}(2-\mu) \end{aligned} \tag{3.328}$$

The difference Δr_{12} between Δr_1 and Δr_2 is therefore

$$\Delta r_{12} = \frac{pD^2}{8E}(2-\mu)\left(\frac{1}{s_1} - \frac{1}{s_2}\right). \tag{3.329}$$

Let us introduce α given by

$$\alpha = s_1/s_2. \tag{3.330}$$

From (3.329) we obtain

$$\Delta r_{12} = \frac{pD^2}{8Es_1}(1-\alpha)(2-\mu). \tag{3.331}$$

Forces F_0 and moments M_0 (see Fig. 3.37) are generated at the edge of the semicylinders to restore congruence. Let us make a few qualitative

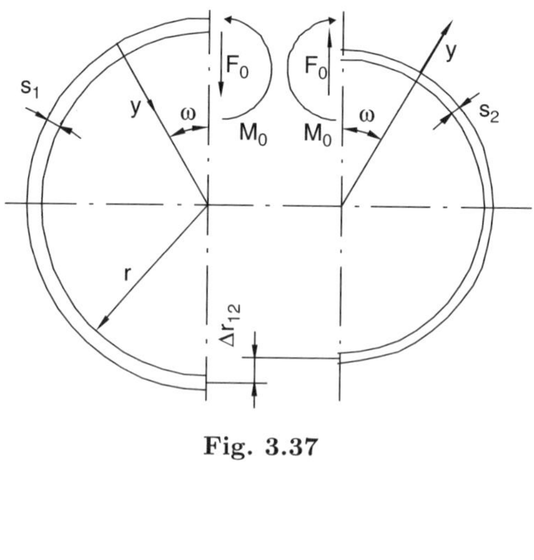

Fig. 3.37

considerations first. The radial deflection in A generated by F_o and M_o that we indicate with y_o must be, following from (3.331),

$$y_0 \equiv \frac{pD^2}{s_1}. \tag{3.332}$$

On the other hand, indicating with M_1 the maximum bending moment in the semicylinder of thickness s_1, from the theory of deflected beams we have

$$y_0 \equiv \frac{M_1 L^2}{I}, \tag{3.333}$$

with L being a characteristic length and I the moment of inertia. Of course, in the present case the characteristic length is the diameter of the cylinder, while the moment of inertia is proportional to s_1^3. Therefore,

$$y_0 \equiv \frac{M_1 D^2}{s_1^3}. \tag{3.334}$$

Comparing (3.334) with (3.332)

$$M_1 \equiv p s_1^2. \tag{3.335}$$

Recalling that $W \equiv s_1^2$ finally that the maximum stress due to moment M_1 is

$$\sigma_1 \equiv \frac{M_1}{s_1^2} \equiv p. \tag{3.336}$$

We determine that the maximum stress generated in the cylinder because of the variation in thickness is proportional to pressure only. It is therefore independent from the thickness of the semicylinder and from the diameter. Let us now look at ways to compute the various parameters. Let ω be the angle between the generic radius and the one at the junction of the two pieces, and y be the radial deflection resulting from F_o and M_o. We know that

$$\frac{d^2 y}{d\omega^2} + y = -\frac{Mr^2\left(1-\mu^2\right)}{EI}, \tag{3.337}$$

where the term $(1-\mu^2)$ takes into account the prevented longitudinal deformations. For the semicylinder of thickness s_1 let us consider the deflections directed towards the center as positive; because the moments that stretch the internal fibers are positive we have

$$M_1 = M_0 - F_0 r \sin\omega. \tag{3.338}$$

Because we set

$$A_1 = -\frac{M_0 r^2\left(1-\mu^2\right)}{EI} = -\frac{12 M_0 r^2\left(1-\mu^2\right)}{E s_1^3}, \tag{3.339}$$

and

$$B_1 = \frac{F_0 r^3 \left(1-\mu^2\right)}{EI} = \frac{12 F_0 r^3 \left(1-\mu^2\right)}{E s_1^3}, \tag{3.340}$$

from (3.337) we have

$$\frac{d^2 y}{d\omega^2} + y = A_1 + B_1 \sin\omega. \tag{3.341}$$

The integral of (3.341), with a zero deflection y for $\omega = \pi/2$, is

$$y = A_1 \left(1 - \sin\omega\right) - \frac{B_1}{2}\omega \cos\omega + C_1 \cos\omega, \tag{3.342}$$

where C_1 is constant. On the other hand, the deflections for $\omega = 0$ and $\omega = \pi$ shall be the same, thus

$$C_1 = \frac{B_1}{2}\pi - C_1, \tag{3.343}$$

then

$$C_1 = \frac{B_1 \pi}{4}. \tag{3.344}$$

Therefore

$$y = A_1 \left(1 - \sin\omega\right) + \frac{B_1}{2}\left(\frac{\pi}{2} - \omega\right)\cos\omega; \tag{3.345}$$

moreover

$$y' = -A_1 \cos\omega - \frac{B_1}{2}\left[\left(\frac{\pi}{2} - \omega\right)\sin\omega + \cos\omega\right]. \tag{3.346}$$

Specifically, at point A with $\omega = 0$

$$y_{01} = A_1 + \frac{\pi}{4} B_1, \tag{3.347}$$

and

$$y'_{01} = -A_1 - \frac{B_1}{2}. \tag{3.348}$$

Let us now examine the semicylinder of thickness s_2, and let us consider the deflections y that are directed towards the outside of the cylinder positive. In this case, the moments that stretch the external fibers are to be considered positive, therefore

$$M_2 = -M_0 - F_0 r \sin\omega. \tag{3.349}$$

Let us assume now

$$A_2 = \frac{M_0 r^2 \left(1-\mu^2\right)}{EI} = \frac{12 M_0 r^2 \left(1-\mu^2\right)}{E s_2^3}, \tag{3.350}$$

and

$$B_2 = \frac{F_0 r^3 \left(1-\mu^2\right)}{EI} = \frac{12 F_0 r^3 \left(1-\mu^3\right)}{E s_2^3}. \tag{3.351}$$

Proceeding in a similar way to what we observed for the semicylinder of thickness s_1, we have

$$y_{02} = A_2 + \frac{\pi}{4} B_2, \tag{3.352}$$

and

$$y'_{02} = -A_2 - \frac{B_2}{2}. \tag{3.353}$$

Congruence requires that

$$y_{01} + y_{02} = \Delta_{12}, \tag{3.354}$$

and

$$y'_{01} = y'_{02}. \tag{3.355}$$

Therefore, recalling previous equations

$$\frac{12r^2\left(1-\mu^2\right)}{E}\left(\frac{-M_0+\frac{\pi}{4}F_0 r}{s_1^3}+\frac{M_0+\frac{\pi}{4}F_0 r}{s_2^3}\right)=\frac{pD^2}{8Es_1}\left(1-\alpha\right)\left(2-\mu\right); \tag{3.356}$$

$$\frac{M_0 - F_0\frac{r}{2}}{s_1^3} = \frac{-M_0 - F_0\frac{r}{2}}{s_2^3}. \tag{3.357}$$

Recalling (3.330) and through various steps we obtain

$$-M_0 + \frac{\pi}{8}F_0 D + \left(M_0 + \frac{\pi}{8}F_0 D\right)\alpha^3 = \frac{ps_1^2}{24}\frac{\left(1-\alpha\right)\left(2-\mu\right)}{1-\mu^2}, \tag{3.358}$$

and

$$M_0 - \frac{F_0 D}{4} = \left(-M_0 - \frac{F_0 D}{4}\right)\alpha^3. \tag{3.359}$$

From (3.359)

$$F_0 = \frac{4M_0}{D}\frac{1+\alpha^3}{1-\alpha^3}, \tag{3.360}$$

and substituting the term F_o in (3.359), we finally obtain

$$M_0 = Cps_1^2, \tag{3.361}$$

with

$$C = \frac{2-\mu}{24\left(1-\mu^2\right)}\frac{\left(1-\alpha^3\right)\left(1-\alpha\right)}{\frac{\pi}{2}\left(1+\alpha^3\right)^2-\left(1-\alpha^3\right)^2} \tag{3.362}$$

Specifically, for $\mu = 0.3$,

$$C = 0.07784\frac{\left(1-\alpha^3\right)\left(1-\alpha\right)}{1.571\left(1+\alpha^3\right)^2-\left(1-\alpha^3\right)^2}. \tag{3.363}$$

Moreover, from (3.360)

$$F_0 = G\frac{ps_1^2}{D}, \tag{3.364}$$

with

$$G = \frac{2-\mu}{6\left(1-\mu^2\right)} \frac{\left(1+\alpha^3\right)\left(1-\alpha\right)}{\frac{\pi}{2}\left(1+\alpha^3\right)^2 - \left(1-\alpha^3\right)^2}. \tag{3.365}$$

Specifically, for $\mu = 0.3$,

$$G = 0.3113 \frac{\left(1+\alpha^3\right)\left(1-\alpha\right)}{1.517\left(1+\alpha^3\right)^2 - \left(1-\alpha^3\right)^2}. \tag{3.366}$$

Figure 3.38 shows the values of C and G obtained from (3.363) and (3.366).

The moment in the semicylinder of thickness s_1 for $\omega = \frac{\pi}{2}$ is given by

$$M_1 = M_0 - F_0 \frac{D}{2} = \left(C - \frac{G}{2}\right) p s_1^2, \tag{3.367}$$

and the maximum stress due to M_1 is

$$\sigma_1 = 6 \left| C - \frac{G}{2} \right| p. \tag{3.368}$$

Assuming that

$$k_1 = 6 \left| C - \frac{G}{2} \right|, \tag{3.369}$$

we have

$$\sigma_1 = k_1 p, \tag{3.370}$$

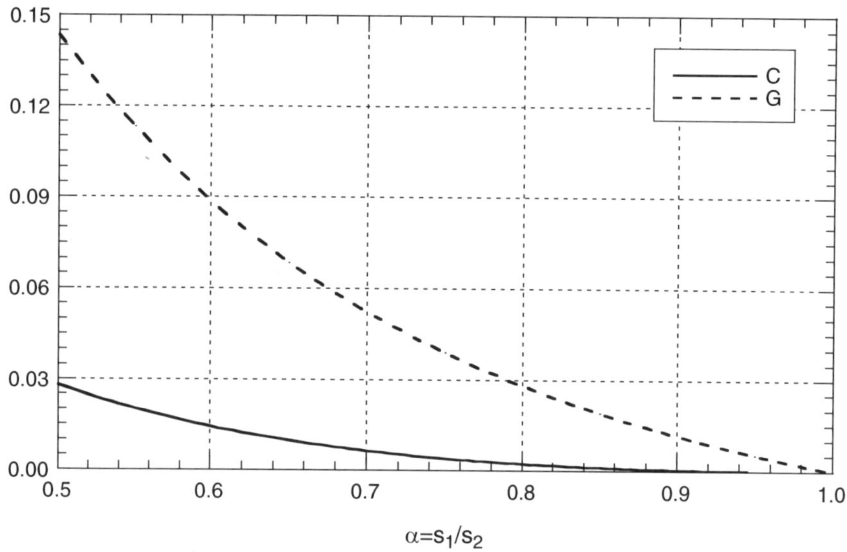

Fig. 3.38

and this confirms the validity of (3.336). As far as the semicylinder of greater thickness we have the following relationship:

$$M_2 = M_0 + F_0 \frac{D}{2} = \left(C + \frac{G}{2} \right) p s_1^2, \tag{3.371}$$

and the corresponding stress σ_2 is

$$\sigma_2 = 6 \left(C + \frac{G}{2} \right) \frac{p s_1^2}{s_2^2} = 6 \left(C + \frac{G}{2} \right) \alpha^2 p. \tag{3.372}$$

With

$$k_2 = 6 \left(C + \frac{G}{2} \right) \alpha^2, \tag{3.373}$$

we finally obtain

$$\sigma_2 = k_2 p. \tag{3.374}$$

Figure 3.39 shows both k_1 and k_2. We notice that the variation in thickness in the cylinder is irrelevant as far as its resistance. Note, e.g., that even if $\alpha = 0.5$, thus $s_2 = 2s_1$, we have $\sigma_1 = 0.2625p$. The hoop membrane stress in the cylinder is equal to $0.5pD/s_1$, and assuming that it is equal to the basic allowable stress f we obtain

$$\sigma_1 = 0.265 \times 2 f s_1/D = 0.525 f s_1/D$$

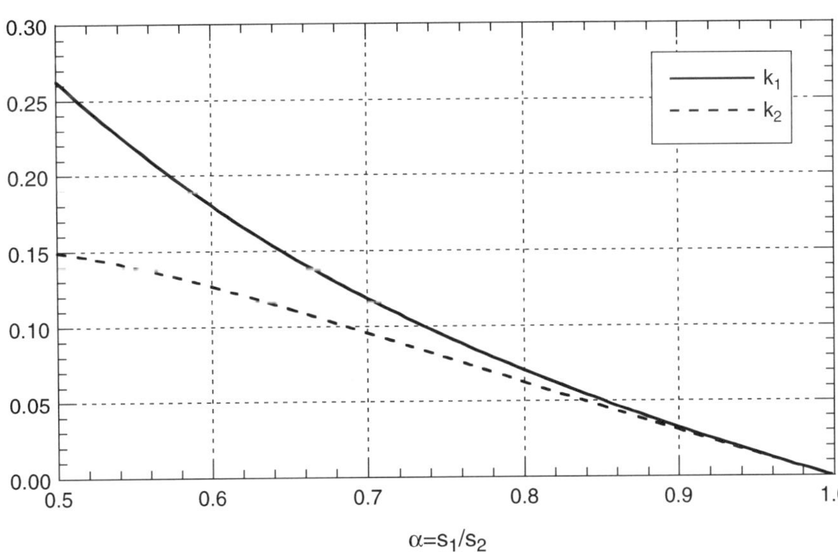

Fig. 3.39

Even if $s_1/D = 0.1$ (an already large value for the structure under investigation) we have

$$\sigma_1 = 0.0525f$$

The maximum stress generated by the moments active in the cylinder because of the variation in thickness is equal to 5% of the allowable stress, and therefore irrelevant as far as the resistance of the cylinder.

At the junction of the two pieces the maximum stress shows an even smaller value, as we determine by comparing the value of C with the absolute value of the term $(C - G/2)$ (see (3.361) and (3.367)). The stresses caused by bending moments may therefore be ignored.

As far as the membrane stresses in the area of the junction are concerned, we have stress peaks the values of which depend on the characteristics of the junction.

With reference to Fig. 3.40, F is the circumferential force per unitary length of the cylinder. The distribution of stresses may be computed with reasonable approximation by looking at the triangle with sides that coincide with those of the junction. For the latter we can write

$$\sigma_r = \frac{F\cos\vartheta}{r\left(\varphi + \dfrac{1}{2}\sin 2\varphi\right)}, \tag{3.375}$$

where σ_r is the stress directed as the radius r that forms the angle ϑ with the center line of the section, whereas φ is the semiangle at the apex of the triangle that corresponds to the tilt of the junction.

The stress σ_x orthogonal to the generic section $A - A$ is not constant and varies instead according to $\cos^2\vartheta$; considering the distance a of the section from the apex of the triangle we have

$$\sigma_x = \sigma_r\cos^2\vartheta = \frac{F\cos^3\vartheta}{r\left(\varphi + \dfrac{1}{2}\sin 2\varphi\right)} = \frac{F\cos^4\vartheta}{a\left(\varphi + \dfrac{1}{2}\sin 2\varphi\right)}. \tag{3.376}$$

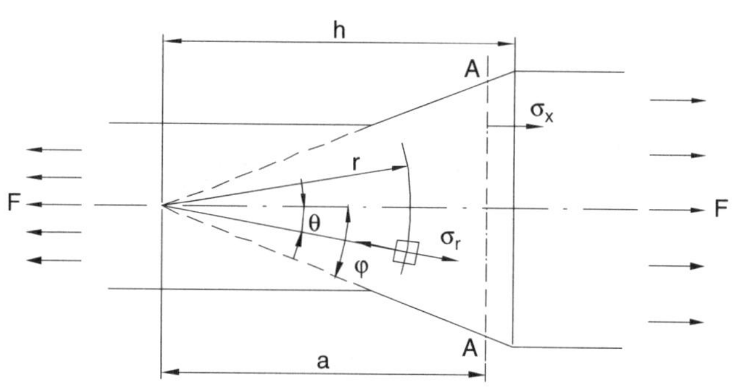

Fig. 3.40

The average value σ_{xm} of the stress σ_x in section $A - A$ is given by

$$\sigma_{xm} = \frac{F}{2a\tan\varphi}; \tag{3.377}$$

therefore

$$\sigma_x = \sigma_{xm} \frac{2\tan\varphi\cos^4\vartheta}{\varphi + \frac{1}{2}\sin 2\varphi}. \tag{3.378}$$

The ratio σ_x/σ_{xm} as a function of ϑ/φ for $\varphi = 10°, 20°$ and $30°$ is shown in Fig. 3.41.

The maximum value of σ_x occurs in correspondence to the center line of the section (for $\vartheta = 0$) and is

$$\sigma_{x\,\max} = k\sigma_{xm}, \tag{3.379}$$

with

$$k = \frac{2\tan\varphi}{\varphi + \frac{1}{2}\sin 2\varphi}. \tag{3.380}$$

The values of k are shown in Fig. 3.42 and highlight the amount of local stress peaks due to the junction. The values of k differ very little from 1 even for relatively large and unusual tilts of the junction. For instance, for a tilt equal to 1/2, we have $k = 1.16$. In the much more common case of tilts equal to 1/4 or 1/3 the increase in value of the stress is no greater than 7%.

Also from this point of view, the variation in thickness of the cylinder is not cause for concern. On the contrary, the stress peaks generated in correspondence to the sharp corners at the beginning and at the end of the junction

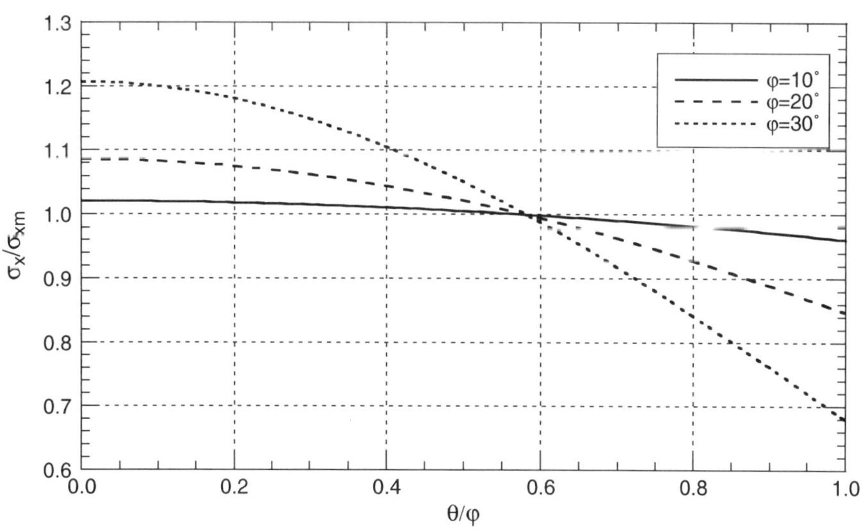

Fig. 3.41

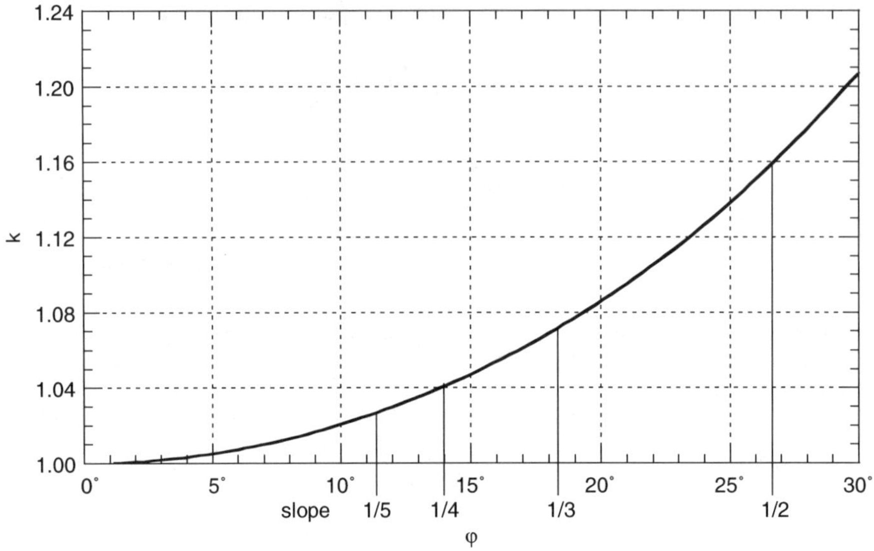

Fig. 3.42

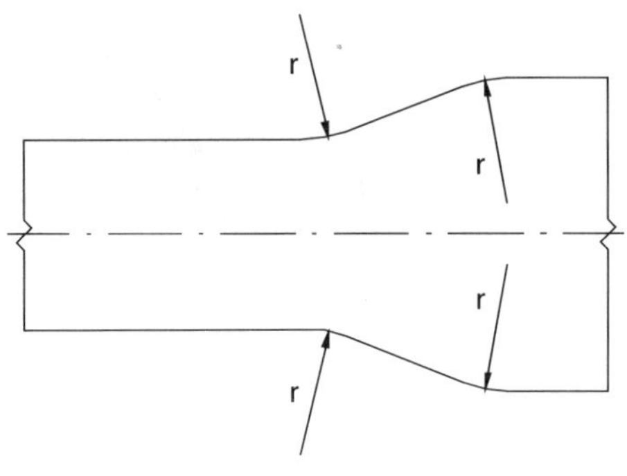

Fig. 3.43

are more important. In order to eliminate, or at least reduce them, it is best to move from constant thickness to variable one by smoothing the corners, as shown in Fig. 3.43.

A different case that may occur is when there is a junction of two cylinder portions of different thickness, as in Fig. 3.44. If we call s_1 the smaller thickness, s_2 the greater thickness, and D the average diameter, the difference Δr_{12} between the variations of radius Δr_1 and Δr_2 of the two portions of the cylinder assumed to be free to undergo deformations is still given by (3.331).

To restore congruence one must apply forces F_o and moments M_o to the edges (see Fig. 3.45). From (2.66) and taking into account the direction of the moments, we have

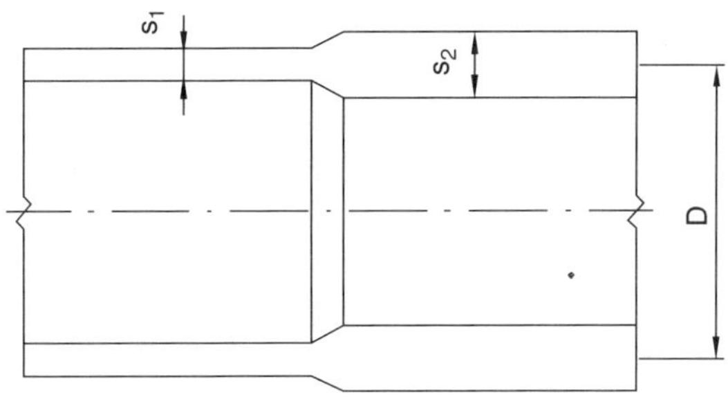

Fig. 3.44

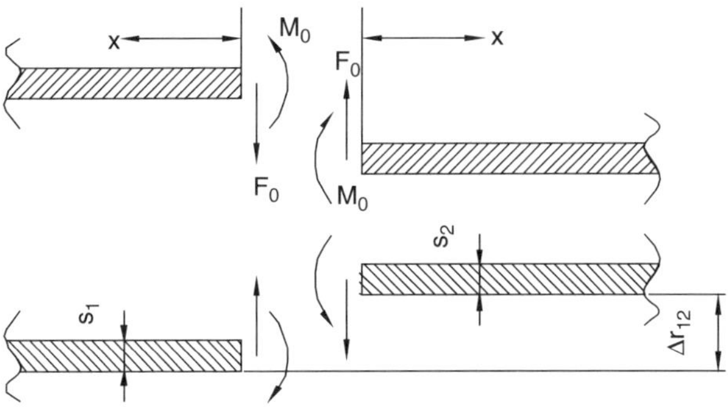

Fig. 3.45

$$0.909\frac{F_0}{E}\left(\frac{D}{s_1}\right)^{3/2} - 1.652\frac{M_0}{Es_1}\frac{D}{s_1} + 0.909\frac{F_0}{E}\left(\frac{D}{s_2}\right)^{3/2} + 1.652\frac{M_0}{Es_2}\frac{D}{s_2}$$
$$= \frac{pD^2}{8E}(2-\mu)\left(\frac{1}{s_1} - \frac{1}{s_2}\right) \quad (3.381)$$

Moreover, from (2.67), to ensure that the rotation of both edges is the same, we have

$$1.652\frac{F_0}{Es_1}\frac{D}{s_1} - 6.007\frac{M_0}{Es_1^2}\sqrt{\frac{D}{s_1}} = 1.652\frac{F_0}{Es_2}\frac{D}{s_2} + 6.007\frac{M_0}{Es_2^2}\sqrt{\frac{D}{s_2}}. \quad (3.382)$$

As in the previous case, assuming that

$$\alpha = \frac{s_1}{s_2}, \quad (3.383)$$

(3.381) and (3.382) may be written as

$$0.909F_0\left(\frac{D}{s_1}\right)^{3/2}\left(1+\alpha^{3/2}\right)-1.652\frac{M_0D}{s_1^2}\left(1-\alpha^2\right)=\frac{pD^2}{8s_1}\left(2-\mu\right)\left(1-\alpha\right); \tag{3.384}$$

$$1.652\frac{F_0D}{s_1^2}\left(1-\alpha^2\right)-6.007\frac{M_0}{s_1^2}\sqrt{\frac{D}{s_1}}\left(1+\alpha^{5/2}\right)=0. \tag{3.385}$$

We find that the determinant of the system with $\mu = 0.3$ is

$$\Delta=-2.73\frac{D^2}{s_1^4}\left[2\left(1+\alpha^{3/2}\right)\left(1+\alpha^{5/2}\right)-\left(1-\alpha^2\right)^2\right]. \tag{3.386}$$

The determinant relative to F_o is

$$\Delta_1=-1.276p\frac{D^2}{s_1^4}\sqrt{Ds_1}\left(1-\alpha\right)\left(1+\alpha^{5/2}\right), \tag{3.387}$$

while the determinant relative to M_o is

$$\Delta_2=-0.352p\frac{D^3}{s_1^3}\left(1-\alpha\right)\left(1-\alpha^2\right). \tag{3.388}$$

Therefore, we obtain

$$F_0=\frac{\Delta_1}{\Delta}=0.467p\sqrt{Ds_1}\frac{\left(1-\alpha\right)\left(1+\alpha^{5/2}\right)}{2\left(1+\alpha^{3/2}\right)\left(1+\alpha^{5/2}\right)-\left(1-\alpha^2\right)^2}, \tag{3.389}$$

and

$$M_0=\frac{\Delta_2}{\Delta}=0.129pDs_1\frac{\left(1-\alpha\right)\left(1-\alpha^2\right)}{2\left(1+\alpha^{3/2}\right)\left(1+\alpha^{5/2}\right)-\left(1-\alpha^2\right)^2}. \tag{3.390}$$

With

$$k_3=0.467\frac{\left(1-\alpha\right)\left(1+\alpha^{5/2}\right)}{2\left(1+\alpha^{3/2}\right)\left(1+\alpha^{5/2}\right)-\left(1-\alpha^2\right)^2}, \tag{3.391}$$

and

$$k_4=0.129\frac{\left(1-\alpha\right)\left(1-\alpha^2\right)}{2\left(1+\alpha^{3/2}\right)\left(1+\alpha^{5/2}\right)-\left(1-\alpha^2\right)^2}. \tag{3.392}$$

force F_o and moment M_o are given by

$$F_0=k_3p\sqrt{Ds_1}, \tag{3.393}$$

and

$$M_0=k_4Ds_1. \tag{3.394}$$

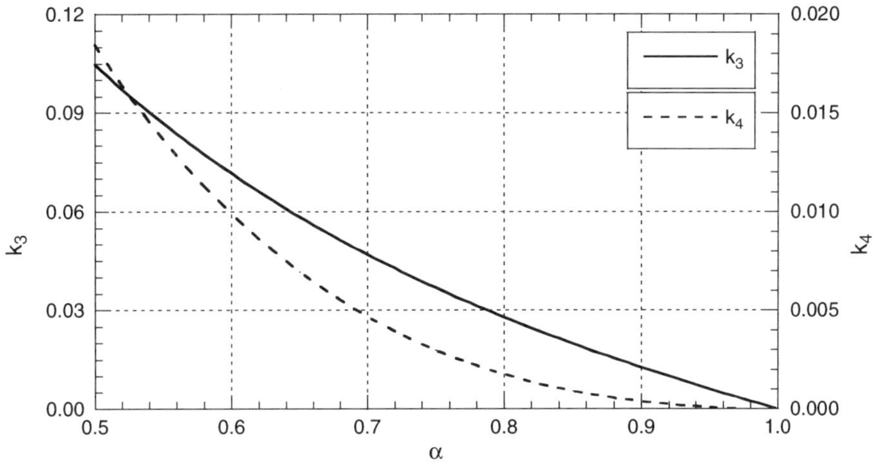

Fig. 3.46

k_3 and k_4 for $\alpha = 0.5 \div 1$ are shown in Fig. 3.46. Equation (3.394) allows one to compute the moment at the edge and consequently the relative stress. The value of the moment, though, is not necessarily the maximum or minimum in correspondence to the junction. To extend the analysis near the edge we have to refer to (2.53) and (2.60).

Keeping in mind the orientation of F_o, the direction of M_o, as well as the meaning of β (2.41), the moment M_x in the cylinder of thickness s_1 at distance x from the edge, recalling (3.393) and (3.394), is

$$M_x = -0.55 k_3 p D s_1 e^{-\beta x} \sin \beta x + k_4 p D s_1 e^{-\beta x} (\sin \beta x + \cos \beta x) . \quad (3.395)$$

The minimum value of M_x is obtained setting the derivative of M_x to zero with respect to x; we observe that since $e^{-\beta x}$ is not zero, we must have

$$k_3 (\sin \beta x - \cos \beta x) = 3.64 k_4 \sin \beta x, \quad (3.396)$$

then

$$\tan \beta x = \frac{1}{1 - 3.64 \dfrac{k_4}{k_3}} . \quad (3.397)$$

By introducing the smaller value of βx in (3.395) that satisfies (3.397), we determine that the minimum value of M_x may be expressed as follows:

$$M_{x(\min)} = -k_5 p D s_1 , \quad (3.398)$$

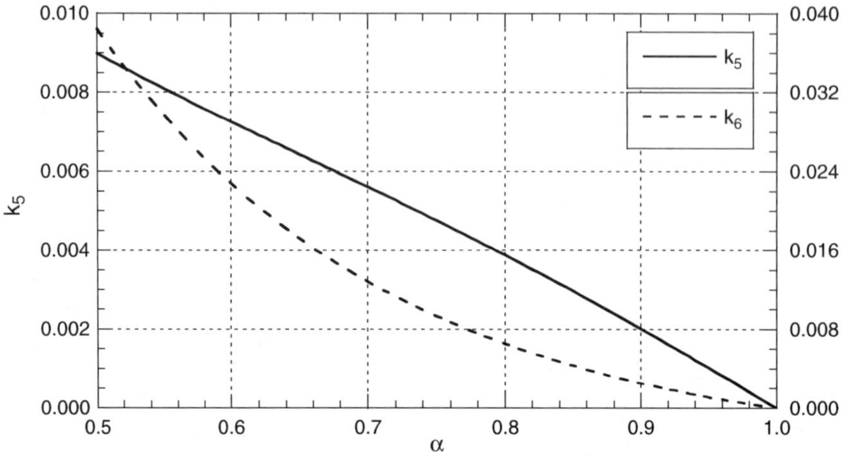

Fig. 3.47

where k_5 is as shown in Fig. 3.47. Similarly, for the cylinder of thickness s_2, taking into account the orientation of F_o and the direction of M_o, we determine that

$$M_x = \frac{k_3}{\sqrt{\alpha}} pDs_1 e^{-\beta x} \sin\beta x + k_4 pDs_1 e^{\beta x} (\sin\beta x + \cos\beta x) , \tag{3.399}$$

where, in this case

$$\beta = \frac{1.817}{\sqrt{Ds_2}} . \tag{3.400}$$

Setting the derivative of M_x to zero with respect to x, we determine that the maximum value of M_x is when

$$\frac{k_3}{\sqrt{\alpha}} (\cos\beta x - \sin\beta x) = 3.64 k_4 \sin\beta x . \tag{3.401}$$

Thus

$$\tan\beta x = \frac{1}{1 + 3.64\sqrt{\alpha}\dfrac{k_4}{k_3}} . \tag{3.402}$$

Introducing the smaller value of βx in (3.399) that satisfies (3.402), we determine that the maximum value of M_x may be expressed as follows:

$$M_{x(\max)} = k_6 pDs_1 , \tag{3.403}$$

where k_6 is shown in Fig. 3.47. We determine that the value of k_4 is higher than the one of k_5 for $\alpha < 0.67$, and viceversa if $\alpha > 0.67$. In absolute terms, the maximum value of the bending moment M_x in the cylinder of thickness s_1 is therefore given by

$$M_x = kpDs_1, \tag{3.404}$$

with $k = k_4$ or $k = k_5$, depending on the situation. Taking into account that the section modulus is

$$W = \frac{s_1^2}{6}, \tag{3.405}$$

the maximum stress σ_{af} generated by M_x becomes

$$\sigma_{af} = 6kp\frac{D}{s_1}. \tag{3.406}$$

If the sizing of the cylinder has been done in such a way that the ideal membrane stress is equal to the basic allowable stress f we have

$$\frac{pD}{2s_1} = f, \tag{3.407}$$

and therefore

$$\sigma_{af} = 12kf. \tag{3.408}$$

Finally,

$$k_7 = \frac{\sigma_{af}}{f} = 12k. \tag{3.409}$$

Recalling the meaning of k, based on Figs. 3.46 and 3.47, the behavior of k_7 is as shown in Fig. 3.48. The maximum value of M_x in the cylinder of thickness s_2 is given by (3.403). Therefore, since

$$W = \frac{s_2^2}{6}, \tag{3.410}$$

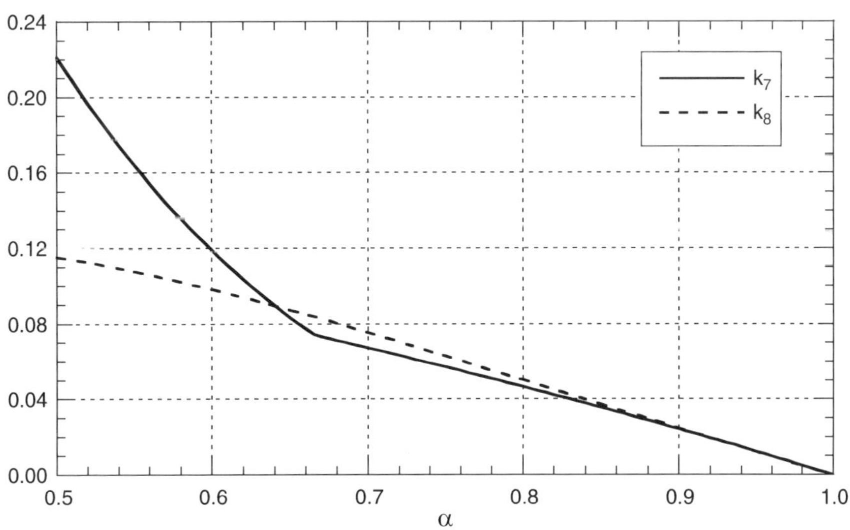

Fig. 3.48

similarly to the previous case, we obtain

$$\sigma_{af} = 6k_6 \frac{pDs_1}{s_2^2} = 6k_6\alpha^2 \frac{pD}{s_1}. \tag{3.411}$$

Recalling (3.407)

$$\sigma_{af} = 12k_6\alpha^2 f, \tag{3.412}$$

and, finally,

$$k_8 = \frac{\sigma_{af}}{f} = 12k_6\alpha^2. \tag{3.413}$$

Based on Fig. 3.47, the behavior of k_8 is as shown in Fig. 3.48.

We notice that the values of both k_7 and k_8 are rather modest. For instance, even in the case of $\alpha = 0.5$ where the thickness s_2 is twice that of s_1, we have $k_7 = 0.22$. Taking into account that, in view of the assumptions made, the longitudinal stress caused by the pressure on the heads is approximately equal to $0.5f$, the total maximum value of the longitudinal stress in correspondence to the junction is equal to $0.72f$, a value quite far from that $3f$ (twice the yield strength) that is allowable in this case, as the stresses due to M_x are secondary. In fact, they result solely from congruence.

As far as the hoop membrane stresses, let us recall one more time (2.66); the deflection y_0 at the edge of the cylinder with thickness s_2 is given by

$$y_0 = 0.909 \frac{F_0}{E} \left(\frac{D}{s_2}\right)^{3/2} + 1.652 \frac{M_0}{Es_2} \frac{D}{s_2}. \tag{3.414}$$

From (3.393) and (3.394) we obtain

$$y_0 = \frac{pD^2}{Es_1} \left(0.909\alpha^{3/2}k_3 + 1.652\alpha^2 k_4\right). \tag{3.415}$$

The hoop stress σ_t is given by

$$\sigma_t = \frac{2Ey_0}{D}, \tag{3.416}$$

then

$$\sigma_t = \frac{2pD}{s_1} \left(0.909\alpha^{3/2}k_3 + 1.652\alpha^2 k_4\right), \tag{3.417}$$

and, from (3.407)

$$k_9 = \frac{\sigma_t}{f} = 4\left(0.909\alpha^{3/2}k_3 + 1.652\alpha^2 k_4\right). \tag{3.418}$$

We have built the curve shown in Fig. 3.49 from Fig. 3.46. We notice that even the values of k_9 are modest. In fact, for $\alpha = 0.5$ we have $k_9 = 0.165$. Under the assumptions made, the general membrane stress is equal to $\alpha f = 0.5f$, and therefore the total membrane stress is equal to $0.665f$. This is a local

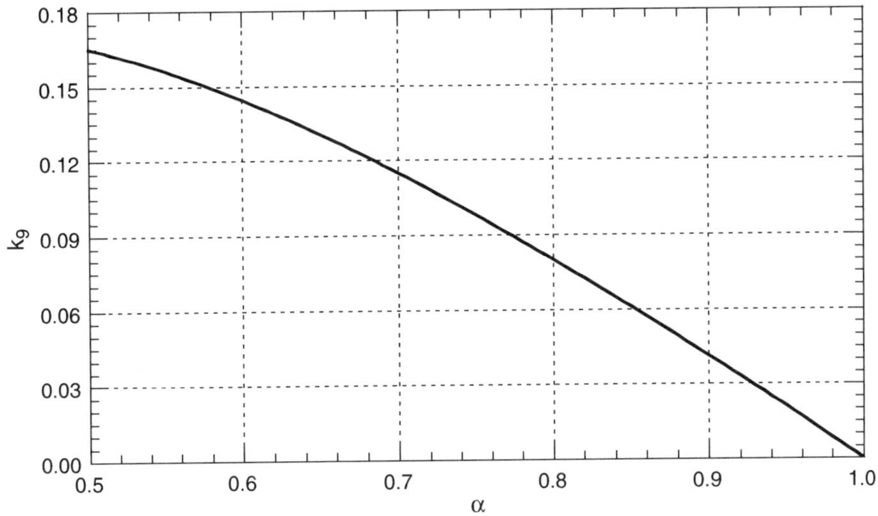

Fig. 3.49

membrane stress, and we can assume an allowable stress equal to $1.5f$, as we discussed in Sect. 1.5. The secondary hoop stresses due to the moment μM_x resulting from M_x for congruence are negligible.

Finally, as far as the cylinder of thickness s_1 is concerned, the deflection at the edge produces negative hoop stresses that cause a local membrane stress of smaller magnitude than the general membrane stress. In conclusion, the junction of two parts of a cylinder, even though they may have greatly different thickness, does not cause a stress state that should concern in terms of the resistance of the cylinder.

4

Cylinders Under External Pressure

4.1 Thick Cylinders

In the case of cylinders that are subject to external pressure the analysis is simpler when they are thick. In fact, in this case danger from buckling can be ruled out (instability-induced critical pressure is considerably higher than the allowable one, based on the mechanical characteristics of the material). Moreover, even possible deviations of the cylinder's profile from perfect roundness (substantial deviations are, after all, quite unlikely) do not influence the cylinder's behavior in any substantial way. This is not the case with thin cylinders, as we shall see later on. From this point of view we can consider thick cylinders those that show s/D_e greater than 0.1, with s being the thickness and D_e the outside diameter. We can apply calculation criteria to these cylinders that are similar to those for cylinders under internal pressure; they can obviously withstand high external pressure.

The equations of equilibrium and congruence are the same, specifically (3.11) and (3.16), and we rewrite them below.

Equation of equilibrium

$$\sigma_t - \sigma_r - r\frac{d\sigma_r}{dr} = 0. \tag{4.1}$$

Equation of congruence

$$\varepsilon_r - \varepsilon_t - r\frac{d\varepsilon_t}{dr} = 0. \tag{4.2}$$

Therefore, the general computation method is identical to that previously shown for cylinders under internal pressure, and we obtain the following equations for radial and hoop stress:

$$\sigma_r = A - \frac{B}{r^2}, \tag{4.3}$$

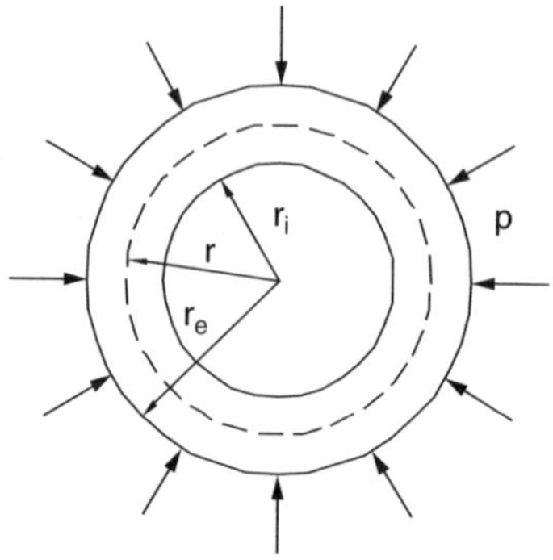

Fig. 4.1

and

$$\sigma_t = A + \frac{B}{r^2}, \tag{4.4}$$

where A and B are constant. For $r = r_i$, where r_i is the inside radius (see Fig. 4.1), we have $\sigma_r = 0$; therefore, from (4.3),

$$A = \frac{B}{r_i^2}, \tag{4.5}$$

then

$$\sigma_r = B\left(\frac{1}{r_i^2} - \frac{1}{r^2}\right). \tag{4.6}$$

On the other hand, for $r = r_e$, where r_e is the outside radius, we must have $\sigma_r = -p$; thus

$$-p = B\left(\frac{1}{r_i^2} - \frac{1}{r_e^2}\right), \tag{4.7}$$

and

$$B = -\frac{pr_e^2}{\frac{r_e^2}{r_i^2} - 1}. \tag{4.8}$$

With the following value for a:

$$a = \frac{r_e}{r_i}, \tag{4.9}$$

we obtain

$$\sigma_r = -p\frac{a^2}{a^2 - 1}\left(1 - \frac{r_i^2}{r^2}\right). \tag{4.10}$$

Finally, from (4.4),

$$\sigma_t = -p\frac{a^2}{a^2 - 1}\left(1 + \frac{r_i^2}{r^2}\right). \tag{4.11}$$

As far as the longitudinal stress caused by pressure on the heads we have

$$\sigma_a = -\frac{\pi p r_e^2}{\pi\left(r_e^2 - r_i^2\right)} = -p\frac{a^2}{a^2-1}. \tag{4.12}$$

Finally, the three principal stresses have the following equations:

$$\begin{aligned}\sigma_t &= -p\frac{a^2}{a^2-1}\left(1+\frac{r_i^2}{r^2}\right)\\ \sigma_r &= -p\frac{a^2}{a^2-1}\left(1-\frac{r_i^2}{r^2}\right)\\ \sigma_a &= -p\frac{a^2}{a^2-1}\end{aligned} \tag{4.13}$$

The behavior of the stresses is shown in Fig. 4.2. The maximum value of the hoop stress occurs in correspondence of the internal fiber and is

$$\sigma_{ti} = -p\frac{2a^2}{a^2-1} = \sigma_{id(i)}. \tag{4.14}$$

This value is also identical to the maximum value of σ_{id} according to Guest–Tresca's failure theory, given that all stresses are negative, and the greatest difference between principal stresses occurs in correspondence of the internal fiber where $\sigma_r = 0$. The sizing in the elastic field would thus lead to the following equation:

$$-p\frac{2a^2}{a^2-1} = -f, \tag{4.15}$$

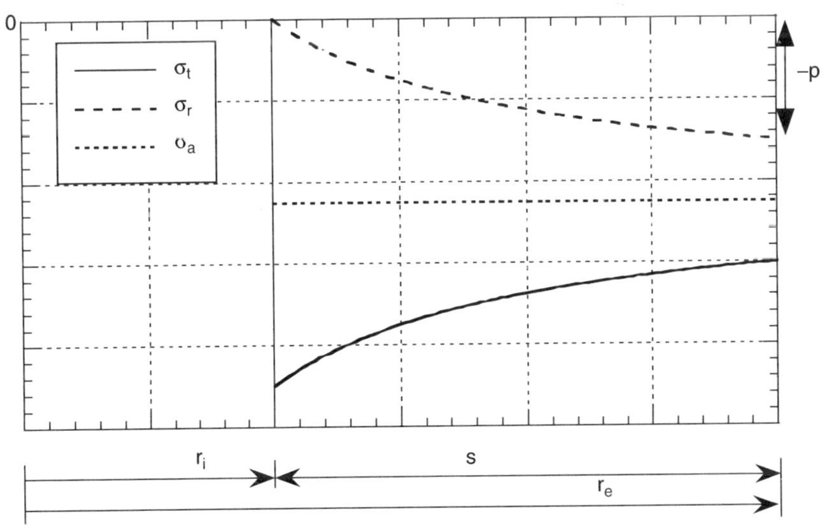

Fig. 4.2

where f is the basic allowable stress. From (4.15) we obtain

$$a = \frac{1}{\sqrt{1 - 2\frac{p}{f}}}; \tag{4.16}$$

then, with D_e being the outside diameter and s the thickness

$$s = \frac{D_e}{2}\left(1 - \sqrt{1 - 2\frac{p}{f}}\right). \tag{4.17}$$

Interestingly, (4.17) is identical to (3.45) relative to a cylinder under internal pressure. Similarly, applying the Huber–Hencky failure theory we obtain the same equation that applied to cylinders under internal pressure (3.46), that is

$$s = \frac{D_e}{2}\left(1 - \sqrt{1 - \sqrt{3}\frac{p}{f}}\right). \tag{4.18}$$

During sizing in the plastic field, as in the case of a cylinder under internal pressure,the condition of danger is represented by

$$\sigma_{id} = \sigma_t - \sigma_r = -\sigma_s; \tag{4.19}$$

reconsidering (4.1) we have

$$-\sigma_s - r\frac{d\sigma_r}{dr} = 0; \tag{4.20}$$

and, through integration

$$\sigma_r = -\sigma_s \log_e r + A, \tag{4.21}$$

where A is constant. Given that, for $r = r_i$, $\sigma_r = 0$ we have

$$A = \sigma_s \log_e r_i. \tag{4.22}$$

Therefore

$$\sigma_r = -\sigma_s \log_e \frac{r}{r_i}. \tag{4.23}$$

On the other hand, for $r = r_e$ we must have $\sigma_r = -p$, that is

$$-p = -\sigma_s \log_e \frac{r_e}{r_i} = -\sigma_s \log_e a; \tag{4.24}$$

it follows that:

$$a = \frac{r_e}{r_i} = e^{p/\sigma_s}. \tag{4.25}$$

Therefore, we rediscovered the same equation relative to cylinders under internal pressure (3.53). The sizing of the cylinder can therefore be done based on the following equation:

$$s = \frac{e^{p/f} - 1}{2e^{p/f}} D_e. \tag{4.26}$$

Figure 4.2 shows that the absolute average value of σ_r is greater than $p/2$; given that based on Guest–Tresca's failure theory the presence of σ_r is favorable, as it shares the same sign of σ_t, we favor resistance by considering $p/2$ as the average value of σ_r. The average value of σ_t is given by

$$\sigma_{tm} = -\frac{pD_e}{2s}. \tag{4.27}$$

The average value of the ideal stress according to Guest–Tresca then becomes

$$\sigma_{id(m)} = -\frac{pD_e}{2s} - \left(-\frac{p}{2}\right) = -\frac{pD_e}{2s} + \frac{p}{2}. \tag{4.28}$$

For $\sigma_{id(m)} = -f$ and from (4.28) we have

$$-p\frac{D_e - s}{2s} = -p\frac{D_m}{2s} = -f, \tag{4.29}$$

and

$$s = \frac{pD_m}{2f}. \tag{4.30}$$

Through a series of steps it becomes

$$s = \frac{pD_e}{2f + p}. \tag{4.31}$$

This is the same sizing equation for cylinders under internal pressure (3.88). We have therefore established that the equations relative to sizing both in the elastic and in the plastic field are the same as those obtained in the case of the cylinder under internal pressure, and all other conclusions that we made maintain their validity, as well.

Specifically, (4.31) may be used within a certain range of values a or, if preferred, of s/D_e or p/f. Based on our past analysis, in general

$$s = \frac{pD_e}{2f + p}. \tag{4.32}$$

Instead of (4.32), the equation

$$s = \left(1 - \sqrt{\frac{f - 1.33p}{f}}\right) \frac{D_e}{2} \tag{4.33}$$

shall be used every time anyone of the following conditions (Fig. 3.9) occurs:

$$p/f > 0.45$$
$$s/D_e > 0.1835$$
$$a > 1.58$$

Unlike what happened with (3.90) and (3.95), the weld joint efficiency z for weldings is missing in (4.32) and (4.33), as the stresses are compressive and z need not be taken into account.

4.2 Thin Cylinders of Infinite Length

It is obvious that the cylinder's infinite length is a pure mathematical abstraction, but it allows us to neglect the influence of the heads or terminal checks (for instance in the case of two coaxial cylinders with interspace pressure) with respect to buckling. If the length of the cylinder is equal to or greater than 10 times the diameter, and if there are no stiffening rings along the cylinder, the behavior of the cylinder can be safely assumed to be identical to that of an infinite length cylinder. Moreover, recalling what observed in Sect. 4.1, thin cylinders are those with $s/D_e \leq 0.1$.

First of all, let us examine the case of a perfectly round cylinder. In this case, lacking bending moments, the stresses in the cylinder can be computed from (4.32). In fact, to simplify sizing and favor safety it is preferable to ignore the positive influence of radial stresses; therefore, if we consider only the average value of the hoop stresses, we have

$$s = \frac{pD_e}{2f}. \tag{4.34}$$

Equations (4.34) and (4.32) are only slightly different in the case of small thickness. With regard to the danger of buckling, let us look at Fig. 4.3. We already know that if y is the hypothetical deviation of the profile from a perfect circle, the internal bending moment for unitary length is given by

$$M_i = -\frac{EI}{r^2}\left(\frac{d^2y}{d\omega^2} + y\right). \tag{4.35}$$

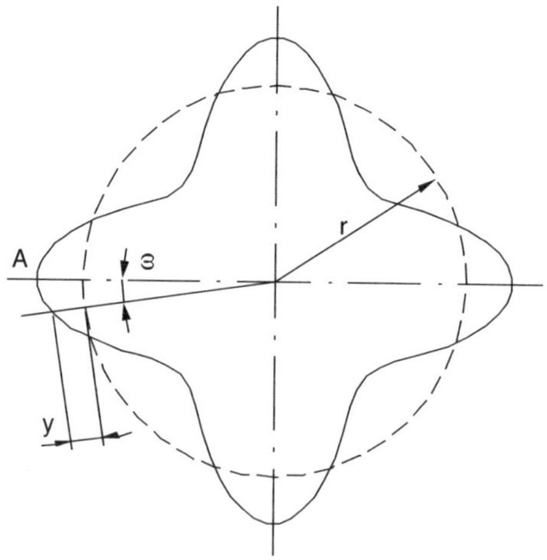

Fig. 4.3

On the other hand, in the case of a circular profile the normal force is

$$N = pr. \tag{4.36}$$

If the deviations from the circular profile are small, the normal force can still be considered equal to the value in (4.36), and the bending moment caused by N and y is given by

$$M = Ny = pry; \tag{4.37}$$

in order to keep the cylinder's sections flat, and with reference to stresses σ_t due to M, we must have

$$\varepsilon_a = \frac{1}{E}(\sigma_a - \mu\sigma_t) = 0; \tag{4.38}$$

then

$$\sigma_a = \mu\sigma_t, \tag{4.39}$$

and therefore

$$\varepsilon_t = \frac{1}{E}(\sigma_t - \mu\sigma_a) = \frac{1}{E}\sigma_t\left(1 - \mu^2\right). \tag{4.40}$$

The circumferential deformations ε_t are not equal to σ_t/E but the term $(1 - \mu^2)$ appears. This term takes into account the requirement that the sections of the cylinder remain flat. If the pressure is critical, the deviation of the profile from the circle is possible; in this case we have

$$M_i = M\left(1 - \mu^2\right). \tag{4.41}$$

If p_{ce} is the critical pressure relative to buckling according to (4.35) and (4.37), then

$$\frac{d^2y}{d\omega^2} + y = -\frac{p_{ce}r^3}{EI}y\left(1 - \mu^2\right), \tag{4.42}$$

and with

$$k^2 - 1 + \frac{p_{ce}r^3}{EI}\left(1 - \mu^2\right) \tag{4.43}$$

we obtain

$$\frac{d^2y}{d\omega^2} + k^2y = 0. \tag{4.44}$$

The general integral of (4.44) is given by

$$y = C_1 \sin k\omega + C_2 \cos k\omega. \tag{4.45}$$

It is not necessary to determine the constants C_1 and C_2, but it suffices to observe that the deformation line is closed, and that for this reason y has the same value if ω is increased by 2π; this requires k to be an integer number n. Therefore

$$k^2 = 1 + \frac{p_{ce}r^3}{EI}\left(1 - \mu^2\right) = n^2. \tag{4.46}$$

Recalling that $I = s^3/12$, one obtains the following value for the critical pressure:

$$p_{ce} = \frac{2}{3} E \frac{n^2 - 1}{1 - \mu^2} \left(\frac{s}{D_e} \right)^3 . \tag{4.47}$$

This is a well-known formula by Bresse that can also be obtained through Von Mises' equation that we will examine in Sect. 4.3 for an infinite length cylinder. The minimum value for p_{ce} is obtained for $n = 2$ (ovalized cylinder); then

$$p_{ce} = \frac{2E}{1 - \mu^2} \left(\frac{s}{D_e} \right)^3 . \tag{4.48}$$

If ν indicates the safety factor

$$s = D_e \sqrt{\nu p \frac{1 - \mu^2}{2E}} , \tag{4.49}$$

and we have, for instance,

$$s = 1.07 D_e \sqrt{\frac{p}{E}} , \tag{4.50}$$

in the case of steel ($\mu = 0.3$) and with a safety factor equal to $\nu = 2.5$. However, rather than using equations that contain the thickness, it is simpler to use equations in which the critical pressure is computed and compared to the actual pressure.

As far as buckling, (4.48) serves this purpose. As for the danger of plastic collapse of the nonovalized cylinder, this occurs when the pressure, that has to be considered critical, is such that the reference stress is equal to the yield strength. Indicating this pressure with p_{cp} we can write

$$\sigma_{tm} = -\frac{p_{cp} D_e}{2s} = -\sigma_s ; \tag{4.51}$$

therefore

$$p_{cp} = 2\sigma_s \frac{s}{D_e} . \tag{4.52}$$

(4.48) and (4.52) show the value of the danger pressure for the two phenomena we examined.

Let us now consider the possibility that the cylinder is not perfectly circular, but that it shows deviations from the circle according to the following equation:

$$\delta = \delta_0 \cos n\omega , \tag{4.53}$$

where n is an integer number (Fig. 4.4).

The bending moment M will have a similar equation, i.e.,

$$M = M_0 \cos n\omega , \tag{4.54}$$

where M_0 is the moment in A with $\omega = 0$.

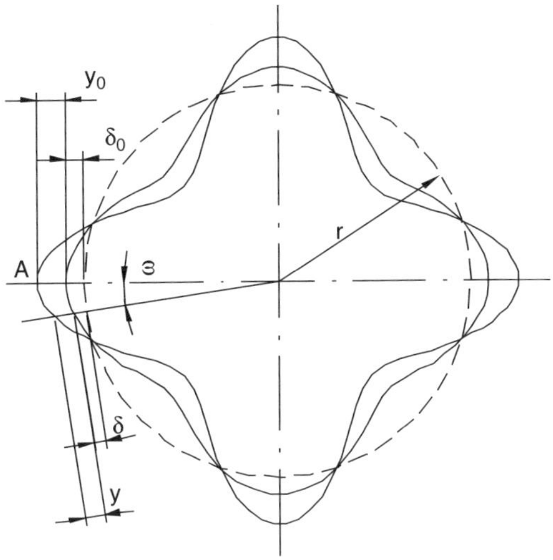

Fig. 4.4

According to (4.35) and (4.41) we can write

$$\frac{d^2y}{d\omega^2} + y = -\frac{M_0}{EI} r^2 \cos n\omega \left(1 - \mu^2\right) . \tag{4.55}$$

Solving (4.55) for y we have

$$y = \frac{M_0}{EI} r^2 \cos n\omega \frac{1 - \mu^2}{n^2 - 1}. \tag{4.56}$$

Note that the moment M_0 in A with $\omega = 0$ is given by

$$M_0 = pr\left(\delta_0 + y_0\right), \tag{4.57}$$

where

$$y_0 = \frac{M_0}{EI} r^2 \frac{1 - \mu^2}{n^2 - 1}. \tag{4.58}$$

Therefore

$$M_0 = pr\left(\delta_0 + \frac{M_0}{EI} r^2 \frac{1 - \mu^2}{n^2 - 1}\right); \tag{4.59}$$

hence

$$M_0 = \frac{pr\delta_0}{1 - \dfrac{pr^3}{EI}\dfrac{1 - \mu^2}{n^2 - 1}}, \tag{4.60}$$

which may also be written as follows:

$$M_0 = \frac{\dfrac{1}{2} p\delta_0 D_e}{1 - \dfrac{3}{2}\dfrac{p}{E}\dfrac{1 - \mu^2}{n^2 - 1}\left(\dfrac{D_e}{s}\right)^3}. \tag{4.61}$$

Recalling that $W = s^2/6$, the maximum stress σ_t due to M_0 is given by

$$\sigma_t = \pm \frac{3p\frac{\delta_0}{D_e}\left(\frac{D_e}{s}\right)^2}{1 - \frac{3}{2}\frac{p}{E}\frac{1-\mu^2}{n^2-1}\left(\frac{D_e}{s}\right)^3}. \tag{4.62}$$

The stress σ_t due to the normal force N is

$$\sigma_t = -\frac{pD_e}{2s}; \tag{4.63}$$

therefore, the maximum total stress in absolute terms is

$$\sigma_t = \frac{pD_e}{2s} + \frac{3p\frac{\delta_0}{D_e}\left(\frac{D_e}{s}\right)^2}{1 - \frac{3}{2}\frac{p}{E}\frac{1-\mu^2}{n^2-1}\left(\frac{D_e}{s}\right)^3}. \tag{4.64}$$

A dangerous condition is created when $\sigma_t = \sigma_s$ (where σ_s is the yield strength), and therefore from (4.64) and indicating the corresponding pressure as critical pressure, we have

$$\frac{p_c D_e}{2s}\left[1 + \frac{6\frac{\delta_0}{D_e}\frac{D_e}{s}}{1 - \frac{3}{2}\frac{p_c}{E}\frac{1-\mu^2}{n^2-1}\left(\frac{D_e}{s}\right)^3}\right] = \sigma_s. \tag{4.65}$$

Finally, based on (4.65), we obtain

$$\frac{p_c D_e}{\sigma_s s} = \frac{2}{1 + \frac{6\frac{\delta_0}{D_e}\frac{D_e}{s}}{1 - \frac{3}{2}\frac{p_c}{\sigma_s}\frac{D_e}{s}\frac{\sigma_s}{E}\frac{1-\mu^2}{n^2-1}\left(\frac{D_e}{s}\right)^3}}. \tag{4.66}$$

If

$$A = \frac{p_c D_e}{\sigma_s s}, \tag{4.67}$$

$$B = 3\frac{\sigma_s}{E}\frac{1-\mu^2}{n^2-1}\left(\frac{D_e}{s}\right)^2, \tag{4.68}$$

and

$$C = 6\frac{\delta_0}{D_e}\frac{D_e}{s}, \tag{4.69}$$

with regard to (4.66) we obtain

$$A = \frac{2}{1 + \dfrac{C}{1 - 0.5AB}}. \tag{4.70}$$

The equation becomes, after a series of steps,

$$A^2B - 2A(1 + B + C) + 4 = 0. \tag{4.71}$$

The roots of equation (4.71) are as follows:

$$A = \frac{1 + B + C \pm \sqrt{(1 + B + C)^2 - 4B}}{B}. \tag{4.72}$$

If the initial deviation δ_0 is equal to zero (perfectly circular profile), $C = 0$ and the two roots of (4.71) are

$$\begin{aligned} A &= 2 \\ A &= 2/B; \end{aligned} \tag{4.73}$$

and using the first of the two above we obtain

$$\frac{p_c D_e}{\sigma_s s} = 2. \tag{4.74}$$

Hence, with p_{cp} being the critical pressure

$$p_{cp} = 2\sigma_s \frac{s}{D_e}, \tag{4.75}$$

which is again (4.52) and determines the critical pressure relative to the danger of the plastic collapse of the cylinder.

From the second one of (4.73) we obtain

$$\frac{p_c D_e}{\sigma_s s} = \frac{2}{3} \frac{E}{\sigma_s} \frac{n^2 - 1}{1 - \mu^2} \left(\frac{s}{D_e}\right)^3; \tag{4.76}$$

it follows that:

$$p_{ce} = \frac{2}{3} E \frac{n^2 \quad 1}{1 - \mu^2} \left(\frac{s}{D_e}\right)^3, \tag{4.77}$$

with p_{ce} as critical pressure. This is again (4.47) and determines the critical pressure caused by buckling. For $\delta_0 \neq 0$ and hence $C \neq 0$, the minimum values of p_c are obtained from the following equation:

$$A = \frac{1 + B + C - \sqrt{(1 + B + C)^2 - 4B}}{B}. \tag{4.78}$$

Figure 4.5 shows (4.58) with dashed lines. The solid lines are not so different from the previous ones and can therefore be replaced, as they correspond to the following simple equations:

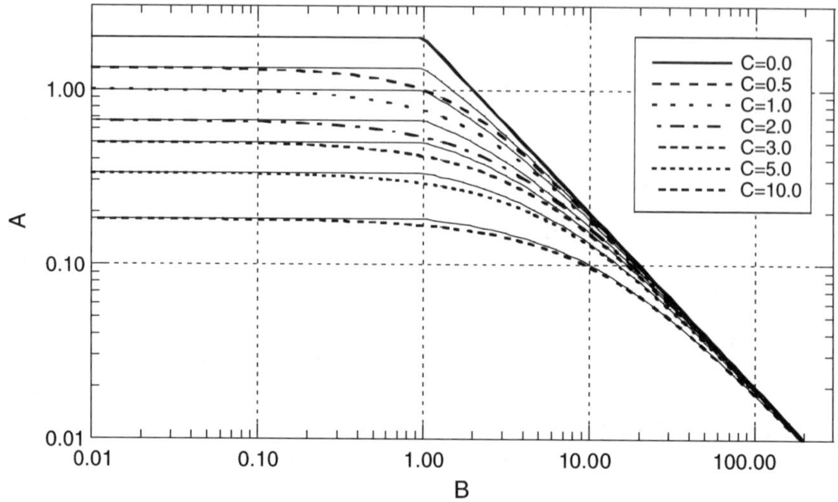

Fig. 4.5

$$A = \frac{2}{1+C} \quad \text{(for } B \leq 1\text{)},$$

and

$$A = \frac{2}{B+C} \quad \text{(for } B \geq 1\text{)}. \tag{4.79}$$

The two equations in (4.79) refer to the danger condition of plastic collapse and buckling, respectively. They are similar to those in (4.73).

Recalling (4.67), (4.68), (4.75), and (4.76) note that

$$A = 2\frac{p_c}{p_{cp}}, \tag{4.80}$$

$$B = \frac{p_{cp}}{p_{ce}}. \tag{4.81}$$

Therefore, in the first case

$$2\frac{p_c}{p_{cp}} = \frac{2}{1+C}, \tag{4.82}$$

and with p'_{cp} as critical pressure corresponding to p_{cp} but for a nonperfectly circular cylinder

$$p'_{cp} = \frac{p_{cp}}{1+C}. \tag{4.83}$$

Moreover, in the second case,

$$\frac{2p_c}{p_{cp}} = \frac{2}{\frac{p_{cp}}{p_{ce}} + C}, \tag{4.84}$$

and with p'_{ce} as critical pressure corresponding to p_{ce} but for a nonperfectly circular cylinder

$$p'_{ce} = \frac{p_{ce}}{1 + C\dfrac{p_{ce}}{p_{cp}}}. \tag{4.85}$$

Based on the equations above, in both cases (with $n = 2$)

$$p'_{cp} = \frac{2\sigma_s \dfrac{s}{D_e}}{1 + 6\dfrac{\delta_0}{D_e}\dfrac{D_e}{s}}; \tag{4.86}$$

$$p'_{ce} = \frac{\dfrac{2E}{1-\mu^2}\left(\dfrac{s}{D_e}\right)^3}{1 + 6\dfrac{\delta_0}{D_e}\dfrac{E/\sigma_s}{1-\mu^2}\dfrac{s}{D_e}}. \tag{4.87}$$

The terms in the denominator are corrective factors that take into account deviations from perfect circularity.

For instance, with $s/D_e = 0.01$, with $\sigma_s = 200\,\mathrm{N\,mm^{-2}}$, with $E = 2,00,000\,\mathrm{N\,mm^{-2}}$, and with $\delta_0/D_e = 0.003$ the denominator of (4.86) is equal to 2.8. Hence, the critical pressure equals 36% of the critical pressure of a nonovalized cylinder.

The denominator of (4.87) (with $\mu = 0.3$) is equal to 1.20, thereby making the critical pressure equal to 83% of the critical pressure of perfectly circular cylinders. Under the same conditions, but with $s/D_e = 0.03$ (i.e., $B \approx 1$) both critical pressures are reduced to 62 % of those for a circular cylinder.

Finally, if there are two waves, as in the present case, the concept of ovalization is usually introduced. Let us indicate with u the ovalization of the cylinder as a percentage; we have

$$u = 2\frac{D_{e(\max)} - D_{e(\min)}}{D_{e(\max)} + D_{e(\min)}} 100(\%) \tag{4.88}$$

where $D_{e(\max)}$ and $D_{e(\min)}$ are the maximum and minimum diameter, respectively. Therefore

$$\frac{\delta_0}{D_e} = \frac{u}{400}. \tag{4.89}$$

4.3 Stiffened Cylinders

As we know, to increase the value of critical pressure relative to buckling, the cylinder can be reinforced with rings. In this case, the critical pressure in a perfectly round and reinforced cylinder can be determined through the well-known von Mises' equation, or from the same equation with the corrective terms suggested by Windenburg and Trilling.

According to von Mises, the critical pressure that we will indicate with p_{ce} (consistently with Sect. 4.2) is given by

$$p_{ce} = \frac{2E}{(n^2-1)\left[1+\left(\frac{2nL}{\pi D_e}\right)^2\right]^2}\frac{s}{D_e} + \frac{2E}{3(1-\mu^2)}\left[n^2-1+\frac{2n^2-1-\mu}{1+\left(\frac{2nL}{\pi D_e}\right)^2}\right]\left(\frac{s}{D_e}\right)^3; \tag{4.90}$$

with the corrective terms of Windenburg and Trilling we have

$$p_{ce} = \frac{2E}{(n^2-1)\left[1+\left(\frac{2nL}{\pi D_e}\right)^2\right]^2}\frac{s}{D_e} + \frac{2E}{3(1-\mu^2)}\left[n^2-1-\frac{2n^2-1-\mu}{1-\left(\frac{2nL}{\pi D_e}\right)^2}\right]\left(\frac{s}{D_e}\right)^3, \tag{4.91}$$

where L is the distance between rings, while n represents an integer number greater than or equal to 2 and that should be chosen to keep p_{ce} at a minimum.

The values for critical pressure that can be obtained from the two equations for practical cases do not substantially differ from one another. At any rate, von Mises' equation is more conservative, and therefore widely adopted in the codes of many countries.

Moreover, we note that for $L = \infty$ we obtain (4.47) both from (4.90) and (4.91). We obtained (4.47) directly in Sect. 4.2 for the cylinder of infinite length.

If the cylinder is not reinforced by rings, L represents the length of the cylinder itself, as the heads or the checks at the ends (in the case of coaxial cylinders with pressure in the inter-space) can themselves be considered reinforcements.

Given the complexity of the equations, whenever the computation has to be performed by hand it may be useful to refer to the curves in Fig. 4.6.

The figure lets us determine the value of the ratio p_{ce}/E as a function of D_e/s and of D_e/L. In addition, Windenburg and Trilling have presented the following approximated equation:

$$p_{ce} = \frac{2.6E\left(\frac{s}{D_e}\right)^{2.5}}{\frac{L}{D_e} - 0.45\left(\frac{s}{D_e}\right)^{0.5}}. \tag{4.92}$$

The equation can be used within the following ranges:

$$0.25 < \frac{L}{D_e} < 2$$

$$0.2 < \frac{100s}{D_e} < 0.7$$

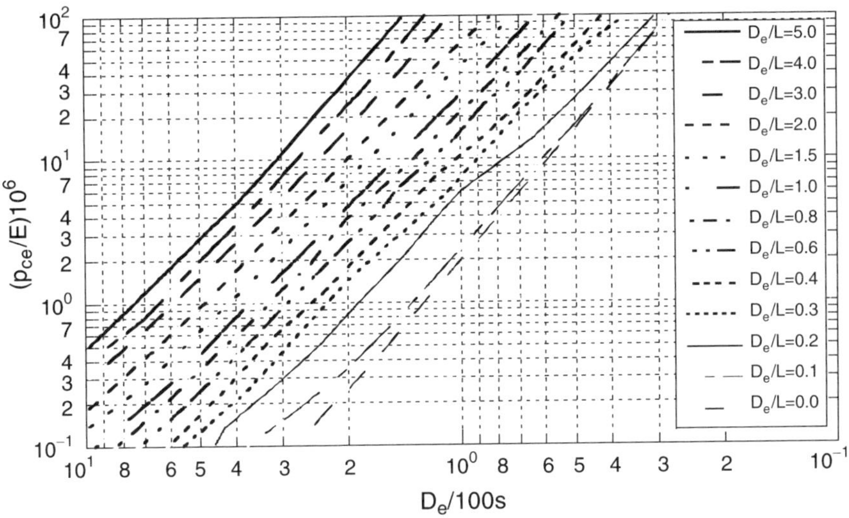

Fig. 4.6

and a mistake in computing p_{ce} not greater than $\pm 2\%$. Equations (4.90) and (4.92) (where applicable) as well as the curves in Fig. 4.6 determine the critical pressure relative to the danger of buckling under the assumption of a perfectly round cylinder.

As far as the danger of plastic collapse of the cylinder, we go back to (4.52), that is

$$p_{cp} = 2\sigma_s \frac{s}{D_e}; \tag{4.93}$$

if the cylinder is not perfectly round, as it is often the case, one has to factor this in, and criteria similar to those discussed for the cylinder of infinite length shall be applied.

Specifically, because of the danger of plastic collapse, and similarly to (4.83), we can write

$$p'_{cr} = \frac{p_{cp}}{1 + C_1}, \tag{4.94}$$

where the term C_1 is a function of the ratio between the maximum deviation from the circle and the outside diameter. Now we want to obtain the equation for this term that is most suitable to reflect the likely reality of the phenomenon.

Evidently, we cannot use (4.69) for C_1, because of the presence of stiffening rings. We note that by replacing (4.78) with (4.79) we practically hypothesized (with regard to plastic collapse) that deformations produced by bending moments can be ignored with respect to the initial deviations from roundness.

In fact, if we ignore such deformations ($y = 0$) the maximum bending moment is given by

$$M = p\frac{D_e}{2}\delta_0, \tag{4.95}$$

to which the following corresponds:

$$\sigma_t = \pm 3p\frac{\delta_0}{D_e}\left(\frac{D_e}{s}\right)^2. \tag{4.96}$$

Recalling (4.63), the maximum hoop stress as an absolute value is given by

$$\sigma_t = \frac{pD_e}{2s} + 3p\frac{\delta_0}{D_e}\left(\frac{D_e}{s}\right)^2 = \frac{pD_e}{2s}\left(1 + 6\frac{\delta_0}{D_e}\frac{D_e}{s}\right); \tag{4.97}$$

with $\sigma_t = \sigma_s$ (where σ_s is the yield strength), from (4.97) we obtain

$$\frac{p_c D_e}{\sigma_s s} = \frac{2}{1 + 6\frac{\delta_0}{D_e}\frac{D_e}{s}}. \tag{4.98}$$

Recalling (4.67) and (4.69), we have

$$A = \frac{2}{1 + C}, \tag{4.99}$$

that is identical to the first equation in (4.79). Based on what was said earlier about edge effects on cylinders (see Sect. 2.2), the ratio $L/\sqrt{D_e s}$ is clearly crucial as far as the impact of rings.

We can reasonably claim that if $L = \sqrt{D_e s}$ we can no longer talk about a reinforced cylinder, but rather about a set of rings placed next to each other, made out of the reinforcement and the portion of metal sheet with a width of $\sqrt{D_e s}$ that collaborates with the ring.

In this case the effect of deviations from roundness is irrelevant, given the large resistance modulus of the ideal rings that constitute the cylinder. Therefore, we believe that we have to assume that $C_1 = 0$. Moreover, if $L = 3\sqrt{D_e s}$, we can actually assume that the deformations caused by bending moments are practically zero for the metal sheet between the rings. In this case, as far as stresses caused by the above moments, we can adopt a plastic flow coefficient of 1.5 without danger.

This is equivalent to saying that, since the adopted safety factor is generally equal to 1.5, if the stresses caused by the moments acted alone, they could reach the yield strength. Introducing the coefficient of plastic collaboration (for which $W = s^2/4$) modifies (4.96) as follows:

$$\sigma_t = \pm 2p\frac{\delta_0}{D_e}\left(\frac{D_e}{s}\right)^2, \tag{4.100}$$

and (4.98) changes as follows:

$$\frac{p_c D_e}{\sigma_s s} = \frac{2}{1 + 4\dfrac{\delta_0}{D_e}\dfrac{D_e}{s}}. \tag{4.101}$$

Therefore,

$$C_1 = 4\frac{\delta_0}{D_e}\frac{D_e}{s}. \tag{4.102}$$

Recalling that for $L = \infty$ we must have

$$C_1 = 6\frac{\delta_0}{D_e}\frac{D_e}{s}, \tag{4.103}$$

and we conclude that we may adopt the following equation for C_1:

$$C_1 = 6\left(1 - \frac{\sqrt{D_e s}}{L}\right)\frac{\delta_0}{D_e}\frac{D_e}{s}. \tag{4.104}$$

In fact, if $L = \sqrt{D_e s}$ we have $C_1 = 0$ regardless of the value of δ_0; if $L = 3\sqrt{D_e s}$ we return to (4.102) again; finally, if $L = \infty$ we return to (4.103) again, as it should be.

Therefore, we can write

$$C_1 = k_1\frac{\delta_0}{D_e}, \tag{4.105}$$

given that

$$k_1 = 6\left(1 - \frac{D_e}{L}\sqrt{\frac{s}{D_e}}\right)\frac{D_e}{s}. \tag{4.106}$$

The term k_1 is shown in Fig. 4.7; of course, we must have $k_1 \geq 0$.

As far as buckling, the discussion in Sect. 4.2 about cylinders that are not perfectly round is based on the assumption that the profile may show the same number of waves that corresponds to the minimum critical pressure. The critical number of waves n_c is a function of the ratios D_e/L and D_e/s, approximately according to the following table:

$\frac{D_e}{L}\sqrt{\frac{D_e}{s}}$	n_c
$= 0 \div 2$	$= 2$
$= 2 \div 4.5$	$= 3$
$= 4.5 \div 7.5$	$= 4$
$= 7.5 \div 10.5$	$= 5$
$= 10.5 \div 14$	$= 6$
$= 14 \div 19$	$= 7$
$= 19 \div 25$	$= 8$

It is quite unlikely that the number of waves of the deformed profile and the number of critical waves be identical when the latter is equal or greater than 6. We therefore recommend to ignore the deviation of the profile from roundness when

$$\frac{D_e}{L}\sqrt{\frac{D_e}{s}} \geq 12.5. \tag{4.107}$$

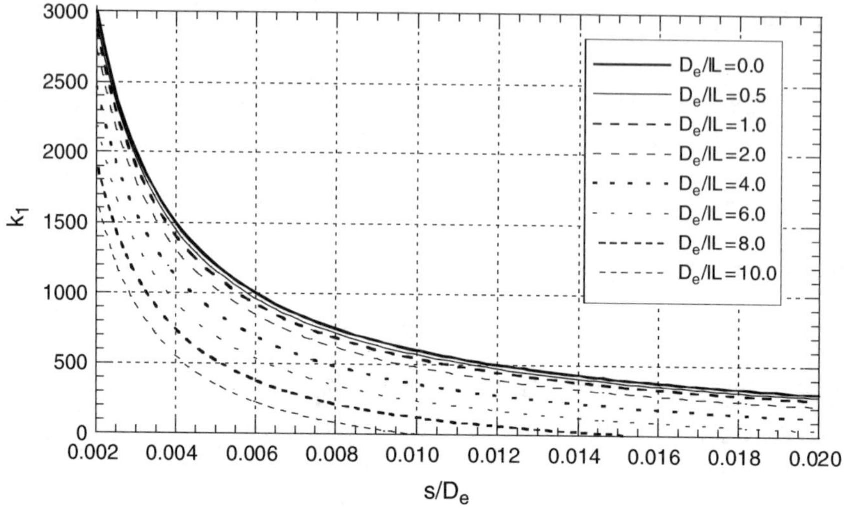

Fig. 4.7

Similarly to (4.85), we use the following equation for the danger of elastic instability:

$$p'_{ce} = \frac{p_{ce}}{1 + C_2 \dfrac{p_{ce}}{p_{cp}}}, \tag{4.108}$$

where C_2 cannot be determined from (4.69), however.

Recalling (4.104) and the analysis leading to (4.107), we recommend the adoption of the following equation for C_2:

$$C_2 = 6\left(1 - \frac{D_e}{L}\sqrt{\frac{s}{D_e}}\right)\left(1 - 0.08\frac{D_e}{L}\sqrt{\frac{D_e}{s}}\right)\frac{D_e}{s}\frac{\delta_0}{D_e}. \tag{4.109}$$

If (4.107) holds, then $C_2 = 0$, as it should be (the negative values for C_2 are, of course, to be ignored). That is,

$$C_2 = k_2 \frac{\delta_0}{D_e}, \tag{4.110}$$

given that

$$k_2 = 6\left(1 - \frac{D_e}{L}\sqrt{\frac{s}{D_e}}\right)\left(1 - 0.08\frac{D_e}{L}\sqrt{\frac{D_e}{s}}\right)\frac{D_e}{s}. \tag{4.111}$$

k_2 is shown in Figs. 4.8 and 4.9.

In summary, the computation steps are as follows.

We compute the pressure p_{cp} from (4.93); then we compute factor k_1 from (4.106) or from the curves in Fig. 4.7; based on the value of δ_0/D_e, we compute

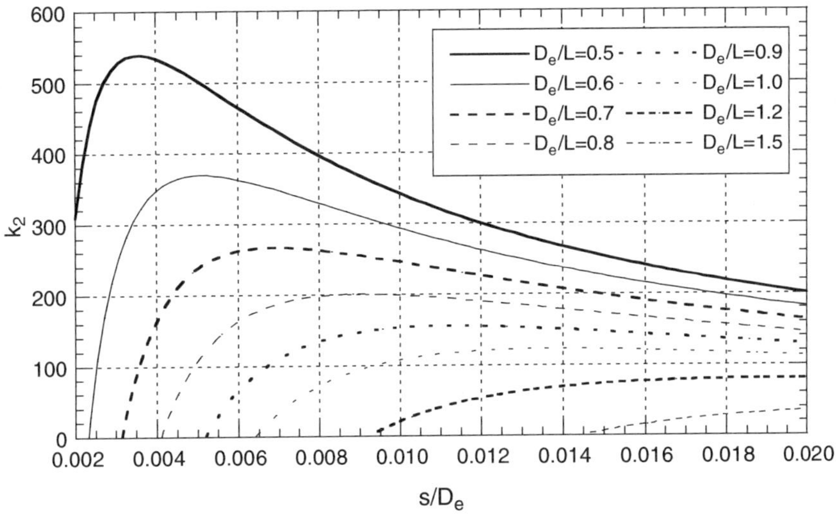

Fig. 4.8

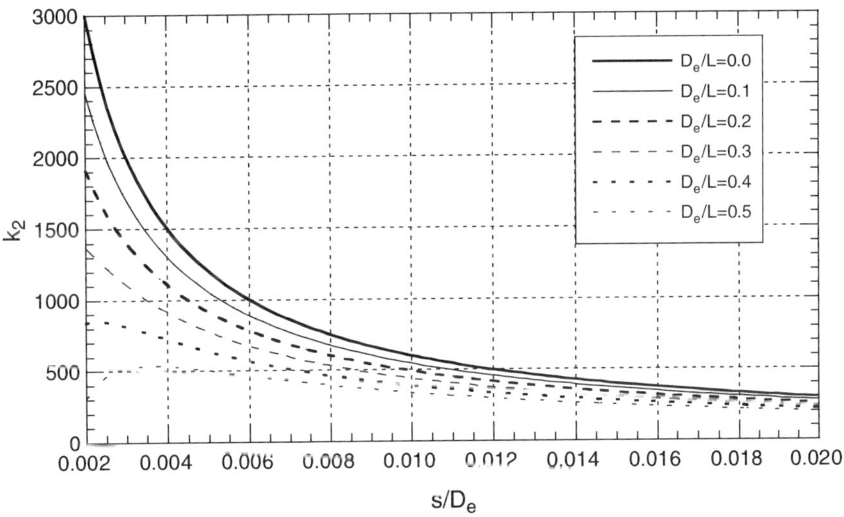

Fig. 4.9

C_1 from (4.105). Finally, we obtain the critical pressure relative to plastic collapse from (4.94). Subsequently, we compute the pressure p_{ce} from (4.90) or from the curves in Fig. 4.6 (or from (4.92) if we are within the range of applicability). Then we compute k_2 from (4.111) or from the curves in Figs. 4.8 and 4.9. Then we compute C_2 from (4.110).

We obtain the critical pressure ${p'}_{ce}$ relative to buckling from (4.108). Table 4.1 shows the different critical pressures for perfectly round cylinders and for those that are not. This highlights the effect of deformations of the cylinder. The table is based on the following assumptions: $\sigma_s = 200\,\mathrm{N\,mm^{-2}}$, $E = 200.000\,\mathrm{N\,mm^{-2}}$ and $\delta_0/D_e = 0.0025$ (which corresponds to an ovalization of 1% in the case of an ovalized cylinder).

When $D_e/L > 2$ we have not considered critical pressures relative to buckling because they are not really significant as the values for p_{ce} are clearly greater than ${p'}_{cp}$, and $k_2 = 0$, i.e., $C_2 = 0$. The last column of the table shows the allowable pressure, indicated with p and obtained through a safety factor equal to 1.5 for ${p'}_{cp}$ and a safety factor equal to 2.5 for ${p'}_{ce}$. The table shows in parenthesis and next to the values for p whether critical pressure for plastic collapse (p) or critical pressure for buckling (e) is crucial.

As one can see, the critical pressure ${p'}_{ce}$ is crucial only for small values of the product of the two dimensionless terms considered here, i.e., for small values of s/L. Note that even an increase of D_e/L, and consequently of the number of rings, does not have a relevant impact on the danger of plastic collapse. It is preferable (within limits, of course) to increase the thickness.

Table 4.1

D_e/L	s/D_e	p_{cp} ($\mathrm{N\,mm^{-2}}$)	k_1	${p'}_{cp}$ ($\mathrm{N\,mm^{-2}}$)	p_{ce} ($\mathrm{N\,mm^{-2}}$)	k_2	${p'}_{ce}$ ($\mathrm{N\,mm^{-2}}$)	p ($\mathrm{N\,mm^{-2}}$)
0.2	0.002	0.8	2973	0.095	0.02	1909	0.018	0.0072 (e)
	0.005	2.0	1183	0.505	0.18	915	0.149	0.0596 (e)
	0.01	4.0	588	1.62	1.26	493	0.908	0.363 (e)
	0.02	8.0	291	4.63	10.4	258	5.66	2.264 (e)
	0.04	16.0	144	11.76	32.7	132	19.52	7.81 (e)
0.5	0.002	0.8	2932	0.096	0.048	309	0.046	0.0184 (e)
	0.005	2.0	1157	0.514	0.46	502	0.357	0.143 (e)
	0.01	4.0	570	1.65	2.64	342	1.57	0.628 (e)
	0.02	8.0	278	4.72	15.6	200	7.90	3.147 (p)
	0.04	16.0	135	11.96	93.2	108	36.22	7.97 (p)
1	0.002	0.8	2865	0.098	0.10	0	0.10	0.040 (e)
	0.005	2.0	1115	0.528	0.96	0	0.96	0.352 (p)
	0.01	4.0	540	1.70	5.60	108	4.06	1.133 (p)
	0.02	8.0	257	4.87	31.91	112	15.07	3.247 (p)
	0.04	16.0	120	12.31	198.1	72	61.36	8.21 (p)
2	0.002	0.8	2731	0.102	0.20	0	0.20	0.068 (p)
	0.005	2.0	1030	0.559	2.08	0	2.08	0.373 (p)
	0.01	4.0	480	1.82	12.6	0	12.6	1.213 (p)
	0.02	8.0	215	5.20	79.9	0	79.9	3.467 (p)
	0.04	16.0	90	13.06	380	18	184	8.71 (p)
4	0.002	0.8	2463	0.112				0.075 (p)
	0.005	2.0	860	0.635				0.423 (p)
	0.01	4.0	360	2.10				1.40 (p)
	0.02	8.0	130	6.04				4.03 (p)
	0.04	16.0	30	14.88				9.92 (p)

As far as safety factors are concerned, we believe that the values used to compute the allowable pressure in Table 4.1 are suitable. In spite of the external pressure being of greater potential danger than the internal one, there is no need to consider a safety factor greater than 1.5 as far as plastic collapse.

However, it is crucial to follow the procedure discussed above to prevent the effects from potential deviations of the cylinder from roundness that do constitute the real source of danger here. In general, one adopts a safety factor greater than 2.5 for the critical pressure relative to buckling, and without taking into account deviations from roundness. Once the latter are factored in there is, no reason to use a value greater than 2.5. This procedure is more rational than an indiscriminate increase of the safety factor with reference to a perfectly round cylinder.

In fact, it suffices to look at Table 4.1 to realize that in certain cases deviations from roundness are irrelevant, while in others they cause a significant decrease in critical pressure. Indeed, if in the latter cases we assign the typical safety factors (3–4) to p_{ce}, we end up with an allowable pressure that, in fact, shows a dangerously low safety factor as far as p'_{ce}.

4.4 Stiffening Rings

Stiffening rings are generally made as shown in Fig. 4.10. In any case, the first problem we face is to determine which portion of the cylinder collaborates with the ring towards reinforcement.

Let us recall Sect. 2.2 and (2.66) and (2.67) that represent the deflection and rotation at the edge of a cylinder under moments M_0 and forces F_0.

If the rings are placed at regular intervals, as is usually the case, the rotation is equal to zero in correspondence of them. This makes it possible to easily examine the behavior of the shell next to the ring. The ring rigidity prevents free deformations of the cylinder under pressure, generating radial deflection y towards the outside. Based on (2.65) and with $\vartheta_o = 0$, we have that

$$C_1 = C_2; \tag{4.112}$$

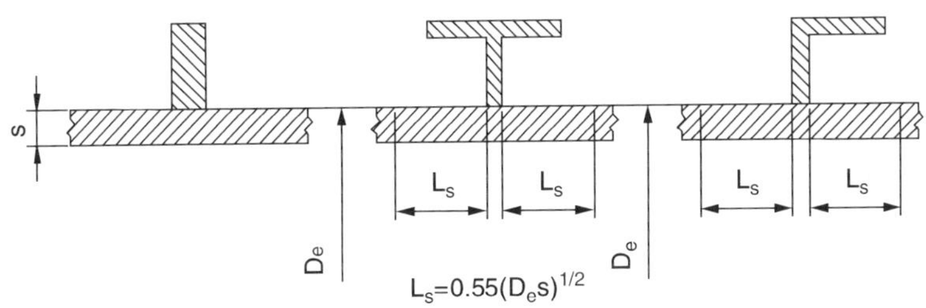

Fig. 4.10

therefore, based on (2.45) the deflections y are given by the following equation:

$$y = C_2 e^{-\beta x} (\sin \beta x + \cos \beta x) ; \tag{4.113}$$

furthermore, based on (2.64),

$$y_0 = C_2, \tag{4.114}$$

and

$$y = y_0 e^{-\beta x} (\sin \beta x + \cos \beta x) . \tag{4.115}$$

The y curve is shown in Fig. 4.11.

If we integrate (4.115) we obtain

$$\int y dx = -\frac{y_0}{\beta} e^{-\beta x} \cos \beta x + C. \tag{4.116}$$

The value of the integral between 0 and ∞ is given by

$$\int_0^\infty y dx = \frac{y_0}{\beta}; \tag{4.117}$$

recalling (2.41), it can be written as follows:

$$\int_0^\infty y dx = \frac{y_0}{1.817} \sqrt{D_e s} = 0.55 y_0 \sqrt{D_e s} \tag{4.118}$$

or

$$\int_0^\infty \frac{y}{y_0} dx = 0.55 \sqrt{D_e s}. \tag{4.119}$$

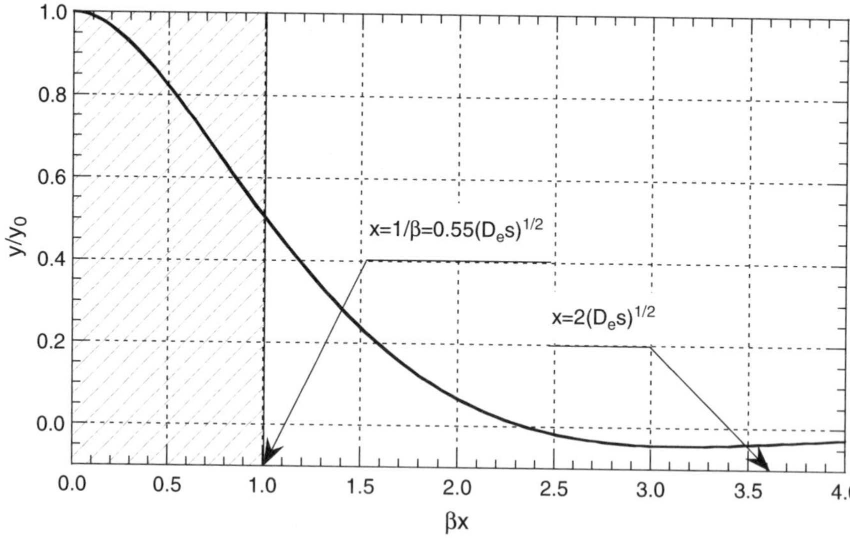

Fig. 4.11

The integral of the curve in Fig. 4.11 for $x = 0 \div \infty$ is therefore equivalent to the dotted area.

Note that even at distance $1.3\sqrt{D_e s}$ from the edge of the ring, the deflection y becomes zero; βx is equal to 2.36; the integral in (4.116) between 0 and 2.36 is

$$\int_0^{2.36} y dx = 0.586 y_0 \sqrt{D_e s}. \tag{4.120}$$

As we can see, it does not substantially differ from the integral between 0 and ∞. Conservatively, one uses the latter to determine the collaborating area of the shell.

The collaboration of the shell decreases as the distance from the ring increases following the curve being considered. Its behavior is equivalent to that of a $0.55\sqrt{D_e s}$ wide metal strip that completely collaborates with the ring, as do the fibers near the ring itself. Therefore, the collaborating shell extends ideally up to $0.55\sqrt{D_e s}$ on both sides of the ring, as shown in Fig. 4.10.

The moment of inertia of the reinforcement has to be computed factoring in the area of the collaborating shell. One could argue that in other instances an extension of the collaborating shell greater than the one indicated may be considered (e.g., in the case of an isolated nozzle on a cylinder), but one should keep in mind that plastic flow phenomena of the material are taken into consideration, thus making it possible to extend the collaborating area.

In this case the computation must be done rigorously in the elastic field, because we have to rule out plastic flow phenomena that may compromise the resistance of the reinforcement with respect to the danger of buckling. Once the procedure to compute the inertia moment of the reinforcement is established, let us now examine how to do the sizing. Let us go back to Sect. 4.2, when we determined that the internal moment in a ring caused by a deviation y from the circle is

$$M_i = -\frac{EI}{r^2}\left(\frac{d^2 y}{d\omega^2} + y\right); \tag{4.121}$$

if N is the normal force active on the ring, the moment is given by

$$M = Ny; \tag{4.122}$$

if the force N is critical (N_c) the deviation y is possible, and both moments are equal. Therefore,

$$\frac{d^2 y}{d\omega^2} + y = -\frac{N_c r^2}{EI} y. \tag{4.123}$$

If we set

$$k^2 = 1 + \frac{N_c r^2}{EI} = 1 + \frac{N_c D_e^2}{4EI}, \tag{4.124}$$

we obtain

$$\frac{d^2y}{d\omega^2} + k^2 y = 0. \tag{4.125}$$

The general integral is given by

$$y = C_1 \sin\omega + C_2 \cos\omega, \tag{4.126}$$

and making the same considerations about the cylinder of infinite length (Sect. 4.2), it necessarily follows that

$$1 + \frac{N_c D_e^2}{4EI} = n^2, \tag{4.127}$$

where n is an integer number.

From (4.127) we obtain

$$N_c = 4\left(n^2 - 1\right)\frac{EI}{D_e^2}, \tag{4.128}$$

that is minimized for $n = 2$, thus

$$N_c = 12\frac{EI}{D_e^2}. \tag{4.129}$$

Introducing the safety factor v we shall have that

$$N \leq \frac{12}{\nu}\frac{EI}{D_e^2}; \tag{4.130}$$

therefore

$$I \geq \frac{\nu}{12E} N D_e^2. \tag{4.131}$$

The force N is given by

$$N = \frac{pD_e}{2} L_r, \tag{4.132}$$

where L_r is a typical length of the cylinder that identifies the area releasing the pressure on the reinforcement, the value of which we will determine later on; therefore, based on (4.132),

$$I \geq \nu \frac{p}{24E} L_r D_e^3. \tag{4.133}$$

The safety factor should be rather large in order to guarantee the resistance of the reinforcement even when it is not perfectly round. A value of $4 \div 5$ is recommended.

From (4.133) we obtain

$$I \geq (0.167 \div 0.208)\,\frac{p}{E} L_r D_e^3. \tag{4.134}$$

Since the modulus of elasticity for steel varies most frequently between 180,000 and $210,000\,\mathrm{N\,mm^{-2}}$ (depending on the temperature), we can use the following equation to compute the moment of inertia:

$$I \geq pL_r D_e^3 \times 10^{-6}, \tag{4.135}$$

with p expressed in $\mathrm{N\,mm^{-2}}$.

As far as the typical length L_r is concerned, let us look again at Fig. 4.11. Conservatively, we may assume that the cylinder releases the forces caused by pressure on the reinforcement along a strip of the cylinder with a width of $2\sqrt{D_e s}$ on each side of the reinforcement. Of course, L_r can never be greater than L, since the latter is the pitch of the reinforcements.

In other words ($L_r \leq L$)

$$L_r = 4\sqrt{D_e s}. \tag{4.136}$$

Note that with $D_e/L = 0.5$ and $s/D_e = 0.01$ we have $\sqrt{D_e s}/L = 0.05$; therefore, $L_r = 0.2L$; the typical length L_r is considerably smaller than the pitch of the reinforcements. With $D_e/L = 1.5$ and $s/D_e = 0.02$ we obtain $\sqrt{D_e s}/L = 0.212$ instead, and $L_r = 0.85L$. The typical length L_r is almost equal to the pitch of the reinforcements. If instead, e.g., $D_e/L = 2$ and $s/D_e = 0.025$, we get $\sqrt{D_e s}/L = 0.316$ and we would obtain $L_r = 1.26L$. In this case one must set $L_r = L$.

If the reinforcements are made of rings with a shape other than the one shown in Fig. 4.10, it is necessary to go through a similar process. As shown in Fig. 4.12, the collaborating area of the metal sheet has a width b given by

$$b = a + 1.1\sqrt{D_e s}. \tag{4.137}$$

It follows that the typical length L_r is

$$L_r = a + 4\sqrt{D_e s}, \tag{4.138}$$

but in any case it is required that

$$L_r \leq L. \tag{4.139}$$

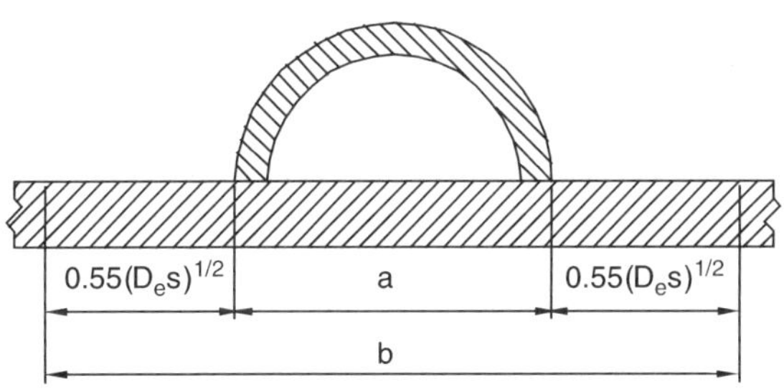

Fig. 4.12

This sizing criterion may be risky if the reinforcements are very far from each other. In fact, even though both the cylinder, as a reinforced cylinder, and the reinforcements are correctly sized in this case, the structure cylinder-reinforcements may not be sufficiently safe with respect to buckling.

In some cases, in fact, it may be advisable to carry out further verifications through the following procedure that we consider quite adequate, even though it is simple and approximate. Be I_1 the moment of inertia of the reinforcement and I_2 the moment of inertia relative to the wall of the cylinder in between reinforcements.

By removing the collaborating wall with the reinforcement from the wall between rings, we have

$$I_2 = \left(L - 1.1\sqrt{D_e s}\right) \frac{s^3}{12}. \tag{4.140}$$

Let us consider the total moment of inertia I, made up by the sum of I_1 and I_2, and compute a virtual thickness s_{id} of the wall of the vessel without reinforcements given by

$$s_{id} = \sqrt[3]{\frac{12I}{L}}. \tag{4.141}$$

We now compute the critical pressure (for p_{cp}, though, we must factor in the actual thickness of the cylinder) from s_{id} and the total length of the cylinder (as long as its ends are connected to very rigid elements, such as heads). If the critical pressure turns out to be too low relative to the working pressure, it will be necessary to increase the moment of inertia of the reinforcements or to increase the number of reinforcements to increase the value of s_{id}.

The need to carry out this further verification becomes evident when the distance L between reinforcements is

$$L > 10\sqrt{D_e s}. \tag{4.142}$$

To avoid this verification, one may adopt the following conservative criterion. Assuming that the moment of inertia of the reinforcement is equal to the minimum allowable value from (4.135), we have that

$$I_1 = pL_r D_e^3 \times 10^{-6}. \tag{4.143}$$

Ignoring the moment of inertia of the wall $I = I_1$ and from (4.141)

$$\left(\frac{s_{id}}{D_e}\right)^3 = 1.2p\frac{L_r}{L} \times 10^{-5}. \tag{4.144}$$

We assume that the cylinder may indeed ideally have infinite length and ignore the possible ovalization. From (4.49) and (4.144) we have that

$$p_{ce} = \frac{2.4E}{1-\mu^2} p \frac{L_r}{L} \times 10^{-5}. \tag{4.145}$$

Recalling that generally $E = 180,000 \div 210,000\,\mathrm{N\,mm^{-2}}$ and that $\mu = 0.3$

$$\frac{p_{ce}}{p} = (4.75 \div 5.53)\,\frac{L_r}{L}, \tag{4.146}$$

and to obtain $p_{ce} \geq 2.5p$ based on (4.146) approximately we must have

$$L_r \geq 0.5L. \tag{4.147}$$

This added condition for the value of L_r beyond those from (4.136) or from (4.138) and (4.139) makes it possible to avoid the additional verification discussed above.

An example will show that this criterion is quite conservative. We assume a vessel with a cylinder 6,000 mm long where

$$\begin{aligned} p &= 0.14\,\mathrm{N\,mm^{-2}}(1.4\,\mathrm{bar}) \\ D_e &= 1,000\,\mathrm{mm} \\ s &= 5\,\mathrm{mm} \\ L &= 2,000\,\mathrm{mm} \\ \text{Ovalization} &= 1\%(\delta_0 = 0.0025) \end{aligned}$$

Since

$$\begin{aligned} D_e/L &= 0.5 \\ s/D_e &= 0.005 \end{aligned}$$

from Table 4.1 we see that the critical pressure is equal to $0.357\,\mathrm{N\,mm^{-2}}$ and that the allowable pressure computed with a safety factor of 2.5 is equal to $0.143\,\mathrm{N\,mm^{-2}}$, therefore greater than the one under working conditions.

Moreover, we have

$$\begin{aligned} \sqrt{D_e s} &= 71\,\mathrm{mm} \\ 1.1\sqrt{D_e s} &= 78\,\mathrm{mm} \\ 4\sqrt{D_e s} &= 284\,\mathrm{mm} \\ L/\sqrt{D_e s} &= 28.2 > 10 \end{aligned}$$

Thus, the typical length L_r will be 284 mm. From (4.143) we have

$$I_1 = 0.14 \times 284 \times 1000^3 \times 10^{-6} = 39760\,\mathrm{mm}^4.$$

Then

$$I_2 = (2000 - 78)5^3/12 = 20020\,\mathrm{mm}^4.$$

Therefore

$$I = 39760 + 20020 = 59780\,\mathrm{mm}^4,$$

$$s_{id} = \sqrt[3]{\frac{12 \times 59780}{2000}} = 7.1\,\mathrm{mm}.$$

Considering the length of the cylinder of 6,000 mm and the ideal thickness determined through (4.91), we observe that the critical number of waves of the cylinder-reinforcements structure is 2 (in the areas of the cylinder in-between reinforcements it is 4 instead), and that the critical pressure p_{ce} with $E = 200,000\,\mathrm{N\,mm^{-2}}$, as in Table 4.1, is

$$p_{ce} = 0.431\,\mathrm{N\,mm^{-2}}.$$

From (4.93) with $\sigma_s = 200\,\mathrm{N\,mm^{-2}}$ and with the actual thickness of the cylinder, we obtain

$$p_{cp} = 2\,\mathrm{N\,mm^{-2}}.$$

Finally, from (4.111)

$$k_2 = 700,$$
$$C_2 = 700 \times 0.0025 = 1.75.$$

Also, from (4.51)

$$p'_{ce} = \frac{0.431}{1 + 1.75\frac{0.431}{2}} = 0.313\,\mathrm{N\,mm^{-2}}.$$

The safety factor is

$$\nu = 0.313/0.14 = 2.23 < 2.5.$$

We observe that the value of L_r is insufficient, but we can demonstrate that with $L_r = 400\,\mathrm{mm}$ the safety factor is $2.54 > 2.5$. In that case

$$L_r = 0.2\,L,$$

which is clearly lower than the value required by (4.147).

In summary, it is preferable to conduct the verification discussed above for the danger of buckling in the cylinder-reinforcements structure.

5

Spherical Vessels

5.1 Spheres Under Internal Pressure

The internal pressure generates three principal stresses, i.e., a circumferential stress σ_t, a meridian stress σ_m, and a radial stress σ_r. The circumferential and meridian stresses are, of course, identical since in a sphere the meridian is a circumference, as well. The values of σ_t, σ_m and σ_r vary across the wall of the sphere, as we shall see later on.

As in the case of the cylinder, in order to do the sizing of the sphere we have to examine the triaxial status of stresses, and determine an ideal stress through one of the theories of failure. Setting the ideal stress at most equal to the allowable stress, we obtain an equation to calculate the minimum required thickness.

Let us look at the basic element shown in Fig. 5.1. The location of the element is given by radius r. Its dimensions are defined by dr, and the angle $d\varphi$. The circumferential stress σ_t, that is constant along the circumference, is active on the sides $A-B$ and $C-D$. Similarly, the meridian stress σ_m, that is constant along the meridian, is active on the faces $A-D$ and $B-C$.

The equilibrium along the circumference and the meridian is enforced by σ_t and σ_m being constant along their respective circumferences. As far as the radial equilibrium is concerned, the force F_{ri} is active on the internal side of the element given by

$$F_{ri} = \sigma_r r^2 d\varphi^2; \tag{5.1}$$

the force F_{re} is active on the external side and is given by

$$F_{re} = -\left(\sigma_r + d\sigma_r\right)\left(r + dr\right)^2 d\varphi^2. \tag{5.2}$$

Neglecting the terms of higher order from (5.2) we have

$$F_{re} = -\left(\sigma_r r^2 + 2\sigma_r r dr + r^2 d\sigma_r\right) d\varphi^2; \tag{5.3}$$

therefore

$$F_{ri} + F_{re} = -\left(2\sigma_r r dr + r^2 d\sigma_r\right) d\varphi^2. \tag{5.4}$$

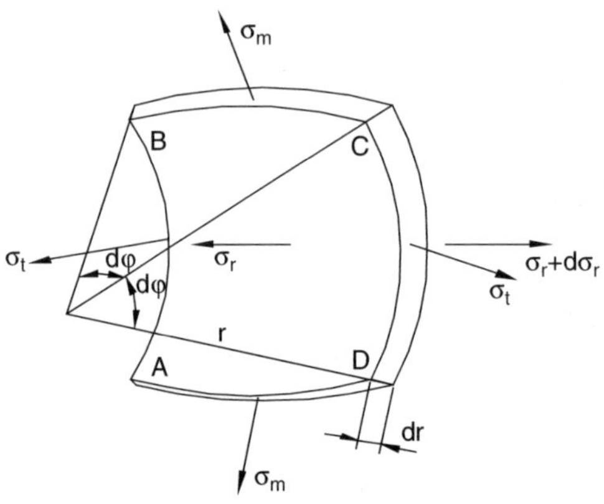

Fig. 5.1

The forces F_t and F_m active on the sides of thickness dr are given by

$$F_t = F_m = \sigma_t r dr d\varphi. \tag{5.5}$$

The component along the radius of these forces is

$$F_{tr} = F_{mr} = 2F_t \frac{d\varphi}{2} = \sigma_t r dr d\varphi^2, \tag{5.6}$$

whereas for the equilibrium we must have

$$F_{ri} + F_{re} + F_{tr} + F_{mr} = 0. \tag{5.7}$$

Therefore, from (5.4) and (5.6) we have

$$-2\sigma_r r dr - r^2 d\sigma_r + 2\sigma_t r dr = 0. \tag{5.8}$$

Dividing (5.8) by $2rdr$ we finally obtain

$$\sigma_t - \sigma_r - \frac{1}{2} r \frac{d\sigma_r}{dr} = 0. \tag{5.9}$$

Equation (5.9) is the equation of equilibrium of the sphere.

As far as congruence of deformations, let us consider the round crown of thickness dr shown in Fig. 5.2. As a consequence of the circumferential deformations ε_t, the radius of the circle α is subject to an elongation

$$\Delta r_\alpha = \varepsilon_t r. \tag{5.10}$$

The radius of circle β in turn undergoes the following elongation:

$$\Delta r_\beta = (\varepsilon_t + d\varepsilon_t)(r + dr). \tag{5.11}$$

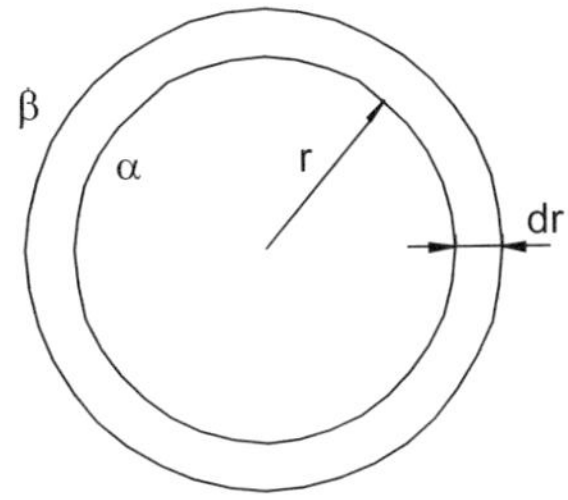

Fig. 5.2

According to congruence, the difference between these elongations has to correspond to an increase in thickness of the crown of

$$\Delta s = \varepsilon_r dr; \tag{5.12}$$

in other words,

$$\Delta r_\beta - \Delta r_\alpha = \Delta s. \tag{5.13}$$

Recalling (5.10), (5.11), and (5.12), we have

$$\varepsilon_r - \varepsilon_t - r\frac{d\varepsilon_t}{dr} = 0. \tag{5.14}$$

Equation (5.14) is the equation of congruence for the sphere, and it is identical to the one already obtained for the cylinder.

Recalling that

$$\begin{aligned} \varepsilon_t &= \frac{1}{E}[\sigma_t - \mu(\sigma_r + \sigma_m)] = \frac{1}{E}[\sigma_t(1-\mu) - \mu\sigma_r] \\ \varepsilon_r &= \frac{1}{E}[\sigma_r - \mu(\sigma_t + \sigma_m)] = \frac{1}{E}[\sigma_r - 2\mu\sigma_t] \end{aligned} \tag{5.15}$$

Based on the equation of congruence

$$\sigma_r - 2\mu\sigma_t - \sigma_t(1-\mu) + \mu\sigma_r - r(1-\mu)\frac{d\sigma_t}{dr} + r\mu\frac{d\sigma_r}{dr} = 0, \tag{5.16}$$

which can also be written as follows:

$$(1+\mu)\left(\sigma_t - \sigma_r - \frac{1}{2}r\frac{d\sigma_r}{dr}\right) + r(1-\mu)\frac{d\sigma_t}{dr} + \frac{1}{2}r(1-\mu)\frac{d\sigma_r}{dr} = 0. \tag{5.17}$$

Recalling (5.9) and (5.17), we finally have

$$\frac{d\sigma_t}{dr} + \frac{1}{2}\frac{d\sigma_r}{dr} = 0. \tag{5.18}$$

If we now derive (5.9), we obtain

$$\frac{d\sigma_t}{dr} - \frac{d\sigma_r}{dr} - \frac{1}{2}\frac{d\sigma_r}{dr} - \frac{1}{2}r\frac{d^2\sigma_r}{dr^2} = 0, \tag{5.19}$$

that is,

$$\frac{d\sigma_t}{dr} - \frac{3}{2}\frac{d\sigma_r}{dr} - \frac{1}{2}r\frac{d^2\sigma_r}{dr^2} = 0, \tag{5.20}$$

and recalling (5.18)

$$\frac{d^2\sigma_r}{dr^2} = -\frac{4}{r}\frac{d\sigma_r}{dr}. \tag{5.21}$$

Now

$$y = \frac{d\sigma_r}{dr}; \tag{5.22}$$

from (5.21) we obtain

$$y' = -\frac{4}{r}y, \tag{5.23}$$

which is an equation with separable variables; the general integral is given by

$$\int \frac{dy}{y} = \int -\frac{4}{r}dr + Z; \tag{5.24}$$

that is,

$$\log_e y = -4\log_e r + Z = \log_e \frac{K}{r^4}, \tag{5.25}$$

where Z and K are constant. Recalling (5.22) we have

$$\frac{d\sigma_r}{dr} = \frac{K}{r^4}, \tag{5.26}$$

after integration we have

$$\sigma_r = A - \frac{B}{r^3}, \tag{5.27}$$

where A and B are constant. Based on (5.9) and (5.27) we also have that

$$\sigma_t = A + \frac{1}{2}\frac{B}{r^3}. \tag{5.28}$$

Note that for $r = r_e$, with r_e being the outside radius, we have $\sigma_r = 0$; therefore,

$$A = \frac{B}{r_e^3}; \tag{5.29}$$

then

$$\sigma_r = B\left(\frac{1}{r_e^3} - \frac{1}{r^3}\right). \tag{5.30}$$

On the other hand, for $r = r_i$, with r_i being the internal radius, we have $\sigma_r = -p$; therefore

$$-p = B\left(\frac{1}{r_e^3} - \frac{1}{r_i^3}\right); \tag{5.31}$$

then

$$B = \frac{pr_e^3}{\dfrac{r_e^3}{r_i^3} - 1}. \tag{5.32}$$

If we set

$$a = \frac{r_e}{r_i}, \tag{5.33}$$

we obtain

$$\sigma_r = \frac{p}{a^3 - 1}\left[1 - \left(\frac{r_e}{r}\right)^3\right], \tag{5.34}$$

and, similarly from (5.28)

$$\sigma_t = \sigma_m = \frac{p}{a^3 - 1}\left[1 + \frac{1}{2}\left(\frac{r_e}{r}\right)^3\right]. \tag{5.35}$$

The three principal stresses are shown in Fig. 5.3, while Fig. 5.4 shows the deformations obtained from (5.15). As one can see, σ_t and σ_m show a maximum value in correspondence of the internal fiber; the same is true for the absolute value of σ_r.

The maximum value of the ideal stress, according to Guest's failure theory, occurs in correspondence of the internal fiber; therefore

$$\sigma_{ti} = p\frac{\dfrac{a^3}{2} + 1}{a^3 - 1}. \tag{5.36}$$

$$\sigma_{ri} = -p$$

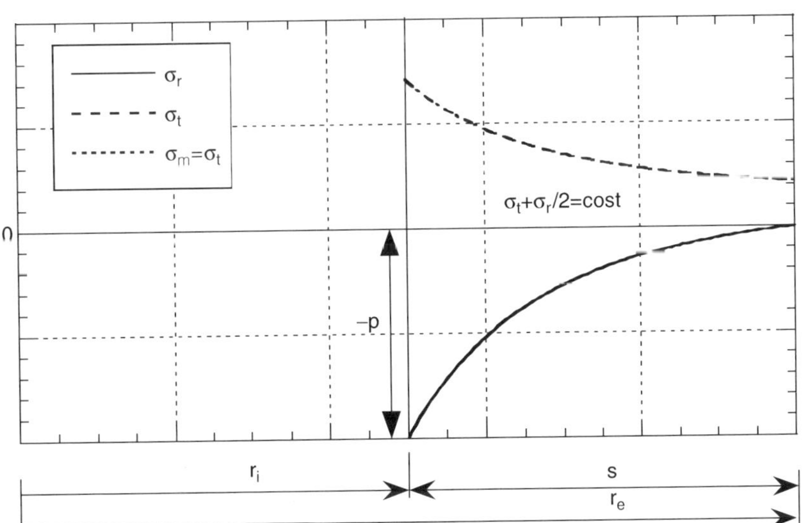

Fig. 5.3

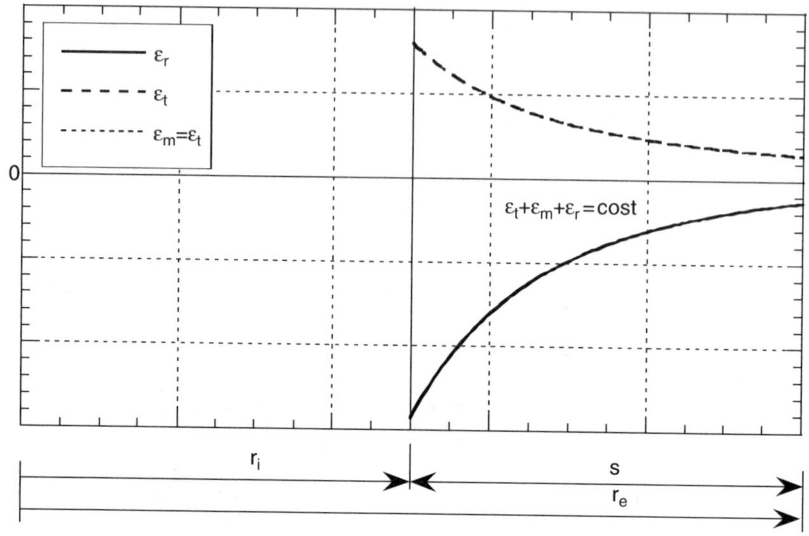

Fig. 5.4

Then

$$\sigma_{id(i)} = \sigma_{ti} - \sigma_{ri} = \frac{3}{2}p\frac{a^3}{a^3-1}. \tag{5.37}$$

If the sphere were to be sized according to Guest's theory, following the traditional criteria of resistance in the elastic field we would set

$$\frac{3}{2}p\frac{a^3}{a^3-1} = f, \tag{5.38}$$

where f is the basic allowable stress. From (5.38) we obtain

$$a = \sqrt[3]{\frac{1}{1-\frac{3}{2}\frac{p}{f}}}, \tag{5.39}$$

and, since D_e is the outside diameter of the sphere and s the thickness

$$s = \frac{D_e}{2}\left(1-\sqrt[3]{1-\frac{3}{2}\frac{p}{f}}\right). \tag{5.40}$$

It may be interesting to note that applying the theory of Huber–Hencky we obtain the same equation. Equation (5.40) is not used to size the sphere. As in the case of the cylinder, we consider the possibility of collaboration of the fibers that are less under stress.

Since we do not consider the yield strength of the most stressed fiber as an indication of danger but rather the plastic flow of the entire sphere, we observe that, according to Guest's theory, we face this condition when in every point

$$\sigma_{id} = \sigma_t - \sigma_r = \sigma_s, \tag{5.41}$$

where σ_s is the yield strength. From (5.19) we have

$$\sigma_s - \frac{1}{2} r \frac{d\sigma_r}{dr} = 0, \tag{5.42}$$

and after integration

$$\sigma_r = 2\sigma_s \log_e r + A, \tag{5.43}$$

where A is constant.

Once again, recalling that for $r = r_e$ we must have $\sigma_r = 0$, we obtain

$$A = -2\sigma_s \log_e r_e; \tag{5.44}$$

therefore

$$\sigma_r = 2\sigma_s \log_e \frac{r}{r_e}. \tag{5.45}$$

On the other hand, for $r = r_i$ we must have $\sigma_r = -p$, thus

$$p = 2\sigma_s \log_e \frac{r_e}{r_i} = 2\sigma_s \log_e a. \tag{5.46}$$

Finally,

$$a = \frac{r_e}{r_i} = e^{p/2\sigma_s}. \tag{5.47}$$

Equation (5.47) means that the sphere shows plastic flow at a given pressure p with a material with yield strength equal to σ_s, if the ratio between outside and inside radius follows equation (5.47).

Needless to say, an adequate safety factor shall be considered. To this end, it suffices to substitute the basic allowable stress f to σ_s to obtain the following equation:

$$a = \frac{r_e}{r_i} = e^{p/2f}. \tag{5.48}$$

From (5.48) we obtain the following:

$$s = \frac{e^{p/2f} \quad 1}{e^{p/2f}} \frac{D_e}{2} \tag{5.49}$$

or

$$s = \frac{D_i}{2} \left(e^{p/2f} - 1 \right), \tag{5.50}$$

where D_i is the inside diameter.

When the sphere shows plastic flow the radial stress is as in (5.45). Recalling (5.41), we also have

$$\sigma_t = \sigma_s \left(1 - 2 \log_e \frac{r_e}{r} \right). \tag{5.51}$$

Following what was said about the cylinder and the deviator of stresses, the deformations are

$$\begin{aligned}\varepsilon_t =& f(r)\left(\sigma_t - \frac{\sigma_r + \sigma_m}{2}\right)\\ \varepsilon_r =& f(r)\left(\sigma_r - \frac{\sigma_t + \sigma_m}{2}\right)\end{aligned} \tag{5.52}$$

with $f(r)$ a function of the radius. Recalling that $\sigma_m = \sigma_t$, we have

$$\begin{aligned}\varepsilon_t =& \varepsilon_m = f(r)\frac{\sigma_t - \sigma_r}{2} = \frac{\sigma_s}{2}f(r)\\ \varepsilon_r =& f(r)(\sigma_r - \sigma_t) = -\sigma_s f(r)\end{aligned}. \tag{5.53}$$

Recalling the equation of congruence (5.14), from (5.53) we obtain

$$-\frac{3}{2}f(r) - \frac{r}{2}f'(r) = 0, \tag{5.54}$$

from which

$$f'(r) = -\frac{3}{r}f(r). \tag{5.55}$$

By integrating the differential equation (5.55) we have

$$\log_e f(r) = -3\log_e r + Z = \log_e \frac{B}{r^3}, \tag{5.56}$$

where Z and B are both constant.

Finally, from (5.56),

$$f(r) = \frac{B}{r^3}. \tag{5.57}$$

Based on (5.53) and setting

$$C = \frac{B\sigma_s}{2}, \tag{5.58}$$

we obtain

$$\begin{aligned}\varepsilon_t =& \varepsilon_m = \frac{C}{r^3}\\ \varepsilon_r =& -2\frac{C}{r^3}\end{aligned} \tag{5.59}$$

where C is constant.

In summary, when the sphere is completely under plastic flow:

$$\begin{aligned}\sigma_t =& \sigma_m = \sigma_s\left(1 - 2\log_e \frac{r_e}{r}\right)\\ \sigma_r =& -2\sigma_s \log_e \frac{r_e}{r}\\ \varepsilon_t =& \varepsilon_m = \frac{C}{r^3}\\ \varepsilon_r =& -2\frac{C}{r^3}\end{aligned} \tag{5.60}$$

Figures 5.5 and 5.6 show the stresses and deformations from (5.60). The behavior of σ_t and σ_m for plastic flow is completely different from that in the elastic field (Fig. 5.3). The behavior of deformations is instead "qualitatively" similar to that for elastic behavior (Fig. 5.4). Equations (5.49) and (5.50) may be substituted by the following ones:

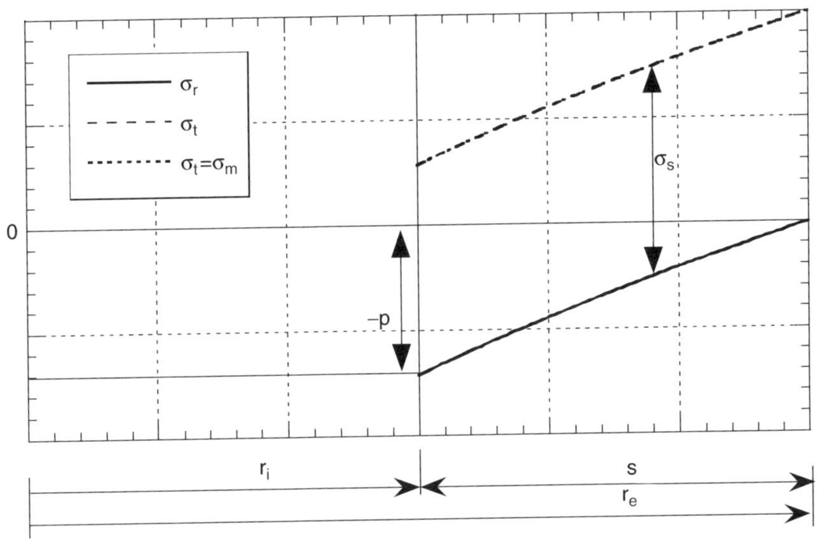

Fig. 5.5

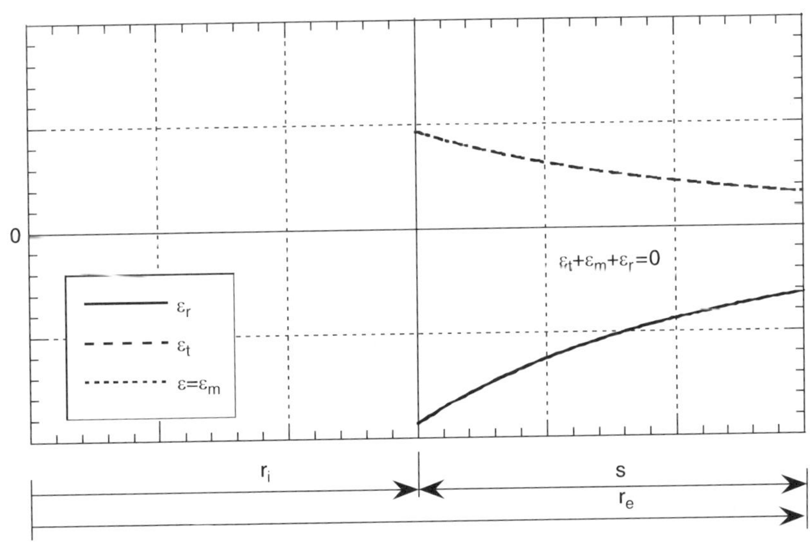

Fig. 5.6

$$s = \frac{pD_e}{4f + \left(1 + 0.08\frac{p}{f}\right)p}, \qquad (5.61)$$

$$s = \frac{pD_i}{4f - \left(1 - 0.08\frac{p}{f}\right)p}. \qquad (5.62)$$

They basically coincide with the exact equations. The mistakes in excess when computing s are: for $p/f = 0.2$ the error is 0.003%; for $p/f = 0.4$ the error is 0.014%; for $p/f = 0.6$ the error is 0.034%.

In favor of resistance, one can also ignore the term between parenthesis in the denominator, obtaining the following:

$$s = \frac{pD_e}{4f + p}, \qquad (5.63)$$

$$s = \frac{pD_i}{4f - p}. \qquad (5.64)$$

Figure 5.7 shows the different equations for both plastic flow and elastic field. Finally, welding may have a value for the allowable stress lower than that of the basic metal, that is,

$$f' = fz, \qquad (5.65)$$

with f' being the allowable stress of the welding and z a coefficient smaller than or equal to the unit called weld joint factor. In this case the (5.63) and (5.64) can be written as follows, replacing f with f':

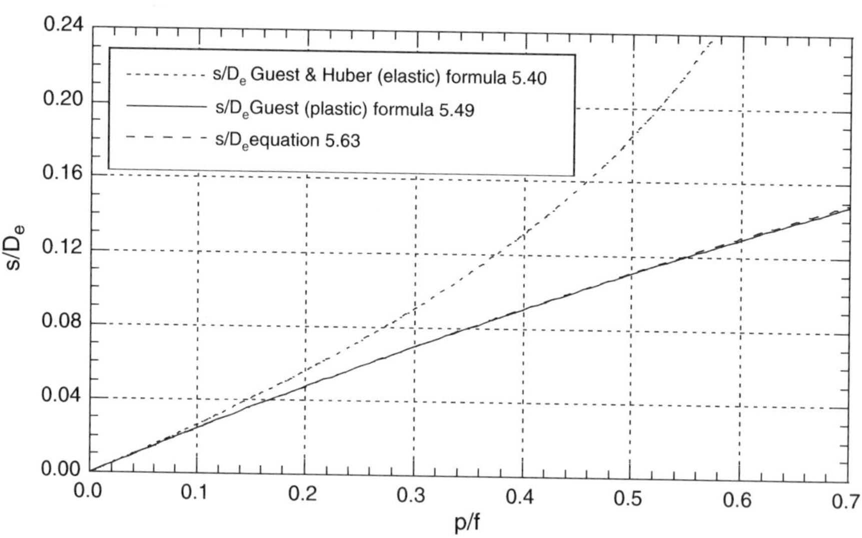

Fig. 5.7

$$s = \frac{pD_e}{4fz + p}, \tag{5.66}$$

$$s = \frac{pD_i}{4fz - p}. \tag{5.67}$$

5.2 Thick Spheres

Equations (5.66) and (5.67) cannot be used in the entire range of values of r_e/r_i. In fact, the greater the value of such ratio, the greater the stress peaks in correspondence of the internal fiber. The use of (5.63) is equivalent to considering $\sigma_{id(m)}$ the ideal reference stress that we obtain from (5.63) by setting $\sigma_{id(m)} = f$.

We note that

$$\sigma_{id(m)} = \frac{p}{4}\frac{D_e - s}{s}. \tag{5.68}$$

As usual, setting

$$a = \frac{r_e}{r_i}, \tag{5.69}$$

from (5.68), through a series of steps, we obtain

$$\sigma_{id(m)} = \frac{p}{4}\frac{a + 1}{a - 1}. \tag{5.70}$$

Recalling (5.37), the ratio γ between the ideal maximum ideal stress on the internal fiber and the ideal reference stress as in (5.63) is given by

$$\gamma = 6\frac{a^3 (a - 1)}{(a^3 - 1)(a + 1)}. \tag{5.71}$$

For instance, for $a = 1.1$ we have $\gamma = 1.15$; for $a = 1.3$ we get $\gamma = 1.44$; for $a = 1.6$ we get $\gamma = 1.83$. Based on what was discussed before, if we use (5.63) the ideal stress on the internal fiber is expressed by

$$\sigma_{id(i)} = \gamma f. \tag{5.72}$$

On the other hand, we already know that the basic allowable stress f is equal to the yield strength (or to the rupture stress under creep at 1,00,000 h) divided by the safety factor of 1.5. Therefore, if $\gamma > 1.5$, the yield strength or creep rupture stress at 1,00,000 h is exceeded in the internal fiber.

On the other hand, we know that σ_s cannot actually be exceeded; the sphere goes partially into plastic flow. The phenomenon may not be of concern, if the working conditions of the sphere and its geometric characteristics are such to rule out further stress peaks (e.g., absence of holes or holes that are largely compensated for by reinforecements, lack or limited presence of stresses caused by thermal flux, no external forces, and so on).

In Sect. 5.4, we will come back to this topic and show how it is possible to use (5.63) even in the case of very large ratios between outside and inside radius, as long as the sphere does not exhibit peaks other than those caused by the value of the above ratio, and if the stresses due to thermal flux are either absent or lower than certain values that depend on such ratio and on the direction of the flux.

Generally, in the case of thick spheres it is necessary or at least advisable to perform a direct analysis of the stresses either by finite element analysis or through experimentation. Without either one, it is at least recommended to limit the maximum value of the ideal stress caused by pressure in correspondence of the internal fiber, so that either σ_s or $\sigma_{R/1000000/t}$ is not exceeded.

Recalling (5.37), and assuming a safety factor of 1.5 (consistently with what discussed in Sect. 1.2) we must write that

$$\frac{3}{2}p\frac{a^3}{a^3-1} = 1.5f, \tag{5.73}$$

where $1.5f$ is equal to the yield strength.

Through a series of steps

$$\frac{1}{a} = \sqrt[3]{1-\frac{p}{f}}. \tag{5.74}$$

We obtain the following equations that refer to the external and internal diameter of the sphere, respectively

$$s = \left(1-\sqrt[3]{1-\frac{p}{f}}\right)\frac{D_e}{2}, \tag{5.75}$$

and

$$s = \left(\frac{1}{\sqrt[3]{1-\frac{p}{f}}}-1\right)\frac{D_i}{2}. \tag{5.76}$$

Again, if the weld joint factor is lower than 1, it must be introduced into the equations that are modified as follows:

$$s = \left(1-\sqrt[3]{1-\frac{p}{fz}}\right)\frac{D_e}{2}, \tag{5.77}$$

and

$$s = \left(\frac{1}{\sqrt[3]{1-\frac{p}{fz}}}-1\right)\frac{D_i}{2}. \tag{5.78}$$

Of course, both (5.77) and (5.78) are operational, if they result in thickness greater than that obtained through (5.66) or (5.67).

Therefore, (5.77) and (5.78) shall be used whenever any one of the following conditions occurs:

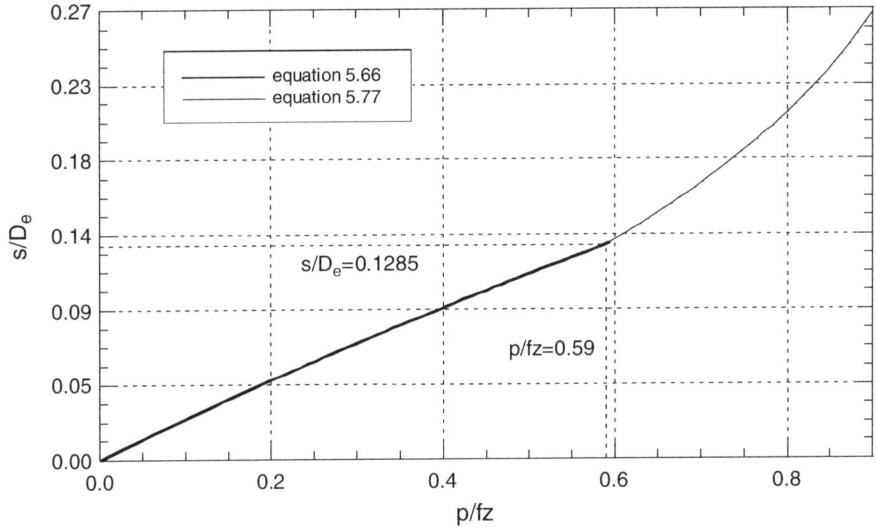

Fig. 5.8

$$\frac{p}{fz} > 0.59 \quad \frac{s}{D_e} > 0.1285$$
$$\frac{s}{D_i} > 0.173 \quad a > 1.346$$

Figure 5.8 shows the behavior of s/D_e with respect to p/fz; the values come from (5.66) and (5.77), respectively.

5.3 Thermal Stresses

As we discussed in Sect. 3.3 about cylinders, the presence of a centrifugal or centripetal thermal flux generates stresses that re-establish the congruence of deformations that would not take place because of the different temperatures present across the thickness, and the resulting different thermal dilations.

The behavior of the temperature through the thickness can be determined with the help of the following considerations. Because of the heat transmission q we have

$$q = S\lambda\frac{dt}{dr}, \tag{5.79}$$

where dt/dr is the temperature gradient, whereas the general surface traversed by heat is

$$S = 4\pi r^2. \tag{5.80}$$

Therefore,

$$\frac{dt}{dr} = \frac{q}{4\pi\lambda r^2}, \tag{5.81}$$

and integrating between r_i and r_e

$$\Delta t = t_e - t_i = \frac{q}{4\pi\lambda}\left(\frac{1}{r_i} - \frac{1}{r_e}\right) = \frac{q}{4\pi r_e \lambda}(a-1). \tag{5.82}$$

Again, we indicate the ratio between outside and inside radius of the sphere with a. From (5.81) we obtain

$$\frac{dt}{dr} = \frac{\Delta t r_e}{(a-1)\, r^2}. \tag{5.83}$$

Equation (5.83) for the temperature gradient will be useful below.

Let us rewrite the equations of equilibrium and congruence obtained in Sect. 5.1

$$\sigma_t - \sigma_r - \frac{1}{2} r \frac{d\sigma_r}{dr} = 0, \tag{5.84}$$

$$\varepsilon_r - \varepsilon_t - r\frac{d\varepsilon_t}{dr} = 0. \tag{5.85}$$

Moreover, recalling that the meridian stress σ_m is equal to the circumferential stress σ_t we have

$$\begin{aligned} \varepsilon_r &= \varepsilon_m = \frac{1}{E}[\sigma_t(1-\mu) - \mu\sigma_r] + \alpha t \\ \varepsilon_r &= \frac{1}{E}(\sigma_r - 2\mu\sigma_t) + \alpha t \end{aligned} \tag{5.86}$$

where α is the thermal expansion coefficient. From (5.85) we obtain

$$\sigma_r - 2\mu\sigma_t - \sigma_t(1-\mu) + \mu\sigma_r - r\frac{d\sigma_t}{dr}(1-\mu) + \mu r\frac{d\sigma_r}{dr} - E\alpha r\frac{dt}{dr} = 0; \tag{5.87}$$

that can be written as

$$-(1+\mu)\left(\sigma_t - \sigma_r - \frac{1}{2}r\frac{d\sigma_r}{dr}\right) - (1-\mu)\left(\frac{r}{2}\frac{d\sigma_r}{dr} + r\frac{d\sigma_t}{dr}\right) - E\alpha r\frac{dt}{dr} = 0. \tag{5.88}$$

Recalling (5.84)

$$\frac{d\sigma_t}{dr} + \frac{1}{2}\frac{d\sigma_r}{dr} = -\frac{E\alpha}{1-\mu}\frac{dt}{dr}, \tag{5.89}$$

we now compute the first derivative of (5.86) to obtain

$$\frac{d\varepsilon_m}{dr} = \frac{1}{E}\left[(1-\mu)\frac{d\sigma_t}{dr} - \mu\frac{d\sigma_r}{dr}\right] + \alpha\frac{dt}{dr}. \tag{5.90}$$

The equatorial sections of the sphere must, of course, remain flat, i.e., $d\varepsilon_m/dr = 0$; thus, from (5.90) we obtain

$$\frac{d\sigma_r}{dr} = \frac{1-\mu}{\mu}\frac{d\sigma_t}{dr} + \frac{E\alpha}{\mu}\frac{dt}{dr}; \tag{5.91}$$

and from (5.89) we have

$$\frac{1+\mu}{2\mu}\frac{d\sigma_t}{dr} = -\frac{1+\mu}{2\mu\,(1-\mu)}E\alpha\frac{dt}{dr}; \tag{5.92}$$

finally,

$$\frac{d\sigma_t}{dr} = -\frac{E\alpha}{1-\mu}\frac{dt}{dr}. \tag{5.93}$$

Based on (5.89), $d\sigma_r/dr = 0$; we conclude that the radial stress shall always be zero, in fact if $r = r_i$ and $r = r_e$ then $\sigma_r = 0$.

Let us recall (5.83). From (5.93) we have that

$$\frac{d\sigma_t}{dr} = -\frac{E\alpha\Delta t}{1-\mu}\frac{r_e}{(a-1)\,r^2}. \tag{5.94}$$

We introduce the quantity K given by

$$K = \frac{E\alpha\Delta t}{1-\mu}, \tag{5.95}$$

and by integrating (5.95) we have that

$$\sigma_t = \sigma_m = \frac{Kr_e}{(a-1)\,r} + C, \tag{5.96}$$

where C is constant.

Given the lack of external forces we must have

$$2\pi\int_{r_i}^{r_e} \sigma_m r dr = 0. \tag{5.97}$$

Therefore,

$$\frac{Kr_e}{a-1}\left(r_e - r_i\right) + \frac{C}{2}\left(r_e^2 - r_i^2\right) = 0; \tag{5.98}$$

then

$$C = -\frac{2Kr_e}{a-1}\frac{r_e - r_i}{r_e^2 - r_i^2} = -\frac{2Ka}{a^2-1}, \tag{5.99}$$

and

$$\sigma_m = K\left[\frac{r_e}{(a-1)\,r} - \frac{2a}{a^2-1}\right]. \tag{5.100}$$

Recalling the meaning of K, we finally have

$$\begin{aligned} \sigma_t &= \sigma_m = \frac{E\alpha\Delta t}{1-\mu}\left[\frac{r_e}{(a-1)r} - \frac{2a}{a^2-1}\right]; \\ \sigma_r &= 0 \end{aligned} \tag{5.101}$$

for $r = r_i$ we have

$$\sigma_{ti} = \sigma_{mi} = \frac{E\alpha\Delta t}{1-\mu}\frac{a}{a+1}, \tag{5.102}$$

whereas for $r = r_e$ we have that

$$\sigma_{te} = \sigma_{me} = \frac{E\alpha\Delta t}{1-\mu}\frac{1}{a+1}. \tag{5.103}$$

σ_t (or σ_m) is shown in Fig. 5.9 for a positive Δt and centripetal flux. If the latter is centrifugal instead, the stresses will obviously change sign.

Given the simultaneous presence of pressure and thermal flux, the maximum values of the stresses generally occur in the internal fiber with centripetal flux, and in the external fiber with centrifugal flux.

Recalling Sect. 5.1, we have, respectively

$$\sigma_{ti} = p\frac{\frac{a^3}{2}+1}{a^3-1} + \frac{E\alpha\Delta t}{1-\mu}\frac{a}{a+1}, \tag{5.104}$$

and

$$\sigma_{te} = p\frac{3/2}{a^3-1} - \frac{E\alpha\Delta t}{1-\mu}\frac{1}{a+1}. \tag{5.105}$$

We introduce the quantity γ

$$\gamma = \pm\frac{E\alpha\Delta t}{2p\,(1-\mu)}, \tag{5.106}$$

with a positive sign for centripetal flux and a negative sign for centrifugal flux.

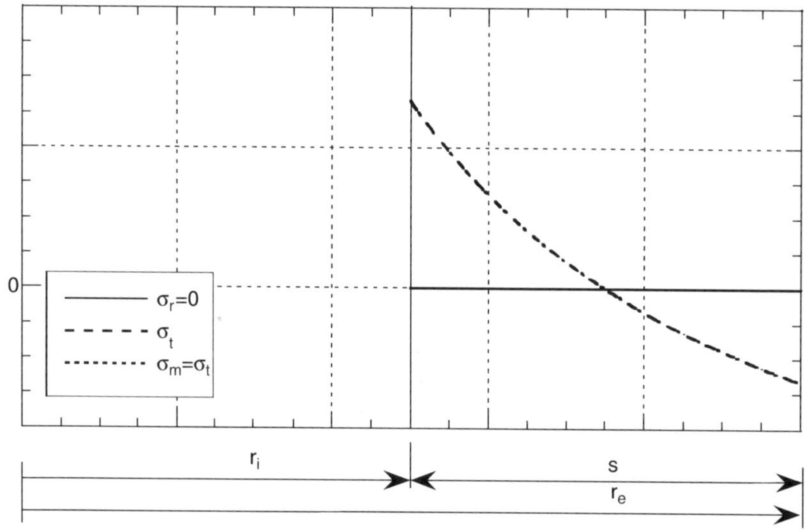

Fig. 5.9

In the case of centripetal flux we obtain

$$\frac{\sigma_{ti}}{p} = \frac{\frac{a^3}{2} + 1}{a^3 - 1} + \gamma \frac{2a}{a+1}. \tag{5.107}$$

In the case of centrifugal flux we obtain

$$\frac{\sigma_{te}}{p} = \frac{3/2}{a^3 - 1} + \gamma \frac{2}{a+1}. \tag{5.108}$$

These ratios are shown in Figs. 5.10 and 5.11. As one can see, the biggest differences between the two ratios above occur for large values of a (i.e., of the ratio between outside and inside radius), and of γ, and then of Δt. The conclusion to draw is that centripetal thermal flux is more dangerous than centrifugal one for the same Δt.

Note that portions of some curves in Fig. 5.11 are dashed because the stress σ_{te} is not always crucial in the case of centrifugal flux. The stress σ_{ti} may reach an absolute value higher than that of σ_{te}, even though the circumferential stress caused by pressure has a different sign than the one caused by flux.

Let us now verify when this is the case.

In the internal fiber

$$\frac{\sigma_{ti}}{p} = \frac{\frac{a^3}{2} + 1}{a^3 - 1} - \gamma \frac{2a}{a+1}. \tag{5.109}$$

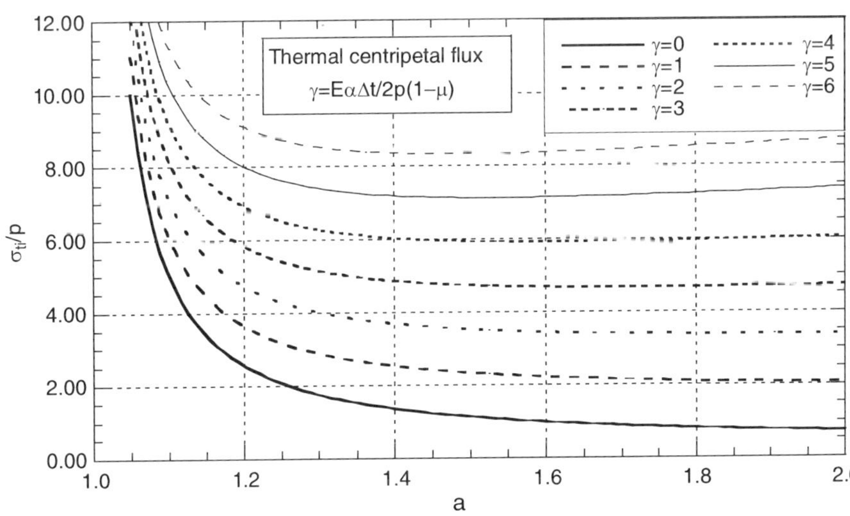

Fig. 5.10

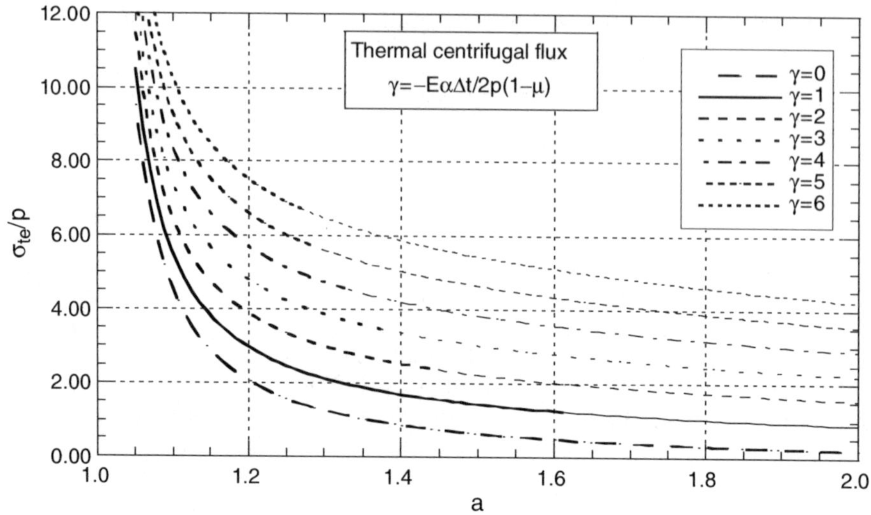

Fig. 5.11

We are interested in the negative values, hence the absolute value is

$$\left|\frac{\sigma_{ti}}{p}\right| = -\frac{\frac{a^3}{2}+1}{a^3-1} + \gamma\frac{2a}{a+1}. \tag{5.110}$$

The condition we are examining occurs when

$$-\frac{\frac{a^3}{2}+1}{a^3-1} + \gamma\frac{2a}{a+1} > \frac{3/2}{a^3-1} + \gamma\frac{2}{a+1}; \tag{5.111}$$

therefore,

$$\gamma > \frac{1}{4}\frac{a^3+5}{a^3-1}\frac{a+1}{a-1}. \tag{5.112}$$

Figure 5.12 shows the curve that represents the right-hand side of (5.112); if the value for γ is greater than the one obtained from this curve, the absolute value of σ_{ti} is greater than σ_{te}. Figure 5.13 shows the value of σ_{ti}/p from (5.109) instead, and confirms previous statements.

If the pressure drops to zero inside the sphere, the absolute value of σ_{ti} is even greater. The condition for the latter value to be greater than σ_{te} is determined by the following inequality:

$$\gamma > \frac{3}{4}\frac{a+1}{(a^3-1)(a-1)}. \tag{5.113}$$

The right-hand side of (5.113) is shown in Fig. 5.12, too.

If we now want to generalize the analysis, as we did in Sect. 3.3 for the cylinder and for any distribution of temperatures, we can write (see (5.93))

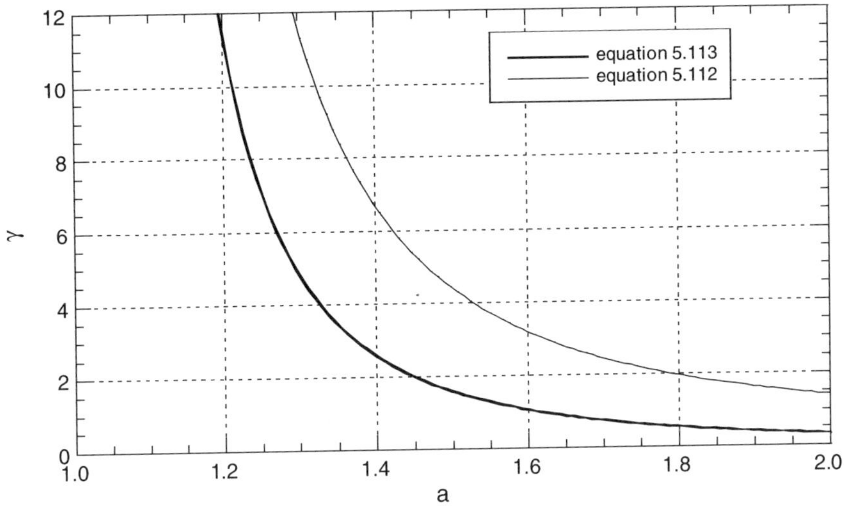

Fig. 5.12

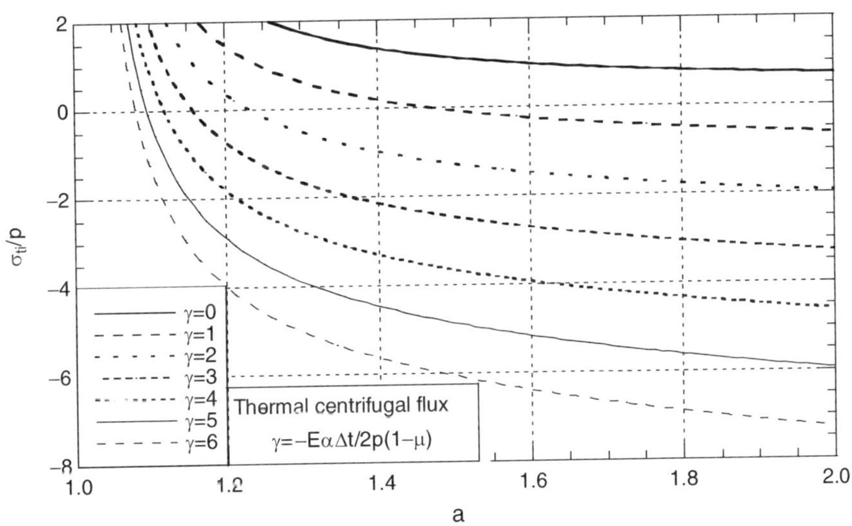

Fig. 5.13

$$\sigma_m - \sigma_{mi} = -\frac{E\alpha}{1-\mu}(t - t_i), \tag{5.114}$$

where t_i is the temperature that corresponds to r_i.

Based on (5.97)

$$\int_{r_i}^{r_e} \sigma_m r dr = 0; \tag{5.115}$$

thus

$$\sigma_{mi} \int_{r_i}^{r_e} r dr - \frac{E\alpha}{1-\mu} \int_{r_i}^{r_e} t r dr + \frac{E\alpha t_i}{1-\mu} \int_{r_i}^{r_e} r dr = 0. \tag{5.116}$$

Solving the integrals that do not contain the temperature we obtain

$$\sigma_{mi} = \frac{2E\alpha}{(1-\mu)\left(r_e^2 - r_i^2\right)} \int_{r_i}^{r_e} t r dr - \frac{E\alpha}{1-\mu} t_i. \tag{5.117}$$

Based on (5.117), (5.114) may be written as follows:

$$\sigma_m = \frac{E\alpha}{1-\mu} \left(\frac{2}{r_e^2 - r_i^2} \int_{r_i}^{r_e} t r dr - t \right). \tag{5.118}$$

The first term in parenthesis may be written as

$$\frac{2}{r_e^2 - r_i^2} \int_{r_i}^{r_e} t r dr = \frac{\int_{r_i}^{r_e} 2\pi t r dr}{\pi\left(r_e^2 - r_i^2\right)}. \tag{5.119}$$

This is the average temperature in the wall of the cylinder and we call it t_m. Therefore, based on (5.118) we have

$$\sigma_t = \sigma_m = \frac{E\alpha}{1-\mu}\left(t_m - t\right), \tag{5.120}$$

and this equation is analogous to (3.177), the one relative to cylinders.

While (3.177) is approximate, (5.120) is not, as one can realize by looking at Fig. 5.9. The behavior of the temperature is qualitatively represented in Fig. 3.16, if we assume that during a transient thermal condition there is a sudden temperature increase, e.g., on the internal wall of the sphere. This is analogous to what we have seen in the case of cylinders.

At the onset of a temperature increase we may assume that the average temperature is about equal to the temperature on the external side. Therefore, based on (5.120)

$$\sigma_{ti} = \sigma_{mi} \cong \frac{E\alpha\Delta t}{1-\mu}, \tag{5.121}$$

and

$$\sigma_{te} = \sigma_{me} \cong 0, \tag{5.122}$$

where Δt is the difference between external and internal side.

Note that under steady-state conditions we have that

$$\sigma_{ti} = \sigma_{mi} \cong \frac{E\alpha\Delta t}{2(1-\mu)}, \tag{5.123}$$

and

$$\sigma_{te} = \sigma_{me} \cong -\frac{E\alpha\Delta t}{2(1-\mu)}. \tag{5.124}$$

As far as the verification criteria applicable to stresses caused by thermal flux, the same considerations made at the end of Sect. 3.3 for cylinders apply here. Specifically, Sect. 5.4 analyzes a sphere solely under internal pressure and possibly thermal flux but without stress peaks from other sources and in absence of fatigue phenomena.

5.4 Partially Plastic Deformed Spheres

In Sect. 5.2, we described a sizing criterion for thick spheres that rules out plastic flow of the internal fibers, limiting the maximum value of the ideal stress to the yield strength. We also pointed out that this limitation implicitly considered the possibility that stress peaks, other than those caused by pressure in a cylinder without holes, existed.

In fact, a partial plastic flow of the sphere may be allowed under certain conditions, provided, of course, that the average value of the ideal stress is not greater than the basic allowable stress of the material.

In other words, one must assess whether the fundamental equation for the sizing of the sphere shown in Sect. 5.1 may also be used in the case of very large values of the ratio between outside and inside radius, and under what conditions. First of all, the equation in question is

$$s = \frac{pD_e}{4f + p}, \tag{5.125}$$

where s is the thickness and f the basic allowable stress. It can also be written as follows:

$$\frac{a-1}{a+1} = \frac{p}{4f}, \tag{5.126}$$

where

$$a = \frac{r_e}{r_i}. \tag{5.127}$$

As we have seen in Sect. 5.1, the maximum value of the ideal stress in the internal fiber, according to Guest–Tresca's failure theory, is given by

$$\sigma_{id(i)} = \frac{3}{2}p\frac{a^3}{a^3 - 1}. \tag{5.128}$$

Therefore, based on (5.126) we have that

$$\sigma_{id(i)} = \frac{6a^3\,(a-1)}{(a+1)\,(a^3-1)}f. \tag{5.129}$$

The nature of this peak is such that, based on the criteria in Sect. 1.5, it is possible for its value to be twice the yield strength. Since the basic allowable stress f is equal to the yield strength divided by the safety factor 1.5, the value of $\sigma_{id(i)}$ can even reach $3f$. Therefore, from (5.129) we have

$$\frac{6a^3(a-1)}{(a+1)(a^3-1)}f = 3f; \tag{5.130}$$

then

$$a^4 - 3a^3 + a + 1 = 0, \tag{5.131}$$

and, finally,

$$a = 2.83.$$

We obtain a very large allowable value for α, one that is considerably greater than the value (1.346) above which we recommended the use of (5.75) instead of (5.125) (see Sect. 5.2).

If the value of a is greater than 1.346, the sphere shows signs of partial plastic flow, and residual stresses are generated at release that we will discuss later on.

To this end, let us examine the behavior of the sphere in the elastic and plastic field in view of Sect. 5.1.

From (5.27) the radial stress σ_r in the area displaying elastic behavior is given by

$$\sigma_r = A - \frac{B}{r^3}, \tag{5.132}$$

where A and B are constant.

On the other hand, for $r = r_e$ we must have $\sigma_r = 0$; thus

$$A = \frac{B}{r_e^3}; \tag{5.133}$$

then

$$\sigma_r = B\left(\frac{1}{r_e^3} - \frac{1}{r^3}\right). \tag{5.134}$$

In the area with plastic flow, F being a constant and based on (5.43), we have

$$\sigma_r = 2\sigma_s \log_e r + F, \tag{5.135}$$

where σ_s is the yield strength.

For $r = r_i$ we must have $\sigma_r = -p$, therefore

$$F = -p - 2\sigma_s \log_e r_i; \tag{5.136}$$

then

$$\sigma_r = 2\sigma_s \log_e \frac{r}{r_i} - p. \tag{5.137}$$

Let r_0 be the radius corresponding to the circumference that separates the area of the sphere showing signs of plastic flow from the area displaying elastic behavior. In addition, we introduce the following dimensionless quantity:

$$a_0 = \frac{r_0}{r_i}. \tag{5.138}$$

For $r = r_o$, the two values of σ_r must coincide, thus

$$B\left(\frac{1}{r_e^3} - \frac{1}{r_0^3}\right) = 2\sigma_s \log_e \frac{r_0}{r_i} - p; \tag{5.139}$$

then

$$B = \frac{2\sigma_s \log_e \dfrac{r_0}{r_i} - p}{\dfrac{1}{r_e^3} - \dfrac{1}{r_0^3}}. \tag{5.140}$$

From (5.134) the radial stress in the area displaying elastic behavior is given by

$$\sigma_r = \left(p - 2\sigma_s \log_e \frac{r_0}{r_i}\right) \frac{\dfrac{1}{r_e^3} - \dfrac{1}{r^3}}{\dfrac{1}{r_0^3} - \frac{1}{r_e^3}}, \tag{5.141}$$

and

$$\sigma_r = (p - 2\sigma_s \log_e a_0) \frac{1 - \dfrac{r_e^3}{r^3}}{\dfrac{a^3}{a_0^3} - 1}. \tag{5.142}$$

From (5.28) and recalling (5.133) and (5.140) we also obtain the following equation of σ_t in the elastic field:

$$\sigma_t = (p - 2\sigma_s \log_e a_0) \frac{1 + \dfrac{r_e^3}{2r^3}}{\dfrac{a^3}{a_0^3} - 1}. \tag{5.143}$$

On the other hand, in the area with plastic flow we know that

$$\sigma_t - \sigma_r = \sigma_s; \tag{5.144}$$

and, from (5.137)

$$\sigma_t = \sigma_s \left(1 + 2\log_e \frac{r}{r_i}\right) - p. \tag{5.145}$$

For $r = r_o$, the two values of σ_t must coincide, thus

$$(p - 2\sigma_s \log_e a_0) \frac{\dfrac{a^3}{2a_0^3} + 1}{\dfrac{a^3}{a_0^3} - 1} = \sigma_s \left(1 - 2\log_e \frac{r}{r_i}\right) - p. \tag{5.146}$$

We obtain

$$\frac{3}{2}p\frac{\dfrac{a^3}{a_0^3}}{\dfrac{a^3}{a_0^3}-1} = \sigma_s\left(\frac{\dfrac{3a^3}{a_0^3}}{\dfrac{a^3}{a_0^3}-1}\log_e a_0 + 1\right). \tag{5.147}$$

Recalling (5.126) and that $\sigma_s = 1.5f$, we finally have

$$\frac{\dfrac{3a^3}{a_0^3}}{\dfrac{a^3}{a_0^3}-1}\left(2\frac{a-1}{a+1} - 1.5\log_e a_0\right) = 1.5. \tag{5.148}$$

Equation (5.148) allows us to obtain the value of a_0 as a function of a by trial-and-error. These values are shown in Fig. 5.14; we considered the values of a between 1.345 and 2.845 that correspond approximately to the onset of plastic flow of the sphere and the value of a from (5.131), respectively. Always taking into account (5.126), the equations for the stresses σ_t, σ_r, and σ_{id} are as follows:

Area with plastic flow ((5.137) and (5.145))

$$\begin{aligned}\sigma_t &= f\left(1.5 + 3\log_e\frac{r}{r_i} - 4\frac{a-1}{a+1}\right)\\ \sigma_r &= f\left(3\log_e\frac{r}{r_i} - 4\frac{a-1}{a+1}\right)\\ \sigma_{id} &= 1.5f = \sigma_s\end{aligned} \tag{5.149}$$

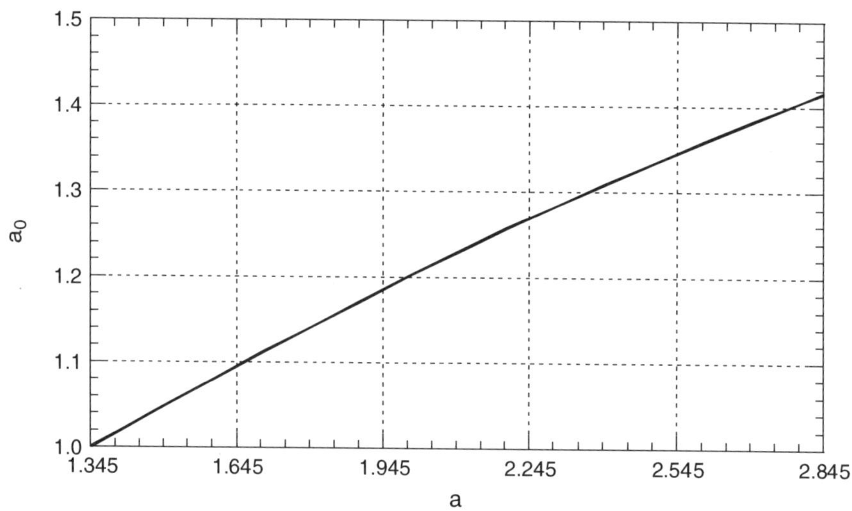

Fig. 5.14

Elastic area ((5.142) and (5.143))

$$\sigma_t = f\left(4\frac{a-1}{a+1} - 3\log_e a_0\right)\frac{1+\dfrac{r_e^3}{2r^3}}{\dfrac{a^3}{a_0^3}-1}$$

$$\sigma_r = f\left(4\frac{a-1}{a+1} - 3\log_e a_0\right)\frac{1-\dfrac{r_e^3}{r^3}}{\dfrac{a^3}{a_0^3}-1} \tag{5.150}$$

$$\sigma_{id} = \frac{3}{2}f\left(4\frac{a-1}{a+1} - 3\log_e a_0\right)\frac{\dfrac{r_e^3}{r^3}}{\dfrac{a^3}{a_0^3}-1}$$

Eliminating the pressure, the release occurs according to the laws of elasticity. Recalling (5.34), (5.35), and (5.126), the corresponding stresses are given by

$$\sigma_t = -4f\frac{a-1}{(a+1)(a^3-1)}\left[1+\frac{1}{2}\left(\frac{r_e}{r}\right)^3\right]$$

$$\sigma_r = -4f\frac{a-1}{(a+1)(a^3-1)}\left[1-\left(\frac{r_e}{r}\right)^3\right] \tag{5.151}$$

$$\sigma_{id} = -6f\frac{a-1}{(a+1)(a^3-1)}\left(\frac{r_e}{r}\right)^3$$

The residual stresses generated during release are the sum of the values from (5.149) or (5.150) and those from (5.151).

Figure 5.15 shows the stresses σ_t and σ_r for the sphere under pressure and during release. In addition, it shows the residual stresses generated during release. Based on (5.148) and Fig. 5.14 we set $a = 2.4$, hence $a_0 = 1.31$.

Figure 5.16 shows the ideal stresses σ_{id}. The absolute maximum value of the ideal stress corresponding to the release is smaller than $3f$ given that $a < 2.83$; as a result, the residual ideal stress as an absolute value is smaller than $1.5f$, i.e., smaller than the yield strength.

Even though we assumed a very high value for a, the area displaying plastic flow is small, equal to about 22% of the thickness. As far as deformations, the difference between ε_t and ε_r, that is proportional to the ideal deformation, is in turn inversely proportional to the cube of the radius in the area displaying plastic flow. This is consistent with the analysis in Sect. 5.1.

Indicating with $\varepsilon_{id(i)}$ and $\varepsilon_{id(o)}$ the ideal deformations in correspondence of the internal fiber and the radius r_o, we can write

$$\varepsilon_{id(i)} = \varepsilon_{id(0)}\frac{r_0^3}{r_i^3}. \tag{5.152}$$

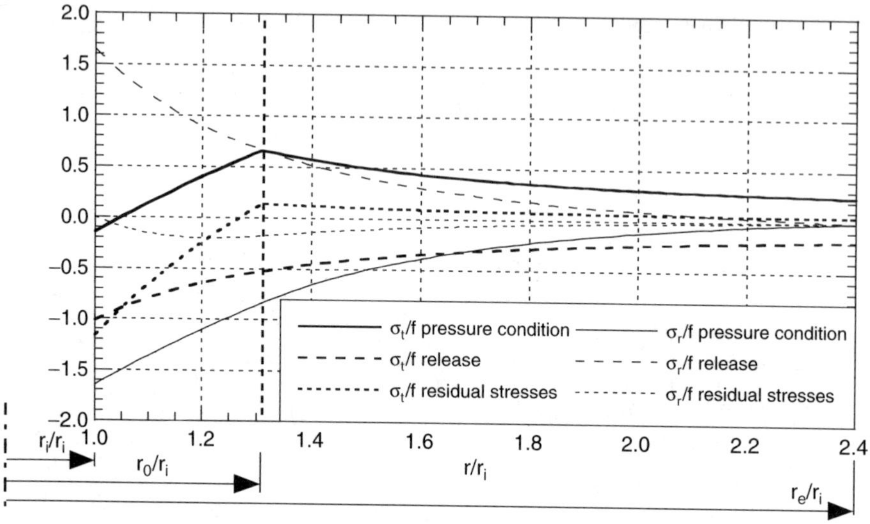

Fig. 5.15

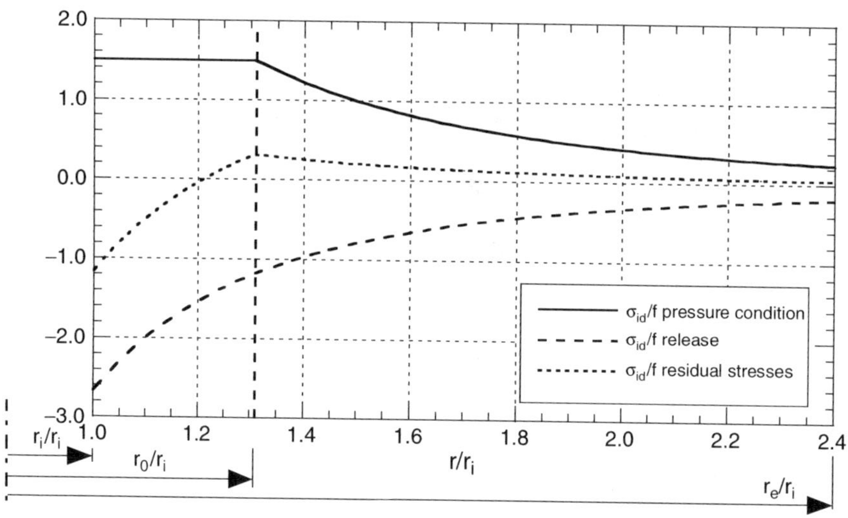

Fig. 5.16

Recalling (5.138), we have

$$\frac{\varepsilon_{id(i)}}{\varepsilon_{id(0)}} = a_0^3. \tag{5.153}$$

Based on Fig. 5.14, it is therefore possible to calculate the ratio between the two deformations discussed above and computed through (5.153). The ratio is shown in Fig. 5.17. This highlights the influence of a on maximum deformations in correspondence of the internal fiber.

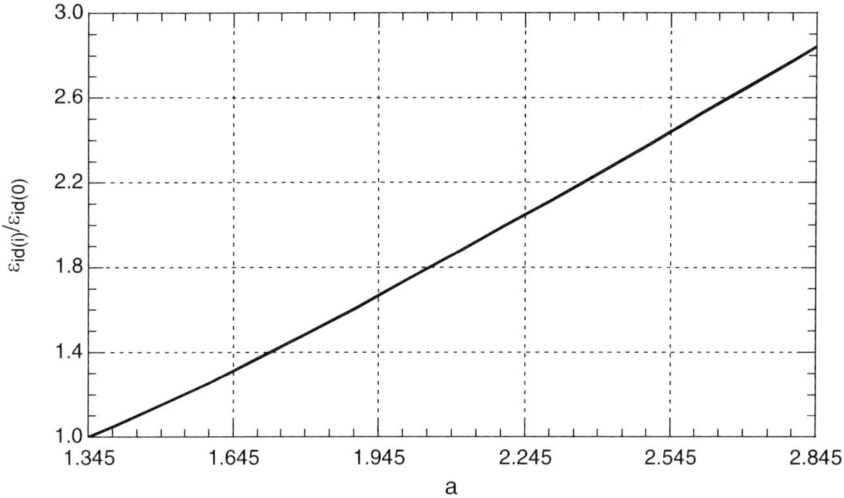

Fig. 5.17

We notice that $\varepsilon_{id(o)}$ corresponds to the yield strength. It is better to limit the deformations in the area of plastic flow, and we recommend that their value never be twice the deformation that corresponds to the yield strength. Consequently, according to Fig. 5.17 the maximum allowable value of a is $a = 2.2$ Therefore, this value replaces the limit 2.83, previously obtained from (5.131), as a condition for the validity of (5.125). Because of this conservative measure we have $p = 1.5f$ (identical to the yield strength).

Moreover, the ideal residual stress at release practically coincides with the allowable stress of the material. In other words, this is equivalent to assuming an allowable stress equal to $2.5f$ instead of $3f$ (as in (5.130)), for the calculation in the elastic field.

Let us now look at the influence of thermal flux on the allowable value a. Recalling Sect. 5.3, if the flux is centripetal the worst situation occurs in the internal fiber. The ideal stress is given by

$$\sigma_{id(i)} = \frac{3}{2}p\frac{a^3}{a^3-1} + \frac{E\alpha\Delta t}{1-\mu}\frac{a}{a+1}. \tag{5.154}$$

We now introduce the dimensionless quantity β

$$\beta = \frac{E\alpha\Delta t}{2f\,(1-\mu)}. \tag{5.155}$$

Recalling (5.126) and setting $\sigma_{id(i)} = 3f$, from (5.154) we obtain

$$6\frac{a-1}{a+1}\frac{a^3}{a^3-1} + \frac{2\beta a}{a+1} = 3, \tag{5.156}$$

and

$$\beta = 1.5\frac{a+1}{a} - 3\frac{a^2(a-1)}{a^3-1}. \tag{5.157}$$

Equation (5.157) lets us determine the maximum allowable value of β, and therefore of Δt as a function of a.

In the case of centrifugal flux, we must first consider the value of the circumferential stress σ_{te} for the external fiber. The latter coincides with the ideal stress ($\sigma_{re} = 0$) and is given by

$$\sigma_{id(e)} = \sigma_{te} = \frac{3}{2}\frac{p}{a^3-1} - \frac{E\alpha\Delta t}{1-\mu}\frac{1}{a+1}. \tag{5.158}$$

In this case both Δt and β are negative; similarly to what we did above we have

$$6\frac{a-1}{(a+1)(a^3-1)} - \frac{2\beta}{a+1} = 3, \tag{5.159}$$

and

$$\beta = 3\frac{a-1}{a^3-1} - 1.5(a+1). \tag{5.160}$$

Equation (5.157) lets us compute the values of β for $\sigma_{te} = 3f$; the values of β that are greater (smaller in absolute value) than those obtained from (5.160) are allowable.

As we said in Sect. 5.3, one cannot neglect the value that the circumferential stress σ_t reaches in the internal fiber. Conservatively factoring in a case with no pressure, such (negative) stress is given by

$$\sigma_{ti} = \frac{E\alpha\Delta t}{1-\mu}\frac{a}{a+1}. \tag{5.161}$$

We set $\sigma_{ti} = -3f$, and through a similar process we obtain

$$\beta = -1.5\frac{a+1}{a}. \tag{5.162}$$

Taking into account the limit of a as far as pressure only ($a = 2.2$) from (5.157), (5.160), and (5.162), the conditions where (5.125) is applicable to size the sphere are identified. This is shown in Fig. 5.18. Again, the values with a positive sign of β refer to a centripetal thermal flux, while the values with a negative sign refer to a centrifugal flux. In particular, the centripetal flux can considerably reduce the allowable a (even below 1.346 considered in Sect. 5.2). One should also keep in mind that modest values of a generally correspond to modest values of the thickness (unless the diameter is very large), which typically means equivalent small values of Δt and β.

The criteria presented in this section can be used for the sizing of thick spheres, if stress peaks other than those under study can be ruled out. As we have seen, the use of (5.125) is justifiable insofar the self-stresses generated during release are utilized. Because of this, when the sphere is again put

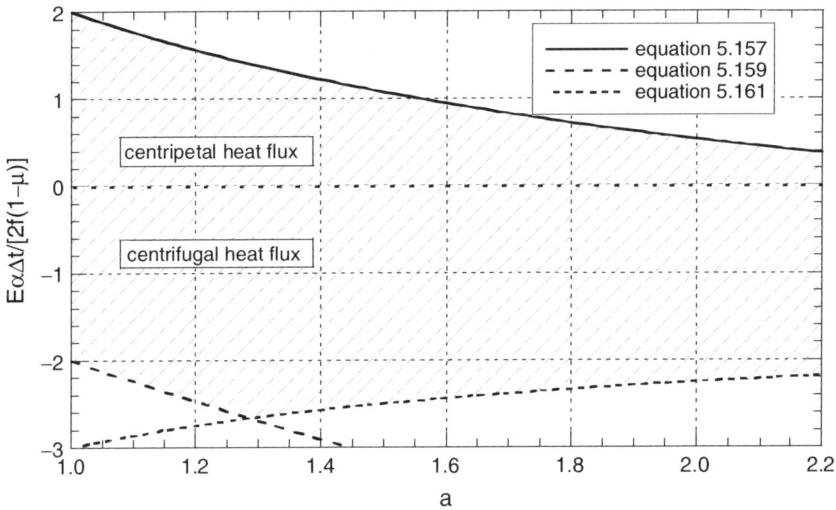

Fig. 5.18

under pressure and temperature, the sphere ends up working elastically and at maximum between stresses $-\sigma_s$ and σ_s.

One should not forget, however, that this situation occurs only in a limited area of the sphere, and that, at any rate, the average ideal stress is not greater than the basic allowable stress of the material. It may be worth pointing out that in the case of large values of a and using (5.125), the average ideal stress is in fact lower than the allowable stress of the material, as discussed in Sect. 5.1.

Specifically, from (5.48) it should be as follows:

$$a = e^{p/2f}. \tag{5.163}$$

From (5.126) we have instead that

$$a = \frac{1 + \dfrac{p}{4f}}{1 - \dfrac{p}{4f}}. \tag{5.164}$$

From (5.164) for $\alpha = 2.2$ we obtain $p/f = 1.5$, whereas from (5.163) we get $p/f = 1.577$. This means that the average ideal stress is not equal to f but to $1.5f/1.577 = 0.95f$.

The design choice of tolerating even large local stress peaks appears to find a solid justification. If the jump in temperature Δt is not related to steady-state conditions, but to a sudden increase in temperature on the internal or external side of the sphere instead, we can still use Fig. 5.18 as long as we assume a conventional and conservative value for Δt, which is twice the actual one, and the number of cycles does not call for a fatigue analysis. To this extent, let us recall the end of Sect. 5.3, specifically (5.121) and (5.123).

5.5 Spheres Under External Pressure

To study the phenomenon of elastic instability of spheres, first we choose to consider the case of a cylinder loaded at the ends and along the vertical axis by an evenly distributed load q (see Fig. 5.19).

Let us assume that the generatrixes have an ondulatory pattern, as shown in the figure, and let us also call y the generic deflection. The deflection corresponds to the variation Δr of the radius of the circumference caused by the axial deformations of the cylinder. As a result of these variations along the circumference,

$$\varepsilon_t = \frac{\Delta r}{r} = \frac{y}{r}, \tag{5.165}$$

and

$$\sigma_t = E\varepsilon_t = \frac{E}{r}y. \tag{5.166}$$

The following force F_t corresponds to the stress for unity of length of the generatrix:

$$F_t = \sigma_t s = \frac{Es}{r}y. \tag{5.167}$$

If we now consider the cylindrical element of unitary width along the circumference, the angle φ corresponding to this element is given by

$$\varphi = \frac{1}{r}. \tag{5.168}$$

The forces F_t exerted on the element form an angle φ between them, and generate a radial resultant active on an element of unitary area. It is, in fact, a pressure that we indicate with q_r; it is given by

$$q_r = 2F_t\frac{\varphi}{2} = \frac{Es}{r^2}y. \tag{5.169}$$

Let us now consider a strip along the generatrix of the cylinder of unitary width loaded along the axis by the load q (Fig. 5.19). The bending moment active in any given point is

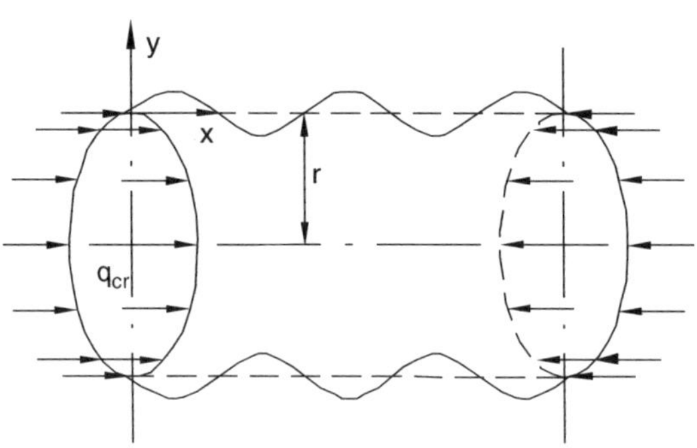

Fig. 5.19

$$M = qy - M_0, \tag{5.170}$$

with M_o as the moment caused by the reactions q_r. The axial moments M generate moments equal to μM along the circumference, since for congruence one cannot have variable deformations through the thickness.

If σ_{af} and σ_{sf} are the stresses generated by the moments M and μM, in fact, we have

$$\begin{aligned} \varepsilon_{af} &= \frac{1}{E}(\sigma_{af} - \mu\sigma_{tf}) = \frac{1}{E}\sigma_{af}(1-\mu^2) \\ \varepsilon_{tf} &= \frac{1}{E}(\sigma_{tf} - \mu\sigma_{af}) = 0 \end{aligned} \tag{5.171}$$

From the first equation in (5.171) and recalling the theory of bent beams, we have

$$\frac{EI}{1-\mu^2}\frac{d^2y}{dx^2} = -M = -qy + M_0. \tag{5.172}$$

Computing the second derivative in (5.172), we obtain

$$\frac{EI}{1-\mu^2}\frac{d^4y}{dx^4} = -q\frac{d^2y}{dx^2} + \frac{d^2M_0}{dx^2}. \tag{5.173}$$

The second derivative of M_o is equivalent to the distributed load q_r with the opposite sign, hence

$$\frac{EI}{1-\mu^2}\frac{d^4y}{dx^4} = -q\frac{d^2y}{dx^2} - \frac{Es}{r^2}y. \tag{5.174}$$

Remembering that $I = s^3/12$ and setting

$$\alpha = \sqrt{\frac{12q\,(1-\mu^2)}{Es^3}}, \tag{5.175}$$

and

$$\beta = \frac{12\,(1-\mu^2)}{r^2 s^2}, \tag{5.176}$$

we have

$$\frac{d^4y}{dx^4} + \alpha^2\frac{d^2y}{dx^2} + \beta y = 0. \tag{5.177}$$

If we consider that the ends of the strip under study are able to rotate, we can write

$$y = A\sin\omega x, \tag{5.178}$$

where A and ω are constant. Therefore, we have

$$\frac{d^2y}{dx^2} = -A\omega^2\sin\omega x, \tag{5.179}$$

and

$$\frac{d^4y}{dx^4} = A\omega^4 \sin \omega x. \tag{5.180}$$

By substituting these equations in (5.177) and eliminating $\sin \omega x$ we obtain

$$\omega^4 - \alpha^2\omega^2 + \beta = 0, \tag{5.181}$$

which has both roots with a positive sign

$$\omega = \sqrt{\frac{\alpha^2}{2} \pm \sqrt{\frac{\alpha^4}{4} - \beta}}. \tag{5.182}$$

The minimum value of α and hence the critical value of q that we indicate with q_{cr} is the one for which ω is still real, that is

$$\frac{\alpha^4}{4} - \beta = 0; \tag{5.183}$$

and, based on (5.176),

$$\alpha^2 = 2\sqrt{\beta} = \frac{2\sqrt{12\,(1-\mu^2)}}{rs}. \tag{5.184}$$

Therefore, from (5.175)

$$\frac{12q_{cr}\left(1-\mu^2\right)}{Es^3} = 4\frac{\sqrt{3\,(1-\mu^2)}}{rs}; \tag{5.185}$$

then

$$q_{cr} = \frac{1}{\sqrt{3\,(1-\mu^2)}}\frac{Es^2}{r}. \tag{5.186}$$

Finally,

$$\sigma_{cr} = \frac{q_{cr}}{s} = \frac{1}{\sqrt{3\,(1-\mu^2)}}\frac{Es}{r}. \tag{5.187}$$

Let us now compare this case with that for the sphere.

Both cylinder and sphere share the characteristic that the bending of the meridians (the generatrixes for the cylinder) is hampered by the rigidity of the parallels (the circumferences for the cylinder) that are subject to deformation caused by the deformation of the meridians themselves. The difference consists of the double curvature for the sphere, and yet this does not change the stiffening effect of the parallels.

In fact, if the tangent to the sphere forms an angle ϑ with the equatorial plane in a generic point, and if the radial deflection is called y, the variation in radius of the parallel is $y \sin \vartheta$. Therefore, by indicating the radius of the parallel with r_p, by analogy with (5.165) and (5.166) we have

$$\varepsilon_t = \frac{y}{r_p} \sin \vartheta, \tag{5.188}$$

and

$$\sigma_t = \frac{Ey}{r_p} \sin\vartheta. \tag{5.189}$$

On the other hand, if r is the radius of the sphere

$$r_p = r\sin\vartheta; \tag{5.190}$$

from (5.149) we have

$$\sigma_t = \frac{Ey}{r}, \tag{5.191}$$

which coincides with (5.166).

Consequently, the forces F_t and the radial reactions q_r are independent from the position of the parallel and coincide with those relative to the cylinder with same radius. Therefore, the way we analyzed the cylinder is applicable to the sphere as well, and is reflected in (5.187).

Let us recall that for the sphere considered as a membrane

$$\sigma = \frac{pr}{2s}; \tag{5.192}$$

thus, the critical pressure is given by

$$p_{cr} = 2\sigma_{cr}\frac{s}{r} = \frac{2Es^2}{r^2\sqrt{3\left(1-\mu^2\right)}}. \tag{5.193}$$

By introducing the external radius r_e instead of the generic radius r of the membrane, we have

$$p_{cr} = \frac{2Es^2}{r_e^2\sqrt{3\left(1-\mu^2\right)}}; \tag{5.194}$$

and finally, for steel with $\mu = 0.3$,

$$p_{cr} = 1.21E\left(\frac{s}{r_e}\right)^2. \tag{5.195}$$

Karman and Tsien recommend the following equation for the allowable pressure:

$$p = \frac{E}{16}\left(\frac{s}{r_e}\right)^2. \tag{5.196}$$

As one can see, by comparing (5.196) with (5.195), a particularly high safety factor is adopted. It is important, though, to consider that (5.196) does not take the out-of-roundness of the sphere into account. For this reason, we prefer the following equation:

$$p = (1-k)\frac{E}{9+0.006\frac{r_e}{s}}\left(\frac{s}{r_e}\right)^2, \tag{5.197}$$

where the value k depends on manufacturing out-of-roundness, according to the following criteria.

Let u be the ovalization of the sphere in percentage

$$u = \frac{D_{i(\max)} - D_{i(\min)}}{D_{i(m)}} 100 = 2\frac{D_{i(\max)} - D_{i(\min)}}{D_{i(\max)} + D_{i(\min)}} 100, \tag{5.198}$$

with $D_{i(m)}$, $D_{i(\max)}$ and $D_{i(\min)}$ equal to the average, maximum, and minimum inside diameter, respectively, k is zero if $u = 0 \div 0.5$ and is equal to 0.2 if $u = 0.5 \div 1$.

In the event of local flattening, one sets

$$u = 400\frac{y_0}{D_e}, \tag{5.199}$$

where y_o is the height of the imperfection measured along the radius.

As far as the verification of the sphere in the plastic field, recalling (5.192) and introducing the external radius r_e, we can write

$$\frac{pr_e}{2s} = f, \tag{5.200}$$

where f is as usual the basic allowable stress.

Then

$$s = \frac{pr_e}{2f} = \frac{pD_e}{4f}; \tag{5.201}$$

or, if we want to highlight the allowable pressure, for comparison with the values in (5.196) or (5.197), we can adopt the following equation:

$$p = 2f\frac{s}{r_e}. \tag{5.202}$$

Equation (5.202) is more conservative than (5.63) in Sect. 5.1.

Nonetheless, given the undoubtedly greater danger from external pressure of the collapse of the sphere, and also taking into account potential out-of-roundness, it is recommended to use the following equation:

$$p = 1.5\,(1-k)\,f\frac{s}{r_e}, \tag{5.203}$$

where k has the same meaning as in (5.197).

Naturally, the allowable pressure is given by the smaller of two values obtained from (5.197) and (5.203), respectively.

6

Heads

6.1 Hemispherical Heads

The analysis of the state of stress at the knuckle between cylinder and hemispherical head is not particularly difficult, as long as one considers both cylinder and semisphere as membranes and one keeps in mind what was said in Sect. 2.2 about edge effects.

Let D be the diameter of both cylinder and head, and s_c and s_f their respective thickness. As we know from Sect. 2.1 ((2.16)), the membrane stresses in the cylinder are

$$\begin{aligned} \sigma_{tc} &= \frac{pD}{2s_c} \\ \sigma_a &= \frac{pD}{4s_c} \end{aligned} \tag{6.1}$$

The circumferential deformation is then

$$\varepsilon_{tc} = \frac{1}{E}\left(\sigma_{tc} - \mu\sigma_a\right) = \frac{1}{E}\frac{pD}{4s_c}\left(2 - \mu\right); \tag{6.2}$$

for the head (see (2.17)) we have instead

$$\sigma_{tf} = \sigma_m = \frac{pD}{4s_f}; \tag{6.3}$$

therefore,

$$\varepsilon_{tf} = \frac{1}{E}\left(\sigma_{tf} - \mu\sigma_m\right) = \frac{1}{E}\frac{pD}{4s_f}\left(1 - \mu\right). \tag{6.4}$$

The elongation Δ_{rc} of the radius of the cylinder is given by

$$\Delta r_c = \varepsilon_{tc}\frac{D}{2} = \frac{pD^2}{8Es_c}\left(2 - \mu\right), \tag{6.5}$$

while the equivalent one Δr_f of the radius of the head is

$$\Delta r_f = \varepsilon_{tf} \frac{D}{2} = \frac{pD^2}{8Es_f}(1-\mu)\,. \tag{6.6}$$

The difference Δr between the elongations is therefore as follows:

$$\Delta r = \Delta r_c - \Delta r_f = \frac{pD^2}{8E}\left(\frac{2-\mu}{s_c} - \frac{1-\mu}{s_f}\right). \tag{6.7}$$

Let us introduce the dimensionless quantity

$$\alpha = \frac{s_c}{s_f}, \tag{6.8}$$

and rewrite (6.7) as follows:

$$\Delta r = \frac{pD^2}{8Es_c}\left[2-\mu-(1-\mu)\,\alpha\right]; \tag{6.9}$$

or, with $\mu = 0.3$,

$$\Delta r = \frac{pD^2}{Es_c}\left[0.2125 - 0.0875\alpha\right]. \tag{6.10}$$

We observe that Δr is zero if

$$\alpha = \frac{2-\mu}{1-\mu} = \frac{1.7}{0.7} = 2.428. \tag{6.11}$$

If the ratio between cylinder and head is 2.428, the radial deformations of both components are congruent, and no stresses are produced at the knuckle between cylinder and head. In this case, the only stresses present are those from (6.1) and (6.3).

In general, Δr is not zero, and radial forces F_0 and moments M_0 are produced at the edge of head and cylinder, as shown in Fig. 6.1.

Because of F_0 and M_0 the edge of the cylinder warps; the radial deflection y_{oc} is given from (2.66) by

$$y_{oc} = 0.909\frac{F_0}{E}\left(\frac{D}{s_c}\right)^{3/2} - 1.652\frac{M_0 D}{Es_c^2}. \tag{6.12}$$

Keeping in mind the orientation of F_0 and the direction of M_0 at the edge of the head, we have the following deflection:

$$y_{of} = 0.909\frac{F_0}{E}\left(\frac{D}{s_f}\right)^{3/2} + 1.652\frac{M_0 D}{Es_f^2}. \tag{6.13}$$

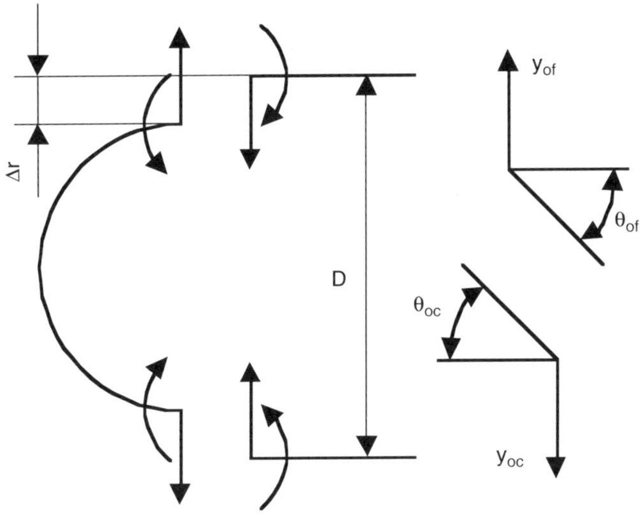

Fig. 6.1

As far as rotations, from (2.67) we have

$$\vartheta_{oc} = -1.652\frac{F_0 D}{E s_c^2} + 6.007\frac{M_0}{E s_c^2}\sqrt{\frac{D}{s_c}}$$
$$\vartheta_{of} = -1.652\frac{F_0 D}{E s_f^2} - 6.007\frac{M_0}{E s_f^2}\sqrt{\frac{D}{s_f}}. \qquad (6.14)$$

Congruence requires that

$$\Delta r = y_{oc} + y_{of}$$
$$\vartheta_{oc} = \vartheta_{of}. \qquad (6.15)$$

In case $s_c = s_f (\alpha = 1)$, we indicate the common thickness with s; based on (6.14) $M_0 = 0$, therefore, from (6.12) and (6.13)

$$\Delta r = 1.817\frac{F_0}{E}\left(\frac{D}{s}\right)^{3/2}. \qquad (6.16)$$

Based on (6.10) and with $\alpha = 1$, we obtain

$$F_0 = 0.0688 p\sqrt{Ds}. \qquad (6.17)$$

Recalling (2.53) and (2.41), we obtain the following equation of the bending moment along the axis of the cylinder or along the meridian of the head

$$M_x = \pm 0.0379 pDs e^{-\beta x}\sin\beta x. \qquad (6.18)$$

Considering the moments that pull the internal fibers of the cylinder and the head to be positive, the plus sign refers to the sphere and the minus sign to the cylinder (see Fig. 6.1).

Taking into account that the moment M_x refers to the unit of length of the circumference, and that the section modulus is therefore $s^2/6$, the longitudinal stress in the cylinder and the meridian stress on the head due to M_x are

$$\sigma_a = \sigma_m = \pm 0.227 \frac{pD}{s} e^{-\beta x} \sin \beta x, \tag{6.19}$$

where the plus sign refers to the external fiber of the cylinder and the internal fiber of the head. In (6.19) x is the distance of the generic fiber from the edge measured along the axis of the cylinder or along the meridian of the head, respectively. As we discussed in Sect. 2.2, the maximum value of M_x, i.e., of σ_a and σ_m, is obtained for $\beta x = \pi/4$, thus

$$|\sigma_a|_{\max} = |\sigma_m|_{\max} = 0.0732 \frac{pD}{s}. \tag{6.20}$$

Moreover, from (6.12) and (6.13), $y_{oc} = y_{of}$; therefore, recalling (6.9) and the first equation of (6.15)

$$y_{oc} = y_{of} = \frac{pD^2}{16Es}. \tag{6.21}$$

A hoop membrane stress is generated at the edge of the cylinder and the head given by

$$\sigma_{to} = \pm E \frac{2y_{oc}}{D} = \pm \frac{pD}{8s}, \tag{6.22}$$

with the plus sign for the head and the minus sign for the cylinder.

In general, from (2.45), and remembering that in the case under study $C_1 = 0$

$$\sigma_t = \pm 0.125 \frac{pD}{s} e^{-\beta x} \sin \beta x. \tag{6.23}$$

Because of M_x, a moment μM_x is generated in the circumferential direction. The hoop stresses on the external and internal fibers of the cylinder and head corresponding to this moment are

$$\sigma_t = \pm 0.0684 \frac{pD}{s} e^{-\beta x} \sin \beta x, \tag{6.24}$$

with the plus sign for the external fiber of the cylinder and the internal fiber of the head.

Again, the maximum value is obtained for $\beta x = \pi/4$ and is

$$|\sigma_t|_{\max} = 0.022 \frac{pD}{s}. \tag{6.25}$$

In view of what discussed thus far, the behavior of the stresses at the junction between cylinder and head is shown in Fig. 6.2.

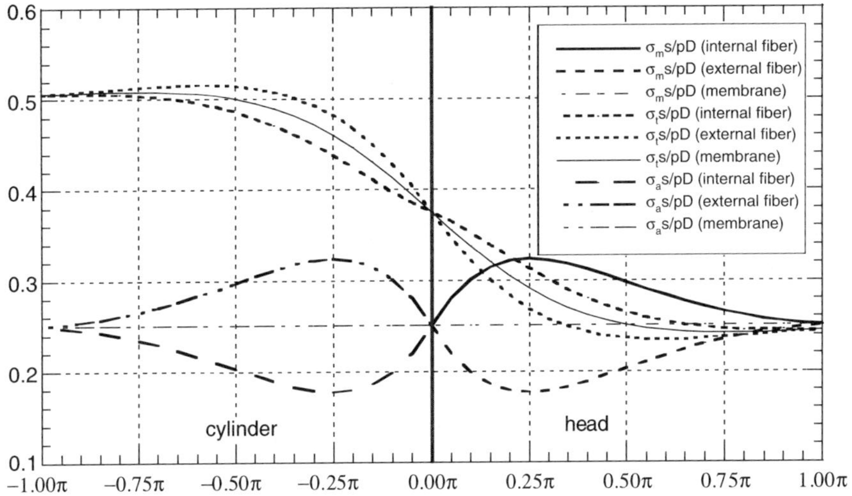

Fig. 6.2

As far as the cylinder is concerned, the maximum value of the hoop membrane stress is obtained for $\beta x = 3\mu/4$ and is

$$\sigma_t = 0.5084\frac{pD}{s}.$$

The maximum value of σ_t is approximately equal to:

$$\sigma_t = 0.515\frac{pD}{s}.$$

As one can see, the membrane value of σ_t locally increases by 1.7%, while an increase of 50% is allowable, based on the criteria presented in Sect. 1.5.

Moreover, we register a small peak of about 3% of the basic membrane stress of the cylinder that is absolutely negligible. The maximum value of σ_a along the axis is equal to

$$\sigma_a = 0.3238\frac{pD}{s}.$$

This value is smaller than the membrane stress σ_t, and thus much smaller than the allowable value which is at least three times σ_t (if the thickness of the cylinder is the minimum allowable), according to the criteria in Sect. 1.5.

In fact, in the case of stress generated to guarantee congruence, a value equal to three times the basic allowable stress of the material is allowed (twice the yield strength). As for the head, if the material is the same used for the cylinder, it is oversized, and the considerations made about the cylinder hold true even more so, because the meridian stresses have the same values than the longitudinal stresses of the cylinder and the hoop stresses are smaller.

Let us consider now the case where the thickness of the head (equal to the thickness of the cylinder) is the minimum allowable because the material used has mechanical characteristics that are inferior to those used for the cylinder. The maximum value of the hoop membrane stress is equal to

$$\sigma_t = 0.375\frac{pD}{s}.$$

with an increase of 50% compared to the membrane stress of the sphere, hence in agreement with the criteria presented in Sect. 1.5. The maximum value of σ_m is

$$\sigma_m = 0.3232\frac{pD}{s},$$

which is quite far from the allowable value equal to three times the basic allowable stress of the material used for the head. We determine that the junction cylinder–head does not generate stresses that require an increase in thickness of the cylinder and the head that are sized independently.

Let us now examine the more general case where the thickness of cylinder and head are different. We introduce the following dimensionless quantities:

$$A = \frac{F_0}{1.817p\sqrt{Ds_c}}, \tag{6.26}$$

and

$$B = \frac{M_0}{pDs_c}. \tag{6.27}$$

Recalling (6.8), (6.10), (6.12), (6.13), (6.14), and (6.15), through a series of steps we obtain the following equations:

$$-A + 2B + A\alpha^2 + 2B\alpha^{5/2} = 0, \tag{6.28}$$

$$A - B + A\alpha^{3/2} + B\alpha^2 = 0.1287 - 0.053\alpha, \tag{6.29}$$

$$A\left(\alpha^2 + 1\right) + 2B\left(\alpha^{5/2} + 1\right) = 0, \tag{6.30}$$

and

$$A\left(\alpha^{3/2} + 1\right) + B\left(\alpha^2 - 1\right) = 0.1287 - 0.053\alpha. \tag{6.31}$$

Then, solving the system we obtain

$$A = \frac{2\left(\alpha^{5/2} + 1\right)(0.1287 - 0.053\alpha)}{2\left(\alpha^{5/2} + 1\right)\left(\alpha^{3/2} + 1\right) - \left(\alpha^2 - 1\right)^2}, \tag{6.32}$$

and

$$B = -\frac{\left(\alpha^2 - 1\right)(0.1287 - 0.053\alpha)}{2\left(\alpha^{5/2} + 1\right)\left(\alpha^{3/2} + 1\right) - \left(\alpha^2 - 1\right)^2}. \tag{6.33}$$

A and B are plotted in Fig. 6.3. We only considered the values of α greater than one, since we can rule out the possibility that the thickness of the head will be greater than that of the cylinder.

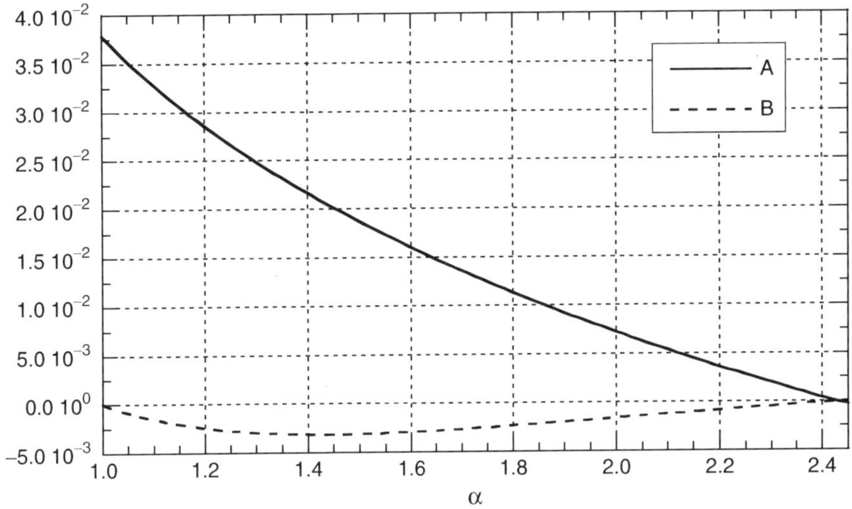

Fig. 6.3

Recalling (2.51), (2.53), (2.60), (6.26), and (6.27), in the cylinder we have

$$M_x = -pDs_c e^{-\beta x} \left[A \sin \beta x - B \left(\sin \beta x + \cos \beta x\right)\right]. \tag{6.34}$$

By integrating (6.34) and making it equal to 0, we obtain

$$\frac{dM_x}{dx} = pDs_c \beta e^{-\beta x} \left[(A - 2B) \sin \beta x - A \cos \beta x\right] = 0; \tag{6.35}$$

then

$$\tan \beta x = \frac{A}{A - 2B}. \tag{6.36}$$

The values of βx that maximize M_x can be obtained as a function of α from (6.36).

As far as the head, recalling (6.8) and keeping in mind the orientation of F_0 and the direction of M_0, we have

$$M_x = -pDs_f e^{-\beta x} \left[A\sqrt{\alpha} \sin \beta x + B\alpha \left(\sin \beta x + \cos \beta x\right)\right]. \tag{6.37}$$

Similarly to (6.36), by computing the derivative of (6.37) and making it equal to 0, we obtain

$$\tan \beta x = \frac{A}{A + 2B\sqrt{\alpha}}. \tag{6.38}$$

The values of βx that satisfy both (6.36) and (6.38) are plotted in Fig. 6.4 (consider that if $s_f \neq s_c$ the value of β is different for the cylinder and the head). The longitudinal stress on the internal or external fibers for the cylinder, due to M_x is given by

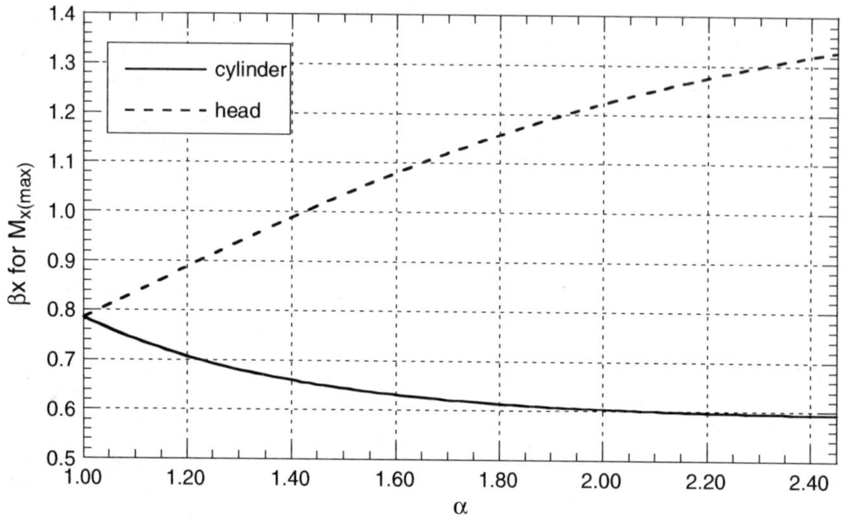

Fig. 6.4

$$\sigma_a = \mp \frac{6pD}{s_c} e^{-\beta x} \left[A \sin \beta x - B \left(\sin \beta x + \cos \beta x \right) \right] . \tag{6.39}$$

Based on the values of βx that satisfy (6.36), one can define the absolute maximum value of σ_a that may be expressed as follows:

$$|\sigma_a|_{\max} = k_a \frac{pD}{s_c} . \tag{6.40}$$

The values of k_a are shown in Fig. 6.5. As for the head, the meridian stress on the internal or external fibers due to M_x is given by

$$\sigma_m = \pm \frac{pD}{s_f} e^{-\beta x} \left[A\sqrt{\alpha} \sin \beta x + B\alpha \left(\sin \beta x + \cos \beta x \right) \right] ; \tag{6.41}$$

by analogy with the cylinder, we can write

$$|\sigma_m|_{\max} = k_m \frac{pD}{s_f} . \tag{6.42}$$

The values of k_m are shown in Fig. 6.5. We now consider the hoop membrane stresses due to the deflections y_{oc} and y_{of}: from (6.12), (6.26) and (6.27) we have

$$y_{oc} = 1.652 \left(A - B \right) \frac{pD^2}{Es_c} . \tag{6.43}$$

The corresponding compressive stress σ_{toc} is given by

$$\sigma_{toc} = -\frac{2Ey_{oc}}{D} = -3.304 \left(A - B \right) \frac{pD}{s_c} , \tag{6.44}$$

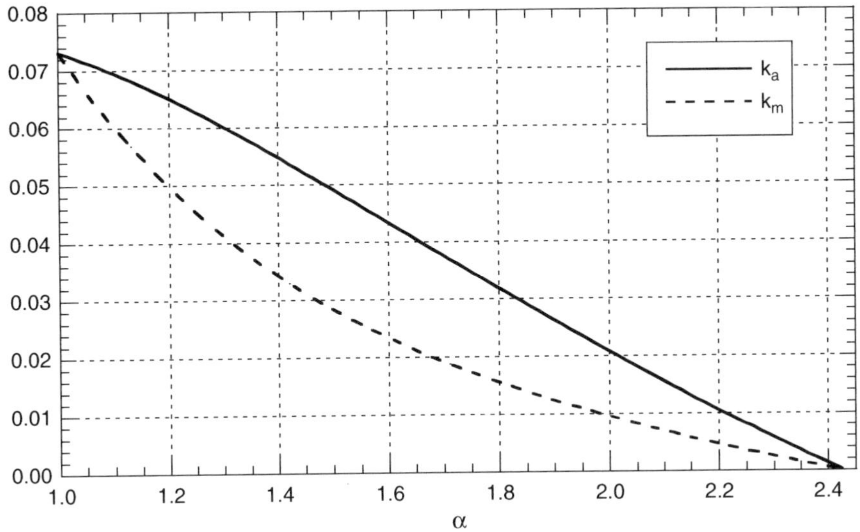

Fig. 6.5

that we may write as follows:

$$\sigma_{toc} = -k_{tc}\frac{pD}{s_c}. \tag{6.45}$$

The values of k_{tc} are shown in Fig. 6.6. For the head we have

$$y_{of} = 1.652\left(A\sqrt{\alpha} + B\alpha\right)\frac{pD^2}{Es_f}. \tag{6.46}$$

The membrane stress at the junction is

$$\sigma_{tof} = 3.304\left(A\sqrt{\alpha} + B\alpha\right)\frac{pD}{s_f}, \tag{6.47}$$

that we rewrite as

$$\sigma_{tof} = k_{tf}\frac{pD}{s_f}. \tag{6.48}$$

The values of k_{tf} are shown in Fig. 6.6. Note that the following condition should be checked at the junction, remembering the equations for the membrane values of σ_t in the cylinder and the sphere, if considered separately

$$\frac{pD}{2s_c} - k_{tc}\frac{pD}{s_c} = \frac{pD}{4s_f} + k_{tf}\frac{pD}{s_f}; \tag{6.49}$$

then

$$0.5 - k_{tc} = (0.25 + k_{tc})\,\alpha. \tag{6.50}$$

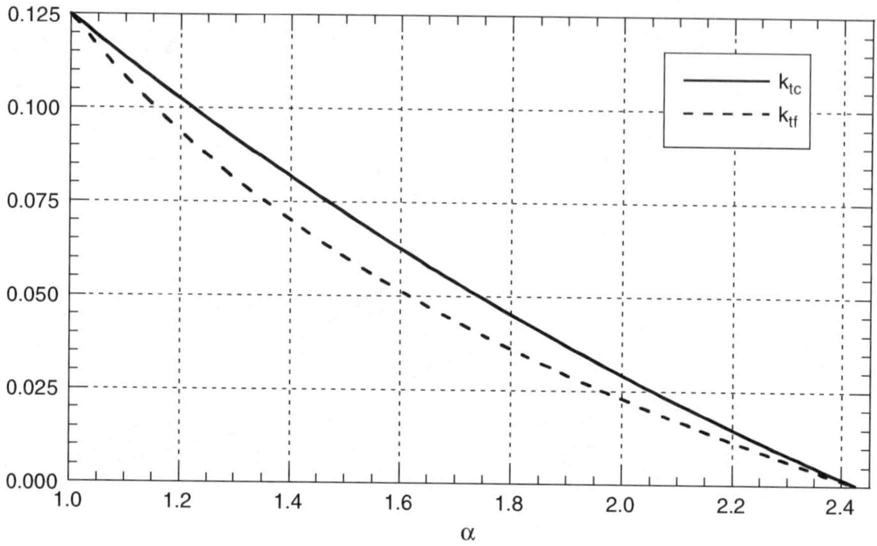

Fig. 6.6

Indeed, this condition is verified for $\alpha = 1$ only, as we can easily see in Fig. 6.2, as well. We should also not forget that if $s_c \neq s_f$, there is an area of knuckle between cylinder and head where the values of σ_t vary between the value on the left-hand side in (6.49) in correspondence of the cylinder, and the value on the right-hand side in (6.49) in correspondence of the head. Moreover, the cylinder exhibits the following hoop stresses at the junction because of the moment μM_0:

$$\sigma_{toc} = \pm 1.8B\frac{pD}{s_c}, \tag{6.51}$$

while for the head we have

$$\sigma_{tof} = \pm 1.8B\frac{pD}{s_f}. \tag{6.52}$$

The positive sign refers to the internal fiber and the negative sign to the external fiber both in (6.51) and (6.52).

Analyzing k_a, k_m, k_{te} and k_{tf} as a function of α, we observe that if the thickness of the cylinder is greater than that of the head (as it is often the case), the stresses at the edge decrease, making the conclusions drawn for $s_c = s_f$ all the more true.

In other words, the fact that the head is connected to the cylinder does not require any greater thickness of the cylinder or the head beyond that obtained considering both components separately. Specifically, if the pressure is external for hemispherical heads, one must apply the sizing criteria discussed in Sect. 5.5 about spheres.

6.2 Dished Heads

The stresses in a dished head are quite complex, and can neither be analyzed in a comprehensive way, nor can they be studied in-depth to point out all required aspects in this book.

If we think of a classic head with three centers, in other words a paraelliptical or torospherical head, there are three structural elements, i.e., a spherical central portion, a knuckle and a cylinder. If they were able to dilate under pressure independently from one another, they would exhibit different deformations at the edges. The study of the stresses in the three elements considered independently shall be integrated by the study of the stresses generated by edge effects. As far as the spherical portion and the cylinder are concerned, the situation is considerably simpler, as we have seen in Sect. 6.1. Finally, the study of the knuckle is quite complex.

Moreover, the study should include the analysis of a large number of cases stemming from the variability of the radius of the sphere and of the knuckle, and from the influence of the ratio between the thickness of the head and that of the cylinder. If, on the other hand, the head is elliptical, one has the problem of evaluating the entity of the deformations generated by the stresses at the edge of a component for which a straightforward analysis is necessary. Evidently, there are no simple solutions, and the problem can be faced either through advanced numerical techniques (e.g., finite element analysis) or through experimental research.

We shall limit ourselves here to compute membrane stresses at the knuckle of the paraelliptical head so that we can compare them with those of the sphere and the cylinder and show the onset of edge effects among the three above components. This study, by its very nature incomplete, will lead us however to make a number of interesting considerations. Most important, they are also applicable to the actual behavior of the paraelliptical head. Finally, we will discuss the criteria for the sizing of the heads in actual designs.

Let us consider the paraelliptical head in Fig. 6.7 where r and r' are the radii of the sphere and of the knuckle, respectively.

As usual, the membrane stresses of the sphere are

$$\sigma_m = \sigma_t = \frac{pr}{2s}, \tag{6.53}$$

where s is the thickness. For the knuckle based on Sect. 2.1 we have

$$\begin{aligned} R_1 &= r' \\ R_2 &= \frac{D/2 - r'}{\sin\alpha} + r'. \end{aligned} \tag{6.54}$$

Naturally, the geometry of the knuckle must be such that for $\alpha = \alpha_0$

$$\frac{D/2 - r'}{\sin\alpha_0} + r' = r; \tag{6.55}$$

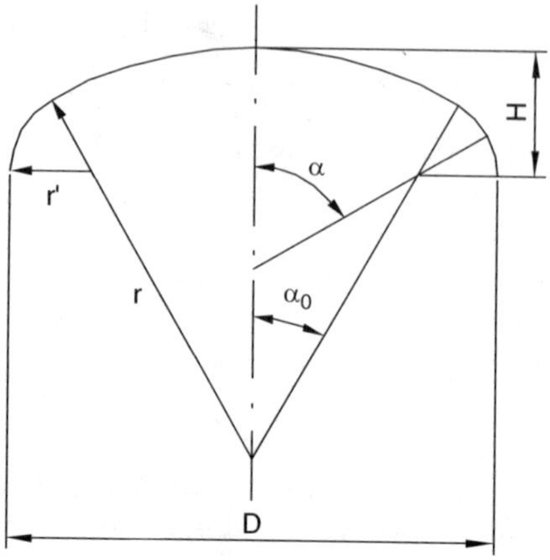

Fig. 6.7

hence,

$$\alpha_0 = \arcsin\left(\frac{D/2 - r'}{r - r'}\right). \tag{6.56}$$

Let us examine now a specific and rather frequent case. We assume that $r = 0.8D$ and $H = 0.25D$. To verify these conditions, $\alpha_0 = 33°$ and $r' = 0.143D$. Let us recall (2.14) and (2.15) in Sect. 2.1; it is possible to compute from (6.54) the dimensionless quantities $\sigma_t s/pD$ and $\sigma_m s/pD$ for the knuckle. For the sphere they are equal to 0.4, as it can be easily determined from (6.53), whereas for the cylinder they are equal to 0.5 and 0.25, respectively. The result is plotted in Fig. 6.8.

While σ_m is a smooth curve moving from the sphere to the cylinder (this is to be expected because σ_m balances the entire thrust on the different portions of the head), the circumferential stress σ_t is rather abrupt at the junctions sphere–knuckle and knuckle–cylinder. Moreover, the knuckle exhibits a negative value of σ_t.

Thus, the deformations are not congruent and come from stresses at the edges of the three elements. They considerably modify σ_m, which has different values outside and inside the head. Bending stresses overlap the meridian membrane stress; the former are typical secondary stresses, according to the classification in Sect. 1.5, because they are present solely to reestablish congruence.

The stresses σ_t undergo substantial change as well. Bending stresses overlap the membrane stress that, differently from what shown in Fig. 6.8, is continuous. The bending stresses are secondary stresses, while the peak of the membrane stress σ_t is typical of a local membrane stress, as long as the conditions discussed in Sect. 1.5 hold.

The stresses are qualitatively shown in Fig. 6.9. As one can see, the negative peak of the hoop stresses is noticeably reduced (compared to that in Fig. 6.8)

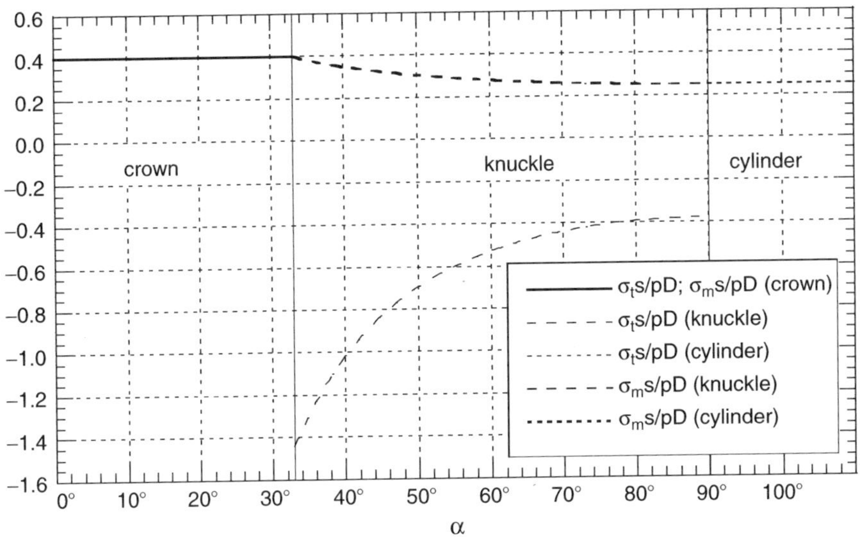

Fig. 6.8

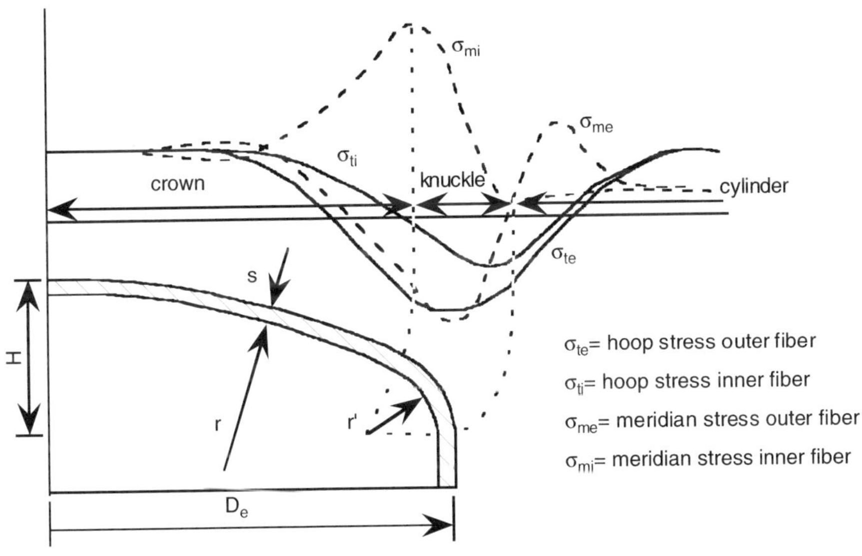

Fig. 6.9

due to the effects at the edges. Nonetheless, especially if the head is almost flat, the value of the local ideal membrane stress may be large and critical as far as the sizing of the head. It will be so in any case, if the ratio between thickness and diameter is small. In fact, the presence of negative σ_t in the knuckle causes the potential for collapse due to buckling. This can be addressed appropriately by increasing the thickness and hence reducing the value of these stresses. We will see later on how this is actually accomplished.

The sizing of the heads can be done by introducing a characteristic quantity called form factor, the values of which are obtained from a diagram. The values are the result of theoretical studies and experimental research, and therefore there are slight differences across various proposals. To explain the nature of the form factor, we refer to a conceptually quite simple method that can be adopted.

The largest values of the deformations occur in the internal fiber. Once the values of the circumferential and the meridian deformations causing the maximum ideal deformation on a head under pressure p are identified, the values for ε_{id} are put on the x-axis and the pressure on the y-axis (Fig. 6.10). As long as the material does not yield, there is proportionality between ε_{id} and p. Once the yield strength is reached in the most stressed fiber, the deformations increase more than proportionally to p.

It is best to consider a danger condition when the plastic component of the ideal deformation is equal to 0.2%. Figure 6.10 highlights a condition of danger at point C in correspondence of plastic deformation BC and pressure p_2.

Let us now consider the meridian (or circumferential) stress of a spherical membrane with a radius r_s equal to the outside diameter D_e of the head under examination, while the thickness is the same as the head. For the sphere we have

$$\sigma_m = \frac{pr_s}{2s} = \frac{pD_e}{2s}. \tag{6.57}$$

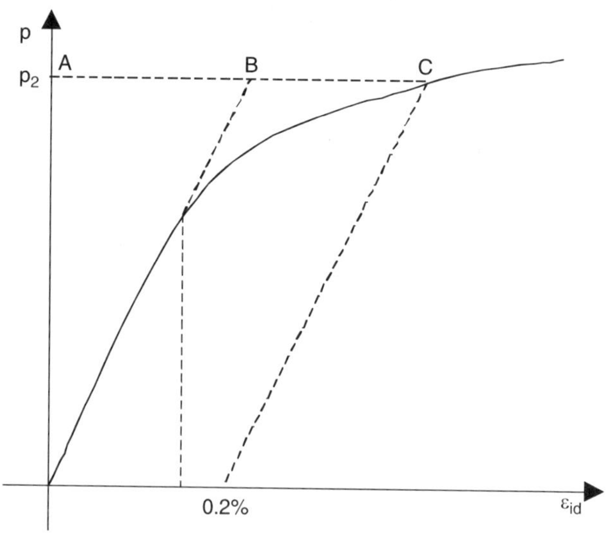

Fig. 6.10

The dangerous pressure p_1 for this ideal sphere is obtained from (6.57) by setting $\sigma_m = \sigma_s$, where σ_s is the yield strength, that is,

$$p_1 = \frac{2s}{D_e}\sigma_s. \tag{6.58}$$

On the other hand, the minimum thickness of the sphere is obtained by setting $\sigma_m = f$, where f is the basic allowable stress. From (6.57) we obtain

$$s = \frac{pD_e}{2f}. \tag{6.59}$$

We set the head to be such that

$$s = \frac{pD_e}{2f}C, \tag{6.60}$$

where C is a dimensionless quantity called form factor, the value of which must be determined.

The dangerous pressure of the head that we know to be equal to p_2 is obtained by substituting σ_s to f in (6.60). Therefore, we have

$$p_2 = \frac{1}{C}\frac{2s}{D_e}\sigma_s, \tag{6.61}$$

and by comparing (6.61) with (6.58) we obtain

$$C = \frac{p_1}{p_2}. \tag{6.62}$$

Therefore, it is necessary to experimentally identify the value of p_2. Then the value of p_1 is computed based on the thickness, the external diameter of the head, the yield strength of the material, and (6.58). Finally, C was computed from (6.62). This made it possible to compute a series of values for C that are subsequently plotted, by varying the geometrical characteristics of the head.

Equation (6.60) can be used for the sizing of dished heads and is included in different national codes. The problem is therefore how to determine C. We described one methodology based on an experimental investigation. In any case, according to modern verification criteria discussed in Sect. 1.5, identifying C is conceptually easy, once the values of the different stresses and their nature (general membrane stresses, local membrane stresses, and secondary ones) have been identified through an appropriate theoretical or experimental investigation. What remains to carry out, though, is the danger of buckling in the knuckle, which only experimental analysis may be able to address.

Figure 6.11 shows a diagram that can be adopted; it allows us to compute C as a function of the ratio H/D_e between the external height of the head and the outside diameter and, for small values of H/D_e, of the ratio s/D_e between the thickness and the outside diameter. Note that for $H/D_e = 0.5$, i.e., for the hemispherical head, we have $C = 0.5$ (see Sect. 6.1).

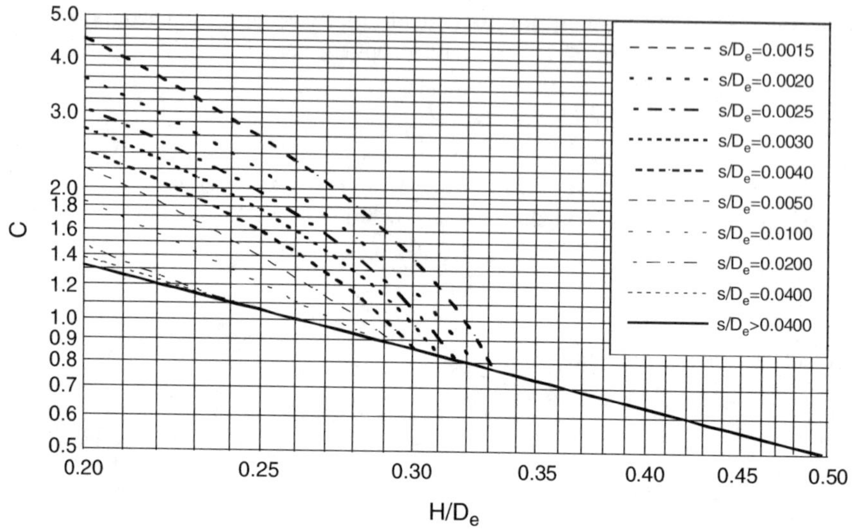

Fig. 6.11

The series of curves on the left side accounts for the danger of buckling in the knuckle. It is not surprising that the value of C becomes larger as the value of the ratio s/D_e becomes smaller. It may be surprising instead that the value of C depends on H/D_e, regardless of the shape of the head. Indeed, Fig. 6.11 may be used for hemispherical, paraelliptical and elliptical heads, albeit with precise limitations for its range of applicability. Specifically, the importance of the radius of the knuckle for paraelliptical heads is quite evident. Obviously, C refers to that head that results in the least favorable state of stress (within the range of applicability of the figure). The drawback of the simplicity of this approach is that it may result in the oversizing of some heads. The limitations mentioned above that also define the range of applicability are as follows:

Hemispherical heads

$$0.003D_e \le s \le 0.16D_e.$$

Elliptical heads

$$0.003D_e \le s \le 0.08D_e; H \ge 0.18D_e.$$

Paraelliptical heads

$$0.003D_e \le s \le 0.08D_e; H \ge 0.18D_e$$
$$r' \ge 0.1D_e; r' \ge 2s$$

or

$$0.01D_e \le s \le 0.03D_e; H \ge 0.18D_e$$
$$r' \ge 0.06D_e$$

or

$$0.02D_e \leq s \leq 0.03D_e; 0.18D_e \leq H \leq 0.22D_e$$
$$r' \geq 0.06D_e.$$

In the above inequalities r is the radius of the spherical portion and r' the radius of the knuckle.

Of course, there is no reason not to use heads outside the conditions shown above. In this case, though, Fig. 6.11 is not valid anymore, and a specific investigation of the state of stress is necessary to determine the required thickness. If the head is under pressure from the outside, it is necessary to perform a double verification. In the first one it is recommended to use the following equation, instead of (6.59) because of the increased danger caused by compressive stresses:

$$s = \frac{pD_e}{1.6f}C. \tag{6.63}$$

One then shall consider the central spherical portion of the head (if it is paraelliptical), or the ideal spherical portion of the radius equal to the radius of the ellipsis in correspondence of the top.

The same verifications, already discussed in Sect. 5.5, are carried out in this case by computing the allowable pressure. Specifically, (5.197) and (5.203) are used with a value for u coming from (5.199).

6.3 Conical Heads and Truncated Cones

Determining the fundamental equation for the sizing of the conical heads is not difficult in view of Sect. 2.1 and following the calculation criteria already discussed for cylinders and spheres.

In fact, from (2.19) and introducing the inside diameter D_i at the base of the cone, the maximum hoop membrane stress is

$$\sigma_{tm} = \frac{pD_i}{2s\cos\alpha}, \tag{6.64}$$

where s is the thickness.

As for the cylinder and the sphere, we set the average between the values on the internal and internal fiber of the cone's wall as the average radial stress, that is

$$\sigma_{rm} = -\frac{p}{2}. \tag{6.65}$$

The average ideal stress that practically corresponds to the general membrane stress, according to Guest's failure theory, is given by

$$\sigma_{id(m)} = \frac{pD_i}{2s\cos\alpha} + \frac{p}{2}, \tag{6.66}$$

where f is the basic allowable stress. Setting $\sigma_{id(m)} = f$ we obtain

$$s = \frac{pD_i}{2f - p} \frac{1}{\cos\alpha}. \tag{6.67}$$

If we consider the weld joint efficiency z to be smaller than one, (6.67) changes into

$$s = \frac{pD_i}{2fz - p} \frac{1}{\cos\alpha}. \tag{6.68}$$

Equation (6.67) (or (6.68)) is fundamental for the sizing of the conical head. The sizing of the conical head or the truncated cone is not simply addressed by this equation. Clearly, the stresses in correspondence of the junction of the cone with the cylinder or cylinders (knuckle as a truncated cone) are quite important. The theoretical analysis of these stresses is equally or even more difficult than that of the junction between a dished head and a cylinder that we already discussed in Sect. 6.2. Therefore, we limit our analysis to some criteria that are generally accepted.

First of all, (6.66) is applicable only if $s/D_e \leq 1/6$ (where D_e is the outside diameter of the cylinder connected to the cone), and if the semiangle at the apex α is lower or equal to 70°. If the angle is greater than that, it is recommended to use the following equation conceptually reminiscent of flat heads, as we shall see in Sect. 6.4, together with (6.66):

$$s = 0,5\,(D_e - r)\,\frac{\alpha}{90}\sqrt{\frac{p}{fz}}, \tag{6.69}$$

where r is the radius of the knuckle (see Fig. 6.13). We must select the smaller thickness between those obtained from (6.67) and (6.69). The junction can be done without a knuckle (see Fig. 6.12). In this case there are two scenarios: first, if the angle α is not greater than the one obtained from Fig. 6.14 (with a maximum of 30°), neither the cylinder nor the cone require an increase in

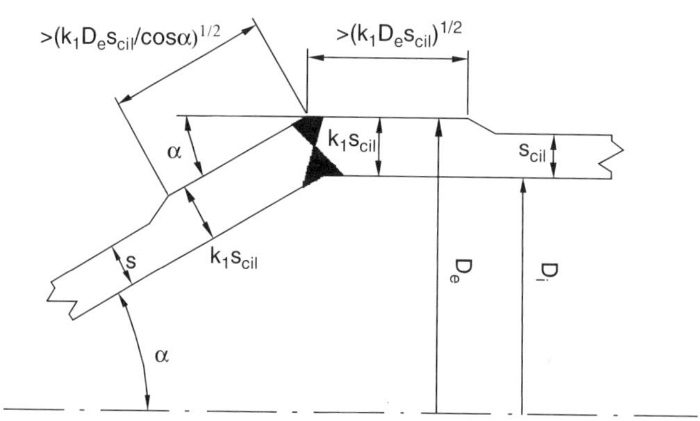

Fig. 6.12

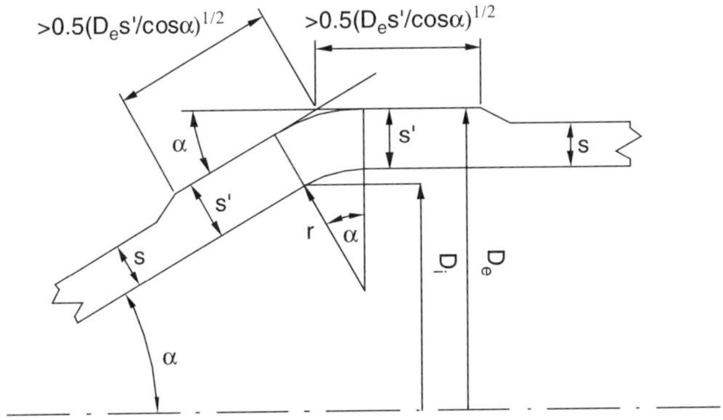

Fig. 6.13

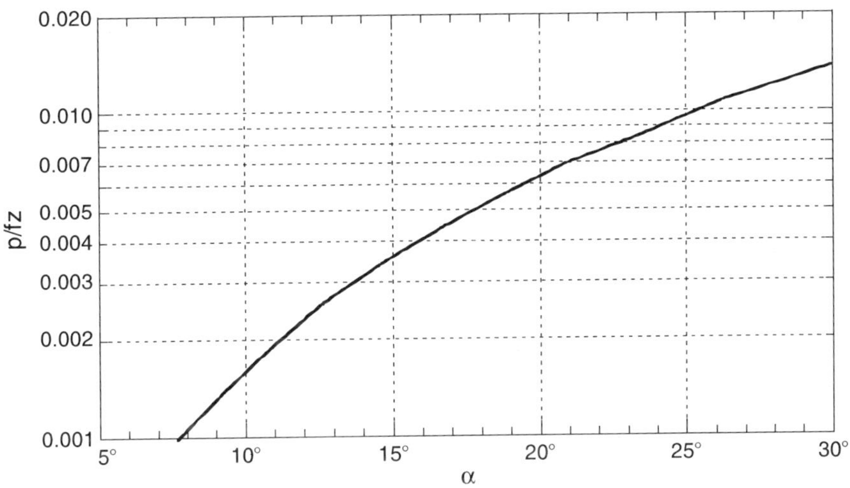

Fig. 6.14

thickness, compared to the value computed through the respective fundamental sizing equations, provided the junction cylinder–cone has a distance from another potential junction at least equal to

$$L_g = 2\sqrt{D_e s_{cil}}, \tag{6.70}$$

where s_{cil} is the minimum required thickness of the cylinder. Second, if α is greater than the one obtained from Fig. 6.14 instead, one shall introduce quantity k_1, obtained from Fig. 6.15. The minimum required thickness of the cylinder must be multiplied by this quantity; the resulting thickness must cover an area of the cylinder with a length at least equal to

$$L_{cil} = \sqrt{k_1 D_e s_{cil}}, \tag{6.71}$$

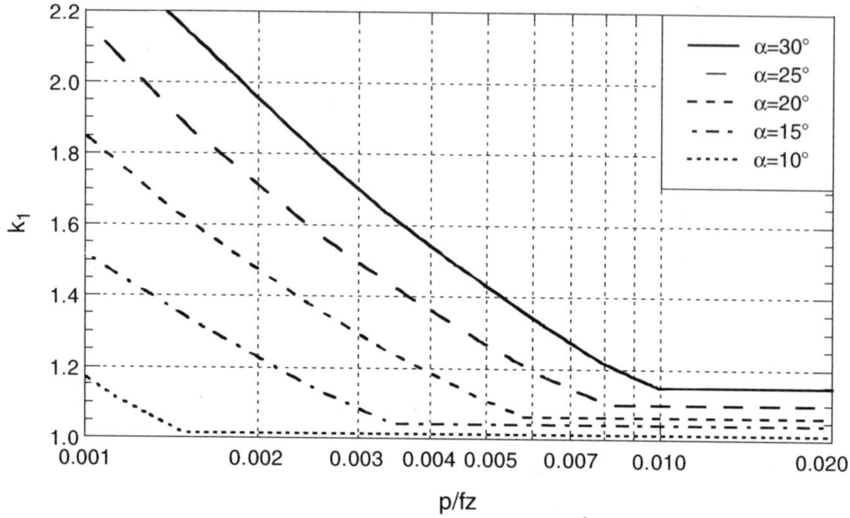

Fig. 6.15

and an area of the cone with a length at least equal to

$$L_{con} = \sqrt{\frac{k_1 D_e s_{cil}}{\cos\alpha}}. \tag{6.72}$$

In addition, the junction cylinder–cone must be at a distance from another potential junction at least equal to

$$L_g = 2\sqrt{k_1 D_e s_{cil}}. \tag{6.73}$$

Examine Fig. 6.12 for further clarifications.

Let us now consider the case of a junction between a cylinder and a cone that is knuckled and where r is the radius of the knuckle. (see Fig. 6.13). The minimum required thickness of the knuckle is given by

$$s' = \frac{pD_e}{2f}C_1, \tag{6.74}$$

with C_1 being a form factor obtained from Fig. 6.16. In any case, the thickness of the knuckle cannot be smaller than the minimum one of the cone ((6.67) or (6.69)). The resulting thickness from (6.74) shall be extended to an area of the cylinder and the cone with a length at least equal to

$$L_{cil} = L_{con} = 0.5\sqrt{\frac{D_e s'}{\cos\alpha}}. \tag{6.75}$$

The same equation applies to the minimum distance L_g from another junction. In addition, the radius of the knuckle cannot be smaller than $0.06D_e$

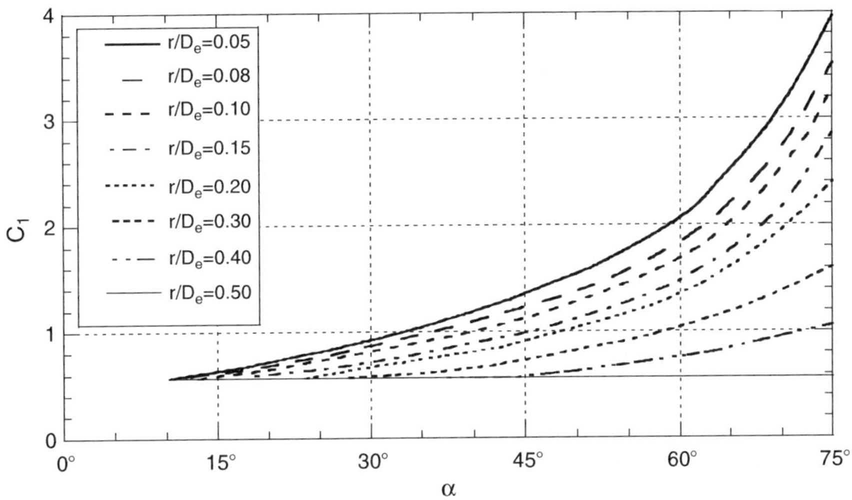

Fig. 6.16

if α is greater than 30°. If this condition is not met, the thickness of the knuckle must be computed according to the criteria discussed about the junction without the knuckle. In other words, it must disregard the presence of the knuckle. In the case of conical reductions, one must also consider the situation in correspondence of the smaller base. The sizing criteria are similar to those presented earlier except for the following changes.

For the junction cone–cylinder without knuckle, the possibility to leave the thickness of both elements unchanged depends on whether the angle α is not greater than the values in Fig. 6.17. If this is the case, (6.70) is valid, too. If α is greater than the one obtained from the figure mentioned above, then an increase in thickness will be necessary through k_2 (Fig. 6.18) as we have done for the greater base of the conical reduction. Equations (6.71), (6.72), and (6.73) hold as well, provided that k_2 replaces k_1.

The same criteria discussed earlier, including (6.74) and (6.75), are true for the junction with knuckle.

If the cone or the conical reduction are under external pressure, it is possible to apply the calculation criteria for the cylinders (Sect. 4.3), including (4.94) and (4.108). The sizing criteria relative to internal pressure discussed earlier must still be verified in any case.

If junctions can be considered efficient reinforcements, in the sense that we shall determine shortly (in any case this is true for the apex of the head), the length L to be included in calculations is that of the generatrix of the cone or of the reduction. As far as the diameter, it is set to $D_{em}/\cos\alpha$, with D_{em} being the average value of the outside diameter.

If the junctions cannot instead be considered efficient reinforcements, L will be the length between the first reinforcement on the cylinder and the

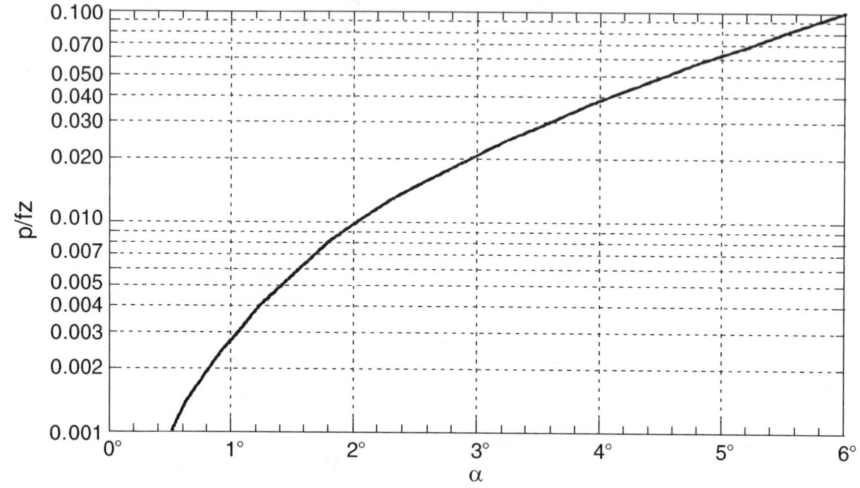

Fig. 6.17

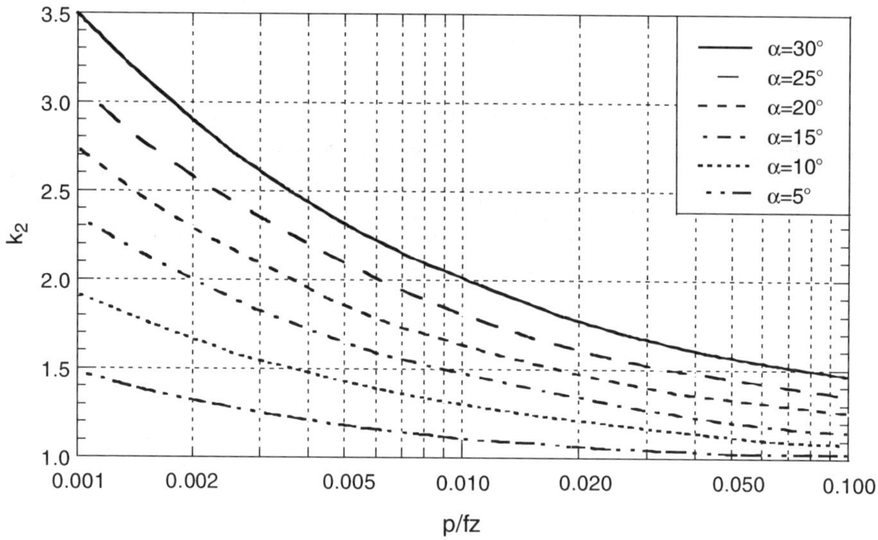

Fig. 6.18

apex of the head, or the distance between the reinforcements of the cylinder connected to the reduction. Moreover, the diameter should be the outside one of the cylinder (the one with the largest value in the case of a truncated reduction). To ensure that the junctions are efficient reinforcements, they must satisfy (4.135).

In our specific case, the moment of inertia is the one corresponding to the potential knuckle and to two areas of the cylinder and the cone to the side of the knuckle. The length will be equal to $0.55\sqrt{D_e s}$, with D_e the external diameter of the cylinder. The moment of inertia must be computed considering the neutral axis to be parallel to the axis of the cone. As far as the length L_r, for the cone we set

$$L_r = 4\sqrt{D_e s}, \tag{6.76}$$

where D_e is the outside diameter of the cylinder, provided, of course, that L_r is not greater than half the distance between the reinforcements next to the one considered. The same criterion holds for the greater base in the conical reduction, while for the smaller one we set

$$L_r = 4\sqrt{D_{em} s}, \tag{6.77}$$

with the same limits of applicability mentioned above.

6.4 Flat Heads

Bending moments are generated under pressure in flat heads, and they cause stresses in the radial and circumferential direction that, in turn, have an influence on the sizing of the head. The generic bending moment is, of course, proportional to the pressure. Moreover, when it refers to the unity of length it is dimensionally a force, and it is proportional to the square of a characteristic dimension of the head (in the case of a round head, it is of course the radius or the diameter).

Therefore, we may write

$$M \equiv pL^2, \tag{6.78}$$

with L being the characteristic dimension. The section modulus of the ideal beam of unitary length and thickness s loaded by the moment M is equal to

$$W = \frac{1}{6}s^2. \tag{6.79}$$

Thus, the maximum stress due to M is given by

$$\sigma = \frac{M}{W} \equiv \frac{pL^2}{s^2}. \tag{6.80}$$

We need to set this stress equal to the allowable stress, which in turn is proportional through eventual multiplicative factors (as we shall see) to the basic allowable stress f typical of the material. Thus, it must be

$$\frac{pL^2}{s^2} \equiv f; \tag{6.81}$$

then

$$s \equiv L\sqrt{\frac{p}{f}}. \tag{6.82}$$

Indicating with C the coefficient of proportionality, called form factor, as for dished heads, we have

$$s = CL\sqrt{\frac{p}{f}}. \tag{6.83}$$

In general, the value C depends on both the shape of the head as well as the type of constraint (only from the latter if the head is round).

For round heads we can write

$$s = CD_i\sqrt{\frac{p}{f}}, \tag{6.84}$$

with D_i being the inside diameter of the vessel that coincides with the diameter of the head. The problem to address has to do with C, therefore.

Let us start by looking at Fig. 6.19, that shows a plate subject to bending in two orthogonal directions.

M_x and M_y are the bending moments per unit of length.

If the generic deflection dz occurring on the plate is small, the middle plane of the element shown on the right in the figure will not warp (i.e., the plane that has the neutral axis in both directions). The deflection is a function of both x and y; therefore, the curvatures $1/r_x$ and $1/r_y$ in both directions are given by

$$\begin{aligned} \frac{1}{r_x} &= -\frac{\delta^2 z}{\delta x^2} \\ \frac{1}{r_y} &= -\frac{\delta^2 z}{\delta y^2} \end{aligned} \tag{6.85}$$

where r_x refers to plane xz and r_y to plane yz.

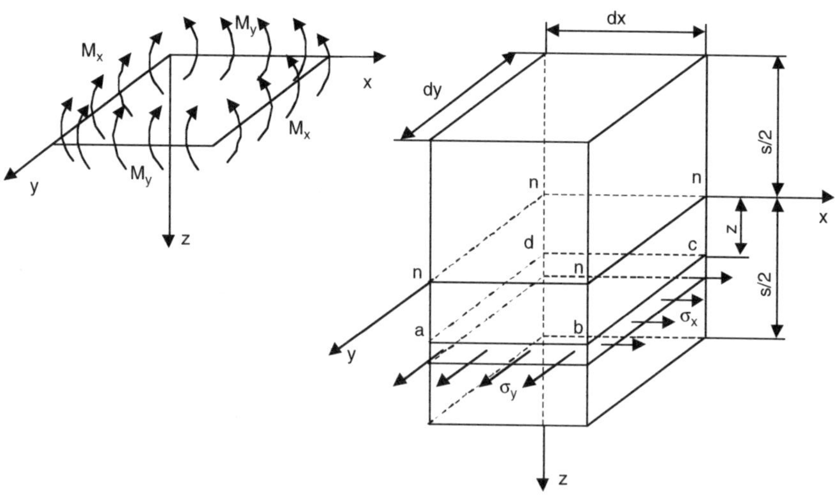

Fig. 6.19

In correspondence of a generic plane *abcd* the deformations ε_x and ε_y are given by

$$\begin{aligned} \varepsilon_x &= \frac{z}{r_x} \\ \varepsilon_y &= \frac{z}{r_y} \end{aligned} \tag{6.86}$$

Thus, we can write

$$\begin{aligned} \frac{1}{E}\left(\sigma_x - \mu\sigma_y\right) &= \frac{z}{r_x} \\ \frac{1}{E}\left(\sigma_y - \mu\sigma_x\right) &= \frac{z}{r_y} \end{aligned} \tag{6.87}$$

and from it

$$\sigma_x\left(1-\mu^2\right) = E\left(\frac{z}{r_x} + \mu\frac{z}{r_y}\right); \tag{6.88}$$

finally

$$\sigma_x = \frac{zE}{1-\mu^2}\left(\frac{1}{r_x} + \mu\frac{1}{r_y}\right). \tag{6.89}$$

On the other hand

$$\sigma_x = \frac{M_x z}{I}, \tag{6.90}$$

with I being the moment of inertia.

Thus

$$M_x = \frac{EI}{1-\mu^2}\left(\frac{1}{r_x} + \mu\frac{1}{r_y}\right); \tag{6.91}$$

similarly,

$$M_y = \frac{EI}{1-\mu^2}\left(\frac{1}{r_y} + \mu\frac{1}{r_x}\right). \tag{6.92}$$

We introduce the quantity K given by

$$K = \frac{EI}{1-\mu^2} \tag{6.93}$$

Therefore, we have

$$\begin{aligned} M_x &= K\left(\frac{1}{r_x} + \mu\frac{1}{r_y}\right) \\ M_y &= K\left(\frac{1}{r_y} + \mu\frac{1}{r_x}\right) \end{aligned} \tag{6.94}$$

Now, let us consider the round head shown in Fig. 6.20. The deformations of a head of this type under pressure depend, of course, uniquely on the distance

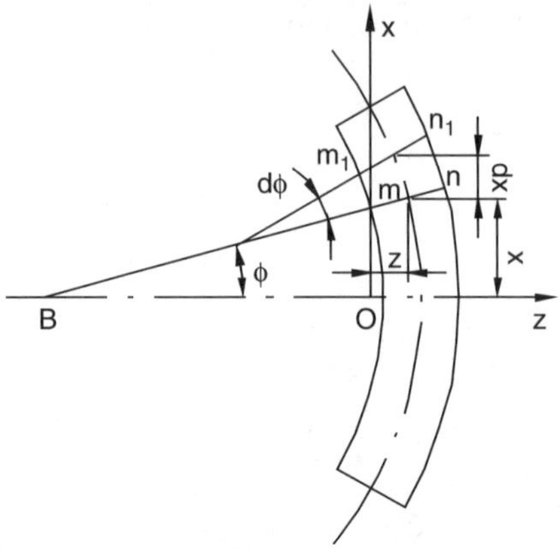

Fig. 6.20

from the center that we indicate with x. The curvature of the head on the diametrical plane is only a function of x, and more precisely

$$\frac{1}{r_x} = -\frac{d^2 z}{dx^2} = \frac{d\varphi}{dx}, \tag{6.95}$$

where φ is the angle in the figure.

The curvature radius orthogonal to the plane xz may be obtained by observing that the cylindrical surface identified by the segment mn changes into a conical surface with the apex at point B as a result of the deformation of the head. Thus, AB represents the radius r_y. Always assuming that the deformations are small, we have

$$\varphi = \frac{x}{r_y}; \tag{6.96}$$

then

$$\frac{1}{r_y} = \frac{\varphi}{x}, \tag{6.97}$$

and based on (6.94) we have

$$\begin{aligned} M_x &= K\left(\frac{d\varphi}{dx} + \mu\frac{\varphi}{x}\right) \\ M_y &= K\left(\frac{\varphi}{x} + \mu\frac{d\varphi}{dx}\right) \end{aligned} \tag{6.98}$$

We observe that M_x produces radial stresses, while M_y causes circumferential stresses. In what follows, and in order to better highlight the significance of the different symbols, we shall use the subscript r for all radial parameters, and the subscript c for the circumferential ones. We also replace x with the radius r.

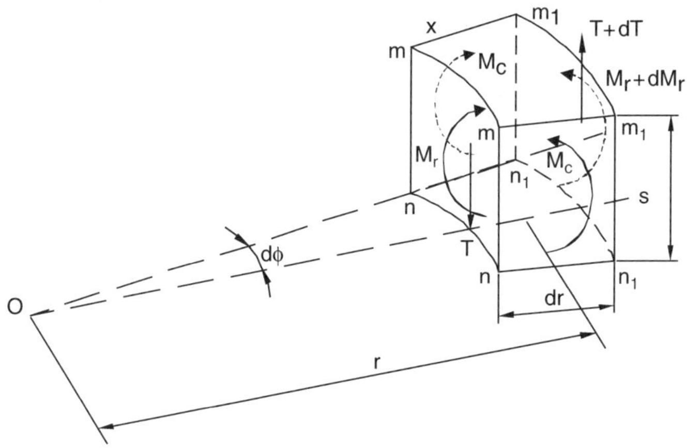

Fig. 6.21

Therefore, we write (6.98) as follows:

$$\begin{aligned} M_r &= K\left(\frac{d\varphi}{dr} + \mu\frac{\varphi}{r}\right) \\ M_c &= K\left(\frac{\varphi}{r} + \mu\frac{d\varphi}{dr}\right) \end{aligned} \tag{6.99}$$

Now let us look at Fig. 6.21. The total moment on the side $mmnn$ is

$$M_r r d\vartheta, \tag{6.100}$$

whereas on the side $m_1m_1n_1n_1$ the moment is

$$(M_r + dM_r)(r + dr)\, d\vartheta. \tag{6.101}$$

On the sides mnm_1n_1 the moment is

$$M_c dr. \tag{6.102}$$

The resultant of the moments active on the sides above in the direction of M_r is

$$2M_c dr\frac{d\vartheta}{2} = M_c dr d\vartheta. \tag{6.103}$$

If T is the shear force per unity of length, the shear force on the $mmnn$ side is

$$T r d\vartheta, \tag{6.104}$$

whereas on the $m_1m_1n_1n_1$ side it is

$$(T + dT)(r + dr)\, d\vartheta. \tag{6.105}$$

The couple generated by these forces is

$$Trd\vartheta\frac{dr}{2} + (T + dT)(r + dr)\,d\vartheta\frac{dr}{2}. \tag{6.106}$$

Neglecting the terms of higher order, this couple may be written as $Trdrd\vartheta$.
For equilibrium it must be

$$(M_r + dM_r)(r + dr)\,d\vartheta - M_r rd\vartheta - M_c drd\vartheta + Trdrd\vartheta = 0, \tag{6.107}$$

that, neglecting the terms of higher order, is reduced to

$$M_r + \frac{dM_r}{dr}r - M_c + Tr = 0. \tag{6.108}$$

Recalling (6.99), we obtain the following

$$\frac{d^2\varphi}{dr^2} + \frac{1}{r}\frac{d\varphi}{dr} - \frac{\varphi}{r^2} = -\frac{T}{K}. \tag{6.109}$$

We observe that in a round head under pressure the value T is

$$T = \frac{\pi r^2 p}{2\pi r} = \frac{pr}{2}. \tag{6.110}$$

Then

$$\frac{d^2\varphi}{dr^2} + \frac{1}{r}\frac{d\varphi}{dr} - \frac{\varphi}{r^2} = -\frac{pr}{2K}, \tag{6.111}$$

and may be written like

$$\frac{d}{dr}\left[\frac{1}{r}\frac{d}{dr}(r\varphi)\right] = -\frac{pr}{2K}. \tag{6.112}$$

Integrating a first time we have

$$\frac{1}{r}\frac{d}{dr}(r\varphi) = -\frac{pr^2}{4K} + C_1; \tag{6.113}$$

integrating a second time we have

$$r\varphi = -\frac{pr^4}{16K} + C_1\frac{r^2}{2} + C_2, \tag{6.114}$$

where C_1 and C_2 are two integration constants.
Finally,

$$\varphi = -\frac{pr^3}{16K} + C_1\frac{r}{2} + \frac{C_2}{r}. \tag{6.115}$$

Let us now look at how one may get to the stress state of the head from (6.115).

First, let us carry out an approximate study and assume that the head may simply be supported on the edge. For $r = 0$ it follows that $\varphi = 0$. Therefore, from (6.115) it must be $C_2 = 0$ and therefore

$$\varphi = -\frac{pr^3}{16K} + C_1 \frac{r}{2}. \tag{6.116}$$

Recalling the first equation in (6.99) we have

$$M_r = K\left[-\frac{3}{16}\frac{pr^2}{K} + \frac{C_1}{2} + \mu\left(-\frac{1}{16}\frac{pr^2}{K} + \frac{C_1}{2}\right)\right], \tag{6.117}$$

which can also be written as

$$M_r = K\left[-\frac{1}{16}\frac{pr^2}{K}\left(3+\mu\right) + \frac{C_1}{2}\left(1+\mu\right)\right]. \tag{6.118}$$

If R is the radius of the head for $r = R$ we have $M_r = 0$.

Thus

$$\frac{C_1}{2}\left(1+\mu\right) = \frac{1}{16}\frac{pR^2}{K}\left(3+\mu\right); \tag{6.119}$$

then

$$M_r = \frac{3+\mu}{16} p\left(R^2 - r^2\right). \tag{6.120}$$

The maximum value of M_r is for $r = 0$, i.e., at the center of the head, and is

$$M_{r0} = \frac{3+\mu}{16} pR^2. \tag{6.121}$$

As far as M_c, recalling (6.99), (6.115), and (6.119) we have

$$M_c = \frac{p}{16}\left[\left(3+\mu\right)R^2 - \left(1+3\mu\right)r^2\right]. \tag{6.122}$$

The maximum value of M_c and of M_r is for $r = 0$ and we observe that

$$M_{c0} = \frac{3+\mu}{16} pR^2, \tag{6.123}$$

which not unexpectedly coincides with the value of M_r; for $r = R$, i.e., at the edge of the head, we have

$$M_{cR} = \frac{1-\mu}{8} pR^2; \tag{6.124}$$

M_r and M_c are plotted in Fig. 6.22 when the head is simply supported at the edge. Let us now consider the opposite case, i.e., with a head that is perfectly clamped at the edge; thus it must be $\varphi = 0$ both for $r = 0$ and $r = R$.

The first condition that we already considered for the simply supported head, results in $C_2 = 0$. Moreover, based on (6.116) we write

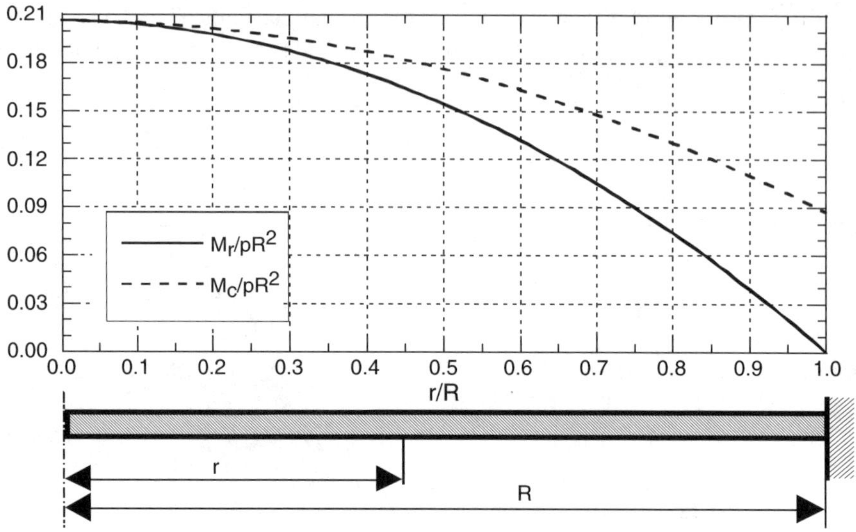

Fig. 6.22

$$-\frac{pR^3}{16K}+C_1\frac{R}{2}=0; \tag{6.125}$$

then

$$C_1=\frac{pR^2}{8K}. \tag{6.126}$$

Introducing C_1 in (6.118) we have

$$M_r=\frac{p}{16}\left[(1+\mu)R^2-(3+\mu)r^2\right]; \tag{6.127}$$

for $r=0$ we have

$$M_{r0}=\frac{1+\mu}{16}pR^2, \tag{6.128}$$

and for $r=R$

$$M_{rR}=-\frac{1}{8}pR^2. \tag{6.129}$$

As far as M_c, proceeding as usual, we obtain

$$M_c=\frac{p}{16}\left[(1+\mu)R^2-(1+3\mu)r^2\right]; \tag{6.130}$$

for $r=0$

$$M_{c0}=\frac{1+\mu}{16}pR^2=M_{r0}; \tag{6.131}$$

and for $r=R$

$$M_{cR}=-\frac{\mu}{8}pR^2. \tag{6.132}$$

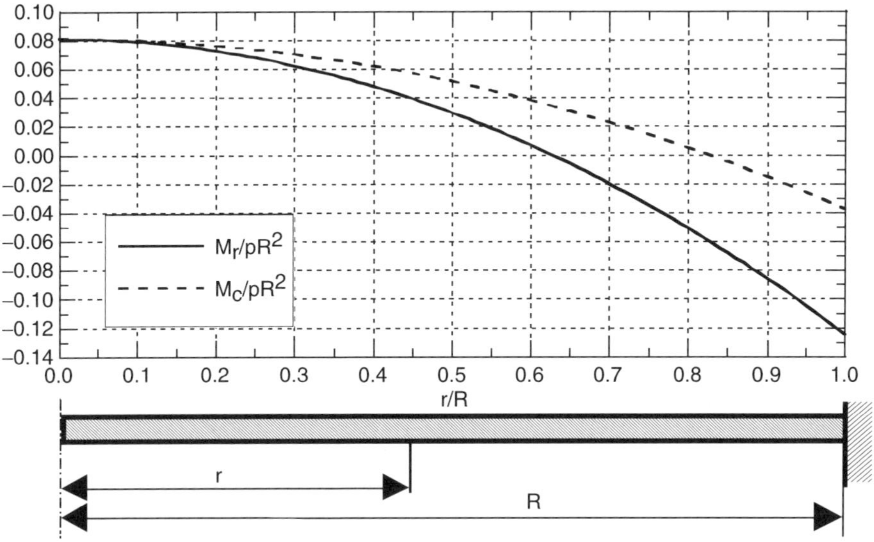

Fig. 6.23

Figure 6.23 shows M_r and M_c when the head is perfectly clamped at the edge. Let us now look at the values of the form factor C that we end up obtaining in the two cases under study. If the head is simply supported, the maximum value of the two identical moments M_r and M_c is at the center of the head. They produce compressive stresses on the fibers corresponding to the side under pressure, and tensile stresses on the opposite fibers.

In the first case we have three principal stresses with a negative sign (the third one is direct like the pressure and equal to $-p$); in the second case we have two principal stresses with a positive sign (the third one is zero). The maximum value of the ideal stress according to Guest's failure theory is therefore equal to one of the two principal stresses.

From (6.121) and keeping in mind that $W = s^2/6$ we have

$$\sigma_{id} = \frac{3\,(3+\mu)}{8}\frac{pR^2}{s^2} \tag{6.133}$$

This is a primary bending stress. Based on Sect. 1.5, the allowable stress is equal to $1.5f$, where f is the basic allowable stress of the material and $1.5f$ corresponds to the yield strength. Therefore

$$\frac{3\,(3+\mu)}{8}\frac{pR^2}{s^2} = 1.5f; \tag{6.134}$$

then

$$s = \frac{\sqrt{3+\mu}}{2}R\sqrt{\frac{p}{f}}, \tag{6.135}$$

or, introducing the diameter of the head D_1,

$$s = \frac{\sqrt{3+\mu}}{4} D_i \sqrt{\frac{p}{f}}. \tag{6.136}$$

Recalling (6.84), we imply that

$$C = \frac{\sqrt{3+\mu}}{4} = 0.454,$$

since $\mu = 0.3$ for steel.

If the head is perfectly clamped, from (6.128) we observe that in the center the ideal stress is given by

$$\sigma_{id} = \frac{3\,(1+\mu)}{8} \frac{pR^2}{s^2}. \tag{6.137}$$

Proceeding as before, we have

$$\frac{3\,(1+\mu)}{8} \frac{pR^2}{s^2} = 1.5f; \tag{6.138}$$

then

$$s = \frac{\sqrt{1+\mu}}{2} R \sqrt{\frac{p}{f}}, \tag{6.139}$$

or, introducing the diameter D_1,

$$s = \frac{\sqrt{1+\mu}}{4} D_i \sqrt{\frac{p}{f}}. \tag{6.140}$$

Therefore C = 0.285 under the assumption that $\mu = 0.3$.

One could argue that the value of the moment is higher at the clamped edge; one should not forget that based on Sect. 1.5 a stress equal to $3f$ is allowable there. Adopting (6.140) for the sizing of the head that is perfectly clamped, the value of σ_{id} at the edge is equal to about $2.31f$. These extreme situations lead only to a preliminary understanding of what the values of C may be. It is evident that the behavior of a head junction with a cylinder will be somewhere in the middle compared to the ones examined. Moreover, one cannot ignore the behavior of the cylinder at the junction. As we shall see, in some cases the behavior of the cylinder is crucial for the sizing of the head that must be so rigid to prevent unacceptable stresses in the cylinder itself.

Let us consider now Fig. 6.24, that shows the wall of the cylinder and half the head. In correspondence of point A there are the force F_0 and the moment M_0 (referred to the unity of length of the average circumference of the cylinder) to reestablish congruence. In the section A–B of the head the moment M'_0 is

$$M'_0 = M_0 - F_0 \frac{s_f}{2}, \tag{6.141}$$

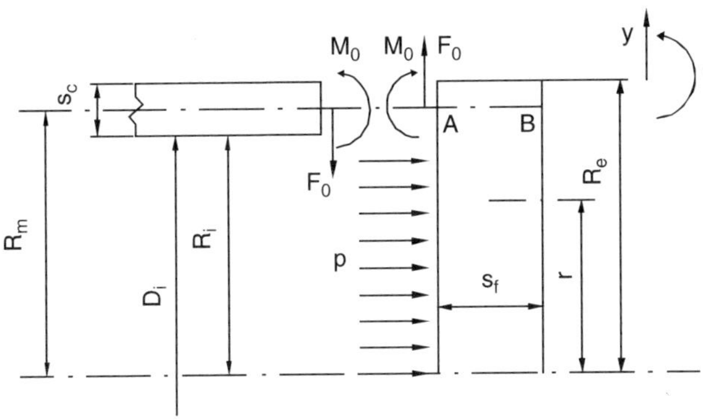

Fig. 6.24

where s_c and s_f are the thickness of the cylinder and the head, respectively; R_m is the average radius of the cylinder, r the generic radius of the head, and a the ratio between outside and inside radius of the cylinder.

In addition, let us introduce the following dimensionless quantities:

$$\alpha = \frac{s_c}{R_m}, \tag{6.142}$$

$$\beta = \frac{s_f}{R_m}, \tag{6.143}$$

$$\gamma = \frac{r}{R_m}, \tag{6.144}$$

$$Z = \frac{F_0}{pR_m}, \tag{6.145}$$

and

$$S = \frac{M_0}{pR_m^2}. \tag{6.146}$$

We consider the impact of the pressure and the stresses F_0 and M_0 separately for the head and the cylinder.

If the cylinder could be considered a membrane (without thickness), as a result of pressure only (the head is simply supported) the rotation φ'_f in A, based on (6.116), (6.119), and (6.93), and recalling that $I = s_f^3/12$, would be

$$\varphi'_f = \frac{3}{2}\frac{pR_m^3}{Es_f^3}(1-\mu). \tag{6.147}$$

Because of the nonzero thickness of the cylinder we have that: (1) the head is not under pressure on the crown between the radii R_i and R_m; (2) the head is more rigid than the head with radius R_m because of the presence of the crown between the radii R_m and R_e.

The first phenomenon is taken into account by the corrective coefficient k_2 given by

$$k_2 = 1 - \alpha^2 + \frac{\alpha^3}{2} - \frac{\alpha^4}{16}. \tag{6.148}$$

The coefficient k_3 accounts for the second phenomenon, and is given by

$$k_3 = 1 - \frac{\alpha\,(4+\alpha)}{2\,(2+\alpha)^2}\,(1-\mu)\,. \tag{6.149}$$

Assuming $\mu = 0.3$, as we shall do from now on, we have

$$k_3 = 1 - 0.35\frac{\alpha\,(4+\alpha)}{2\,(2+\alpha)^2}. \tag{6.150}$$

Introducing the coefficient k_1 given by

$$k_1 = k_2 k_3. \tag{6.151}$$

We finally have

$$\varphi'_f = \frac{3}{2}\frac{k_1 p R_m^3}{E s_f^3}\,(1-\mu) = 1.05\frac{k_1 p R_m^3}{E s_f^3}, \tag{6.152}$$

and k_1, k_2 and k_3 have been plotted in Fig. 6.25. Of course, the three coefficients are equal to one for $\alpha = 0$, i.e., for the cylinder with zero thickness.

The shift y'_f of point A corresponding to φ'_f is given by

$$y'_f = -\frac{\varphi'_f s_f}{2} = -0.525\frac{k_1 p R_m^3}{E s_f^2}. \tag{6.153}$$

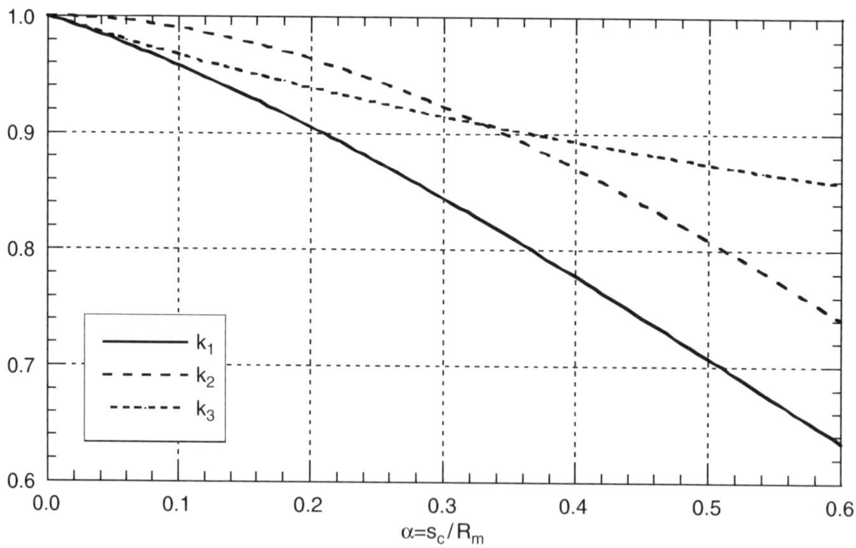

Fig. 6.25

Thus, based on the equations above we have

$$\frac{E\varphi'_f}{p} = 1.05k_1 \left(\frac{R_m}{s_f}\right)^3 = 1.05\frac{k_1}{\beta^3}; \tag{6.154}$$

$$\frac{Ey'_f}{pR_m} = -0.525k_1 \left(\frac{R_m}{s_f}\right)^2 = -0.525\frac{k_1}{\beta^2}. \tag{6.155}$$

Through (6.116) and (6.118) with $p = 0$, $M_r = M'_0$ and $r = R_m$ we observe that the rotation φ''_f due to M'_0 would be given by

$$\varphi''_f = \frac{12M'_0R_m}{Es_f^3}(1-\mu) = 8.4\frac{M'_0R_m}{Es_f^3}. \tag{6.156}$$

This value must be corrected through the coefficient k_3; also, recalling (6.141) we have

$$\varphi''_f = -4.2\frac{k_3F_0R_m}{Es_f^2} + 8.4\frac{k_3M_0R_m}{Es_f^3}. \tag{6.157}$$

The shifting of point A caused by the moment M'_0, similarly to (6.153), is given by

$$y''_f = -\frac{\varphi''_f s_f}{2} = 2.1\frac{k_3F_0R_m}{Es_f} - 4.2\frac{k_3M_0R_m}{Es_f^2}. \tag{6.158}$$

The force F_0 is responsible for a radial shift of the section $A-B$ that occurs as follows. Ignoring for the time being the external crown, the force F_0 corresponds to a negative pressure F_0/s_f applied to a cylinder with external radius R_m and internal radius zero.

From the first equation in (4.13) we obtain

$$\sigma_t = \sigma_r = \frac{F_0}{s_f}; \tag{6.159}$$

then

$$\varepsilon_t = \frac{F_0}{Es_f}(1-\mu) \tag{6.160}$$

and the shift y''_f is

$$y''_f = \varepsilon_t R_m = \frac{F_0R_m}{Es_f}(1-\mu). \tag{6.161}$$

Even in this case one must introduce the coefficient k_3; thus, from (6.161)

$$y''_f = 0.7\frac{k_3F_0R_m}{Es_f}. \tag{6.162}$$

Finally, recalling (6.158)

$$y''_f = 2.8\frac{k_3F_0R_m}{Es_f} - 4.2\frac{k_3M_0R_m}{Es_f^2}. \tag{6.163}$$

Recalling (6.144), (6.145), and (6.146)

$$\frac{E{\varphi''}_f}{p} = -4.2k_3\frac{Z}{\beta^2} + 8.4k_3\frac{S}{\beta^3}; \tag{6.164}$$

$$\frac{E{y''}_f}{pR_m} = 2.8k_3\frac{Z}{\beta} - 4.2k_3\frac{S}{\beta^2}. \tag{6.165}$$

In summary, at point A we have

$$\frac{E\varphi_f}{p} = 1.05\frac{k_1}{\beta^3} - 4.2k_3\frac{Z}{\beta^2} + 8.4k_3\frac{S}{\beta^3}; \tag{6.166}$$

$$\frac{Ey_f}{pR_m} = -0.525\frac{k_1}{\beta^2} + 2.8k_3\frac{Z}{\beta} - 4.2k_3\frac{S}{\beta^2}. \tag{6.167}$$

Let us now examine the behavior of the cylinder at the junction with the head. Because of the pressure in correspondence with radius R_m, and based on Sect. 3.1. we have

$$\begin{aligned}\sigma_t &= \frac{p}{a^2-1}\left[1+\left(\frac{R_e}{R_m}\right)^2\right]\\ \sigma_r &= \frac{p}{a^2-1}\left[1-\left(\frac{R_e}{R_m}\right)^2\right]\\ \sigma_a &= \frac{p}{a^2-1}\end{aligned} \tag{6.168}$$

Calculating ε_t based on σ_t, σ_r and σ_a and remembering that the deflection ${y'}_c$ is equal to $\varepsilon_t R_m$ through a series of steps we obtain

$${y'}_c = \frac{pR_m}{8E\alpha}\left(4-4\alpha+\alpha^2\right)\left[(2-\mu)+(1+\mu)\,\alpha+\frac{1+\mu}{4}\alpha^2\right]. \tag{6.169}$$

We introduce the coefficient k_4 shown in Fig. 6.26 and given by

$$k_4 = \frac{1}{8\alpha}\left(4-4\alpha+\alpha^2\right)\left(1.7+1.3\alpha+0.325\alpha^2\right). \tag{6.170}$$

We obtain

$$\frac{E{y'}_c}{pR_m} = k_4. \tag{6.171}$$

As far as F_0 and M_0, we refer back to Sect. 2.2 and more specifically to (2.66) and (2.67) (where the rotation is indicated with the symbol ϑ).

Introducing the radius R_m instead of D ($R_m = D/2$) and taking into account the direction of M_0 and y''_c as well as the direction of F_0 and y''_c, we obtain

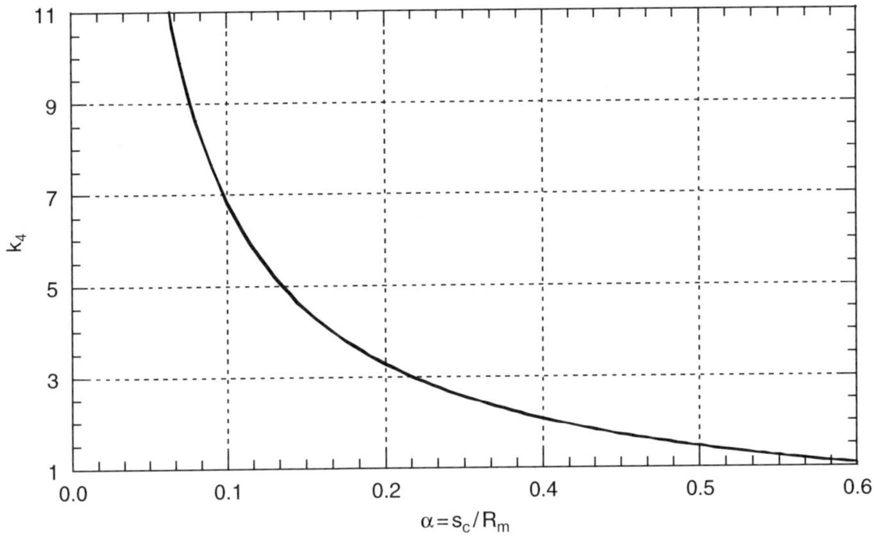

Fig. 6.26

$$\varphi''_c = -3.304\frac{F_0 R_m}{E s_c^2} - 8.495\frac{M_0}{E s_c^2}\sqrt{\frac{R_m}{s_c}}; \tag{6.172}$$

$$y''_c = -2.57\frac{F_0}{E}\left(\frac{R_m}{s_c}\right)^{3/2} - 3.304\frac{M_0 R_m}{E s_c^2}. \tag{6.173}$$

Factoring in (6.171), in summary we have

$$\frac{E\varphi_c}{p} = -3.304\frac{Z}{\alpha^2} - 8.495\frac{S}{\alpha^{5/2}}; \tag{6.174}$$

$$\frac{E y_c}{p R_m} = k_4 - 2.57\frac{Z}{\alpha^{3/2}} - 3.304\frac{S}{\alpha^2}. \tag{6.175}$$

Let us introduce the following quantities

$$\begin{aligned} H &= -4.2\frac{k_3}{\beta^2} + 3.304\frac{1}{\alpha^2} \\ L &= 8.4\frac{k_3}{\beta^3} + 8.495\frac{1}{\alpha^{5/2}} \\ P &= -1.05\frac{k_1}{\beta^3} \\ Q &= 2.8\frac{k_3}{\beta} + 2.57\frac{1}{\alpha^{3/2}} \\ U &= k_4 + 0.525\frac{k_1}{\beta^2} \end{aligned} \tag{6.176}$$

that are a function of α and β, i.e., of the ratios between the thickness of the cylinder, the head and the average radius R_m.

From (6.166), (6.167), (6.174) and (6.175) we ascertain that to have congruence we must have

$$\begin{aligned} HZ + LS &= P \\ QZ + HS &= U \end{aligned} \tag{6.177}$$

Solving the system then we obtain

$$Z = \frac{PH - LU}{H^2 - LQ}; \tag{6.178}$$

$$S = \frac{HU - PQ}{H^2 - LQ}. \tag{6.179}$$

Through these equations it is possible to compute Z and S once α and β are known. Z and S lead to force F_0 and moment M_0 that are unknown, and through them one obtains the stresses present in the cylinder and the head.

As far as the latter, we have the following. Due to the pressure the radial stress in the head is

$$\sigma_r = \pm \frac{3}{8} \frac{p\,(3+\mu)}{s_f^2} \left(k_5 R_m^2 - r^2\right) = \pm 1.2375 \frac{p}{s_f^2} \left(k_5 R_m^2 - r^2\right), \tag{6.180}$$

where k_5 is a coefficient shown in Fig. 6.27 given by

$$\begin{aligned} k_5 &= \frac{(2-\alpha)^2}{4\,(3+\mu)} \left[(1-\mu)\, \frac{4+4\alpha-\alpha^2}{4+4\alpha+\alpha^2} + 2\,(1+\mu) \left(1 + 2\log_e \frac{2}{2-\alpha}\right)\right] \\ &= 0.07576\,(2-\alpha)^2 \left[0.7 \frac{4+4\alpha-\alpha^2}{4+4\alpha+\alpha^2} + 2.6 \left(1 + 2\log_e \frac{2}{2-\alpha}\right)\right] \end{aligned} \tag{6.181}$$

As usual, we find that for $\alpha = 0$ we have $k_5 = 1$. In that case the radial stress corresponds to the moment M_r from (6.120).

Therefore, we have

$$\frac{\sigma_r}{p} = \pm \frac{1.2375}{\beta^2} \left(k_5 - \gamma^2\right). \tag{6.182}$$

By analogy, for the circumferential stress σ_c

$$\frac{\sigma_c}{p} = \pm \frac{1.2375}{\beta^2} \left(k_5 - 0.5757\gamma^2\right), \tag{6.183}$$

that corresponds to the moment M_c from (6.122) if $k_5 = 1$.

Finally, σ_a/p is equal to -1 for the internal fiber and to 0 for the external one. Because of F_0 and M_0 we have

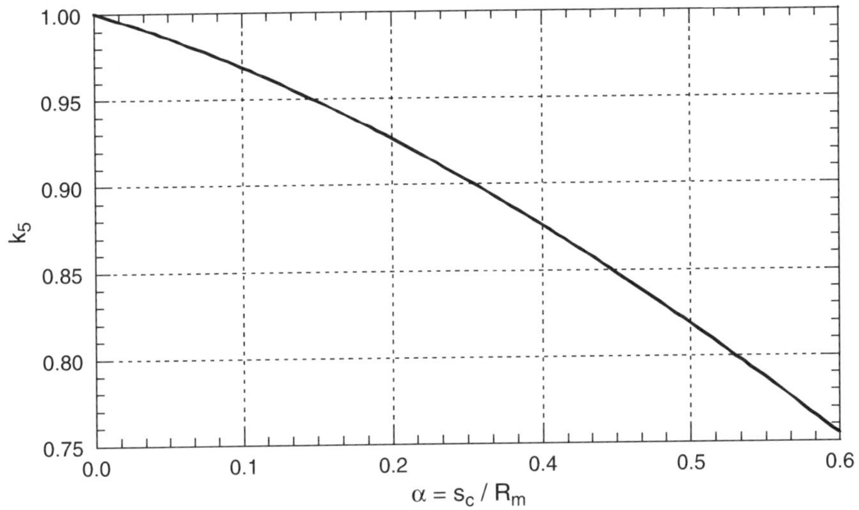

Fig. 6.27

$$\sigma_r = \sigma_c = \frac{k_3 F_0}{s_f} \pm \frac{6k_3}{s_f^2} M'_0. \tag{6.184}$$

Recalling the meaning of M'_0 and factoring in (6.182) and (6.183), through a number of steps we obtain the following equations:

For the fibers of the head inside the vessel

$$\begin{aligned} \frac{\sigma_r}{p} &= -1.2375\frac{k_5 - \gamma^2}{\beta^2} - \frac{6k_3 S}{\beta^2} + \frac{4k_3 Z}{\beta} \\ \frac{\sigma_c}{p} &= -1.2375\frac{k_5 - 0.5757\gamma^2}{\beta^2} - \frac{6k_3 S}{\beta^2} + \frac{4k_3 Z}{\beta} \\ \frac{\sigma_a}{p} &= -1 \end{aligned} \tag{6.185}$$

For the fibers outside

$$\begin{aligned} \frac{\sigma_r}{p} &= 1.2375\frac{k_5 - \gamma^2}{\beta^2} + \frac{6k_3 S}{\beta^2} - \frac{2k_3 Z}{\beta} \\ \frac{\sigma_c}{p} &= 1.2375\frac{k_5 - 0.5757\gamma^2}{\beta^2} + \frac{6k_3 S}{\beta^2} - \frac{2k_3 Z}{\beta} \\ \frac{\sigma_a}{p} &= 0 \end{aligned} \tag{6.186}$$

Let us now examine the stresses in the cylinder at the junction with the head. Because of the pressure in the internal fiber and based on Sect. 3.1 we have

$$\begin{aligned}\frac{\sigma_t}{p} &= \frac{a^2+1}{a^2-1} = \frac{4+\alpha^2}{4\alpha}\\ \frac{\sigma_r}{p} &= -1\\ \frac{\sigma_a}{p} &= \frac{1}{a^2-1} = \frac{(2-\alpha)^2}{8\alpha}\end{aligned} \tag{6.187}$$

In the fiber outside

$$\begin{aligned}\frac{\sigma_t}{p} &= \frac{2}{a^2-1} = \frac{4-4\alpha+\alpha^2}{4\alpha}\\ \frac{\sigma_r}{p} &= 0\\ \frac{\sigma_a}{p} &= \frac{(2-\alpha)^2}{8\alpha}\end{aligned} \tag{6.188}$$

Longitudinal stresses are created as a result of moment M_0, thus

$$\frac{\sigma_a}{p} = \pm\frac{6M_0}{ps_c^2} = \pm\frac{6S}{\alpha^2}. \tag{6.189}$$

Hoop stresses are created because of the deflection y_f'' that are given by

$$\frac{\sigma_t}{p} = \frac{Ey''{}_c}{pR_m} = -2.57\frac{Z}{\alpha^{3/2}} - 3.304\frac{S}{\alpha^2}. \tag{6.190}$$

Finally, because of M_0 the moment μM_0 is created in the circumferential direction, which causes the following hoop stresses:

$$\frac{\sigma_t}{p} = \pm\frac{6\mu M_0}{ps_c^2} = \pm 1.8\frac{S}{\alpha^2}. \tag{6.191}$$

In conclusion, in the internal fiber we have

$$\begin{aligned}\frac{\sigma_t}{p} &= \frac{4+\alpha^2}{4\alpha} - 2.57\frac{Z}{\alpha^{3/2}} - 5.104\frac{S}{\alpha^2}\\ \frac{\sigma_a}{p} &= \frac{(2-\alpha)^2}{8\alpha} - 6\frac{S}{\alpha^2}\\ \frac{\sigma_r}{p} &= -1\end{aligned} \tag{6.192}$$

In the one outside we have

$$\begin{aligned}\frac{\sigma_t}{p} &= \frac{4-4\alpha+\alpha^2}{4\alpha} - 2.57\frac{Z}{\alpha^{3/2}} - 1.504\frac{S}{\alpha^2}\\ \frac{\sigma_a}{p} &= \frac{(2-\alpha)^2}{8\alpha} + 6\frac{S}{\alpha^2}\\ \frac{\sigma_r}{p} &= 0\end{aligned} \tag{6.193}$$

Once the different stresses in the cylinder and the head are computed, we obtain the ideal stress based on the theory of failure of Guest–Tresca. Specifically, based on the previous equations it is possible to compute the ratio σ_{id}/p that leads to

$$\frac{\sigma_{id}}{f} = \frac{\sigma_{id}}{p}\frac{p}{f}. \qquad (6.194)$$

Note that the minimum required thickness of the cylinder s_{co} based on Sect. 3.1 is given by

$$s_{co} = \frac{pR_m}{f}, \qquad (6.195)$$

and

$$\frac{s_c}{s_{co}} = \frac{s_c}{R_m}\frac{R_m}{s_{co}} = \frac{\alpha}{\frac{p}{f}}. \qquad (6.196)$$

The coefficient α that appears several times in the previous equations may be replaced by the product of the ratio between actual thickness of the cylinder and minimum required thickness for the ratio between pressure and basic allowable stress of the material.

In conclusion, the ratio σ_{id}/f is a function of p/f, of s_c/s_{co} and of the quantity β which is the unknown, since the thickness of the head depends on it. Once the value of σ_{id}/f is set, and the ratios p/f and s_c/s_{co} are known, it is possible to determine the value of s_f/R_m. On the other hand, the thickness of the head is given by (6.84) which can be written like this

$$\frac{s_f}{R_m} = C\,(2-\alpha)\sqrt{\frac{p}{f}}, \qquad (6.197)$$

and

$$C = \frac{s_f}{(2-\alpha)\,R_m\sqrt{\frac{p}{f}}}. \qquad (6.198)$$

Therefore, the form factor is only a function of p/f and s_c/s_{co}, once the value of σ_{id}/f has been set. Based on the criteria presented in Sect. 1.5, one must assume $\sigma_{id} = 1.5f$ (corresponding to the yield strength) for the stress at the center of the head, while one can assume $\sigma_{id} = 3f$ (twice the yield strength) at the junction both for the cylinder and the head.

The curves in Fig. 6.28 were developed according to these criteria. Note that limiting σ_{id} not to be greater than $1.5f$ at the center of the head, as well as limiting σ_{id} not to be greater than $3f$ at the edge of the cylinder, $1.55f$ is not exceeded at the junction in the head. This results in an adequate safety margin for the possible presence of welding in this position, as it happens to be the case, in fact, with some types of heads (see Fig. 6.29).

We have also introduced a calculation criterion that we believe to be crucial for safety. If the yield strength in the cylinder is exceeded at the junction (considering it may reach up to $3f$), the type of constraints change, and this has an obvious impact on the stresses at the center of the head. This

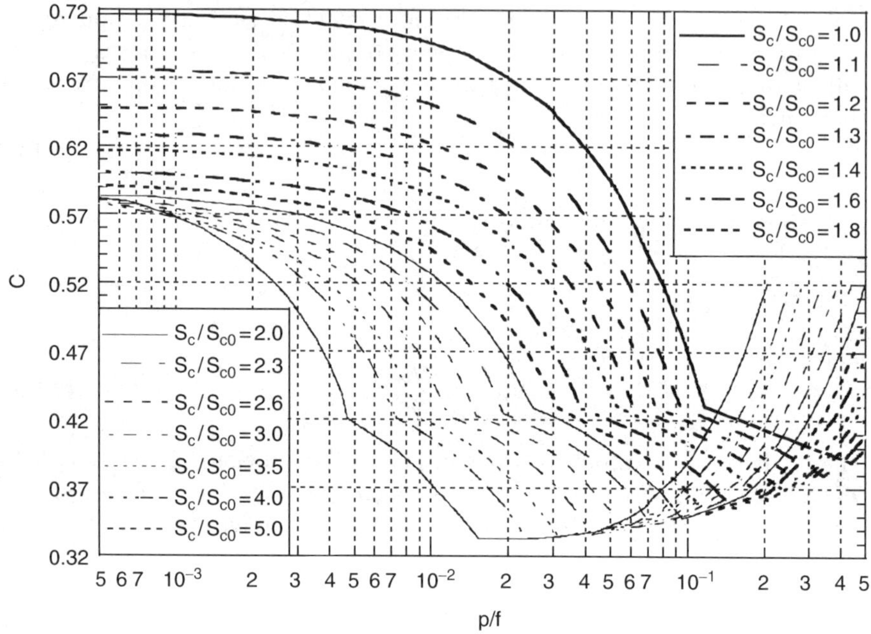

Fig. 6.28

phenomenon was factored in as follows. If the ideal stress in the cylinder exceeds $1.5f$, as far as the stress at the center of the head, reduced values F_0 and M_0 were considered for the ratio between $1.5f$ and the maximum ideal stress in the cylinder. This is equivalent to considering the onset of a plastic hinge when the maximum ideal stress in the cylinder exceeds $1.5f$.

Figure 6.28 shows that under certain conditions the value C is noticeably greater than the one computed earlier in the case of simple supported head. This depends on the fact that if the cylinder is under too great a stress (beyond $3f$), the only way to reduce the stresses is to increase the thickness of the head and consequently its rigidity. This will decrease the values of both F_0 and M_0.

Note that in general every curve shows three distinct portions that correspond to the following situations:

- The left portion corresponds to those instances where the state of stress in the cylinder is crucial. Here the values of C are the greatest because of the mentioned need to stiffen the head and thus reduce the stresses in the cylinder. The maximum value of the ideal stress in the cylinder is equal to $3f$.
- The central portion corresponds to those instances where the state of stress at the center of the head is critical. The maximum value of the ideal stress in the head is equal to $1.5f$. In the cylinder the maximum value of the ideal stress lies between $1.5f$ and $3f$. This means that the onset of a plastic hinge at the junction with the cylinder has been factored in.

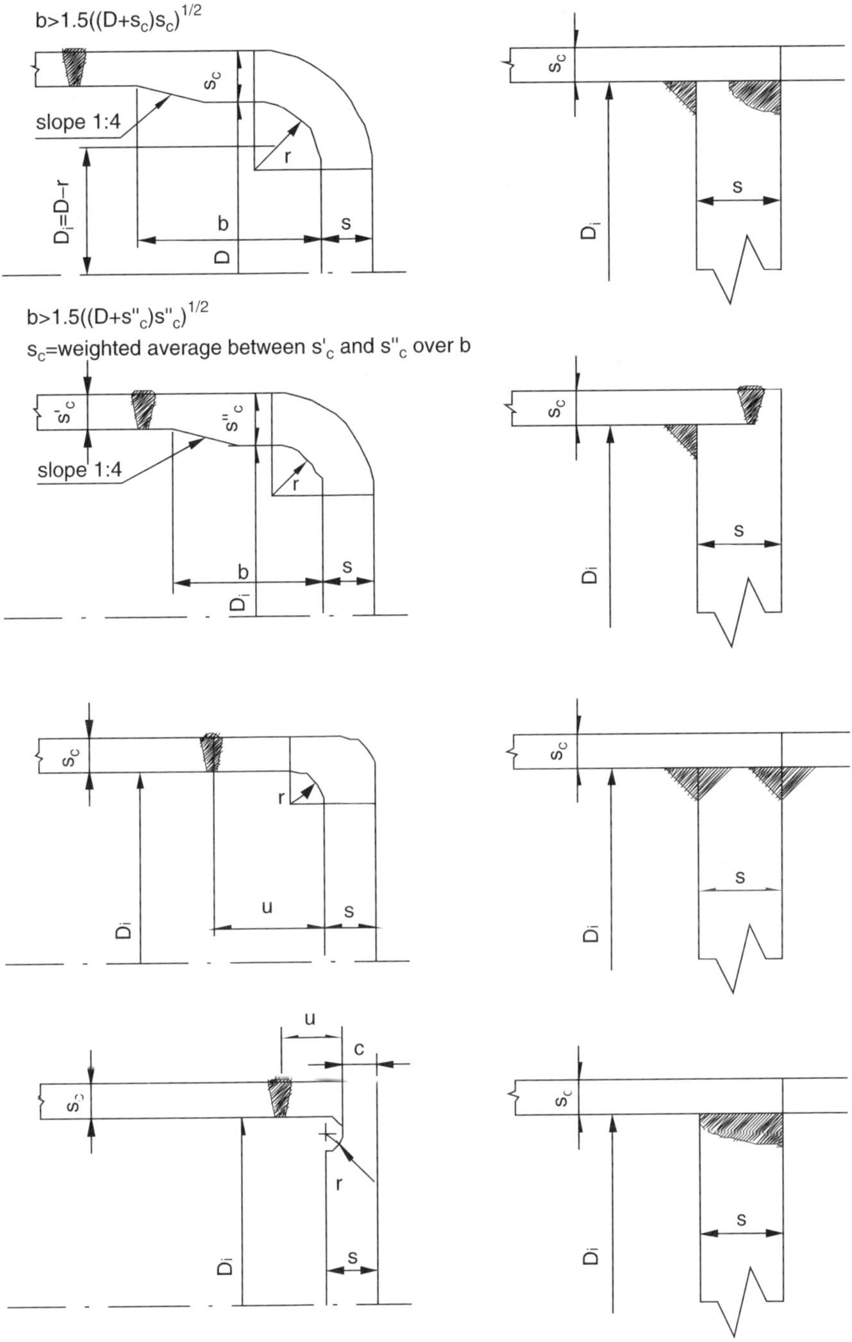
b>1.5((D+s_c)s_c)^1/2
slope 1:4
s_c
r
D_i=D−r
b
s
D
b>1.5((D+s"_c)s"_c)^1/2
s_c=weighted average between s'_c and s"_c over b
s'_c
s"_c
slope 1:4
r
b
s
D_i
s_c
r
u
s
D_i
u
c
s_c
r
s
D_i
s_c
s
D_i
s_c
s
D_i
s_c
s
D_i
s_c
s
D_i

Fig. 6.29

- The right portion corresponds to those instances where the state of stress at the center of the head is crucial, and the maximum value of the ideal stress in the cylinder is smaller than $1.5f$. Thus, there is no yielding in the cylinder or in the head.

In conclusion and with reference to Fig. 6.29 that shows some types of heads, note that the maximum ideal stress in the cylinder goes below $1.5f$, for the entire range of values of p/f and s_c/s_{co} considered in Fig. 6.28, at a distance from the head equal to

$$u \geq 0.26\sqrt{(D_i + s_c)\, s_c}. \tag{6.199}$$

Specifically, in correspondence to the right portion of the curves in Fig. 6.28 this condition holds for any value of u. This may be of interest to determine the best position for the welding in the case of heads with a cylindrical portion. For the head in Fig. 6.30 the minimum section corresponding to the reduction in thickness must be able to withstand the thrust exerted on the head. This is equal to

$$F = \frac{\pi}{4} D^2 p. \tag{6.200}$$

The area of the withstanding section is given by

$$A = \pi D c. \tag{6.201}$$

The average shear stress is

$$\tau_m = \frac{F}{A} = \frac{1}{4}\frac{pD}{c}. \tag{6.202}$$

Recalling that the behavior of τ in a rectangular section is parabolic, the maximum value of τ is

$$\tau_{\max} = \frac{3}{2}\tau_m = \frac{3}{8}\frac{pD}{c}. \tag{6.203}$$

For pure shear the ideal stress according to Huber–Hencky's failure theory is given by

$$\sigma_{id} = \sqrt{3}\tau; \tag{6.204}$$

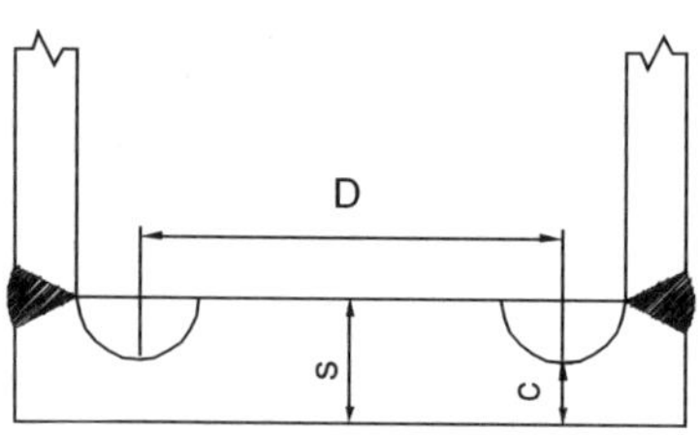

Fig. 6.30

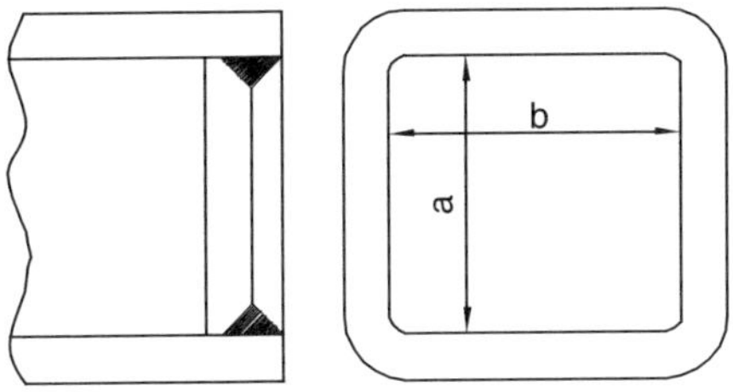

Fig. 6.31

setting $\sigma_{id} = f$, we obtain

$$\frac{3}{8}\sqrt{3}\frac{pD}{c} = f. \tag{6.205}$$

Therefore it must be

$$c \geq 0.65\frac{pD}{f}. \tag{6.206}$$

A comparable study of rectangular heads (see Fig. 6.31) would be particularly difficult. To examine the problem from a theoretical point of view, it is necessary to adopt simplified calculation criteria that inevitably lead to an oversizing of the head.

Generally, this type of head involves smaller vessels at low pressure, so that the oversizing has a modest economical impact. Leaving aside the behavior of the vessel at the junction for now, the worst condition for the head is evidently the one with simply supported edges. Under this type of constraint and in agreement with Timoshenko and Lessels (1925) the maximum moment at the center of the head is given by

$$M = \frac{1}{8}\frac{pa^2}{1 + 1.61\alpha^3}, \tag{6.207}$$

where α is the ratio between shorter and longer side of the head, that is,

$$\alpha = \frac{a}{b}. \tag{6.208}$$

Note that $\alpha = 0$, i.e., for $b = \infty$, since the influence of the leaning shorter sides is no longer present, the head behaves like a beam of length equal to a.

The maximum stress that coincides with the ideal stress, as we have seen for round heads, is given by

$$\sigma_{id} = \sigma = \frac{6M}{s^2} = \frac{3}{4}\frac{pa^2}{s^2\left(1 + 1.61\alpha^3\right)}. \tag{6.209}$$

As usual, setting $\sigma_{id} = 1.5f$ from (6.209) we obtain

$$s = \frac{0.707}{\sqrt{1 + 1.61\alpha^3}}a\sqrt{\frac{p}{f}}, \tag{6.210}$$

that may be written as follows:

$$s = Ca\sqrt{\frac{p}{f}}, \tag{6.211}$$

where C is the form factor given by

$$C = \frac{0.707}{\sqrt{1 + 1.61\alpha^3}}. \tag{6.212}$$

With (6.211) we go back to equation (6.83), but having set the shorter side of the head as the characteristic dimension. The value of the form factor in this case that does not depend on this type of constraint that was set in advance, but is instead a function of the geometry of the head through the ratio a/b.

The values of C are shown in Fig. 6.32. At this point it would be legitimate to doubt the suitability of the vessel to absorb the moments at the clamped edges that, contrary to previous assumptions, are actually present at the junction.

A simple and conservative calculation allows us to dismiss these doubts. We anticipate some of the results that will be developed in Sect. 7.3 for the study of quadrangular vessels. As we shall see, the thickness of the vessel may be computed through the following equation:

$$s_1 = \frac{pm}{3f} + \sqrt{\frac{4}{3}\frac{m^3 + n^3}{m + n}\frac{p}{f}} \tag{6.213}$$

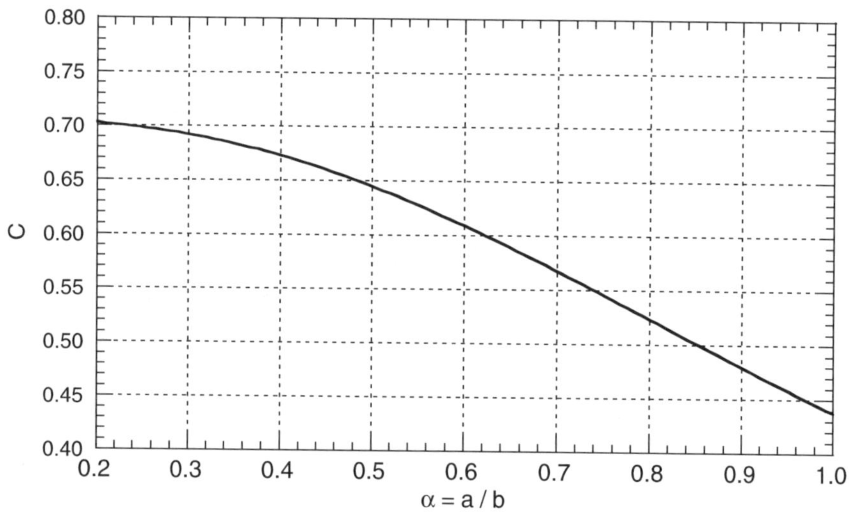

Fig. 6.32

where

$$m = \frac{b}{2}$$
$$n = \frac{a}{2} \qquad (6.214)$$

We ignore the first term of equation (6.213) that is also very small compared to the second. This way we assume that the thickness of the vessel is less than the actual one in favor of resistance.

Through a series of steps and recalling (6.214), we obtain

$$s_1 = \frac{b}{\sqrt{3}}\sqrt{\frac{1+\alpha^3}{1+\alpha}}\sqrt{\frac{p}{f}}. \qquad (6.215)$$

At the center of the longer side b the bending moment in the vessel is given by (and with reference to Sect. 7.3)

$$M_1 = \frac{pb^2}{24}\frac{1+3\alpha-2\alpha^3}{1+\alpha}. \qquad (6.216)$$

Note that if $\alpha = 1$, i.e., if the sides are equal, we have

$$M_1 = \frac{pb^2}{24} \qquad (6.217)$$

which is a common equation for bended beams that are perfectly clamped at the ends, as is the case if the sides are equal and there are no rotations at the edges.

Recalling (6.215), the stress corresponding to M_1 is given by

$$\sigma_1 = \frac{6M_1}{s_1^2} = \frac{3}{4}\frac{1+3\alpha-2\alpha^3}{1+\alpha^3}f. \qquad (6.218)$$

The moment μM_1 caused by the moment M_1 is generated for congruence along the axis of the vessel; therefore, at the center of the longer side there is an axial stress equal to

$$\sigma_2 = \mu\sigma_1 = 0.225\frac{1+3\alpha-2\alpha^3}{1+\alpha^3}f. \qquad (6.219)$$

The stress caused by the thrust on the heads, though irrelevant compared to σ_2 and thus ignored, should be added to the latter.

Conservatively, we assume that the stress σ_2 is exerted at the edge of the vessel ignoring the stiffening effect of the head that reduces, of course, the value of the moment M_1 and consequently of σ_2, as well. Finally, to favor resistance we assume that the head transmits the moments equivalent to the condition of a perfectly clamped head to the vessel.

Again, according to Timoshenko the bending moment at the center of side b is equal to

$$M_a = \frac{1}{12}\frac{pa^2}{1+0.623\alpha^6} \tag{6.220}$$

that is reduced to $M_a = pa^2/12$ for $\alpha = 0$, as required.

In the vessel the corresponding stress is given by

$$\sigma_3 = \frac{6M_a}{s_1^2} = \frac{1.5\alpha^2\,(1+\alpha)}{(1+0.623\alpha^6)\,(1+\alpha^3)}f. \tag{6.221}$$

Therefore, the total stress at the edge of the vessel is

$$\sigma_a = \sigma_2 + \sigma_3 = \frac{0.225\left(1+3\alpha-2\alpha^3\right) + \dfrac{1.5\alpha^2\,(1+\alpha)}{1+0.623\alpha^6}}{1+\alpha^3}f. \tag{6.222}$$

In Fig. 6.33 we have shown the ratio σ_a/f. As one can see, its value is always smaller than 1.4, while, as shown in Sect. 1.5, under these conditions a stress equal to $3f$ is allowable (twice the yield strength). Therefore, there is no danger at the junction. In conclusion, the sizing of the head may be done through (6.211).

Even though the type of constraint is substantially different from the simple support even when perfectly clamped, this does not put at risk the resistance of the head (the moment at the center of the head is in any case smaller than the one we used to obtain (6.211)), or the vessel, since the maximum stress at the junction is by far smaller than $3f$.

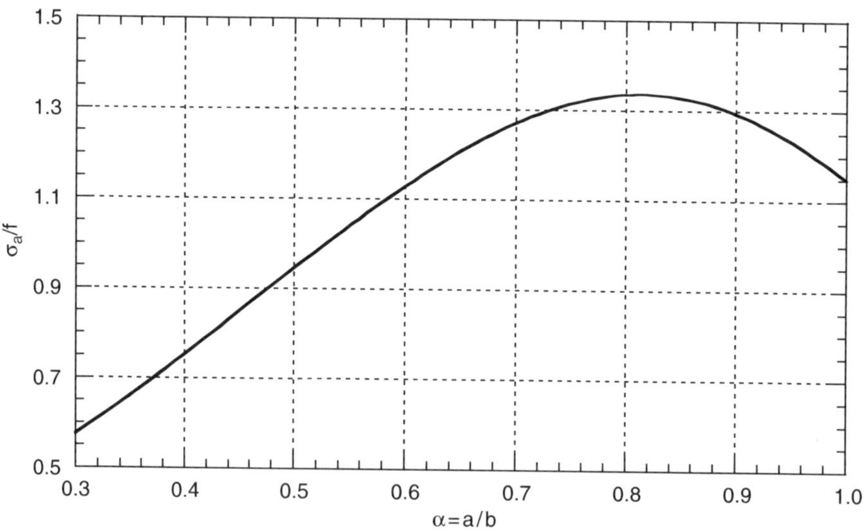

Fig. 6.33

At the end of this presentation we would like to analyze the consequences of the presence of a temperature gradient through the thickness of a round head.

The analysis is simple if the head is modelled as an isolated element, as we did at the beginning of the section. If, on the other hand, the head is considered as an integral part of the cylinder-head structure, as it actually is, the difficulty of a general purpose analysis is significant. In practice, in order to reach a correct evaluation of the stress generated by the temperature gradient, each case must be examined separately, for instance through finite element analysis.

We assume that the different fibers of the head are at different temperatures. More specifically, we assume a linear temperature increase from the internal side of the vessel to the external one as is the case if the head is crossed by a heat flux. Δt is the difference in temperature between outside and inside.

The difference in dilation between the external fiber and the average one is therefore equal to

$$\Delta\varepsilon = \alpha\frac{\Delta t}{2}, \tag{6.223}$$

where α is the thermal expansion coefficient.

The latter involves a constant curvature radius r_x given by

$$r_x = \frac{s}{2\Delta\varepsilon} = \frac{s}{\alpha\Delta t}. \tag{6.224}$$

Thus, the head takes on the shape of a portion of a sphere with radius r_x.

Recalling (6.95) and substituting the x-abscissa with the radius r, the rotation φ is given by

$$\varphi = \frac{r}{r_x} = \frac{\alpha\Delta t}{s}r, \tag{6.225}$$

and at the edge of the head with radius R the corresponding rotation indicated as φ'_R is given by

$$\varphi'_R = \frac{\alpha R\Delta t}{s}. \tag{6.226}$$

If the head is simply supported, it is free to deform, and therefore the temperature gradient Δt does not generate any stress.

If, on the other hand, it is clamped, the moments M_r and M_c are generated. Note that without pressure (6.115) is reduced to

$$\varphi = C_1\frac{r}{2} + \frac{C_2}{r}. \tag{6.227}$$

If φ''_R is the rotation at the edge generated by the moments above and if we recall that at the center $\varphi = 0$

$$C_2 = 0, \tag{6.228}$$

and

$$C_1 = \frac{2\varphi''_R}{R}. \tag{6.229}$$

From (6.99) and factoring in (6.227) and (6.228) we obtain

$$M_r = M_c = \frac{KC_1}{2}(1+\mu), \tag{6.230}$$

and from (6.229)

$$M_r = M_c = \frac{K\varphi''_R}{R}(1+\mu), \tag{6.231}$$

where M_r and M_c are equal and independent from radius r. From (6.231) we obtain

$$\varphi''_R = \frac{M_r R}{K(1+\mu)}, \tag{6.232}$$

and recalling (6.93)

$$\varphi''_R = \frac{M_r R(1-\mu)}{EI} = 12\frac{M_r R(1-\mu)}{Es^3}. \tag{6.233}$$

The total rotation at the edge of head φ_R is therefore given by

$$\varphi_R = \varphi'_R + \varphi''_R = \frac{\alpha R \Delta t}{s} + 12\frac{M_r R(1-\mu)}{Es^3}. \tag{6.234}$$

In the case of perfectly clamped edges, since $\varphi_R = 0$ we obtain

$$M_r = M_c = -\frac{\alpha E \Delta t s^2}{12(1-\mu)}. \tag{6.235}$$

These moments are negative, similarly to those moments when the head is under pressure p.

Of course, if Δt is negative because the temperature is higher on the internal side of the vessel, as it is more frequently the case, the moments are positive. Generally, once the value of φ_r is computed through (6.234), we obtain

$$M_r = M_c = -\frac{Es^2}{12(1-\mu)}\left(\frac{s}{R}\varphi_R - \alpha\Delta t\right). \tag{6.236}$$

The corresponding maximum stress is equal to

$$\sigma_r = \sigma_c = \pm\frac{E}{2(1-\mu)}\left(\frac{s}{R}\varphi_R - \alpha\Delta t\right). \tag{6.237}$$

The absolute value of the moment M_r from (6.235) represents a maximum, and the value of this moment is practically somewhere between zero and this value. In order to identify it, it would be necessary to correlate the rotation φ_R with the moment M_r based on the geometrical characteristics of the cylinder, as was done for the pressure. The state of stress is also influenced by the average temperatures of the cylinder and the head (that are not necessarily the same), as well as by the temperature gradient through the wall of the cylinder. Therefore, there are many parameters at play and many case studies that make a general analysis quite cumbersome. It make sense therefore to examine the specific case, if there is any reason for concern, via finite element analysis.

The value of M_r from (6.235) may be considered for an approximate and conservative analysis. Note that this is relevant at the edge of the head if Δt is positive, since it has the same sign of the corresponding moment due to the pressure, while it is relevant at the center of the head if Δt is negative for the same reason. Finally, note that the stresses caused by moments generated by Δt originate from the respect for congruence,; this way they are self-limiting and may be considered secondary.

7

Special Components and Tubes

7.1 Elliptical Tubes

The sizing of tubes with an elliptical cross-section (Fig. 7.1) under internal pressure may be done according to the following criteria. Note that the nodal circle, i.e., the circle going through the points of the ellipsis with a zero bending moment, almost overlaps the circle with a radius equal to the average of the axis of the ellipsis. The radius ρ of the nodal circle is given by

$$\rho^2 = \frac{\int \delta^2 dl}{l}, \tag{7.1}$$

with l the length developed by the average fiber and δ the distance of the generic fiber from the center of stress corresponding to the center of the nodal circle.

For the properties of the nodal circle the bending moment per unit of length is given anywhere by

$$M = \frac{p}{2}\left(\delta^2 - \rho^2\right). \tag{7.2}$$

If we consider point A of the ellipsis (the extremity of the major axis) we have

$$\delta = \frac{a}{2}, \tag{7.3}$$

$$\rho = \frac{a+b}{4}, \tag{7.4}$$

where a and b are the major and minor axis, respectively.

In A we have

$$M_A = \frac{p}{2}\left[\frac{a^2}{4} - \frac{(a+b)^2}{16}\right]; \tag{7.5}$$

then

$$M_A = \frac{p}{32}\left(3a^2 - 2ab - b^2\right). \tag{7.6}$$

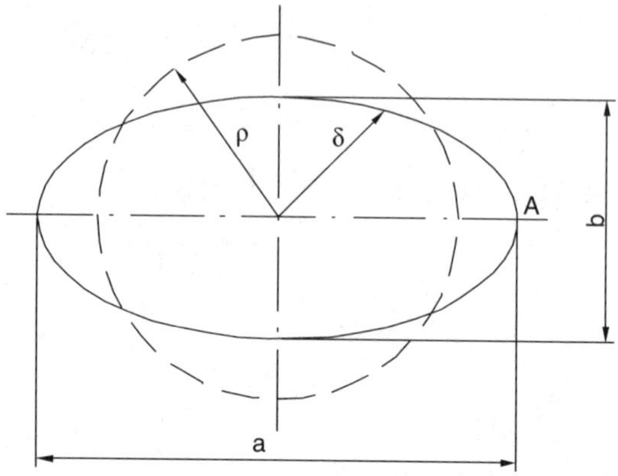

Fig. 7.1

Recalling that $W = s^2/6$, the maximum corresponding stress is

$$\sigma_f = \pm \frac{3}{16} p \left(\frac{a}{s}\right)^2 \left[3 - 2\frac{b}{a} - \left(\frac{b}{a}\right)^2\right] . \tag{7.7}$$

The normal force per unitary length in A is given by

$$N = \frac{pa}{2} , \tag{7.8}$$

thus the corresponding stress is equal to

$$\sigma_n = \frac{p}{2}\frac{a}{s} , \tag{7.9}$$

where s is the thickness. The total is

$$\sigma = \frac{pa}{2s} + \frac{3}{16} p \left(\frac{a}{s}\right)^2 \left[3 - 2\frac{b}{a} - \left(\frac{b}{a}\right)^2\right] . \tag{7.10}$$

The stress σ_n is a general membrane stress. Based on Sect. 1.5 one must set it as

$$\frac{pa}{2s} = f , \tag{7.11}$$

where f is the basic allowable stress of the material; then

$$s = \frac{pa}{2f} . \tag{7.12}$$

The stress σ_f is a primary bending stress. The stress σ, i.e., the sum of σ_n and σ_f, must therefore be limited to $1.5f$, where $1.5f$ corresponds to the yield strength.

We introduce the dimensionless quantity k_1 given by

$$k_1 = 4.5\left[3 - 2\frac{b}{a} - \left(\frac{b}{a}\right)^2\right], \tag{7.13}$$

and (7.10) is reduced to the following:

$$\sigma = \frac{pa}{2s} + \frac{p}{24}\left(\frac{a}{s}\right)^2 k_1. \tag{7.14}$$

With $\sigma = 1.5f$, from (7.14) we obtain

$$\left(\frac{s}{a}\right)^2 - \frac{p}{3f}\frac{s}{a} - \frac{p}{36f}k_1 = 0; \tag{7.15}$$

after solving the equation we have

$$\frac{s}{a} = \frac{p}{6f} + \sqrt{\frac{p^2}{36f^2} + \frac{p}{36f}k_1}; \tag{7.16}$$

and finally

$$s = \frac{pa}{6f}\left(1 + \sqrt{1 + \frac{f}{p}k_1}\right). \tag{7.17}$$

Of course, the value s obtained from (7.17) shall not be smaller than the one in (7.12).

This condition corresponds to the following inequality:

$$1 + \sqrt{1 + \frac{f}{p}k_1} \geq 3; \tag{7.18}$$

then

$$k_1 \geq 3\frac{p}{f}, \tag{7.19}$$

or, based on (7.13) as well

$$\frac{b}{a} \leq 2\sqrt{1 - \frac{p}{6f}} - 1. \tag{7.20}$$

Taking into account that elliptical tubes are used for low pressures, the condition described above is verified for tubes with profiles that differ even slightly from the circular one, too.

For example, if $p = 1\,\mathrm{N\,mm^{-2}}$ and $f = 100\,\mathrm{N\,mm^{-2}}$, from (7.20) we see that, if (7.17) holds, $b/a \leq 0.998$. This means that (7.17) holds true for all elliptical tubes. If the ratio b/a is defined as α, (7.13) may be written as follows:

$$k_1 = 4.3\left(3 - 2\alpha - \alpha^2\right). \tag{7.21}$$

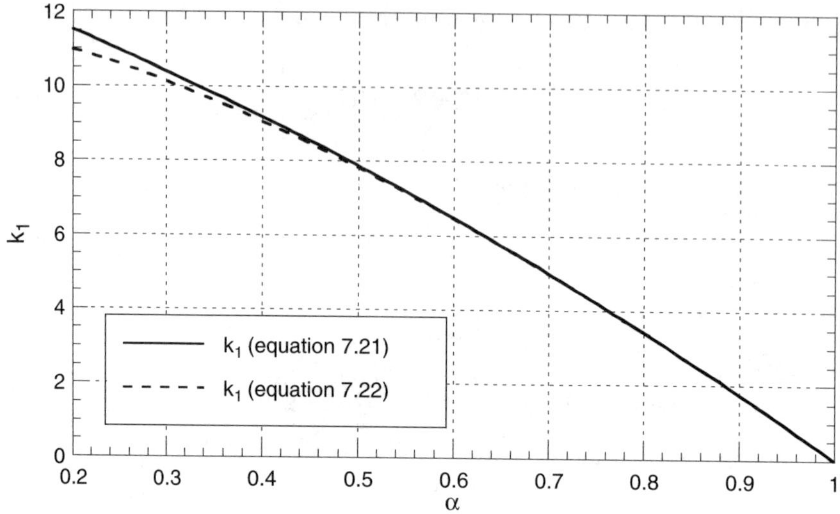

Fig. 7.2

Through more sophisticated calculations we get to (7.17) where quantity k_1, though, is given by

$$k_1 = 9\left(1 - \alpha^2\right)k, \tag{7.22}$$

with

$$k = 1 + \frac{1}{4}\frac{1-\alpha^2}{1+\alpha^2} + \frac{5}{128}\left(\frac{1-\alpha^2}{1+\alpha^2}\right)^3 + \frac{15}{1024}\left(\frac{1-\alpha^2}{1+\alpha^2}\right)^5 . \tag{7.23}$$

Figure 7.2 shows both values of k_1 as a function of α. As one can see, they differ in a negligible way for $\alpha \geq 0.5$. Even if $\alpha < 0.5$, the differences are quite modest, and in any case the value of k_1 from (7.21) is in favor of resistance since it is greater. Assuming for the sake of simplicity that the thickness is proportional to the root of k_1 (thus ignoring the impact of the normal load and considering only that of the bending moment), using (7.21) instead of (7.22), one obtains the following percentual increases in thickness.

For $\alpha = 0.5$ the increase is equal to 0.3%, and for $\alpha = 0.4$ it is 0.6%; for $\alpha = 0.3$ it is 1.28%, and finally for $\alpha = 0.2$ it is equal to 2.4%. These are clearly small values. The increases in thickness are substantial but still within acceptable limits only for very small and unlikely values of α. Given the simplicity of (7.21), its use is preferable to (7.22) for the calculation of k_1.

7.2 Torus and Bended Tubes

To determine the stresses in torus elements and in the curves of the tubes due to internal pressure, and considering that these components are not very thick, one can apply the theory about membranes presented in Sect. 2.1. Generally, if σ_1 and σ_2 are the meridian and the circumferential stress of the membrane, respectively, and R_1 and R_2 are the radii of the meridian and on the normal plane to the meridian, respectively, the equation of equilibrium is

$$\frac{\sigma_1}{R_1} + \frac{\sigma_2}{R_2} = \frac{p}{s}, \tag{7.24}$$

where s is the thickness of the membrane.

In the case at hand (Fig. 7.3) the radius R_1 corresponds to radius r of the tube, and the stress σ_1 is the hoop stress in the tube itself.

The radius R_2 is shown in Fig. 7.3 and has the following equation:

$$R_2 = \frac{r_0 + y}{\sin\vartheta}, \tag{7.25}$$

where r_0 is the radius of the torus referred to the axis of the tube and y the distance of the fiber from the diameter parallel to the axis of the torus. It has a positive sign for the fibers outside such diameter and a negative sign for those inside. In this case the stress σ_2 is the longitudinal stress in the tube. From (7.24) and with σ_t and σ_a the hoop and the longitudinal stress in the tube, respectively, we have

$$\frac{\sigma_t}{r} + \frac{\sigma_a \sin\vartheta}{r_0 + y} = \frac{p}{s}. \tag{7.26}$$

Recalling that for the membrane

$$\sigma_a = \frac{pr}{2s}; \tag{7.27}$$

thus

$$\frac{\sigma_t}{r} + \frac{p}{2s}\frac{r\sin\vartheta}{r_0 + y} = \frac{p}{s}. \tag{7.28}$$

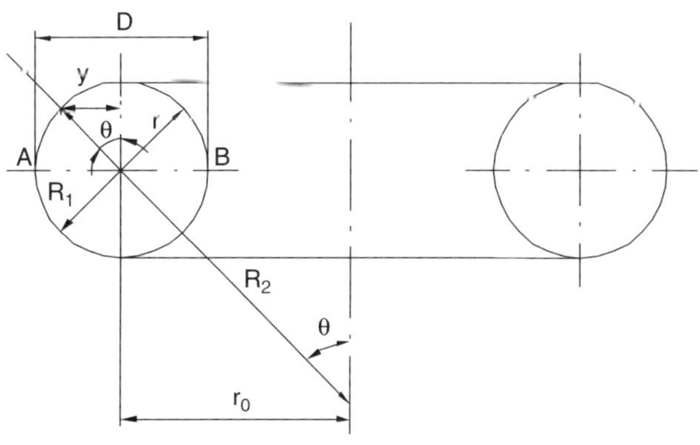

Fig. 7.3

On the other hand

$$r \sin \vartheta = y; \tag{7.29}$$

therefore

$$\frac{\sigma_t}{r} + \frac{p}{2s}\frac{y}{r_0 + y} = \frac{p}{s}. \tag{7.30}$$

From (7.30) and through a series of steps, we obtain

$$\sigma_t = \frac{pD}{2s}\frac{r_0 + \dfrac{y}{2}}{r_0 + y}, \tag{7.31}$$

where D is the diameter of the tube.

Specifically, for $\vartheta = \pi/2$ and $\vartheta = -\pi/2$, i.e., in correspondence of the extrados and the intrados (points A and B of the tube) we have:

Extrados ($y = D/2$)

$$\sigma_t = \frac{pD}{2s}\frac{r_0 + \dfrac{D}{4}}{r_0 + \dfrac{D}{2}}. \tag{7.32}$$

Intrados ($y = -D/2$)

$$\sigma_t = \frac{pD}{2s}\frac{r_0 - \dfrac{D}{4}}{r_0 - \dfrac{D}{2}}. \tag{7.33}$$

Note that $pD/2s$ is none other than the hoop stress occurring in the straight tube that we call σ_{to}. Therefore, generally

$$\sigma_t = \sigma_{t0}\frac{1 + \dfrac{y}{2r_0}}{1 + \dfrac{y}{r_0}} = \sigma_{t0}\frac{1 + \dfrac{D}{4r_0}\sin\vartheta}{1 + \dfrac{D}{2r_0}\sin\vartheta}. \tag{7.34}$$

The ratio σ_t/σ_{to} along the circumference is shown in Fig. 7.4 for $r_0/D = 2$. The double curvature in correspondence of the extrados reduces the hoop stress compared to the one in the straight tube (that corresponds to the stress σ_t for $y = 0$), while it is greater on the intrados.

On the other hand, the construction of the torus or the curve of a tube implies a reduction in thickness on the extrados and an increase in thickness on the intrados. If s' and s'' represent the thickness in the positions indicated above (A and B in Fig. 7.3) and σ'_t and σ''_t the hoop stresses, respectively, we have

$$\sigma'_t = \frac{pD}{2s'}\frac{r_0 + \dfrac{D}{4}}{r_0 + \dfrac{D}{2}}; \tag{7.35}$$

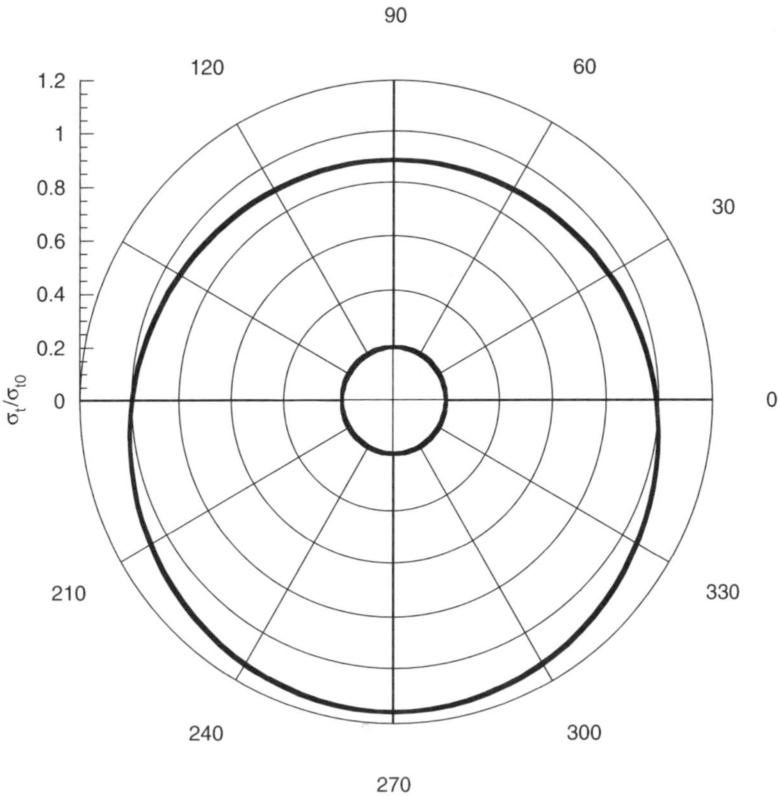

Fig. 7.4

$$\sigma''_t = \frac{pD}{2s''} \frac{r_0 - \dfrac{D}{4}}{r_0 - \dfrac{D}{2}}. \tag{7.36}$$

If we assume that the average fiber does not elongate, the original length of the segment of the tube at 180° through which the curve was obtained is equal to

$$L_0 = \pi r_0. \tag{7.37}$$

The length of the fiber on the extrados after the curvature becomes

$$L' = \pi \left(r_0 + \frac{D}{2} \right), \tag{7.38}$$

whereas the elongation ε'_a in the direction of the axis is given by

$$\varepsilon'_a = \frac{L' - L_0}{L_0} = \frac{1}{2} \frac{D}{r_0}. \tag{7.39}$$

In the plastic field deformations occur without variations in volume, and Poisson's ratio is equal to 0.5. Therefore, deformations ε'_t and ε'_r in the circumferential and radial direction assume the following values:

$$\varepsilon'_t = \varepsilon'_r = -\frac{1}{2}\varepsilon'_a = -\frac{1}{4}\frac{D}{r_0}. \tag{7.40}$$

Since s_0 is the original thickness, the thickness on the extrados becomes

$$s' = \left(1 - \frac{1}{4}\frac{D}{r_0}\right) s_0. \tag{7.41}$$

Based on (7.35), the circumferential stress σ'_t is given by

$$\sigma'_t = \frac{pD}{2\left(1 - \frac{1}{4}\frac{D}{r_0}\right) s_0} \frac{r_0 + \frac{D}{4}}{r_0 + \frac{D}{2}}, \tag{7.42}$$

that may be written as

$$\sigma'_t = A' \frac{pD}{2s_0}. \tag{7.43}$$

As we have already seen, the hoop stress in the straight tube is equal to

$$\sigma_{t0} = \frac{pD}{2s_0}; \tag{7.44}$$

A' indicates the ratio between the value of the stress σ'_t on the extrados and the one in correspondence of the straight tube. Its value is

$$A' = \frac{1 + \frac{1}{4}\frac{D}{r_0}}{\left(1 - \frac{1}{4}\frac{D}{r_0}\right)\left(1 + \frac{1}{2}\frac{D}{r_0}\right)}. \tag{7.45}$$

From (7.45), we obtain the following values of A':

$$\frac{r_0}{D} = 1.5 \quad A' = 1.05$$
$$\frac{r_0}{D} = 2 \quad A' = 1.028$$
$$\frac{r_0}{D} = 3 \quad A' = 1.012$$

As one can see, the hoop stress in correspondence of the extrados would be slightly greater than the one relative to the straight tube. By analogy, for the fiber in correspondence of the intrados we obtain

$$\varepsilon''_t = \varepsilon''_r = -\frac{1}{2}\varepsilon''_a = \frac{1}{4}\frac{D}{r_0}, \tag{7.46}$$

and the stress σ''_t assumes the following value:

$$\sigma''_t = A'' \frac{pD}{2s_0}, \tag{7.47}$$

since

$$A'' = \frac{1 - \dfrac{1}{4}\dfrac{D}{r_0}}{\left(1 + \dfrac{1}{4}\dfrac{D}{r_0}\right)\left(1 - \dfrac{1}{2}\dfrac{D}{r_0}\right)}. \tag{7.48}$$

From (7.48) we obtain

$$\frac{r_0}{D} = 1.5 \quad A'' = 1.07$$
$$\frac{r_0}{D} = 2 \quad A'' = 1.036$$
$$\frac{r_0}{D} = 3 \quad A'' = 1.015$$

Figure 7.5 shows the values for A' and A''. Again, σ_t relative to the extrados would be greater than the one for the straight tube. In fact, if adequate manufacturing practices are taken, one obtains a value of $\varepsilon_a{}'$ greater than the theoretical one given by (7.39) on the extrados, because the fiber that does not become elongated is more on the inside than the average one, and because the elongation is not constant along the curve, but decreases at the ends. Moreover, the absolute value of $\varepsilon_t{}'$ is greater than the theoretical one, but the value of $\varepsilon_r{}'$ is smaller.

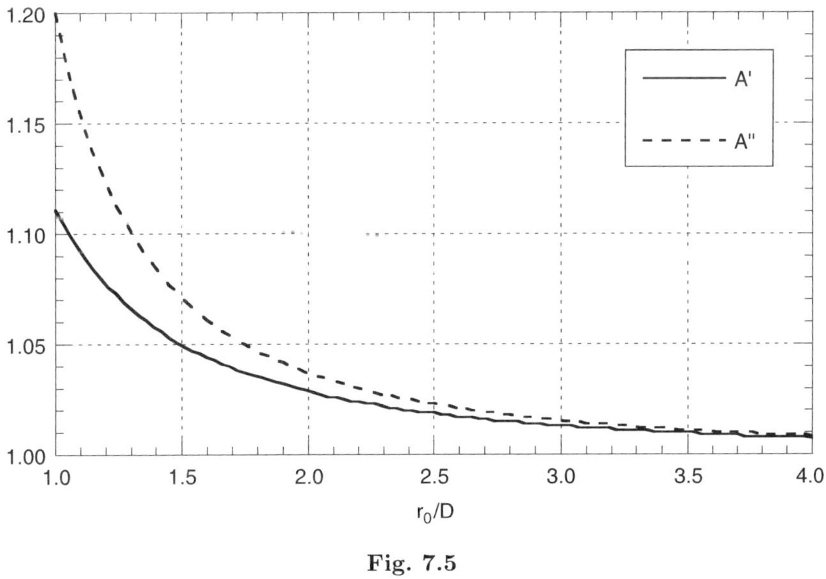

Fig. 7.5

On the intrados the absolute value of ε_a'' is slightly smaller than the theoretical one. In addition, the value of ε_r'' is greater than ε_t'' and greater than the theoretical one.

In practice, the reduction in thickness on the extrados is smaller than the theoretical one, while the increase in thickness on the intrados is greater than the theoretical one. The actual situation is therefore better than the one resulting from calculations, provided the curvature is done correctly. Under these conditions, the curvature does not represent a concern and does not require an increase in thickness of the tube to factor in the stretching along the curve. This is the customary practice.

Thus, it is possible to require the workshop to ensure that variations in thickness do not cause an increase of σ_t with respect to the value for the straight tube. Through a series of steps that are not shown here, one obtains the following equations about the allowable variation in thickness (in percentage and as an absolute value):

For the extrados

$$\frac{\Delta s}{s_0} \leq \frac{25}{\dfrac{r_0}{D} + 0.5} (\%) . \tag{7.49}$$

For the intrados

$$\frac{\Delta s}{s_0} \geq \frac{25}{\dfrac{r_0}{D} - 0.5} (\%) . \tag{7.50}$$

Figure 7.6 shows the values of $\Delta s/s_0$.

These are limitations that one can live with. Recalling the values of A' and A'' that are not far from one (at least for $r_0/D \geq 1.5$), it is even possible

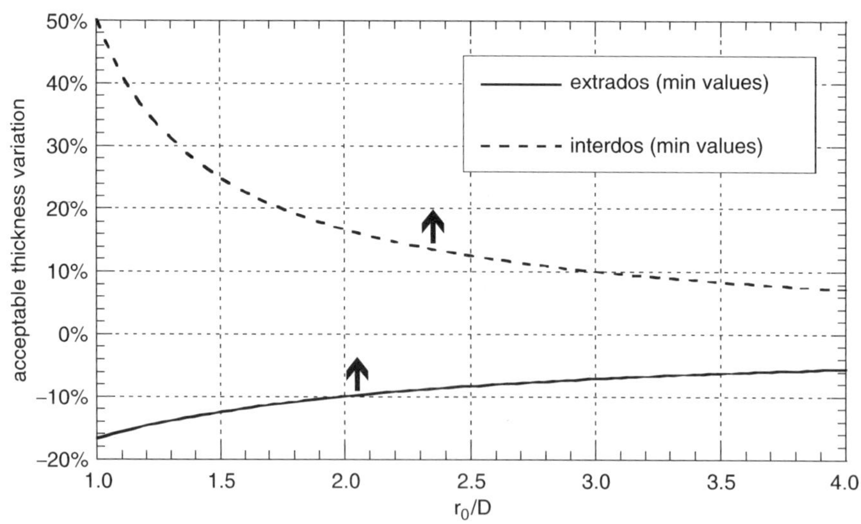

Fig. 7.6

to tolerate variations in thickness equal to the theoretical ones, as long as the ratio between the radius of curvature and the diameter is not smaller than 1.5.

Instead of (7.49) and (7.50), one obtains the following conditions:

For the extrados

$$\frac{\Delta s}{s_0} \leq 25\frac{D}{r_0}\,(\%)\,. \tag{7.51}$$

For the intrados

$$\frac{\Delta s}{s_0} \geq 25\frac{D}{r_0}\,(\%)\,. \tag{7.52}$$

7.3 Quadrangular Vessels

Let us consider the quadrangular vessel with the section shown in Fig. 7.7, where the internal length of the longest side has been set at $2m$, and let us do the calculation with regard to a portion of the vessel of unitary length.

Ignoring the deformations at the corners, we can model the structure by considering it similar to the frame in Fig. 7.8. 1,2,3, and 4 are the knots of the frame, and based on the sign convention used for frames, where the clockwise moments are positive and the counter clockwise moments are negative, the moments of perfectly clamped edge in knot 1 at the end of beam 1–2 and beam 1–3 are, respectively

$$\overline{M}_{12} = \frac{1}{12}p\,(2m)^2 = \frac{1}{3}pm^2; \tag{7.53}$$

$$\overline{M}_{13} = -\frac{1}{12}p\,(2n)^2 = -\frac{1}{3}pn^2. \tag{7.54}$$

It is well known that the moment M_{12} at the end of beam 1–2 in correspondence of knot 1 is

$$M_{12} = \overline{M}_{12} + 2M'_{12} + M'_{21}, \tag{7.55}$$

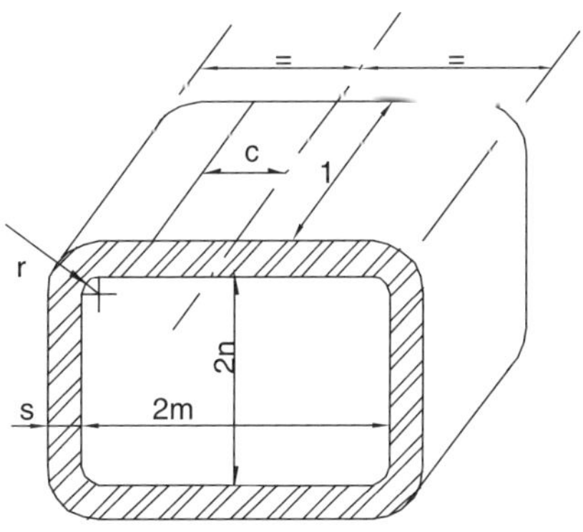

Fig. 7.7

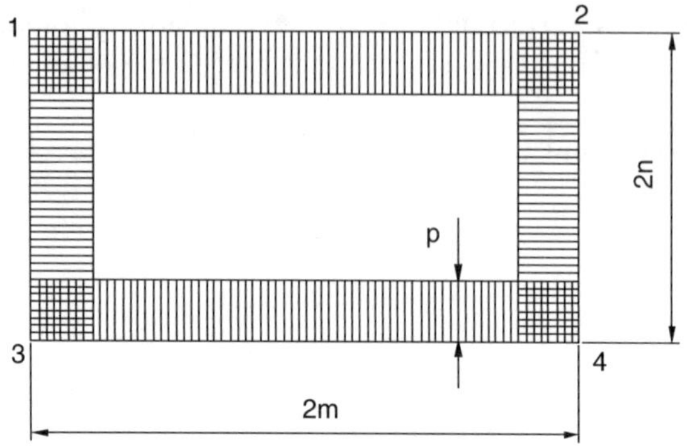

Fig. 7.8

where

$$M'_{12} = 2E\frac{I_{12}}{2m}\vartheta_1, \tag{7.56}$$

and

$$M'_{21} = 2E\frac{I_{12}}{2m}\vartheta_2. \tag{7.57}$$

E is the modulus of elasticity, I_{12} is the moment of inertia of the beam 1–2, and ϑ_1 and ϑ_2 are the rotations of knots 1 and 2. Clearly, for symmetry $\vartheta_2 = -\vartheta_1$, therefore

$$M'_{21} = -M'_{12}. \tag{7.58}$$

Thus, from (7.55)

$$M_{12} = \overline{M}_{12} + M'_{12}. \tag{7.59}$$

By analogy one may write

$$M_{13} = \overline{M}_{13} + M'_{13}. \tag{7.60}$$

On the other hand,

$$M'_{13} = 2E\frac{I_{13}}{2n}\vartheta_1. \tag{7.61}$$

Given that $I_{13} = I_{12}$ because the thickness of the vessel is the same on all sides, from (7.56) and (7.61) we obtain

$$\frac{M'_{13}}{M'_{12}} = \frac{m}{n}. \tag{7.62}$$

For the equilibrium of knot 1 we must have $M_{12} + M_{13} = 0$, therefore

$$M_{12} + M_{13} = \overline{M}_{12} + \overline{M}_{13} + M'_{12}\left(1 + \frac{m}{n}\right) = 0; \tag{7.63}$$

then

$$M'_{12} = -\left(\overline{M}_{12} + \overline{M}_{13}\right)\frac{n}{n+m}; \tag{7.64}$$

and, finally,

$$M_{12} = \overline{M}_{12} - \left(\overline{M}_{12} + \overline{M}_{13}\right) \frac{n}{n+m}. \tag{7.65}$$

Through a series of steps that are not shown here and recalling (7.53) and (7.54), we obtain

$$M_{12} = \frac{1}{3} p \frac{m^3 + n^3}{m+n}. \tag{7.66}$$

The reaction in 1 because of the pressure on the beam 1–2 is equal to

$$R_1 = pm. \tag{7.67}$$

The thrust on the strip of unitary length between corner 1 and the parallel to the axis of the vessel at distance c from the center line of the beam 1–2 (see Fig. 7.7) is equal to

$$F = p\,(m - c)\,. \tag{7.68}$$

In correspondence of this line the moment is given by

$$M_c = M_{12} - R_1\,(m - c) + F \frac{m - c}{2}. \tag{7.69}$$

Recalling (7.66), (7.67), and (7.68), we have

$$M_c = \frac{1}{3} p \frac{m^3 + n^3}{m+n} - pm\,(m - c) + p \frac{(m - c)^2}{2}. \tag{7.70}$$

Through a series of steps that are not shown here we obtain

$$M_c = \frac{1}{3} p \frac{m^3 + n^3}{m+n} - \frac{p}{2} \left(m^2 - c^2\right). \tag{7.71}$$

Setting

$$A = \frac{1}{3} \frac{m^3 + n^3}{m+n}, \tag{7.72}$$

we obtain

$$M_c = \left(A - \frac{m^2 - c^2}{2}\right) p. \tag{7.73}$$

The tensile force N is also given by

$$N = pn. \tag{7.74}$$

According to Guest, the maximum value of the stress present in the fiber stretched at the maximum distance from the neutral axis that coincides with the ideal stress σ_{id}, is therefore given by

$$\sigma_{id} = \frac{N}{s} + \frac{|M_c|}{\frac{s^2}{6}} = \frac{pn}{s} + \frac{6\,|M_c|}{s^2}. \tag{7.75}$$

Based on the criteria presented in Sect. 1.5, since this is about the combination of a general membrane stress with a primary bending stress, the allowable stress is equal to $1.5f$, where $1.5f$ corresponds to the yield strength. Setting $\sigma_{id} = 1.5f$, from (7.26) we obtain

$$s = \frac{\frac{pn}{3} + \sqrt{\left(\frac{pn}{3}\right)^2 + 4\,|M_c|\,f}}{f}. \tag{7.76}$$

Ignoring the first term under the root because it is very small compared to the second, we finally have

$$s = \frac{pn}{3f} + \sqrt{\frac{4\,|M_c|}{f}}. \tag{7.77}$$

In the absence of holes, something we will discuss in Chap. 8, the maximum value of the bending moment occurs at the corner. In this position the verification shall be done considering the end of the shorter side because its equivalent is a much greater value of N given by:

$$N = pm. \tag{7.78}$$

In fact, the sizing of the vessel must be done based on the following equation that takes (7.66) into account, as well

$$s = \frac{pm}{3f} + \sqrt{\frac{4}{3}\,\frac{m^3+n^3}{m+n}\,\frac{p}{f}}, \tag{7.79}$$

where $2m$ is the greatest dimension of the vessel (Fig. 7.7).

Since the vessel has typically equal sides, (7.80) is reduced to the following:

$$s = 2m\left(\frac{p}{6f} + \sqrt{\frac{p}{3f}}\right). \tag{7.80}$$

To refine the calculation versus the simplistic criterion above, it is best to examine the behavior of the vessel at the corners, to highlight the influence of the radius in the corner.

To accomplish this task we look at the vessel with equal sides and consider the portion of circular crown (see Fig. 7.8) at the corner. M_0 is the bending moment, and N_0 and T_0 are the normal force and the shear one at the ends. If r is the radius and s the thickness of the knuckle, in reference to the unitary length of the vessel we obtain

$$N_0 = pm, \tag{7.81}$$

and

$$T_0 = p\,(m - r). \tag{7.82}$$

Let us now consider the thrust caused by pressure that corresponds to a generic angle γ. It is equal to

$$F_1 = pr\sin\gamma; \tag{7.83}$$
$$F_2 = pr\left(1-\cos\gamma\right). \tag{7.84}$$

With reference to the center of the section $C-D$, the lever arms of these forces are

$$a = \left(r+\frac{s}{2}\right)\sin\gamma - \frac{r\sin\gamma}{2} = \frac{r+s}{2}\sin\gamma, \tag{7.85}$$

and

$$b = \left(r+\frac{s}{2}\right)\cos\gamma - \frac{r}{2}\left(1+\cos\gamma\right) = \frac{r+s}{2}\cos\gamma - \frac{r}{2}. \tag{7.86}$$

In correspondence of section $C-D$ the bending moment is equal to

$$M = M_0 + T_0\left(r+\frac{s}{2}\right)\sin\gamma - N_0\left(r+\frac{s}{2}\right)\left(1-\cos\gamma\right) + F_1 a - F_2 b. \tag{7.87}$$

Based on the previous equations and through a series of steps we obtain

$$M = M_0 + p\left(m-r\right)\left(r+\frac{s}{2}\right)\sin\gamma - pm\left(r+\frac{s}{2}\right)\left(1-\cos\gamma\right) + pr\left(r+\frac{s}{2}\right)\left(1-\cos\gamma\right), \tag{7.88}$$

and, finally,

$$M = M_0 + p\left(m-r\right)\left(r+\frac{s}{2}\right)\left(\sin\gamma+\cos\gamma-1\right). \tag{7.89}$$

Now we introduce the following dimensionless quantities

$$\alpha = \frac{r}{m}, \tag{7.90}$$

and

$$\beta = \frac{r}{s}. \tag{7.91}$$

From (7.89) we have

$$M = M_0 + pm^2\alpha(1-\alpha)\left(1+\frac{1}{2\beta}\right)(\sin\gamma+\cos\gamma-1). \tag{7.92}$$

Since the vessel has equal sides, the section of the knuckle corresponding to $\gamma = 45°$ does not rotate. Therefore, it is of interest to determine which rotation is for $A-B$ with respect to the latter. To accomplish this, we do an approximate calculation using the criteria of the bent beam, in spite of the shape of the knuckle and its chunky shape. The rotation ϑ of section $A-B$ is therefore given by

$$\vartheta = -\int_0^{\pi/4}\frac{M}{EI}\left(r+\frac{s}{2}\right)d\vartheta = -\frac{r+\frac{s}{2}}{EI}\int_0^{\pi/4}M\,d\vartheta, \tag{7.93}$$

where K is the integral in (7.93). It is given by

$$K = \frac{\pi}{4} M_0 + pm^2 \alpha (1 - \alpha) \left(1 + \frac{1}{2\beta} \right) \left(1 - \frac{\pi}{4} \right), \tag{7.94}$$

or

$$K = 0.7854 M_0 + 0.2146 pm^2 \alpha (1 - \alpha) \left(1 + \frac{1}{2\beta} \right). \tag{7.95}$$

On the other hand, recalling (7.56) and factoring in that the length of the portion with a straight axis of the vessel is equal to $2(m - r)$, we have

$$M'_0 = 2E \frac{I}{2(m - r)} \vartheta = \frac{EI}{m - r} \vartheta. \tag{7.96}$$

Based on (7.90), (7.91), and (7.93)

$$M'_0 = 2E \frac{I}{2(m - r)} \vartheta = \frac{EI}{(m - r)} \vartheta. \tag{7.97}$$

The moment of perfectly clamped edge for the portion with length $2(m - r)$ is equal to

$$\overline{M}_0 = \frac{1}{3} p (m - r)^2 = \frac{1}{3} pm^2 (1 - \alpha)^2. \tag{7.98}$$

Moreover, based on (7.59)

$$M_0 = \overline{M}_0 - M'_0 = \frac{1}{3} pm^2 (1 - \alpha)^2 - \alpha \frac{1 + \dfrac{1}{2\beta}}{1 - \alpha} K. \tag{7.99}$$

Finally, recalling (7.95)

$$M_0 \left(1 + 0.7854 \alpha \frac{1 + \dfrac{1}{2\beta}}{1 - \alpha} \right) = \frac{1}{3} pm^2 (1 - \alpha)^2 - 0.2146 pm^2 \alpha^2 \left(1 + \frac{1}{2\beta} \right)^2. \tag{7.100}$$

Thus we obtain the value of the moment M_0 given by

$$M_0 = \frac{1}{3} pm^2 \frac{(1 - \alpha)^2 - 0.6438 \alpha^2 \left(1 + \dfrac{1}{2\beta} \right)^2}{\left(1 + 0.7854 \alpha \dfrac{1 + \dfrac{1}{2\beta}}{1 - \alpha} \right)}. \tag{7.101}$$

The maximum moment in the knuckle M_1 occurs for $\gamma = 45°$ and, based on (7.92), is equal to

$$M_1 = M_0 + pm^2\alpha(1-\alpha)\left(1+\frac{1}{2\beta}\right)\left(\sqrt{2}-1\right). \tag{7.102}$$

In conclusion, we have

$$M_1 = \frac{1}{3}pm^2 \left[\frac{(1-\alpha)^2 - 0.6438\alpha^2\left(1+\frac{1}{2\beta}\right)^2}{1+0.7854\alpha\frac{1+\frac{1}{2\beta}}{1-\alpha}} + 1.2426\alpha\,(1-\alpha)\left(1+\frac{1}{2\beta}\right)\right]. \tag{7.103}$$

We now introduce the quantity k_1

$$k_1 = \left[\frac{(1-\alpha)^2 - 0.6438\alpha^2\left(1+\frac{1}{2\beta}\right)^2}{1-0.2146\alpha+0.7854\frac{\alpha}{2\beta}} + 1.2426\alpha\left(1+\frac{1}{2\beta}\right)\right](1-\alpha), \tag{7.104}$$

and we have

$$M_1 = \frac{1}{3}k_1 pm^2. \tag{7.105}$$

Note that based on (7.66) with $m = n$, the quantity k_1 represents the ratio between the moment M_1 computed according to the criteria discussed above as well as the moment according to the simplified method presented at the beginning of the section.

The quantity k_1 is shown in Fig. 7.9. As far as the center line of the vessel is concerned, note that the relative moment indicated with M_2 is given by

$$M_2 = M_0 - \frac{1}{2}p\,(m-r)^2 = M_0 - \frac{pm^2}{2}\,(1-\alpha)^2. \tag{7.106}$$

We introduce the dimensionless quantity k_2 given by

$$k_2 = \left[3\,(1-\alpha) - 2\frac{(1-\alpha)^2 - 0.6438\alpha^2\left(1+\frac{1}{2\beta}\right)^2}{1-0.2146\alpha+0.7854\frac{\alpha}{2\beta}}\right](1-\alpha). \tag{7.107}$$

The moment M_2 may be obtained from the following equation:

$$M_2 = -\frac{1}{6}k_2 pm^2. \tag{7.108}$$

The quantity k_2 represents the ratio between the moment in the center line computed according to the criteria above and the moment M_c obtained from (7.73) with $n = m$ and $c = 0$. This is therefore the moment in the center line

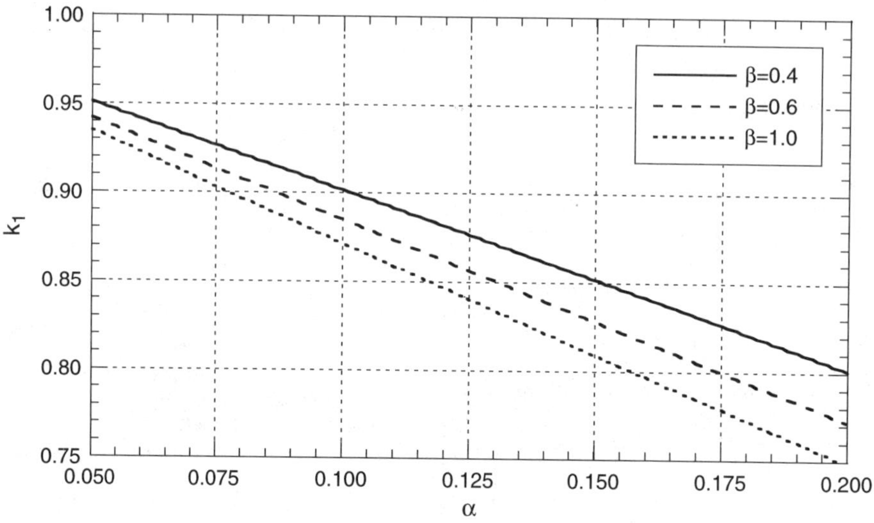

Fig. 7.9

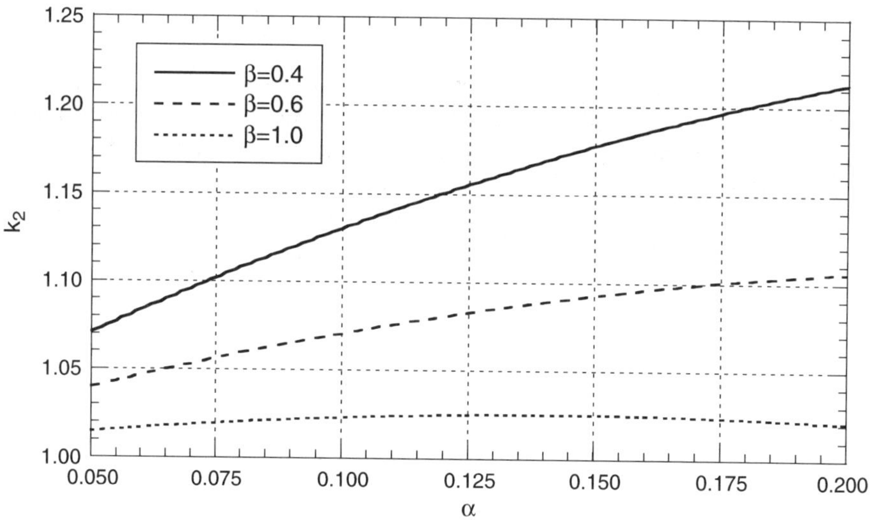

Fig. 7.10

computed according to the simplified method presented at the beginning of the section.

The parameter k_2 is shown in Fig. 7.10. Looking at the figure, we see the beneficial impact of the knuckle on the moment in the corner that is also crucial for verification, since k_1 is smaller than one.

The quantity k_2 is greater than one instead, and one can see that the value of the moment is greater in the center line than the one resulting from the simplified calculation that does not factor in the knuckle. The phenomenon is not worrisome for vessels without holes, since in absolute terms the value is always $M_1 > M_2$. If there are drills at the center of the vessel, special considerations must be made. We will discuss this in Chap. 8.

Figure 7.11 shows the bending moment for a vessel with $\alpha = 0.15$ and $\beta = 0.6$. The dashed line indicates the bending moment that corresponds to the calculation criteria the knuckle notwithstanding. To complete the investigation, one must consider the value of the normal force N at the knuckle for $\gamma = 45°$.

We see that in the knuckle

$$N = (N_0 - F_2)\cos\gamma + (T_0 - F_1)\sin\gamma. \tag{7.109}$$

Recalling (7.81), (7.82), (7.83), and (7.84), we have

$$N = p\,(m - r)\,(\sin\gamma + \cos\gamma) + pr. \tag{7.110}$$

For $\gamma = 45°$ we have

$$N_1 = 1.414p\,(m - r) + pr. \tag{7.111}$$

We introduce the quantity k_3 given by

$$k_3 = 1.414 - 0.414\alpha, \tag{7.112}$$

and we have

$$N_1 = k_3 pm. \tag{7.113}$$

Based on our calculation criterion, the sizing of the vessel with equal sides can therefore be done using the following equation instead of (7.80)

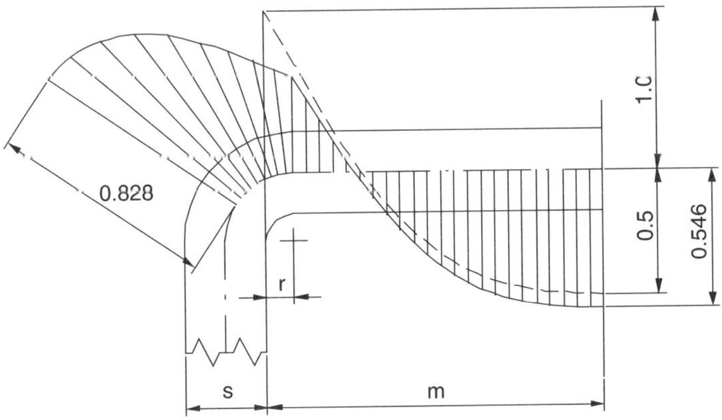

Fig. 7.11

$$s = 2m\left(\frac{pk_3}{6f} + \sqrt{\frac{pk_1}{3f}}\right). \tag{7.114}$$

One may argue that in correspondence of the knuckle there are stress peaks that the suggested simplified calculation, based on the hypothetical proportionality in the knuckle between the stresses generated by the bending moment and the distance from the neutral axis, does not take into account. In fact, these peaks are present and may locally cause plastic flow of limited proportions. One must also keep in mind that these are secondary stresses, according to the classification presented in Sect. 1.5. The phenomenon does not endanger the resistance of the vessel. At the most, it may slightly increase the moment in the center line because of the greater than expected deformation of the knuckle. The quadrangular vessels may have one or more baffles. Besides allowing the distribution of the fluid if it is functionally required, they reduce the value of the stresses along the side with the largest width.

Generally, the pressure is the same on both sides of the baffle. In that case, the latter is subject to the axial force generated by the pressure on the walls that are connected to the baffle. Moreover, the bending moments at the ends generated for congruence by the potential rotation are active. If there is only a single central baffle, this rotation is zero. Moreover, considering that the thickness of the baffle is usually a lot less than that of the walls of the vessel, these moments are of limited magnitude compared to those along the walls. Consequently, it is possible to simplify the calculation by considering the baffles as tie rods of ideal zero rigidity.

Different calculation criteria may be chosen to determine the value of the moments present along the walls of the vessel, e.g., Clapeyron's equation of the three moments. We prefer the calculation methodology typical of frames, of course exploiting the conditions of symmetry to simplify the procedure. Specifically, we present the procedure relative to a vessel with only one baffle in detail.

We refer to Fig. 7.12 that shows the section of the vessel. We examine a portion of it of unitary length, and we consider the clockwise moments as positive. The moment M_{23} in knot 2 relative to beam $2-3$ is given by

$$M_{23} = \overline{M}_{23} + 2M'_{23} + M'_{32}, \tag{7.115}$$

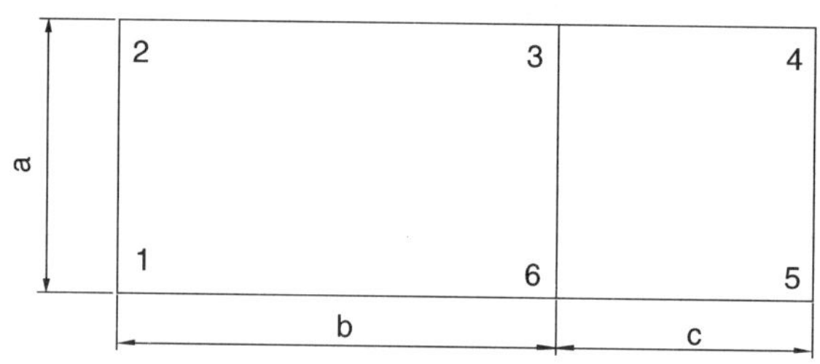

Fig. 7.12

where $\overline{M}_{23}$ is the corresponding moment of perfectly clamped edge, while

$$\begin{aligned} M'_{23} &= 2EK_{23}\vartheta_2 \\ M'_{32} &= 2EK_{23}\vartheta_3 \end{aligned} \tag{7.116}$$

and ϑ_2 and ϑ_3 are the rotations of knots 2 and 3, respectively, while K_{23} is the ratio between the moment of inertia of the beam and its length, that is:

$$K_{23} = \frac{I_{23}}{b} = \frac{s_{23}^3}{12b}, \tag{7.117}$$

with s_{23} being the thickness of the wall.

By analogy

$$M_{21} = \overline{M}_{21} + 2M'_{21} + M'_{12}. \tag{7.118}$$

On the other hand, for the sake of symmetry

$$M'_{12} = -M'_{21}. \tag{7.119}$$

Moreover, since the moments M' are proportional to the value of K of the beam we are examining in any given knot

$$M'_{21} = \frac{K_{12}}{K_{23}} M'_{23}. \tag{7.120}$$

For the equilibrium of knot 2 we must have

$$M_{21} + M_{23} = 0; \tag{7.121}$$

and based on (7.115), (7.118), (7.119), and (7.120) including a series of steps, we obtain

$$\overline{M}_{21} + \overline{M}_{23} + \left(2 + \frac{K_{21}}{K_{23}}\right) M'_{23} + M'_{32} = 0. \tag{7.122}$$

From now on the sum of the moments with perfectly clamped edge relative to the beams converging in a knot will be indicated by the symbol $M*$. Therefore, for knot 2 we have

$$M_2^* = \overline{M}_{21} + \overline{M}_{23}, \tag{7.123}$$

and similarly for the other knots.

We may write

$$\left(2 + \frac{K_{21}}{K_{23}}\right) M'_{23} + M'_{32} = -M_2^*. \tag{7.124}$$

Similarly, for knot 4 we have

$$\left(2 + \frac{K_{45}}{K_{34}}\right) M'_{43} + M'_{34} = -M_4^*. \tag{7.125}$$

For the equilibrium of knot 3 and considering the rigidity of the baffle to be zero, we must have

$$M_3^* + 2M'_{32} + 2M'_{34} + M'_{23} + M'_{43} = 0. \qquad (7.126)$$

On the other hand

$$M'_{34} = \frac{K_{34}}{K_{23}} M'_{32}. \qquad (7.127)$$

Equation (7.126) may also be written as

$$M'_{23} + 2\left(1 + \frac{K_{34}}{K_{23}}\right) M'_{32} + M'_{43} = -M_3^*, \qquad (7.128)$$

and (7.125) may be written as

$$\frac{K_{34}}{K_{23}} M'_{32} + \left(2 + \frac{K_{45}}{K_{34}}\right) M'_{43} = -M_4^*. \qquad (7.129)$$

Equations (7.124), (7.128), and (7.129) constitute a system of three equations in three unknowns (M'_{23}, M'_{32}, and M'_{43}). The determinant of the system is given by

$$\Delta = \left(2 + \frac{K_{45}}{K_{34}}\right)\left(3 + 2\frac{K_{12}}{K_{23}}\right) + \left(2 + \frac{K_{12}}{K_{23}}\right)\left(3\frac{K_{34}}{K_{23}} + 2\frac{K_{45}}{K_{23}}\right). \qquad (7.130)$$

If the thickness is the same for all walls of the vessel, (7.130) may be simplified as follows (see Fig. 7.12)

$$\Delta = \left(2 + \frac{c}{a}\right)\left(3 + 2\frac{b}{a}\right) + \left(2 + \frac{b}{a}\right)\left(3\frac{b}{c} + 2\frac{b}{a}\right). \qquad (7.131)$$

The characteristic determinant of the unknown M'_{23} is given by the following equation:

$$\Delta_1 = -\left(4 + 3\frac{K_{34}}{K_{23}} + 2\frac{K_{45}}{K_{23}} + 2\frac{K_{45}}{K_{34}}\right) M_2^* + \left(2 + \frac{K_{45}}{K_{34}}\right) M_3^* - M_4^*. \qquad (7.132)$$

In case of constant thickness, the latter may be written as follows:

$$\Delta_1 = -\left(4 + 3\frac{b}{c} + 2\frac{b}{a} + 2\frac{c}{a}\right) M_2^* + \left(2 + \frac{c}{a}\right) M_3^* - M_4^*; \qquad (7.133)$$

hence

$$M'_{23} = \frac{\Delta_1}{\Delta}. \qquad (7.134)$$

In addition, based on (7.124)

$$M'_{32} = -M_2^* - \left(2 + \frac{K_{12}}{K_{23}}\right) M'_{23}, \qquad (7.135)$$

or, in case of constant thickness,

$$M'_{32} = -M_2^* - \left(2 + \frac{b}{a}\right) M'_{23}. \tag{7.136}$$

From (7.127) that we rewrite

$$M'_{34} = \frac{K_{34}}{K_{23}} M'_{32}, \tag{7.137}$$

or, in case of constant thickness

$$M'_{34} = \frac{b}{c} M'_{32}. \tag{7.138}$$

From (7.128)

$$M'_{43} = -M_3^* - M'_{23} - 2\left(1 + \frac{K_{34}}{K_{23}}\right) M'_{32}, \tag{7.139}$$

or, in case of constant thickness

$$M'_{43} = -M_3^* - M'_{23} - 2\left(1 + \frac{b}{c}\right) M'_{32}. \tag{7.140}$$

Once M'_{23}, M'_{32}, M'_{34} and M'_{43} are known, one easily obtains the moments on the knots 2, 3, and 4 that, because of symmetry and with the opposite sign, correspond to the moments on the knots 1, 6, and 5.

In fact

$$M_{23} = \overline{M}_{23} + 2M'_{23} + M'_{32} \tag{7.141}$$
$$M_{34} = \overline{M}_{34} + 2M'_{34} + M'_{43} \tag{7.142}$$
$$M_{43} = \overline{M}_{43} + 2M'_{43} + M'_{34} \tag{7.143}$$

Finally, note that referring the moments to a portion of the vessel of unitary width, and indicating the pressure with p, in relation with Fig. 7.12 we have

$$\overline{M}_{45} = -\overline{M}_{21} = \frac{pa^2}{12} \tag{7.144}$$
$$\overline{M}_{23} = -\overline{M}_{32} = \frac{pb^2}{12} \tag{7.145}$$
$$\overline{M}_{34} = -\overline{M}_{43} = \frac{pc^2}{12} \tag{7.146}$$
$$M_2^* = p\frac{b^2 - a^2}{12} \tag{7.147}$$
$$M_3^* = p\frac{c^2 - b^2}{12} \tag{7.148}$$
$$M_4^* = p\frac{a^2 - c^2}{12} \tag{7.149}$$

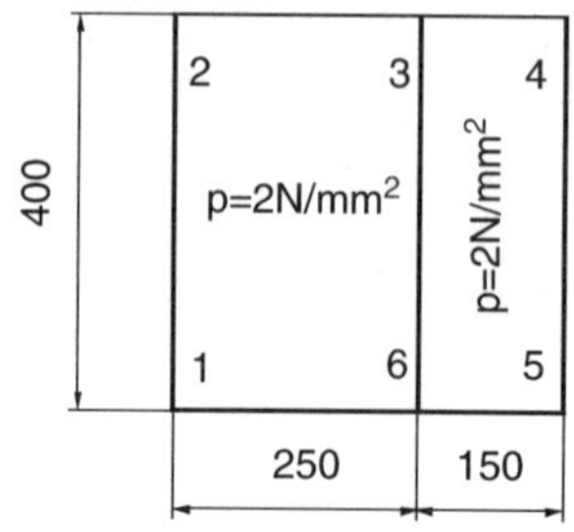

Fig. 7.13

For example, let us examine the vessel with constant thickness and pressure $2\,\mathrm{N\,mm^{-2}}$ (20 bar) shown in Fig. 7.13.

We have

$$a = 400\,\mathrm{mm}$$
$$b = 250\,\mathrm{mm}$$
$$c = 150\,\mathrm{mm}$$
$$\overline{M}_{23} = -\overline{M}_{32} = 10417\,\mathrm{N\,mm\,mm^{-1}}$$
$$\overline{M}_{34} = -\overline{M}_{43} = 3750\,\mathrm{N\,mm\,mm^{-1}}$$
$$M_2^* = -16250\,\mathrm{N\,mm\,mm^{-1}}$$
$$M_3^* = -6667\,\mathrm{N\,mm\,mm^{-1}}$$
$$M_4^* = 22917\,\mathrm{N\,mm\,mm^{-1}}$$

From (7.131) we obtain

$$\Delta = 26.5$$

From (7.133)

$$\Delta_1 = -11\,(-16250) + 2.375\,(-6667) - 22917 = 140000$$

and then

$$M'_{23} = \frac{140000}{26.5} = 5283\,\mathrm{N\,mm\,mm^{-1}}$$
$$M'_{32} = 16250 - 2.625 \times 5283 = 2382\,\mathrm{N\,mm\,mm^{-1}}$$
$$M'_{34} = 1.667 \times 2382 = 3970\,\mathrm{N\,mm\,mm^{-1}}$$
$$M'_{43} = 6667 - 5283 - 5.333 \times 2382 = -11320\,\mathrm{N\,mm\,mm^{-1}}$$
$$M_{23} = 10417 + 2 \times 5283 + 2382 = 23365\,\mathrm{N\,mm\,mm^{-1}}$$
$$M_{34} = 3750 + 2 \times 3970 - 11320 = 370\,\mathrm{N\,mm\,mm^{-1}}$$
$$M_{43} = -3750 + 2(-11320) + 3970 = -22420\,\mathrm{N\,mm\,mm^{-1}}$$

To double-check the results note that

$$\overline{M}_{21} = -2 \times \frac{400^2}{12} = -26667\,\mathrm{N\,mm\,mm^{-1}}$$
$$M'_{21} = 5283 \times \frac{250}{400} = 3301\,\mathrm{N\,mm\,mm^{-1}}$$
$$M'_{12} = -M'_{21} = -3301\,\mathrm{N\,mm\,mm^{-1}}$$
$$M_{21} = -26667 + 2 \times 3301 - 3301 = -23366\,\mathrm{N\,mm\,mm^{-1}}$$

Aside from the sign, this practically coincides with the value of M_{23}. Moreover

$$M_{32} = -10417 + 2 \times 2382 + 5283 = -370\,\mathrm{N\,mm\,mm^{-1}}.$$

Again, aside from the sign, this coincides with the value of M_{34}.

Finally

$$\overline{M}_{45} = 26667\,\mathrm{N\,mm\,mm^{-1}}$$
$$M'_{45} = -11320 \times \frac{150}{400} = -4245\,\mathrm{N\,mm\,mm^{-1}}$$
$$M'_{54} = -M'_{45} = 4245\,\mathrm{N\,mm\,mm^{-1}}$$
$$M_{45} = 26667 - 2 \times 4245 + 4245 = 22422\,\mathrm{N\,mm\,mm^{-1}}$$

Once more, aside from the sign, it coincides with the value of M_{43}.

It is interesting to notice that if the thickness of the baffle is equal to half the one of the walls of the vessel, the bending moment on the ends may be

$$M_{36} = \frac{250}{400}\left(\frac{1}{2}\right)^3 2382 = 186\,\mathrm{N\,mm\,mm^{-1}}.$$

This is clearly an irrelevant value when compared with the value of the moments M_{23} and M_{43}. The fact that it was ignored during calculations (assuming an ideal rigidity of the baffle to be zero) does not influence the validity of the results.

In any case, since $M_{36} = M'_{36}$ because of symmetry, to take into account the rigidity of the baffle one simply has to modify (7.128) as

$$M'_{23} + 2\left(1 + \frac{K_{34}}{K_{23}} + \frac{1}{2}\frac{K_{36}}{K_{23}}\right) M'_{32} + M'_{43} = -M_3^*, \tag{7.150}$$

and, consequently, to modify (7.130), (7.132), and (7.139), as well.

In addition, we have

$$M_{36} = \frac{K_{36}}{K_{23}} M'_{32}. \tag{7.151}$$

Finally, if there is more than one baffle, one can proceed in a way similar to what presented thus far. The computation process is more time-consuming, but not difficult at all. Obviously, it is possible to use one of the well-known calculation methods of successive approximations, for instance the one by Cross and Kani if the calculation is manual. Actually, a number of programs are available to design vessels with a high number of baffles.

7.4 Flanges

The great popularity of flanges used as connectors between pressure vessels and tubes notwithstanding, there is no relatively simple sizing criteria available that experts can agree on. For this reason, flanges used in one country show quite substantial differences in thickness compared to those used in another country. This depends on the fact that the calculation of the state of stress in a flange is rather complex, and only the introduction of drastic oversimplifications makes it possible to set up an easy calculation. The latter, however, may cause a great deal of perplexity as far as its validity, exactly because of the very simplifications introduced.

We would like to suggest a calculation method that is a compromise between an absolutely rigorous criterion and one that is too rudimental. We will introduce unavoidable simplifications, but the calculation procedure will be based on theoretically rigorous criteria already presented in Sect. 2.2 for edge effects, and in Sect. 6.4 for dished heads.

Consider the flange in Fig. 7.14, already showing a few simplifications. The flange is the coupling between a cylinder with constant thickness s_c (at the end of the section we will discuss the possibility to extend the calculation to flanges with a tapered tang) and a circular crown of thickness s_f. Moreover, the external radius of the crown is set to correspond to the radius of the drilling. In other words, we are considering a circular crown without drilling. Force F originating from the bolts is applied to the edge of the crown.

We assume here that the reaction of the gasket is not applied to the crown. This way, the latter is subjected to the force F throughout its extension. We introduce another simplification by assuming that the deformations in the area of the flange, simultaneously belonging to the crown and to the ideal continuation of the cylinder, may be ignored. Both coupled elements may be modeled as shown in Fig. 7.15. Given the congruence of deformations,

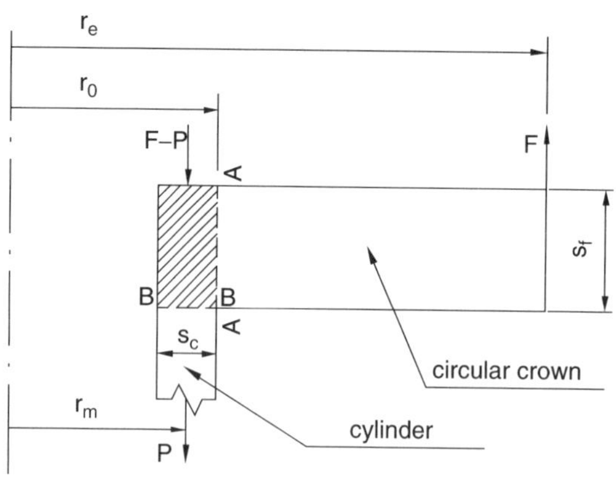

Fig. 7.14

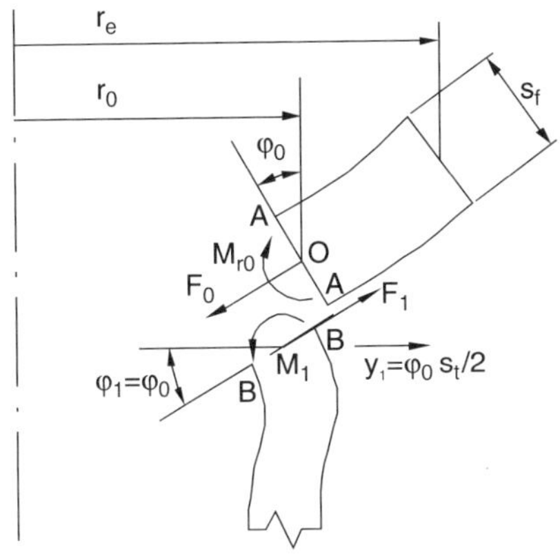

Fig. 7.15

the radial moment M_{ro} and the force F_o are generated in correspondence of section $A - A$ (and, later on, a circumferential moment M_{co}). By ignoring the deformations in the area of the flange limited by the sections $A - A$ and $B - B$, if section $A - A$ is subject to a rotation φ_0, the same rotation is present in section $B - B$. Finally, as part of the simplification, we assume that O (on the neutral axis of section $A - A$) is not subjected to considerable radial displacement, given the great rigidity of the crown.

Under this assumption, section $B - B$ is subjected to a radial displacement equal to

$$y_1 = \varphi_0 \frac{s_f}{2}. \tag{7.152}$$

Let us now examine the circular crown in view of Sect. 6.4 and also of the following differential equation (6.109) that we rewrite as

$$\frac{d^2\varphi}{dr^2} + \frac{1}{r}\frac{d\varphi}{dr} - \frac{\varphi}{r^2} = -\frac{T}{K}; \tag{7.153}$$

in this case we have

$$T = \frac{F}{2\pi r}; \tag{7.154}$$

and, recalling by analogy equation (6.112), we may write

$$\frac{d}{dr}\left[\frac{1}{r}\frac{d}{dr}(r\varphi)\right] = -\frac{F}{2\pi K r}. \tag{7.155}$$

Integrating once, we obtain

$$\frac{d}{dr}(r\varphi) = -\frac{F}{2\pi K} r \log_e r + C_1 r; \tag{7.156}$$

and integrating a second time

$$r\varphi = -\frac{F}{2\pi K}\left(\frac{r^2}{2}\log_e r - \frac{r^2}{4}\right) + C_1\frac{r^2}{2} + C_2, \tag{7.157}$$

where C_1 and C_2 are two constants of integration.

Finally, from (7.157) we have

$$\varphi = -\frac{Fr}{8\pi K}(2\log_e r - 1) + C_1\frac{r}{2} + \frac{C_2}{r}. \tag{7.158}$$

By deriving (7.158), we obtain

$$\frac{d\varphi}{dr} = -\frac{F}{8\pi K}(2\log_e r + 1) + \frac{C_1}{2} - \frac{C_2}{r^2}. \tag{7.159}$$

Recalling (6.99), through a series of steps we obtain the following equations of the radial moment M_r and the circumferential moment M_c

$$M_r = -\frac{F}{8\pi}[2\log_e r(1+\mu) + 1 - \mu] + K\left[\frac{C_1}{2}(1+\mu) - \frac{C_2}{r^2}(1-\mu)\right]; \tag{7.160}$$

$$M_c = -\frac{F}{8\pi}[2\log_e r(1+\mu) - (1-\mu)] + K\left[\frac{C_1}{2}(1+\mu) + \frac{C_2}{r^2}(1-\mu)\right]. \tag{7.161}$$

By adding (7.161) to (7.160) and dividing by 2, we obtain

$$\frac{M_r + M_c}{2} = -\frac{F}{4\pi}(1+\mu)\log_e r + K(1+\mu)\frac{C_1}{2}; \tag{7.162}$$

then

$$\frac{KC_1}{2} = \frac{M_r + M_c}{2(1+\mu)} + \frac{F}{4\pi}\log_e r. \tag{7.163}$$

By subtracting (7.161) from (7.160) and dividing by 2, we obtain

$$\frac{M_r - M_c}{2} = -\frac{F}{8\pi}(1-\mu) - K(1-\mu)\frac{C_2}{r^2}; \tag{7.164}$$

then

$$K\frac{C_2}{r^2} = -\frac{M_r - M_c}{2(1-\mu)} - \frac{F}{8\pi}. \tag{7.165}$$

The moments M_{ro} and M_{co} correspond to section $A-A$ with $r = r_0$. Based on the equations above, one may write

$$\frac{KC_1}{2} = \frac{M_{r0} + M_{c0}}{2(1+\mu)} + \frac{F}{4\pi}\log_e r_0, \tag{7.166}$$

and

$$K\frac{C_2}{r_0^2} = -\frac{M_{r0} - M_{c0}}{2(1-\mu)} - \frac{F}{8\pi}. \tag{7.167}$$

Recalling (7.160) in relation to (7.166) and (7.167), through a series of steps we obtain the following equation:

$$\begin{aligned} M_r =& \frac{F}{8\pi}\left[(1+\mu)\log_e\left(\frac{r_0}{r}\right)^2 + (1-\mu)\left(\frac{r_0^2}{r^2} - 1\right)\right] \\ &+ \frac{M_{r0} + M_{c0}}{2} + \frac{M_{r0} - M_{c0}}{2}\left(\frac{r_0}{r}\right)^2 \end{aligned} \tag{7.168}$$

By analogy, from (7.161) we obtain

$$\begin{aligned} M_c =& \frac{F}{8\pi}\left[(1+\mu)\log_e\left(\frac{r_0}{r}\right)^2 - (1-\mu)\left(\frac{r_0^2}{r^2} - 1\right)\right] \\ &+ \frac{M_{r0} + M_{c0}}{2} - \frac{M_{r0} - M_{c0}}{2}\left(\frac{r_0}{r}\right)^2 . \end{aligned} \tag{7.169}$$

We introduce the dimensionless quantity ρ given by

$$\rho = \frac{r_0}{r}, \tag{7.170}$$

and, assuming that $\mu = 0.3$, we finally have

$$M_r = \frac{M_{r0} + M_{c0}}{2} + \frac{M_{r0} - M_{c0}}{2}\rho^2 - \frac{F}{8\pi}\left[1.3\log_e\rho^2 - 0.7\left(1-\rho^2\right)\right]; \tag{7.171}$$

$$M_c = \frac{M_{r0} + M_{c0}}{2} - \frac{M_{r0} - M_{c0}}{2}\rho^2 - \frac{F}{8\pi}\left[1.3\log_e\rho^2 + 0.7\left(1-\rho^2\right)\right]. \tag{7.172}$$

From (6.99)

$$M_c - \mu M_r = \frac{K\varphi}{r}\left(1-\mu^2\right). \tag{7.173}$$

From (7.171) and (7.172) we obtain

$$\frac{K\varphi}{r} = \frac{M_{r0} + M_{c0}}{2.6} + \frac{M_{r0} - M_{c0}}{1.4}\rho^2 + \frac{F}{8\pi}\left[\log_e\rho^2 + 1 - \rho^2\right]; \tag{7.174}$$

specifically, for $r = r_0$ and given that $\rho = 1$, we obtain

$$\frac{K\varphi_0}{r_0} = \frac{M_{r0} + M_{c0}}{2.6} + \frac{M_{r0} - M_{c0}}{1.4} = 1.099M_{c0} - 0.3297M_{r0}. \tag{7.175}$$

A relationship has been established between the rotation φ_0 of section $A - A$ and the moments M_{ro} and M_{co} that are present in that section.

Let us now examine the cylinder with force F_1 and moment M_1 on the edge of it. From (2.66) and (2.67), introducing the average radius r_m of the cylinder and indicating the rotation at the edge with $\varphi_1(=\varphi_0)$, we have

$$y_1 = 2.57\frac{F_1}{E}\left(\frac{r_m}{s_c}\right)^{3/2} - 3.304\frac{M_1 r_m}{E s_c^2}; \tag{7.176}$$

$$\varphi_1 = \varphi_0 = -3.304\frac{F_1 r_m}{E s_c^2} + 8.495\frac{M_1}{E s_c^2}\sqrt{\frac{r_m}{s_c}}. \tag{7.177}$$

Therefore, from (7.152) we obtain

$$\frac{F_1}{E}\left(\frac{r_m}{s_c}\right)^{3/2}\left(2.57 + 1.652\frac{s_f}{s_c}\sqrt{\frac{s_c}{r_m}}\right) = \frac{M_1 r_m}{E s_c^2}\left(3.304 + 4.247\frac{s_f}{s_c}\sqrt{\frac{s_c}{r_m}}\right). \tag{7.178}$$

Introducing the following dimensionless quantities:

$$\gamma = \frac{s_c}{r_m}, \tag{7.179}$$

and

$$\delta = \frac{s_f}{s_c}. \tag{7.180}$$

From (7.178) we have

$$F_1 = \frac{M_1}{r_m\sqrt{\gamma}}\frac{3.304 + 4.247\delta\sqrt{\gamma}}{2.57 + 1.652\delta\sqrt{\gamma}}. \tag{7.181}$$

Given that forces and moments refer to the length unit, thus factoring in the variation in radius, we have

$$F_1 = F_0\frac{r_0}{r_m} = F_0\left(1 + \frac{\gamma}{2}\right), \tag{7.182}$$

and

$$M_1 = \left(M_{r0} - F_0\frac{s_f}{2}\right)\frac{r_0}{r_m} = \left(M_{r0} - F_0\frac{s_f}{2}\right)\left(1 + \frac{\gamma}{2}\right). \tag{7.183}$$

Then we obtain

$$M_1 = M_{r0}\left(1 + \frac{\gamma}{2}\right) - F_1\frac{s_f}{2}; \tag{7.184}$$

from (7.181), through a series of steps

$$M_1 = M_{r0}\frac{(2.57 + 1.652\delta\sqrt{\gamma})\left(1 + \frac{\gamma}{2}\right)}{2.57 + 3.304\delta\sqrt{\gamma} + 2.123\delta^2\gamma}. \tag{7.185}$$

On the other hand, from (7.177) and (7.181) through different steps we obtain

$$\varphi_0 = \frac{M_1}{E s_c^2\sqrt{\gamma}}\frac{10.916}{2.57 + 1.652\delta\sqrt{\gamma}}; \tag{7.186}$$

then

$$\varphi_0 = \frac{M_{r0}}{Es_c^2\sqrt{\gamma}} \frac{10.916\left(1+\frac{\gamma}{2}\right)}{2.57+3.304\delta\sqrt{\gamma}+2.123\delta^2\gamma}. \quad (7.187)$$

Finally, recalling that

$$K = \frac{EI}{1-\mu^2} = \frac{Es_f^2}{10.916}, \quad (7.188)$$

we obtain

$$\frac{K\varphi_0}{r_0} = \frac{M_{r0}s_f^3}{r_0 s_c^2\sqrt{\gamma}} \frac{1+\frac{\gamma}{2}}{2.57+3.304\delta\sqrt{\gamma}+2.123\delta^2\gamma}. \quad (7.189)$$

In conclusion

$$\frac{K\varphi_0}{r_0} = 0.389M_{r0}\frac{\delta^3\sqrt{\gamma}}{1+1.285\delta\sqrt{\gamma}+0.826\delta^2\gamma}. \quad (7.190)$$

We introduce the following dimensionless quantity:

$$k_1 = 0.354\frac{\delta^3\sqrt{\gamma}}{1+1.285\delta\sqrt{\gamma}+0.826\delta^2\gamma}. \quad (7.191)$$

From (7.175) and (7.190) we obtain

$$1.099M_{c0} - 0.3297M_{r0} = 1.099k_1M_{r0}; \quad (7.192)$$

then

$$M_{c0} = M_{r0}\left(k_1+0.3\right). \quad (7.193)$$

If at the limit $s_f = 0$, the circular crown is perfectly clamped in the cylinder, given its zero rigidity. We have $k_1 = 0$ and $M_{co} = 0.3M_{ro}$, as is to be expected, since $\varphi_0 = 0$ and $M_{co} = \mu M_{ro}$. Finally, from (7.171) and (7.193) we have

$$M_r = M_{r0}\left(\frac{1.3+k_1}{2}+\frac{0.7-k_1}{2}\rho^2\right) - \frac{F}{8\pi}\left[1.3\log_e\rho^2 - 0.7\left(1-\rho^2\right)\right]; \quad (7.194)$$

if we refer to the external radius of the crown r_e and set

$$\alpha = \frac{r_0}{r_e}, \quad (7.195)$$

given that in this position $M_r = 0$, from (7.194) we obtain

$$M_{r0} = \frac{F}{4\pi}\frac{0.7\left(1-\alpha^2\right)-1.3\log_e\alpha^2}{1.3+0.7\alpha^2+k_1\left(1-\alpha^2\right)}. \quad (7.196)$$

Be now

$$k_2 = \frac{1}{8\pi}\frac{0.7\left(1-\alpha^2\right)-1.3\log_e\alpha^2}{1.3+0.7\alpha^2+k_1\left(1-\alpha^2\right)}; \quad (7.197)$$

based on the equations above we obtain

$$M_{r0} = 2k_2F, \tag{7.198}$$

and

$$M_{c0} = 2k_2F\left(k_1 + 0.3\right), \tag{7.199}$$

and, more in general from (7.171) and (7.172)

$$M_r = F\left\{k_2\left[k_1 + 1.3 - \left(k_1 - 0.7\right)\rho^2\right] - \frac{0.7\left(1-\rho^2\right) - 1.3\log_e \rho^2}{8\pi}\right\}; \tag{7.200}$$

$$M_c = F\left\{k_2\left[k_1 + 1.3 + \left(k_1 - 0.7\right)\rho^2\right] + \frac{0.7\left(1-\rho^2\right) + 1.3\log_e \rho^2}{8\pi}\right\}. \tag{7.201}$$

Let us consider the following quantities:

$$k_3 = \frac{\left(1 + 0.643\delta\sqrt{\gamma}\right)\left(1 + \frac{\gamma}{2}\right)}{1 + 1.285\delta\sqrt{\gamma} + 0.826\delta^2\gamma}, \tag{7.202}$$

and

$$k_4 = \frac{1 + 1.285\delta\sqrt{\gamma}}{1 + 0.643\delta\sqrt{\gamma}}; \tag{7.203}$$

from (7.185) we determine that

$$M_1 = k_3 M_{r0} = 2k_2k_3F. \tag{7.204}$$

Moreover, from (7.181) we have

$$F_1 = 1.285k_4\frac{M_1}{\sqrt{r_m s_c}}. \tag{7.205}$$

Recalling (2.53), (2.60), and (2.41), the bending moment in the cylinder at distance x from the edge, is given by

$$M_x = M_1 e^{-\beta x}\left[\cos\beta x - \left(k_4 - 1\right)\sin\beta x\right] \tag{7.206}$$

with

$$\beta = \frac{1.285}{\sqrt{r_m s_c}}. \tag{7.207}$$

The derivative of (7.206) leads to

$$\frac{dM_x}{dx} = -\beta e^{-\beta x}\left[k_4\cos\beta x - \left(k_4 - 2\right)\sin\beta x\right]. \tag{7.208}$$

Setting the derivative to zero, we have

$$\beta x = \frac{k_4}{k_4 - 2}. \tag{7.209}$$

Given that based on (7.203) $k_4 < 2$, the values of $\tan \beta x$ are negative. This means that the maximum value of M_x is at the edge and thus equal to M_1. For $\beta x > \pi/2$ we register a minimum that is generally not of interest for sizing. Also, from (7.206) we determine that the moment becomes zero for

$$\tan \beta x = \frac{1}{k_4 - 1}; \tag{7.210}$$

given that $1 < k_4 < 2$, we have:

$$1.1 < \beta x < \frac{\pi}{2}.$$

Therefore, from (7.207)

$$0.86\sqrt{r_m s_c} < x < 1.22\sqrt{r_m s_c}. \tag{7.211}$$

Practically, we may consider the edge effects negligible at a distance from the edge itself equal to $\sqrt{r_m s_c}$. This conclusion will be useful to suggest an approximate calculation criterion for the flanges with a tapered tang. In summary, we may proceed as follows. Based on the values of γ and δ, one computes the quantities k_1 and k_3. From the values of k_1 and α one computes k_2.

Once k_1 and k_2 are known, one computes the moments M_{ro} and M_{co} that make it possible to verify the circular crown at the junction of the cylinder. Once k_3 is known as well, one computes the moment M_1 (maximum in the cylinder), and based on it and the axial stress from the pressure, the verification at the edge may be completed. Moreover, through (7.200) and (7.201) the stress condition of the entire crown may be checked. The moment M_x can also be computed through k_4 and (7.206) in any given position of the cylinder.

The quantities k_1, k_2, k_3 and k_4 are shown in Figs. 7.16–7.19. If the flange has a tapered tang (see Fig. 7.20), we suggest the use of the following calculation criterion. The radius r_0 corresponds to the junction of the tang; the thickness s_c of the cylinder and the average radius r_m are assumed to be equal to the weighted average of the values along the tang and the cylindrical portion of the flange, as long as the height of the tang and the cylindrical portion is at least equal to $0.8\sqrt{r_m s_c}$; otherwise, one must include a percentage area of the tube to which the flange is connected together with its presumable thickness.

Other than that, one proceeds as explained above but with the following warnings. To verify the tang, one must consider the moment M_1 in reference to the thickness of the tang at the junction of the crown. One must also check that the moment at the beginning of the cylindrical portion (computed through (7.206)) is compatible with the thickness of the same cylindrical portion. In

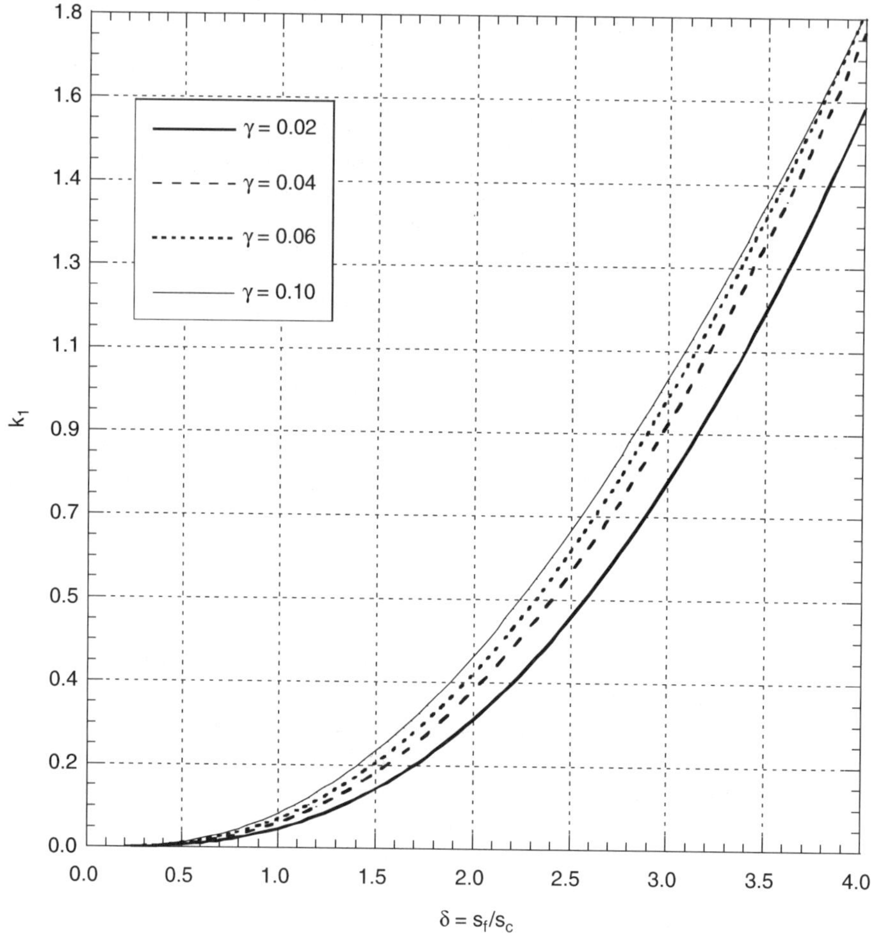

Fig. 7.16

terms of verification criteria, we would like to suggest the following. The hoop stresses in the cylinder (or in the tang) are ignored. In fact, they are relatively modest, given the considerable radial rigidity of the flange, and generally not crucial for the sizing.

For the crown we consider the moments M_{ro} and M_{co}, and compute the corresponding stress based on the greater one. Note that an essential requirement of a flange is that it is not subject to substantial permanent deformations that may compromise the tightness. However, we prefer to rule out plastic deformations in terms of the moments M_{ro} and M_{co}. In other words, as we said before, the calculated stresses must not exceed $1.5f$, i.e., the yield strength. Of course, if this value is reached, given the presence of membrane stresses, the

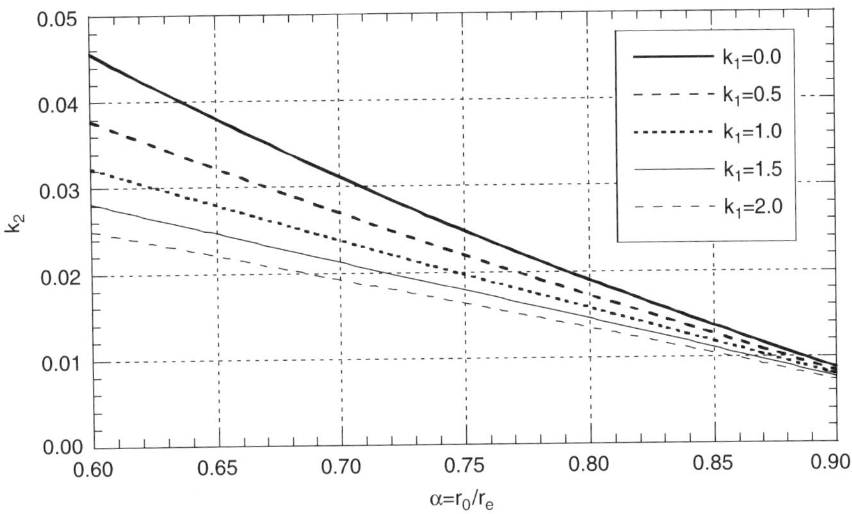

Fig. 7.17

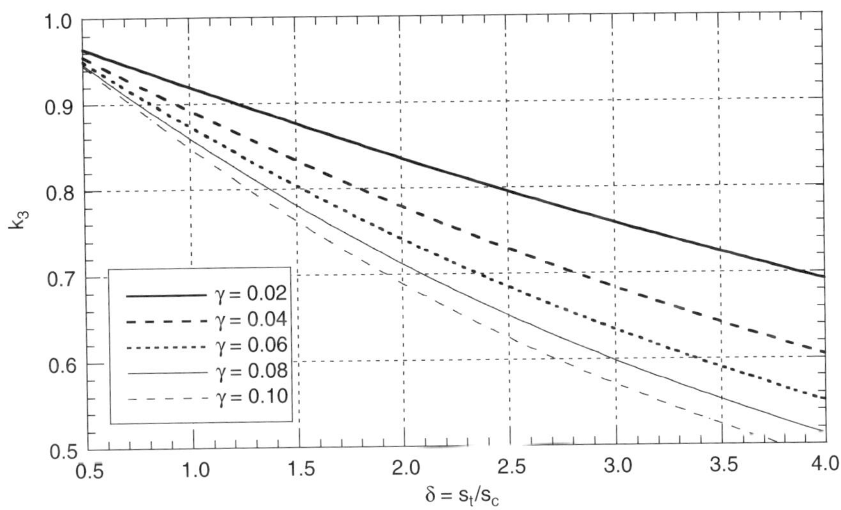

Fig. 7.18

material will be yielded in a few areas. The phenomenon occurs only locally, within a small area, and can be tolerated.

As far as the tang is concerned, based on moment M_1 (in case of tapered tang, the moment that corresponds to the beginning of the cylindrical portion, as well) and the axial stress caused by pressure, we factor in the maximum stress. Even in this case it must not be greater than $1.5f$. About the possibility of limited yielding of the material we refer back to the above. At this point, it

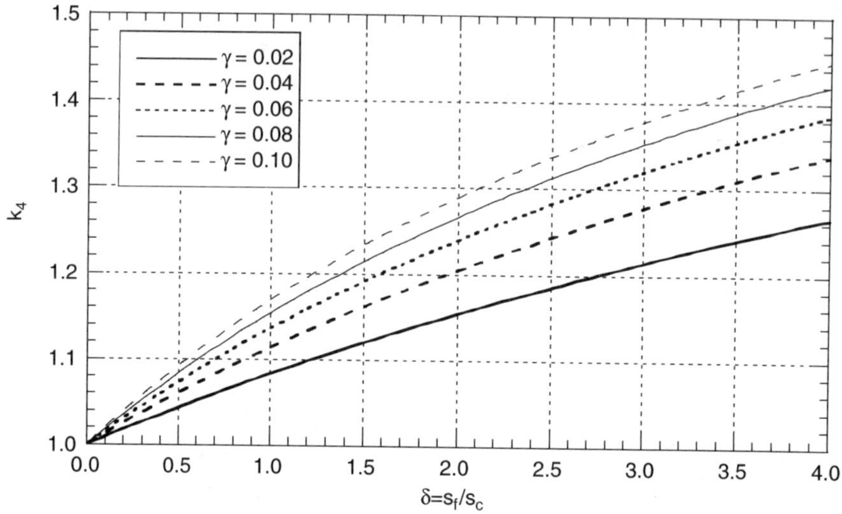

Fig. 7.19

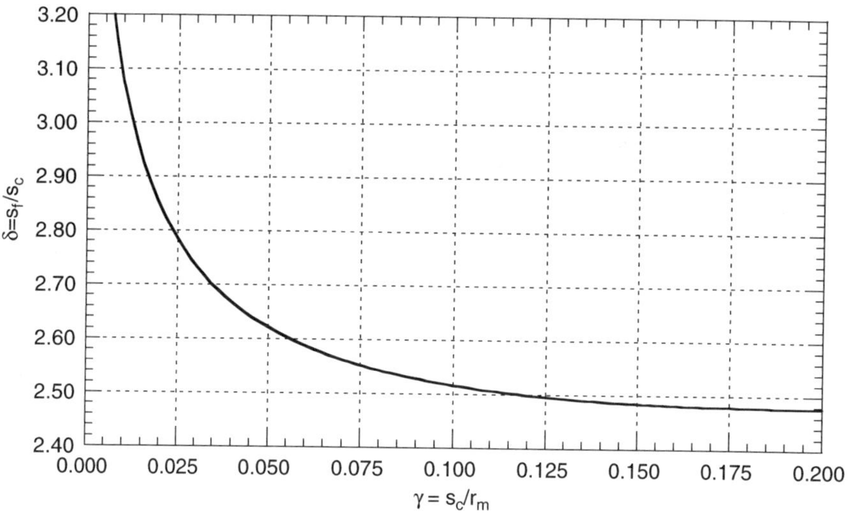

Fig. 7.20

may be interesting to comment on the impact of the different dimensionless quantities on the value of the moments, as well as the impact of potential local plastic flow.

We determine that the value of k_1 is influenced by those of γ and δ. The latter is the ratio between thickness of the crown and thickness of the cylinder, and k_1 rapidly increases as δ grows. Based on (7.197) as k_1 increases, k_2

decreases, and consequently reduces the value of M_{ro}. If the crown is particularly rigid and the rotations are consequently small, the value of the moment at the junction of the cylinder is small, too (including the moment M_1 on the cylinder). We have that for $\delta \to \infty$ also $k_1 \to \infty$; in addition, $k_2 = 0$ and therefore $M_{ro} = 0$.

As far as M_{co}, its value is influenced both by k_1 and k_2; in fact, the influence of k_1 is greater. In other words, if the rigidity of the crown increases, so does the moment M_{co}. As one can see, by increasing the thickness of the crown, the radial moment decreases while the circumferential one increases. Interestingly, both moments coincide if $k_1 = 0.7$. Thus, from (7.191) we can obtain the values of δ as a function of γ according to $M_{co} = M_{ro}$. They are shown in Fig. 7.20 and range between 2.6 and 3 for typical scenarios. Thus, it is possible to obtain a better utilization of the circular crown through a sizing of the flange that corresponds to the values of δ taken from the figure.

Note that one shall also take into account the other phenomenon, i.e., the potential local presence of plastic flow at the junction with the tang caused by circumferential stresses. In this case the rotation φ_0 is greater than the one obtained from (7.190). This corresponds to an ideal increase of δ, even though it is preferable that the actual values of δ are smaller than those obtained from (7.158).

In fact, in doing so if there is a little plastic flow of the flange, the moment M_{ro} decreases in favor of safety for the tang (given that M_1 is a function of M_{ro} through k_3), whereas the moment M_{co} increases. If, however, the value of δ is smaller than the one obtained from Fig. 7.20, it means that $k_1 < 0.7$ and that $M_{co} < M_{ro}$. In other words, there is a certain safety margin against the negative effects of plastic flow on the value M_{co}.

Another criterion to consider for the verification of the flange consists of partially abandoning the principle of elastic behavior of the crown–cylinder structure. More appropriately, one can allow that the material in the cylinder or in the tang of the flange may show signs of plastic flow, the elastic behavior of the crown notwithstanding. In fact, doing the calculations one frequently finds that given a force F, while the stresses in the tang are close to yield strength or ideally greater, in the crown they are indeed considerably less. Therefore, the crown may be more stressed when increasing F, if one accepts the fact that the material in the tang shows signs of plastic flow. Of course, this changes the pattern of deformations in the tang versus the result obtained through calculation in the elastic range. Specifically, the value φ_0 changes, and this must be factored in to define the value of the moments M_{ro} and M_{co}.

For this reason, we plan to illustrate the simplest and yet most appropriate calculation criteria to perform this investigation. The following is the most elementary criterion. We assume that $M_{ro} = 0$. We refer to (7.195) and remember that for $r = r_e$ it must be $M_r = 0$. In that case, from (7.168) we obtain

$$\frac{F}{8\pi}\left[1.3\log_e \alpha^2 + 0.7\left(\alpha^2 - 1\right)\right] + \frac{M_{c0}}{2}\left(1 - \alpha^2\right) = 0; \tag{7.212}$$

then

$$M_{c0} = \frac{F}{4\pi}\left(0.7 - 1.3\frac{\log_e \alpha^2}{1-\alpha^2}\right). \tag{7.213}$$

Setting

$$k_5 = \frac{1}{8\pi}\left(0.7 - 1.3\frac{\log_e \alpha^2}{1-\alpha^2}\right); \tag{7.214}$$

one has

$$M_{c0} = 2k_5F, \tag{7.215}$$

and k_5 is shown in Fig. 7.21. If F is such that the yield strength is exceeded in the tang of the flange (or in the cylinder connected to the crown), it can be considered to be the sum of forces F_A and F_B, where the former induces an ideal stress equal to the yield strength in the most stressed area of the tang.

The same considerations made earlier are applicable in this case. The second force causes the onset of a plastic hinge in the same section. One admits that the tang and the cylinder do not substantially contribute to the rigidity of the flange anymore in favor of resistance. In other words, one allows that the moment M_{ro} caused by force F_B is zero. For this reason, the characteristic moments for the verification of the flange become as follows:

$$M_{r0} = 2k_2F_A, \tag{7.216}$$
$$M_{c0} = 2k_2F_A\,(k_1 + 0.3) + 2k_5F_B, \tag{7.217}$$

and

$$M_1 = 2k_2k_3F_A. \tag{7.218}$$

Incidentally, (7.206) is still valid.

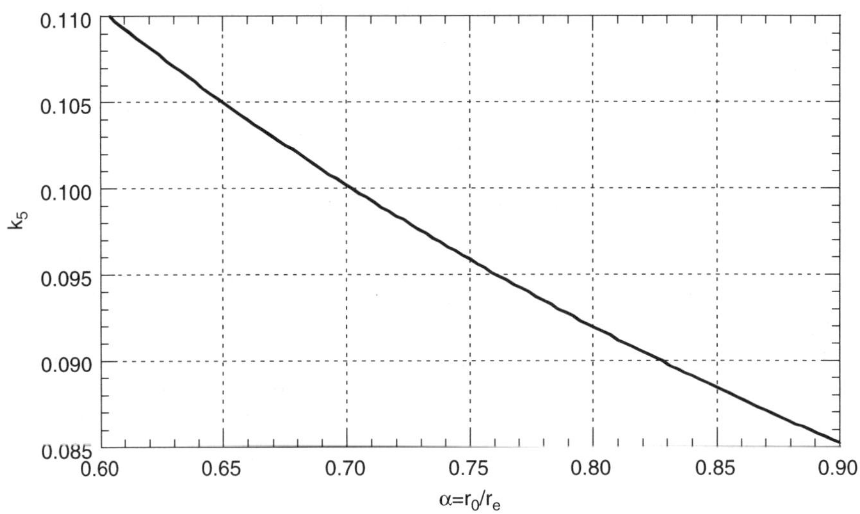

Fig. 7.21

The application of this criterion calls for the determination of F_A; it is also required to check the impact of circumferential stresses, as well. In this context, first of all we observe that the latter have a positive sign, regardless of their origin being related to the deflection y or to internal pressure. As far as Guest–Tresca's failure theory, the stresses do not influence the value σ_{id} in correspondence of the external fibers (of course, if their value is lower than the stresses due to M_x and the axial thrust P, as it is usually the case).

Therefore, verifications should be done as far as internal fibers only. One may neglect the stresses caused by pressure in correspondence of section $B-B$, given the rigidity of the flange, but one cannot ignore the stress due to deflection y_1, as well as the stresses due to the moment μM_1 produced by M_1 out of respect of congruence. As far as the stress due to y_1, note that, through (7.181), it is possible to calculate F_1 from M_1, while the value for y_1 is obtained through (7.176).

The corresponding membrane stress is given by

$$\sigma_t = \frac{E y_1}{r_m}. \tag{7.219}$$

Note that based on (7.205) through a series of steps one obtains the following:

$$\sigma_t = 3.304 \frac{M_1}{s_c^2} \left(k_4 - 1\right). \tag{7.220}$$

For positions other than section $B-B$, based on Sect. 2.2 we establish that the stress σ_t is equal to

$$\sigma_t = 3.304 \frac{M_1}{s_c^2} e^{-\beta x} \left[\left(k_4 - 1\right) \cos \beta x + \sin \beta x\right]. \tag{7.221}$$

It is important, though, to keep in mind that if the section at hand is sufficiently far from section $B-B$, the circumferential stress caused by the pressure may not be neglected. It is safe to assume that the rigidity effect of the circular crown ends at a distance equal to $0.8\sqrt{r_m s_c}$.

At such a distance from section $B-B$, one must consider both σ_t computed through (7.221) and the circumferential stress caused by pressure. Once the maximum ideal stress in the tang is known, the value F_A is equal to computation force F multiplied by the ratio between the yield strength and the maximum stress mentioned previously.

This procedure is particularly conservative since it assumes that once yielding is reached in a section, the moments M_{ro} and M_1 do not increase anymore if F increases. This does not match reality, and we plan to suggest another approximate criterion that better corresponds to the actual phenomenon. Therefore, let us consider a rectangular section showing partial plastic flow (see Fig. 7.22). M' be the moment at the limit of the material's elastic behavior. It is given by

$$M' = \frac{1}{6} s^2 \sigma_s. \tag{7.222}$$

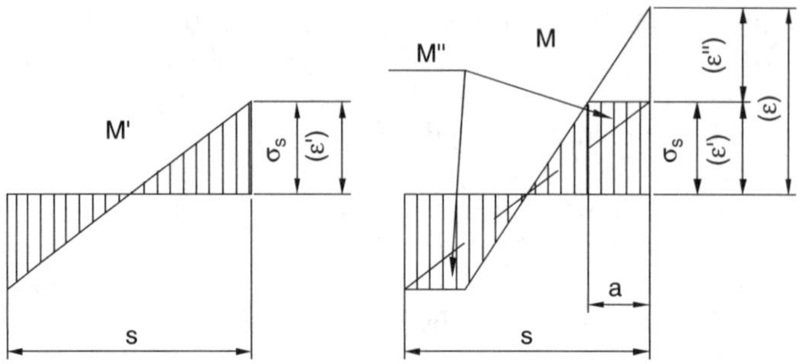

Fig. 7.22

We now consider the generic moment M for a section showing partial plastic flow. We have (see also Fig. 7.22)

$$M = a\,(s-a)\,\sigma_s + \frac{1}{6}\,(s-2a)^2\,\sigma_s. \tag{7.223}$$

Let us consider the moment M as the sum of the elastic moment M' and of a moment M'' caused by plastic flow.

We have

$$M'' = M - M' = \frac{1}{3}s^2\frac{a}{s}\left(1-\frac{a}{s}\right)\sigma_s = 2M'\frac{a}{s}\left(1-\frac{a}{s}\right). \tag{7.224}$$

By analogy, $\varepsilon' = \sigma_s/E$ will be the deformation of the fiber that is most distant from the neutral axis at the limit of elastic behavior.

The deformation ε in case of elastic–plastic behavior is equal to

$$\varepsilon = \varepsilon'\frac{s}{s-2a}.$$

We may consider this deformation as the sum of deformation ε' and a plastic deformation ε'' given by

$$\varepsilon'' = \varepsilon - \varepsilon' = \varepsilon'\frac{\dfrac{2a}{s}}{1-\dfrac{2a}{s}}; \tag{7.225}$$

from (7.224) through a series of steps we obtain

$$\frac{2a}{s} = 1 - \sqrt{1-2\frac{M''}{M'}}; \tag{7.226}$$

and, based on (7.225), we have

$$\frac{\varepsilon''}{\varepsilon'} = \frac{1}{\sqrt{1-2\dfrac{M''}{M'}}} - 1. \tag{7.227}$$

If the material were to behave elastically because of moment M'', we would have

$$\frac{\varepsilon''}{\varepsilon'} = \frac{M''}{M'}. \tag{7.228}$$

We identify a quantity k_6 that has the following equation:

$$k_6 = \frac{1}{\dfrac{M''}{M'}} \left(\frac{1}{\sqrt{1 - 2\dfrac{M''}{M'}}} - 1 \right). \tag{7.229}$$

This quantity would be equal to one if the material's behavior due to moment M'' were elastic. Since this is not the case, k_6 is greater than 1, as shown in Fig. 7.23. It also goes to infinity for $M'' = 0.5M'$. In fact, in that case the total moment is equal to $1.5M'$, the section shows complete plastic flow, and the deformations are ideally infinite.

k_6 clearly shows the impact of plastic flow on deformations.

If we conservatively hypothesize that, once yielding is reached in the tang, an additional increase of the moments introduce plastic flow not just in the section at hand but throughout the entire tang, the rotation φ_0'' in correspondence of section $A - A$ generated by moment M'' (identical to the one in section $B - B$) is given by

$$\varphi_0'' = k_6 \varphi_{0e}'', \tag{7.230}$$

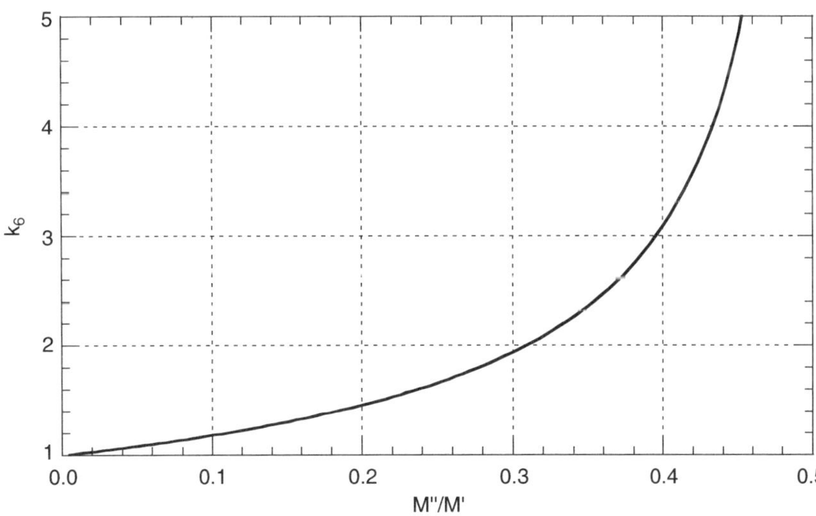

Fig. 7.23

where φ''_{0e} is the computed rotation assuming an elastic behavior of the material.

If one assumes the behavior of the crown to be elastic, (7.175) is still valid even with φ''_0 instead of φ_0.

This is not the case with (7.190). It is necessary to introduce a new quantity k''_1 given by

$$k''_1 = k_6 k_1, \tag{7.231}$$

that replaces k_1 in both (7.193) and (7.197).

In summary, we suggest the following calculation criteria.

One calculates the flange according to the original criteria relative to the elastic behavior of flanges. One then obtains the state of stress in the different sections including the circumferential stresses, as previously discussed. Once established that the yield strength is not exceeded both in the tang and the crown, one considers the maximum value of σ_{id} in the tang. The moment M'_{ro} that causes σ_{id} to increase to σ_s in this section is then computed. There is a corresponding value of F_A that can easily be computed proportionally. Then one adopts a generic value for M''_{ro} and, referring to (7.229) or to Fig. 7.23, the value of k_6 can be established based on the ratio M''_{ro}/M'_{ro} (equal to the generic ratio M''/M').

Once k_6 is known from (7.231), one computes k''_1 introduced to replace k_1 in (7.197) and (7.199). In this way M''_{co} is obtained. By indicating with M'_{ro} and M'_{co} the values for M_{ro} and M_{co} that correspond to F_A, the total is

$$\begin{aligned} M_{r0} &= M'_{r0} + M''_{r0} \\ M_{c0} &= M'_{c0} + M''_{c0} \end{aligned} \tag{7.232}$$

By trial-and-error one calculates the value of M_{ro} or M_{co} so that the relative stress be equal to σ_s.

From the corresponding value of M''_{ro} one obtains F_B given by (see (7.198))

$$F_B = \frac{M''_{r0}}{2k_2}, \tag{7.233}$$

with k_2 computed based on k''_1.

The total force F is given by

$$F = F_A + F_B. \tag{7.234}$$

One may argue that the computation process is based on the assumption that only the moment M_x is active in the tang, whereas both a thrust P and the relative stress are present with the pressure. In other words, there is no pure bending but rather combined tensile force and bending moment. It would be easy to demonstrate that in this case the recommended procedure is conservative. In conclusion, we would like to exemplify a calculation process.

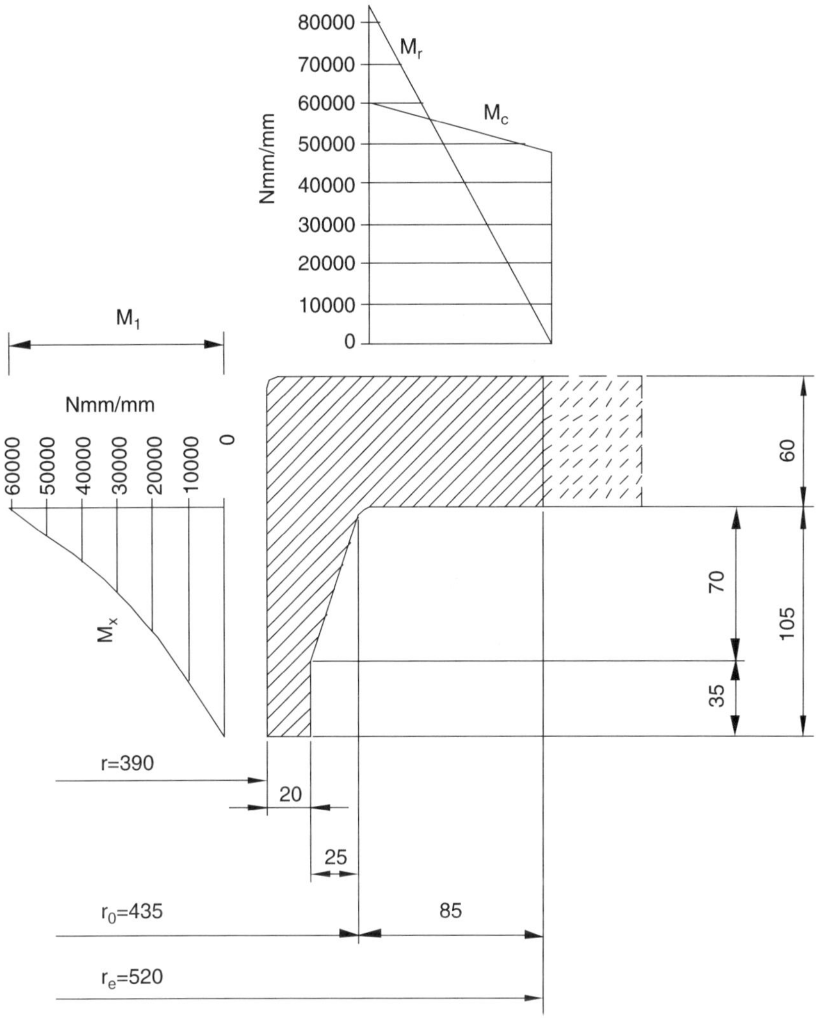

Fig. 7.24

Let us assume a flange with a tapered tang shown in Fig. 7.24 designed for a pressure of $4\,\mathrm{N\,mm^{-2}}$(40 bar). We have

$$r_0 = 435\,\mathrm{mm}$$
$$r_e = 520\,\mathrm{mm}$$
$$\alpha = \frac{435}{520} = 0.836$$

The average thickness of the tang is equal to

$$s_c = \frac{45 + 20}{2} = 32.5\,\mathrm{mm}.$$

The computing thickness is assumed to be as follows (we consider 18 mm of length of the cylindrical portion):

$$s_c = \frac{32.5 \times 70 + 20 \times 18}{88} = 29.9\,\text{mm}.$$

The average radius is therefore equal to

$$r_m = 390 + \frac{29.9}{2} = 404.95\,\text{mm}.$$

We note that

$$0.8\sqrt{r_m s_c} = 0.8\sqrt{404.95 \times 29.9} = 88.$$

This length corresponds to the one considered for the calculation of the weighted average of the thicknesses. The quantities γ and δ are equal to

$$\gamma = \frac{29.9}{404.95} = 0.0738$$
$$\delta = \frac{60}{29.9} = 2.01$$

As you can see, the value of δ is smaller than the one obtained from Fig. 7.20 for $\gamma = 0.0738$. Thus, it will be $M_{co} < M_{ro}$.

In fact, from (7.191) we have

$$k_1 = 0.40.$$

From (7.197) we obtain

$$k_2 = 0.01409,$$

and from (7.202) and (7.203)

$$k_3 = 0.719$$
$$k_4 = 1.259$$

Note that based on (7.210) the moment M_x becomes zero for

$$\tan \beta x = \frac{1}{1.259 - 1} = 3.86,$$

and

$$\beta x = 1.317,$$

in correspondence of which we have

$$x = 1.025\sqrt{r_m s_c} = 113\,\text{mm}.$$

Therefore, the moment is practically zero at the connection of the flange to the tube.

The pressure $4\,\text{N}\,\text{mm}^{-2}$ is matched by a thrust of about 19, 00, 000 N. We assume that the force in the bolts is equal to 30, 00, 000 N (60% greater than the thrust caused by pressure). Thus, based on (7.198) and (7.199) we have

$$M_{r0} = 2 \times 0.01409 \times 3000000 = 84540\,\text{N}\,\text{mm}\,\text{mm}^{-1}$$
$$M_{c0} = (0.40 + 0.30) \times 84540 = 59180\,\text{N}\,\text{mm}\,\text{mm}^{-1}$$

Then, for the crown we have

$$W = \frac{60^2}{6} = 600\,\text{mm}^3$$
$$\sigma = \frac{84540}{600} = 140.9\,\text{N}\,\text{mm}^{-2}$$

Based on (7.204) we have

$$M_1 = 0.719 \times 84540 = 60784\,\text{N}\,\text{mm}\,\text{mm}^{-1}.$$

At the connection of the tang with the crown we have

$$A = 116600\,\text{mm}^2$$
$$W = \frac{45^2}{6} = 337.5\,\text{mm}^3$$

Thus, factoring in the thrust caused by the pressure

$$\sigma = \frac{1900000}{116600} + \frac{60748}{337.5} = 196.3\text{N}\,\text{mm}^{-2}.$$

At the connection of the cylindrical portion we have

$$x = 70\,\text{mm}$$
$$\beta x = \frac{1.285 \times 70}{\sqrt{404.95 \times 29.9}} = 0.817$$

Based on (7.206) we have

$$M_x = 60784 \times 0.441 \times 0.495 = 13268\text{N}\,\text{mm}\,\text{mm}^{-1}.$$

Moreover,

$$A = 50265\,\text{mm}^2$$
$$W = \frac{20^2}{6} = 66.67\,\text{mm}^3$$
$$\sigma = \frac{1900000}{50265} + \frac{13268}{66.67} = 236.8\text{N}\,\text{mm}^{-2}$$

If a material with a yield strength equal to $240\,\text{N}\,\text{mm}^{-2}$ was used for the flange, the verification is positive since the calculated stresses are all smaller than this value (equal to $1.5f$).

In Fig. 7.24 we also show the behavior of the moments using (7.200), (7.201), and (7.206). Note that the maximum value of M_c is considerably smaller than M_{ro} and greatly smaller than the moment corresponding to the yield strength (equal to $144000\,\mathrm{N\,mm\,mm^{-1}}$); potential local plastic flow does not compromise the stability of the flange and its performance in terms of tightness.

Let us now apply the first of the introduced calculation criteria that factor in the possibility of partial plastic flow of the tang. Based on (7.220) in correspondence of section $B-B$ we have

$$\sigma_t = 58.2\,\mathrm{N\,mm^{-2}}.$$

In addition, recalling that $M_1 = 60784\,\mathrm{N\,mm\,mm^{-1}}$ and that $W = 337,5\,\mathrm{mm^3}$ because of the moment μM_1 we have

$$\sigma_t = \pm 0.3\frac{60784}{337.5} = \pm 54\,\mathrm{N\,mm^{-2}}.$$

As a result of the pressure in the longitudinal direction we have

$$\sigma_a = \frac{1900000}{116600} = 16.3\,\mathrm{N\,mm^{-2}}.$$

Finally, as a result of M_1

$$\sigma_a = \pm\frac{60784}{337.5} = \pm 180.1\,\mathrm{N\,mm^{-2}}.$$

On the internal fiber we have

$$\begin{aligned}
\sigma_t &= 58.2 - 54 = 4.2\,\mathrm{N\,mm^{-2}}\\
\sigma_a &= 16.3 - 180.1 = -163.8\,\mathrm{N\,mm^{-2}}\\
\sigma_{id} &= 4.2 - (-163.8) = 168\,\mathrm{N\,mm^{-2}}
\end{aligned}$$

At distance $0.8\sqrt{r_m s_c}$ from section $B-B$, based on (7.221) and given that $\beta x = 1.028$ we have

$$\sigma_t = 224.6 \times 0.3577 \times 0.989 = 79.5\,\mathrm{N\,mm^{-2}}.$$

As a result of moment μM_x, observing that in this position $M_x = 6399\,\mathrm{N\,mm\,mm^{-1}}$ and $W = 66.67\,\mathrm{mm^3}$, we have

$$\sigma_t = \pm 0.3\frac{6399}{66.67} = \pm 28.8\,\mathrm{N\,mm^{-2}}.$$

As far as pressure we have

$$a = \frac{410}{390} = 1.051;$$

and on the internal fiber

$$\sigma_t = 4\frac{1.051^2 + 1}{1.051^2 - 1} = 80.2\,\mathrm{N\,mm^{-2}}.$$

As a result of pressure in the longitudinal direction

$$\sigma_a = \pm\frac{1900000}{50265} = 37.8\,\mathrm{N\,mm^{-2}}.$$

As a result of M_x

$$\sigma_a = \pm\frac{6399}{66.67} = \pm 96\,\mathrm{N\,mm^{-2}}.$$

On the internal fiber we have

$$\sigma_t = 79.5 - 28.8 + 80.2 = 130.9\,\mathrm{N\,mm^{-2}}$$
$$\sigma_a = 37.8 - 96 = -58.2\,\mathrm{N\,mm^{-2}}$$
$$\sigma_{id} = 130.9 - (-58.2) = 189.1\,\mathrm{N\,mm^{-2}}$$

As one can see, there are indeed no yielding phenomena in the flange.

Furthermore, in this case the circumferential stresses do not impact the maximum value of the ideal stress. One can also easily see that the yielding would occur at the connection of the cylindrical portion with a force F equal to 3048000 N. We now assume that the force is equal to 3510000 N. Then we have

$$F_A = 3048000\ \mathrm{N}$$
$$F_B = 462000\ \mathrm{N}$$

Based on (7.216) we obtain

$$M_{r0} = 2 \cdot 0.01409 \times 3048000 = 85893\,\mathrm{N\,mm\,mm^{-1}}.$$

Based on (7.218) we have

$$M_1 = 0.719 \times 85893 = 61757\,\mathrm{N\,mm\,mm^{-1}}.$$

Given that $k_5 = 0.0894$ and based on (7.217) we have

$$M_{c0} = 0.7 \times 85893 + 2 \times 0.0894 \times 462000 = 142730\,\mathrm{N\,mm\,mm^{-1}}.$$

As one can see, the moment M_{co} shows a sudden increase compared to the value obtained through the computation done earlier for $F = 3000000\,N$. Its value is also clearly greater than M_{ro}.

The corresponding stress is equal to

$$\sigma = \frac{142730}{600} = 237.9\,\mathrm{N\,mm^{-2}}.$$

It is smaller than the yield strength of the material so that the flange can be considered as positively verified, even though the yield strength is ideally exceeded in the tang.

Let us now apply the second calculation criterion that better takes into account the partial plastic flow in the tang. Based on previous calculations we have

$$F_A = 3048000\,\mathrm{N},$$

matched by

$$M'_{r0} = 85893\,\mathrm{N\,mm\,mm^{-1}}$$
$$M'_{c0} = 0.7 \times 85893 = 60125\,\mathrm{N\,mm\,mm^{-1}}$$

We shall carry out the computation by trial-and-error.

We have

$$M''_{r0} = 0.3M'_{r0} = 25768\,\mathrm{N\,mm\,mm^{-1}}.$$

Based on (7.228) or Fig. 7.23 it is matched by

$$k_6 = 1.937.$$

Then

$$k''_1 = 1.937 \times 0.4 = 0.7748.$$

And based on (7.197)

$$k_2 = 0.01331.$$

Therefore, based on (7.193) we have

$$M''_{c0} = 25768\,(0.7748 + 0.3) = 27695\,\mathrm{N\,mm\,mm^{-1}}.$$

In summary

$$M_{r0} = 85893 + 25768 = 111661\,\mathrm{N\,mm\,mm^{-1}}$$
$$M_{c0} = 60125 + 27695 = 87820\,\mathrm{N\,mm\,mm^{-1}}$$

Moreover

$$F_B = \frac{25768}{2 \times 0.01331} = 967994\ \mathrm{N}.$$

We try again as follows:

$$M''_{r0} = 0.4M'_{r0} = 34357\,\mathrm{N\,mm\,mm^{-1}}$$
$$k_6 = 3.09$$
$$k''_1 = 3.09 \times 0.4 = 1.236$$
$$k_2 = 0.01246$$
$$M''_{c0} = 34357\,(1.236 + 0.3) = 52772\,\mathrm{N\,mm\,mm^{-1}}$$

In the end

$$M_{r0} = 85893 + 34357 = 120250\,\mathrm{N\,mm\,mm^{-1}}$$
$$M_{c0} = 60125 + 52772 = 112897\,\mathrm{N\,mm\,mm^{-1}}$$
$$F_B = \frac{34357}{2 \times 0.01246} = 1378692\,\mathrm{N}$$

It is possible to further increase the moment M''_{ro} since both M_{ro} and M_{co} are smaller than $144000\,\mathrm{N\,mm\,mm^{-1}}$ that corresponds to the yield strength.

Let us try with

$$M''_{r0} = 0.45M'_{r0} = 38652\,\mathrm{N\,mm\,mm^{-1}}$$
$$k_6 = 4.80$$
$$k''_1 = 4.80 \times 0.4 = 1.92$$
$$k_2 = 0.01137$$
$$M''_{c0} = 38652\,(1.92 + 0.3) = 85807\,\mathrm{N\,mm\,mm^{-1}}$$

The total is

$$M_{r0} = 85893 + 38652 = 124545\,\mathrm{N\,mm\,mm^{-1}}$$
$$M_{c0} = 60125 + 85807 = 145932\,\mathrm{N\,mm\,mm^{-1}}$$
$$F_B = \frac{38652}{2 \times 0.01137} = 1699736\ \mathrm{N}$$

The value M_{co} is slightly greater than $144000\,\mathrm{N\,mm\,mm^{-1}}$. Thus, the value F_B must be slightly smaller than the one found.

One can more or less assume that

$$F_B = 1672000\,\mathrm{N}.$$

Therefore, the total results in

$$F = 3048000 + 1672000 = 4720000\ \mathrm{N}.$$

By comparing the three calculation procedures, one concludes that the allowable force F is slightly greater than 3000000 N when assuming an elastic behavior of the entire flange. Using the simplistic and conservative elastic–plastic calculation shown earlier, we obtain $F = 3510000\,\mathrm{N}$. The last procedure brings the value of F up to 4720000 N.

Note that in the last two cases even a slight increase of F greatly increases the value of M_{co}. We suggest caution in identifying the allowable value for F, especially given the uncertainty regarding the actual value of the tightening force. Therefore, we recommend that the allowable value of F be equal to 85%

of the one obtained through the last calculation procedure we introduced. In that case we would have

$$F = 0.85 \times 4720000 = 4012000\,\text{N}.$$

And this is still clearly greater (by 31.6%) than the value for the elastic behavior of the entire flange.

7.5 Piping with Internal Warm Fluid

If the extremities of pipes used for warm fluids are tied, as they necessarily are, to fixed points, or to points that shift in a predetermined way, stresses in the piping and forces and moments at the anchorage ends will occur. Once the sizing of the piping as far as internal pressure is completed according to the usual criteria relative to cylinders, it is necessary to do a calculation of the piping to identify the internal stresses caused by expansions. In addition, one must consider the stresses caused by the weight of the piping. To this extent, one must obviously factor in the locations of the tie rods, or the saddle supports, the features of possible springs, and so on.

Once the stresses are identified, the verification of the piping may be done according to coded calculation criteria. We refer to the calculation method in the American Code for Pressure Piping edited by the ASME. We consider three Cartesian axis that we indicate with 1, 2, and 3, respectively, so that axis 3 coincides with the one of the piping, or in the case of a curve, with its tangent. We assume that M_1, M_2, and M_3 are the moments on the normal planes to the axis 1, 2, and 3. M_1 and M_2 are bending moments, whereas M_3 is a torsional moment.

Let us compute the resulting bending moment M_f given by

$$M_f = \sqrt{M_1^2 + M_2^2}. \tag{7.235}$$

At this point, the code requires the introduction of a stress intensification factor i that factors in local stress peaks whenever the verification is relative to singular points of the piping, such as curves, branch pipes and flanges. The maximum stress σ_f caused by the bending moment, considering that W is the section modulus, is given by

$$\sigma_f = \frac{iM_f}{W}. \tag{7.236}$$

The values for i may be taken from Table 7.1. It also lists the parameters h that factor in a possible increase in actual flexibility compared with the theoretical one.

As one can see, in the specific case of the curves this parameter is greater than one. This means that if one must assume an increase in stresses in curves during verification through i, on the other hand during the calculation of the

Table 7.1

description	flexibility factor k	stress int. factor i	flexibility character h	sketch
welding elbow[a,b,c,d] or pipe bend	$\frac{1.65}{h}$	$\frac{0.9}{h^{2/3}}$	$\frac{sR_1}{(r_2)^2}$	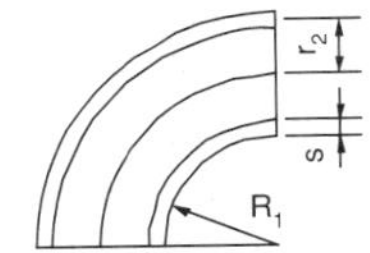
closely spaced miter bend[a,b,c] $a < r_2(1+\tan\theta)$	$\frac{1.52}{h^{5/6}}$	$\frac{0.9}{h^{2/3}}$	$\frac{sa\cot\theta}{2(r_2)^2}$	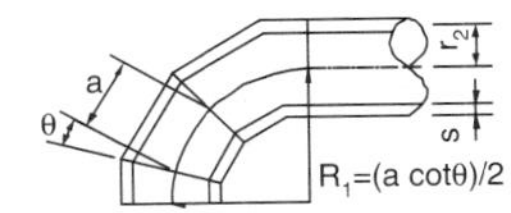
widely spaced miter bend[a b,e] $a \geq r_2(1+\tan\theta)$	$\frac{1.52}{h^{5/6}}$	$\frac{0.9}{h^{2/3}}$	$\frac{1+\cot\theta}{2}\,\frac{s}{r_2}$	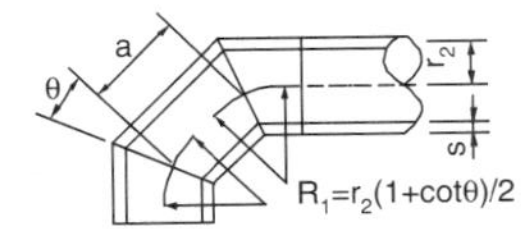
welding tee[a,b,f] per ASA B16.9	1	$\frac{0.9}{h^{2/3}}$	$4.4\frac{s}{r_2}$	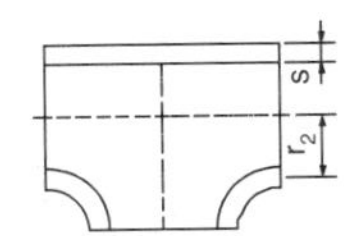
reinforced fabricated tee[a,b,f,g] with pad or saddle	1	$\frac{0.9}{h^{2/3}}$	$\frac{(s+0.5s_e)^{5/2}}{s^{3/2}r_2}$	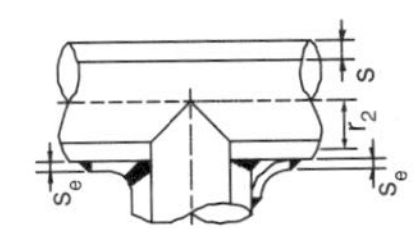

Table 7.1. *Continued*

description	flexibility factor k	stress int. factor i	flexibility character h	sketch
unreinforced fabricated tee[a,b,f]	1	$\frac{0.9}{h^{2/3}}$	$\frac{s}{r_2}$	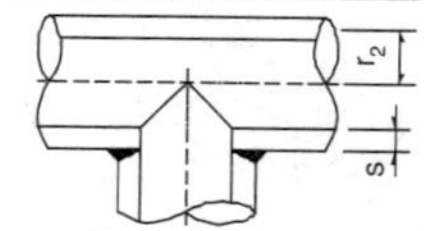
butt-welded joint, reducer, or welding neck flange	1	1.0		
double welded slip-on flange	1	1.2		
fillet welded joint (single welded), socket welded flange or single welded slip-on flange	1	1.3		
lap joint flange (with ASA B16.9 lap joint stub)	1	1.6		
screwed pipe joint or screwed flange	1	2.3		
corrugated straight pipe or corrugated or creased bend[h]	5	2.5		

[a] For fittings and miter bends the flexibility factor (k) and stress intensification factor (i) in the table apply to bending in any plane and shall not be less than unity; factors for torsion equal unity. Both factors apply over the effective arc length (shown by heavy center lines in the sketches) for curved and miter elbows, and to the intersection point for tees

[b] The values of k and i can be calculate by entering with the characteristic h computed from the formulas given where, r_2 is the mean radius of matching pipe $s =$ for elbows and miter bends the nominal wall thickness of the fitting (see note 7); s for tees, the nominal wall thickness of matching pipe; R_1 the bend radius of welding elbow or pipe bend; θ, one-half angle between adjacent miter axes; a the miter spacing at center line; s_e the pad or saddle thickness

[c] Where flanges are attached to one or both ends, the values of k and i in the table shall be corrected by the factor C_1 given below, entering with the computed h One end flanged : $C_1 = h^{1/6}$. Both ends flanged : $C_1 = h^{1/3}$

[d] The engineer is cautioned that cast butt-welding elbows may have considerably heavier walls than that of the pipe with which they are used. Large errors may be introduced unless the effect of these greater thicknesses is considered

[e] Also includes single-miter joint

[f] The stress intensification factors i in the table were obtained from tests on full size outlet connections. For less than full size outlets, the full size values should be used until more applicable values are developed

[g] When $s_e > 1.5s$ use $h = 4.05s/r_2$ [h] Factors shown apply to bending; flexibility factor for torsion equals 0.9

piping, one may consider an increased flexibility of the curve through the corrective parameter h. As we all know, the polar section modulus of a pipe is double than that of a bending one. Therefore, the torsional stress caused by M_3 is given by

$$\tau = \frac{M_3}{2W}. \tag{7.237}$$

Because of internal pressure, the weight of the piping and the fluid, as well as potential external forces, axial stresses are also generated. We indicate their maximum value in the section under study with σ_a. If the impact of the weight can be ignored, σ_a is reduced to the longitudinal stress caused by the pressure that we indicate with σ_{ap}. It is equal to

$$\sigma_{ap} = \frac{p}{a^2 - 1} = p\frac{D_i^2}{D_e^2 - D_i^2}. \tag{7.238}$$

In favor of safety it may be computed more simply with the following equation:

$$\sigma_{ap} = \frac{pD_e}{4s}. \tag{7.239}$$

Therefore, factoring in the bending moment M_f the maximum axial stress is equal to

$$\sigma'_a = \sigma_f + \sigma_a; \tag{7.240}$$

and, applying Guest's failure theory, the ideal stress is given by

$$\sigma_{id} = \sqrt{(\sigma_f + \sigma_a)^2 + 4\tau^2}. \tag{7.241}$$

According to the code in question, the ideal stress is computed based on the following simpler and cautionary, albeit less rational criteria, instead.

One calculates an ideal stress referred only to the stresses caused by prevented dilations. Indicated with σ_d, it is given by

$$\sigma_d = \sqrt{{\sigma_f}^2 + 4\tau^2}. \tag{7.242}$$

Recalling (7.236) and (7.237), it may also be written as follows:

$$\sigma_d = \frac{\sqrt{(iM_f)^2 + M_3^2}}{W}. \tag{7.243}$$

Moreover, we consider a global ideal stress given by

$$\sigma_{id} = \sigma_d + \sigma_a. \tag{7.244}$$

By comparing (7.244) with (7.241) and recalling (7.242), it is evident that the adopted criterion is conservative. The ideal stress must be compared to the allowable stress f_p. In the absence of substantial fatigue phenomena it is assumed to be equal to

$$f_p = 1.25\left(f_c + f_h\right), \tag{7.245}$$

where f_c and f_h are the basic allowable stresses for the material at room temperature (cold) and working temperature (hot) of the piping, respectively.

In order for the verification to be positive, we must have

$$\sigma_d + \sigma_a \leq 1.25\,(f_c + f_h)\,, \tag{7.246}$$

or

$$\sigma_d \leq 1.25\,(f_c + f_h) - \sigma_a. \tag{7.247}$$

The code also requires the introduction of a stress-range reduction factor that accounts for the impact of fatigue. By indicating it with F, (7.247) must be written as follows:

$$\sigma_d \leq F\,[1.25\,(f_c + f_h) - \sigma_a]\,; \tag{7.248}$$

and this is, in fact, the equation obtained by applying calculation criteria of the code, except for the different symbolism.

The values of F may be obtained from Table 7.2. As one can see, if the number of cycles is smaller than 7000, one assumes $F = 1$. The stress σ_a must not be greater than f_h. In the case of $\sigma_a = f_h$, or if for sake of resistance and in spite of $\sigma_a < f_h$, one does not take it into account, (7.248) may be written more simply as follows:

$$\sigma_d \leq F\,(1.25 f_c + 0.25 f_h)\,. \tag{7.249}$$

The adoption of the allowable stress present in (7.245) finds justification in the criteria presented in Sect. 1.5 relative to self-limiting stresses.

As far as the piping we are examining, one must remember that the internal stresses originate from prevented thermal expansion, and not from applied external forces. Clearly, if the material yields in an area of the piping, the ensuing plastic deformations rapidly lead the piping itself to the respect of congruence, including a self-limitation of internal stresses and without any danger of rupture for the material. Therefore, it is possible to adopt a very high allowable stress in relation to the computation of stresses within the elastic field.

More precisely, we recall that in Sect. 1.5 we described how it is possible to adopt an allowable stress equal to twice the yield strength. In our specific

Table 7.2

Cycles	F
7,000 and less	1.0
7,000 to 14,000	0.9
14,000 to 22,000	0.8
22,000 to 45,000	0.7
45,000 to 100,000	0.6
Over 100,000	0.5

case, we observe that the piping alternates between room temperature and the operational maximum temperature. Both σ_{sc} and σ_{sh} indicate the yield strength under cold and hot working conditions, respectively. As usual, we schematize the curve characteristic of the material, by hypothesizing its elastic–plastic behavior. Moreover, we adopt the modulus of elasticity E_h under hot conditions during the entire warm-up phase, and the modulus of elasticity E_c under cold conditions during the entire next phase of cooling down (it can be demonstrated that this simplification works in favor of resistance).

If the material is exploited to the fullest, the behavior of stresses as a function of deformations is shown in Fig. 7.25, that represents the absence or presence of cold spring. After the first warm-up phase during which the material yields, in the following phase of cooling down (and during the next cycles, as well) the stress varies between σ_{sh} and $-\sigma_{sc}$ (or, of course, between $-\sigma_{sc}$ and σ_{sh}). Through an adequate cold spring it is possible, at least in theory, to eliminate the yielding at hot temperature. With the exception of the first warm-up, the material behaves elastically with the stress varying between the yield strength under cold and hot conditions. Thus, computing the stresses according to the laws of elasticity, in order to prevent the material from behaving elastically during the cycles subsequent to the first one, it is necessary and sufficient that the ideal stress does not exceed the sum of both the yield strength under warm and cold working conditions. In other words, the allowable stress f_p is equal to

$$f_p = \sigma_{sc} + \sigma_{sh}. \tag{7.250}$$

In view of the considerations above, it is also possible to understand why the code requires the internal stresses to be calculated based on E_c and still referring to the expansions occurring under hot working conditions (the value of the modulus of elasticity is crucial in terms of the entity of the stresses

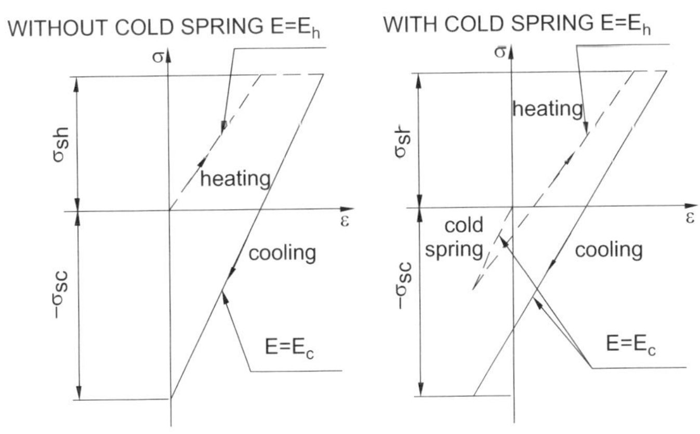

Fig. 7.25

considering their origin from prevented expansions). In fact, this way one refers to the cooling down phase.

Based on Fig. 7.25 one also establishes that the cold spring does not impact the entity of the final stresses, and its potential presence is actually not considered during the computation of stresses. One notices, though, that the cold spring obviously reduces the deformations caused by heat. This has repercussions on the forces and moments present at fixed points of the piping. Thus, the only purpose of the cold spring is to reduce these reaction forces, as we shall see later on.

Reconsidering (7.250), note that the allowable stresses under cold and hot conditions (ruling out the impact of creep) are generally based on the yield strength divided by the safety factor assumed to be equal to 1.5. Thus, (7.250) may be written as follows:

$$f_p = 1.5\,(f_c + f_h)\,. \tag{7.251}$$

As one can see, the code adopts a conservative value of f_p smaller than the latter by substituting the coefficient 1.25–1.5.

In summary, the calculation process consists of computing the moments M_1, M_2 and M_3 caused by the prevented expansions according to the modulus of elasticity E_c under cold conditions. Once M_f is computed through (7.235), one obtains σ_d through (7.243). At this point one must compute the values for f_c and f_h based on the mechanical characteristics of the steel, the maximum working temperature, and the required safety factors. Then, one must compute σ_a based on the pressure, the weight of the piping and the fluid, and verify that $\sigma_a \leq f_h$. After identifying the value of F, one must check for every significant section of the piping that (7.248) is fulfilled (or (7.249), for simplicity purposes and in favor of resistance).

As far as the forces and moments on the anchorages that must be calculated to size their matching structures, the code in question delivers the following calculation criteria. We assume S is the generic force or moment obtained through the calculation of the piping based on E_c. By analogy, S_c and S_h are the force or moment on the anchorages under cold and hot conditions. With C indicating the cold spring factor (variable from 0 to 1 for cold spring zero or equal to 100%), one calculates the following quantity:

$$C_h = \left(1 - \frac{2}{3}C\right)\frac{E_h}{E_c}\,. \tag{7.252}$$

The coefficient 2/3 accounts for the impossibility to practically fully assure the specified cold spring during the erection. The forces or moments under hot conditions are given by

$$S_h = C_h S\,. \tag{7.253}$$

Now we compute C_1 given by

$$C_1 = 1 - \frac{f_h}{\sigma^*}\frac{E_c}{E_h}\,, \tag{7.254}$$

where f_{h_c} has the usual meaning while σ^* represents the maximum value of σ_d in the entire piping. If C_1 is negative it must be set to zero. By indicating with C_f the largest of C and C_1, the forces or moments under cold conditions are given by

$$S_c = -C_c S. \tag{7.255}$$

In view of previous considerations, C_1 takes into account the onset of self-stresses at cold temperatures caused by plastic flow of the material under hot temperatures in the most stressed section. As far as the calculation of the internal stresses in the piping, the manual procedures are generally extremely time-consuming, especially if the piping has offtakes. Fortunately, there are several computer programs to quickly identify these stresses or the ones caused by the weight of the piping or the fluid.

Of course, these programs are also able to directly perform the verifications and to determine whether the piping is correctly sized or not without further manual calculations. Computers also assist in rapidly examining different solutions for the layout of piping. This allows the designer to adopt a suitable solution from a static point of view, and also to select among different scenarios the one at lower cost, thus optimizing the project.

7.6 Expansion Compensators

The rigorous analysis of the behavior of an expansion compensator in the presence of internal pressure and deformations caused by the expansion of the piping to which it is connected is quite difficult because of its geometry. We would like to introduce a calculation criterion by approximation which is acceptable in terms of practical requirements and leads to a conservative sizing without unnecessary complications. As we shall see, the equations are based on dimensionless quantities with a precise analytical expression. Thus, they are quite suitable for running on a computer. Since manual calculations are not excluded but would be cumbersome, we have decided to show these parameters in a number of figures.

The reader will see that given the complexity of the topic, the presentation will not be elementary, the methodology based on approximations notwithstanding. First of all, we shall examine the behavior of one circular crown. With reference to Fig. 7.26, we consider the circular crown with inside radius r_1, outside radius r_2, and thickness s. We assume a deformation that shows a deflection δ_1 in correspondence of the internal radius, while rotations are prevented in correspondence of both r_1 and r_2.

Let us consider an elementary strip of this crown with unitary width in correspondence of the internal radius. If I_1 is the moment of inertia for $r = r_1$, the generic moment of inertia is

$$I = I_1 \frac{r}{r_1}. \tag{7.256}$$

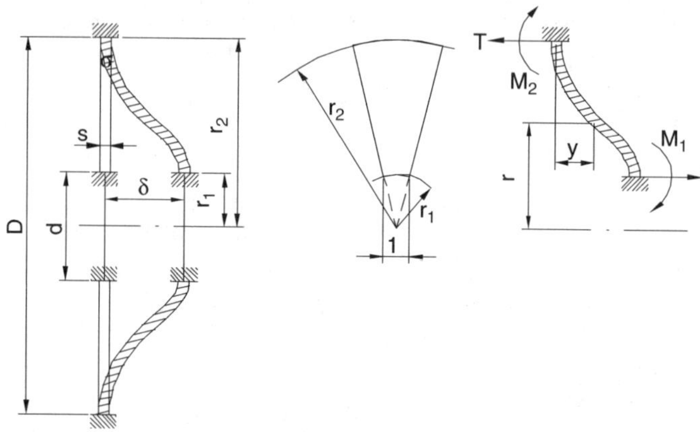

Fig. 7.26

The generic bending moment M is given by

$$M = M_1 - T\,(r - r_1)\,, \tag{7.257}$$

and we introduce K

$$K = \frac{E}{1 - \mu^2}, \tag{7.258}$$

where E is the modulus of elasticity and μ Poissons's factor.

As a result of prevented transversal deformations, we have

$$KIy'' = -M = -M_1 + T\,(r - r_1)\,. \tag{7.259}$$

By integrating once and recalling (7.256), we obtain

$$\frac{KI_1 y'}{r_1} = -M_1 \log_e r + T\,(r - r_1 \log_e r) + C_1, \tag{7.260}$$

where C_1 is a constant of integration.

After integrating a second time we have

$$\frac{KI_1 y}{r_1} = -M_1\,(r \log_e r - r) + T\left[\frac{r^2}{2} - r_1\,(r_1 \log_e r - r)\right] + C_1 r + C_2, \tag{7.261}$$

where C_2 is the second constant of integration.

On the other hand, for $r = r_1$ and $r = r_2$ we have $y' = 0$. At this point, we introduce α

$$\alpha = \frac{r_2}{r_1} = \frac{D}{d}. \tag{7.262}$$

Based on (7.260) we obtain

$$M_1 = Tr_1 \frac{\alpha - 1 - \log_e \alpha}{\log_e \alpha}, \tag{7.263}$$

and

$$C_1 = Tr_1 \left(\frac{\alpha - 1 - \log_e \alpha}{\log_e \alpha} \log_e r_1 - 1 + \log_e r_1 \right). \tag{7.264}$$

For $r = r_2$ we also have $y = 0$.

Based on (7.260) we obtain the following equation of the constant C_2:

$$C_2 = Tr_1^2 \alpha \left[\frac{\alpha - 1 - \log_e \alpha}{\log_e \alpha} (\log_e \alpha - 1) + \log_e \alpha - \frac{\alpha}{2} \right]. \tag{7.265}$$

Once the equations of constant C_1 and C_2 are known, from (7.261) and (7.263) and setting $r = r_1$, through a series of steps we obtain

$$KI_1\delta_1 = Tr_1^3 \left[\frac{\alpha - 1 - \log_e \alpha}{\log_e \alpha} (1 - \alpha + \alpha \log_e \alpha) + \frac{1}{2} - \frac{\alpha^2}{2} + \alpha \log_e \alpha \right]. \tag{7.266}$$

Introducing k_1 given by

$$k_1 = \frac{\alpha - 1 - \log_e \alpha}{\log_e \alpha} (1 - \alpha + \alpha \log_e \alpha) + \frac{1}{2} - \frac{\alpha^2}{2} + \alpha \log_e \alpha, \tag{7.267}$$

we can write

$$T = \frac{KI_1}{k_1 r_1^3} \delta_1. \tag{7.268}$$

The total force F required to provoke the deflection δ_1 in the circular crown, and recalling that $I_1 = s^3/12$, is given by

$$F = 2\pi r_1 T = \frac{\pi}{6} \frac{K}{k_1} \frac{s^3}{r_1^2} \delta_1, \tag{7.269}$$

where k_1 is shown in Fig. 7.27. Let us now consider an ideal strip of constant width undergoing the same deformation δ_1 (Fig. 7.28). Its width is equal to the average width of the previously analyzed strip. In that case the bending moment is zero in correspondence of the center line, and considering half a strip, one can write

$$[t]KI\frac{\delta_1}{2} = \frac{1}{3} T \left(\frac{l}{2} \right)^3, \tag{7.270}$$

where l is the length of the strip that is equal to $r_2 - r_1$, if it is considered to belong to the circular crown; therefore, we can write

$$T = \frac{12KI}{r_1^3 (\alpha - 1)^3} \delta_1. \tag{7.271}$$

We observe that in this case

$$I = \frac{\alpha + 1}{2} \frac{s^3}{12}. \tag{7.272}$$

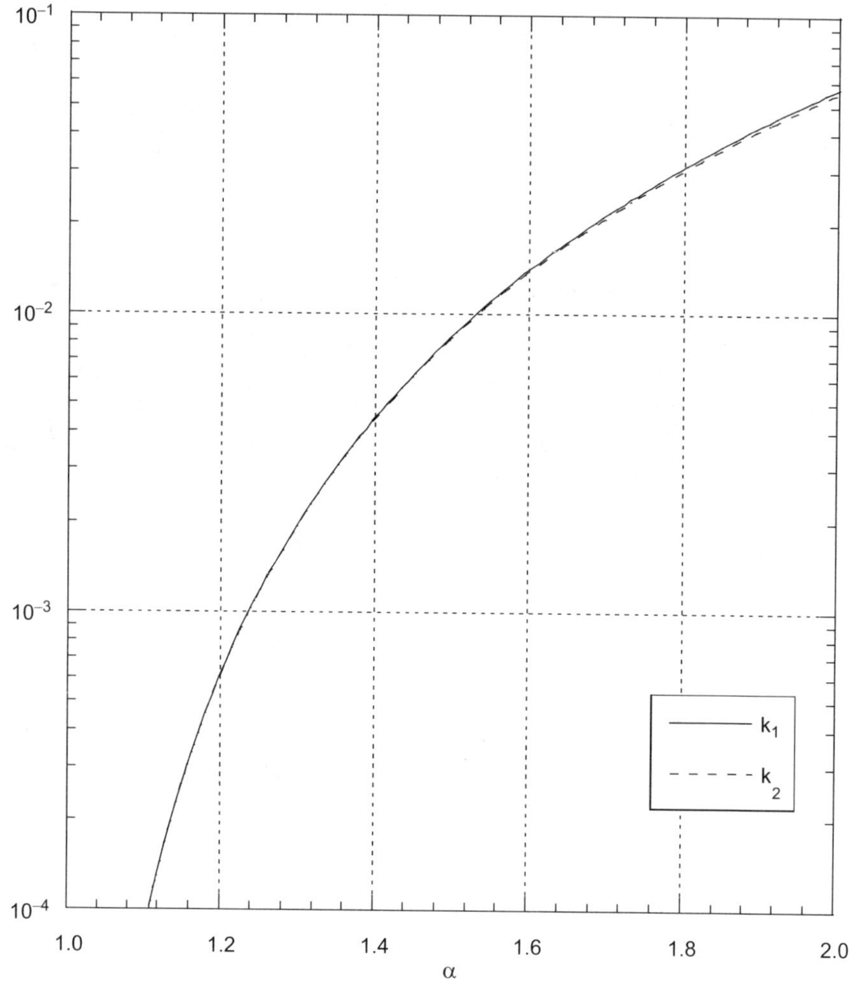

Fig. 7.27

By transferring the result to the entire crown, the force F that generates the deflection δ_1 is obtained by multiplying T by $2\pi r_1$, and similarly to (7.269), it follows that:

$$F = 2\pi r_1 T = \pi K \frac{\alpha + 1}{(\alpha - 1)^3} \frac{s^3}{r_1^2} \delta_1 \tag{7.273}$$

that we may also write as

$$F = \frac{\pi}{6} \frac{K}{k_2} \frac{s^3}{r_1^2} \delta_1, \tag{7.274}$$

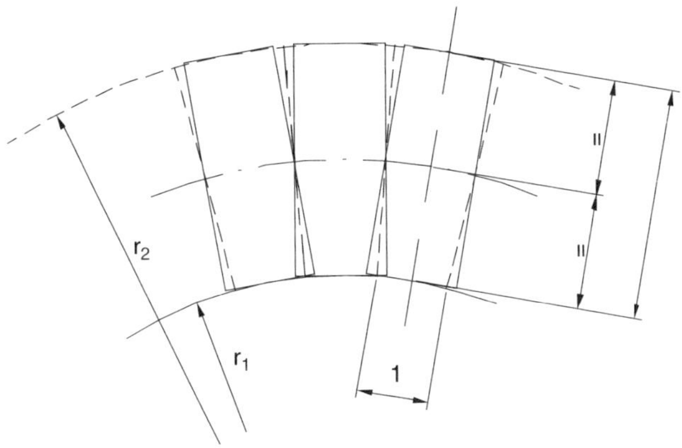

Fig. 7.28

given that

$$k_2 = \frac{(\alpha - 1)^3}{6\,(\alpha + 1)}; \tag{7.275}$$

k_2 is shown in Fig. 7.27. By comparing (7.274) with (7.269), we see that the two equations differ for the presence of k_2 instead of k_1. On the other hand, observing Fig. 7.27, we note that the two parameters almost coincide. For example, for $\alpha = 2$ the difference is 3%; for $\alpha < 2$ the difference is smaller. Basically, (7.274) may therefore substitute (7.269), and the values for T may be considered practically the same in both cases.

Let us now consider (7.263) and observe that the maximum corresponding stress, given that $W = s^2/6$, is equal to

$$\sigma_1 = 6k_3 \frac{Tr_1}{s^2}. \tag{7.276}$$

Assuming that

$$k_3 = \frac{\alpha - 1 - \alpha \log_e \alpha}{\log_e \alpha}, \tag{7.277}$$

by analogy, from (7.257) for $r = r_2$

$$M_2 = M_1 - Tr_1\,(\alpha - 1)\,; \tag{7.278}$$

M_2 is matched by a width of the strip equal to α, thus

$$W = \frac{\alpha s^2}{6}. \tag{7.279}$$

The maximum corresponding stress σ_2, considering that M_2 is negative, is equal to

$$\sigma_2 = 6k_4 \frac{Tr_1}{s^2}, \tag{7.280}$$

given that

$$k_4 = 1 - \frac{1}{\alpha} - \frac{\alpha - 1 - \log_e \alpha}{\alpha \log_e \alpha}. \tag{7.281}$$

Let us now consider the ideal strip of constant length equal to the average width of the previous strip. We have

$$M_1 = -M_2 = T\frac{r_1 r_2}{2} = \frac{T r_1}{2}(\alpha - 1); \tag{7.282}$$

moreover,

$$W = \frac{\alpha + 1}{2}\frac{s^2}{6}; \tag{7.283}$$

thus

$$\sigma_1 = \sigma_2 = 6k_5\frac{T r_1}{s^2}, \tag{7.284}$$

since

$$k_5 = \frac{\alpha - 1}{\alpha + 1}. \tag{7.285}$$

k_3, k_4 and k_5 are shown in Fig. 7.29. As you can see, the value of k_5 is between k_3 and k_4. The most interesting parameter is k_3, which is greater than k_5.

One can see that the ratio between k_3 and k_5 may be assumed to be with sufficient approximation equal to

$$k_6 = \frac{k_3}{k_5} \cong \frac{2 + \alpha}{3}. \tag{7.286}$$

The conclusions that may be drawn from this investigation are the following.

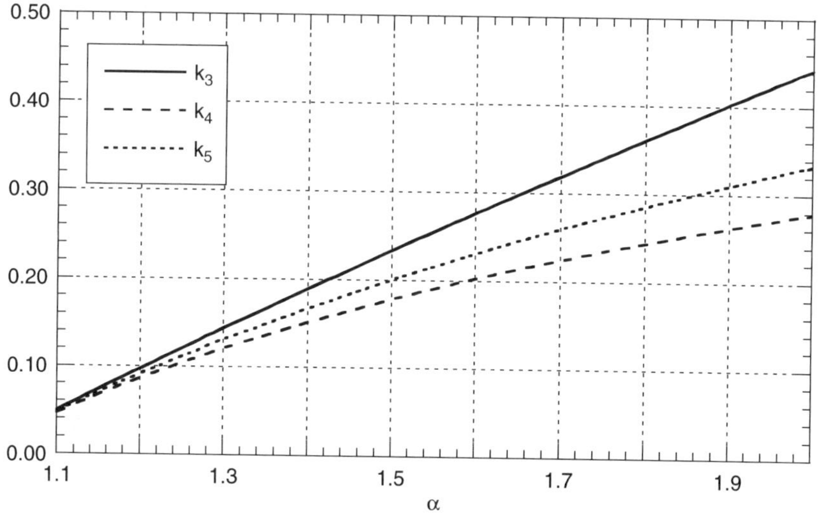

Fig. 7.29

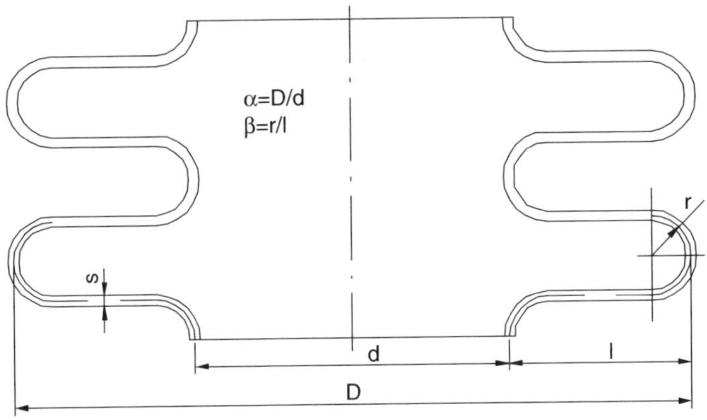

Fig. 7.30

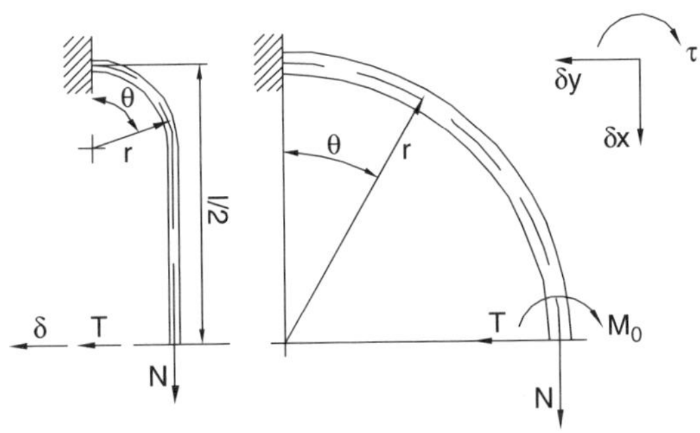

Fig. 7.31

The examined crown may be considered as a set of elementary strips of constant width in relation to the average circumference of the crown, and with regard to the force F that is necessary to generate the deflection δ_1. As far as the maximum stress caused by the bending moments on the crown, one can refer to the set of strips of constant width, simply remembering to multiply the resulting maximum stress by k_6.

Let us now consider a fourth of a wave of a compensator (see Figs. 7.30 and 7.31), and specifically the portion of the torus. To that extent, we examine the behavior of a curved beam corresponding to a fourth of circumference of constant and unitary width clamped at one end. The forces T and N and the moment M_0 are present at the opposite end.

By applying the principle of virtual works, one obtains the following equations for δ_x, δ_y and τ relative to the force T:

$$\delta_x = \frac{1}{2}\frac{Tr^3}{KI}$$

$$\delta_y = \frac{\pi}{4}\frac{Tr^3}{KI} \tag{7.287}$$

$$\tau = \frac{Tr^2}{KI} \tag{7.288}$$

As far as N

$$\delta_x = \left(\frac{3}{4}\pi - 2\right)\frac{Nr^3}{KI} \tag{7.289}$$

$$\delta_y = \frac{1}{2}\frac{Nr^3}{KI} \tag{7.290}$$

$$\tau = \left(\frac{\pi}{2} - 1\right)\frac{Nr^2}{KI} \tag{7.291}$$

Finally, as far as M_0

$$\delta_x = \left(\frac{\pi}{2} - 1\right)\frac{M_0 r^2}{KI} \tag{7.292}$$

$$\delta_y = \frac{M_0 r^2}{KI} \tag{7.293}$$

$$\tau = \frac{\pi}{2}\frac{M_0 r}{KI} \tag{7.294}$$

Note that since the bending moment goes to zero at the end of the quarter of wave for symmetry, the moment M_0 is given by

$$M_0 = T\left(\frac{l}{2} - r\right). \tag{7.295}$$

Setting

$$\beta = \frac{r}{l}, \tag{7.296}$$

and recalling the equations above, we obtain

$$KI\delta_x = \left[1 + \left(\frac{\pi}{2} - 1\right)\left(\frac{1}{\beta} - 2\right)\right]\frac{Tr^3}{2} + \left(\frac{3}{4}\pi - 2\right)Nr^3, \tag{7.297}$$

$$KI\delta_y = \left(\frac{\pi}{2} + \frac{1}{\beta} - 2\right)\frac{Tr^3}{2} + \frac{1}{2}Nr^3, \tag{7.298}$$

and

$$KI\tau = \left[2 + \frac{\pi}{2}\left(\frac{1}{\beta} - 2\right)\right]\frac{Tr^2}{2} + \left(\frac{\pi}{2} - 1\right)Nr^2. \tag{7.299}$$

Because of the considerable rigidity of the torus' portion of the compensator in the radial direction, it is reasonable to assume that the deflection δ_x is almost zero. Setting $\delta_x = 0$ from (7.297), through a series of steps we obtain

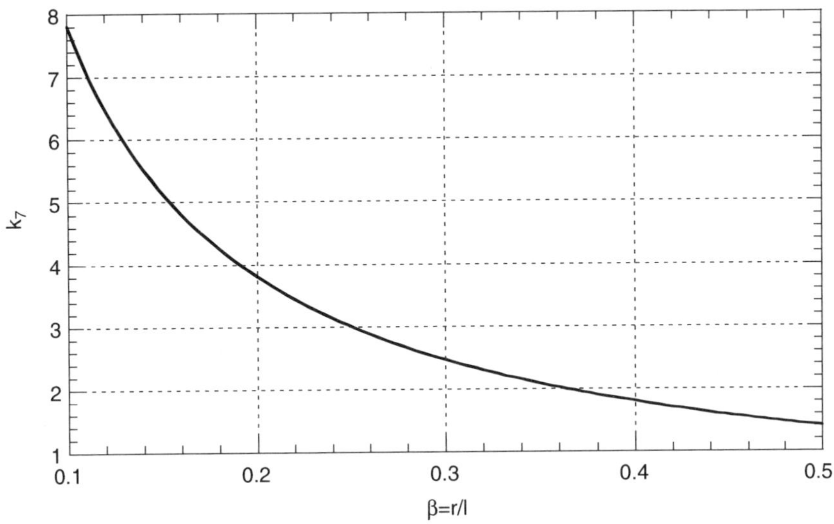

Fig. 7.32

$$N = -k_7 T, \tag{7.300}$$

given that

$$k_7 = \frac{6 - 2\pi + \dfrac{\pi - 2}{\beta}}{3\pi - 8}. \tag{7.301}$$

k_7 is shown in Fig. 7.32. Moreover, based on (7.298), (7.299), and (7.300) we obtain

$$KI\delta_y = (\alpha - 0.429 - k_1)\,\frac{Tr^3}{2}; \tag{7.302}$$

$$KI\delta_y = 0.571\,(1.376\alpha - 1 - k_7)\,Tr^2. \tag{7.303}$$

The deflection δ corresponding to the end of the quarter of wave can be obtained from the following equation:

$$KI\delta = KI\delta_y + KI\tau\left(\frac{l}{2} - r\right) + \frac{T}{3}\left(\frac{l}{2} - r\right)^3. \tag{7.304}$$

From (7.302) and (7.303), and recalling (7.296), through a series of steps we obtain the following equation:

$$KI\delta = k_8 T l^3, \tag{7.305}$$

given that

$$\begin{aligned} k_8 =& 0.5\beta^2 - 0.2146\beta^3 - 0.5\beta^3 k_7 + 0.04166\,(1 - 2\beta)^3 \\ &+ (0.3925 - 0.2855\beta - 0.2855\beta k_7)\,(1 - 2\beta)\,\beta \end{aligned}. \tag{7.306}$$

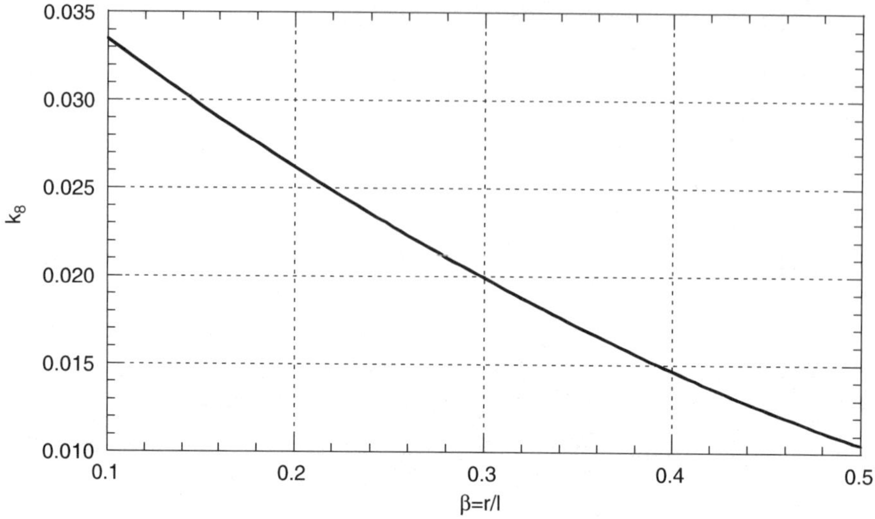

Fig. 7.33

k_8 is shown in Fig. 7.33.

Now we see that in the portion of the torus we have

$$M = T\left\lfloor \frac{l}{2} - r(1-\cos\vartheta) \right\rfloor + Nr(1-\sin\vartheta). \tag{7.307}$$

From both (7.296) and (7.300) we obtain

$$M = Tl\left\lfloor \frac{l}{2} - \beta(1-\cos\vartheta) - k_7\beta(1-\sin\vartheta) \right\rfloor. \tag{7.308}$$

The maximum value for M can be obtained by remembering that

$$\frac{dM}{d\vartheta} = Tl\beta\left[-\sin\vartheta + k_7\cos\vartheta\right]. \tag{7.309}$$

Setting the derivative to zero, we obtain the following value for ϑ:

$$\vartheta_0 = \arctan k_7. \tag{7.310}$$

Therefore, it is possible to obtain the maximum value for M from (7.310) that may be expressed as follows:

$$M_{\max} = k'_9 Tl. \tag{7.311}$$

The value for k'_9 may be obtained from Fig. 7.34.

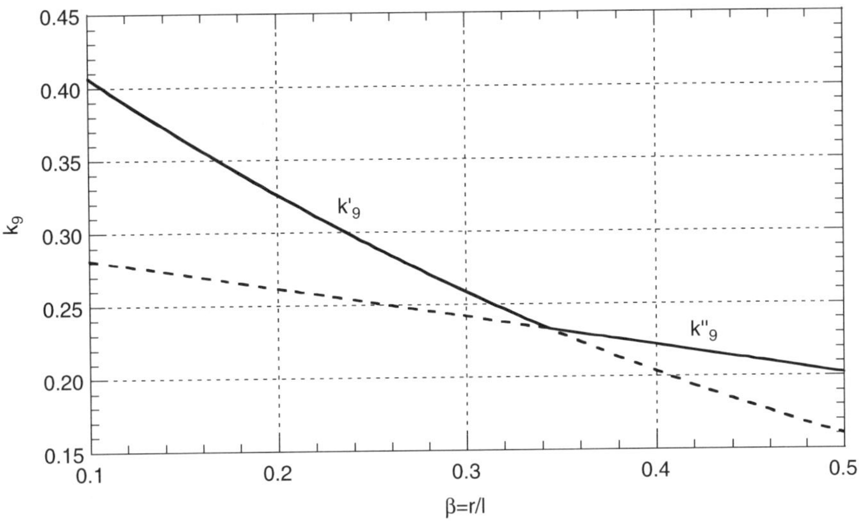

Fig. 7.34

One should also consider the minimum value for M that results when $\vartheta = 0$, and is given by

$$M_{\min} = -k''_9 Tl, \tag{7.312}$$

given that

$$k''_9 = \frac{1}{2} - k_7 \beta. \tag{7.313}$$

k''_9 is shown in Fig. 7.34, as well.

As we can see, both k'_9 and k''_9 may be crucial, depending on the case. For the resistance verification, as far as k_9 the greater of the two will have to be assumed. Given that $W = s^2/6$, the maximum stress is given by

$$\sigma_{\max} = 6k_9 \frac{Tl}{s^2}. \tag{7.314}$$

Recalling the meaning of parameter k_6 (7.286), we conservatively write for the wave compensator

$$\sigma_{\max} = 6k_9 k_6 \frac{Tl}{s^2}. \tag{7.315}$$

From (7.305) and indicating the deformation of a wave with Δ, we have

$$KI\Delta = 4k_8 Tl^3, \tag{7.316}$$

and recalling (7.258) we have

$$\Delta_{\max} = 8\frac{k_8 \left(1-\mu^2\right)}{k_6 k_9} \frac{l^2}{Es} \sigma_{\max}. \tag{7.317}$$

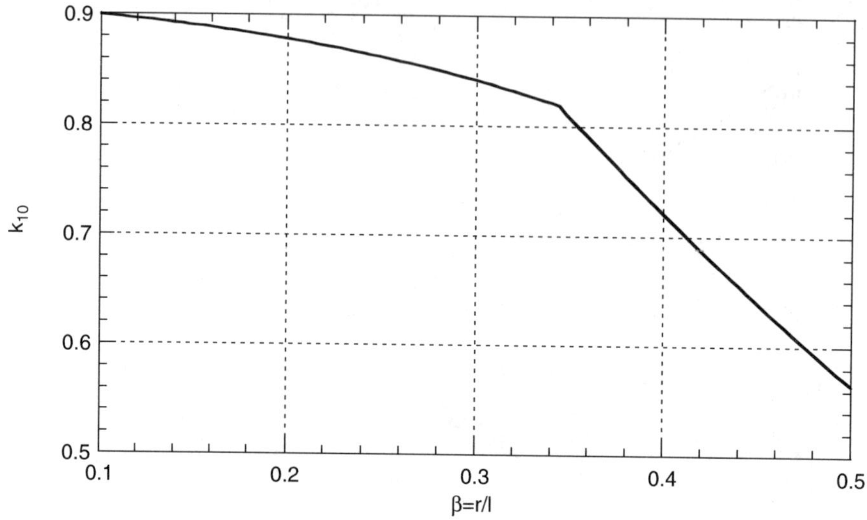

Fig. 7.35

Based on the criteria presented in Sect. 1.5 and 7.5 about piping, we may assume that

$$\sigma_{\max} = 1.5\,(f_c + f_h)\,, \tag{7.318}$$

where f_c and f_h are the allowable stresses under cold and hot conditions.

Therefore, from (7.317) we obtain

$$\Delta_{\max} = \frac{k_{10}}{k_6}\,(f_c + f_h)\,\frac{l^2}{Es}\,, \tag{7.319}$$

given that

$$k_{10} = 12\left(1 - \mu^2\right)\frac{k_8}{k_9}\,. \tag{7.320}$$

k_{10} is shown in Fig. 7.35.

Of course, according to the criteria considered for the values f_c and f_h the modulus of elasticity at room temperature is the one to be adopted (Sect. 7.5). The value $\Delta_{\max}$ obtained from (7.319) represents the maximum value for the expansion that any wave may be able to absorb.

Let us now compute the value for the reaction F_h exerted by the compensator on the piping as a result of the expansions. With reference to Fig. 7.30 and recalling the observations about the circular crown, we can write

$$F_h = \pi\frac{D + d}{2}T\,. \tag{7.321}$$

In this case T is referred to a strip of unitary width. Recalling (7.316) we have:

$$F_h = \frac{0.036}{k_8} E\,(D + d) \left(\frac{s}{l}\right)^3 \Delta, \tag{7.322}$$

where Δ is the actual expansion absorbed by a wave, given that $\Delta \le \Delta_{\max}$. It is worth noting that F_h can never be greater than the force that causes the total plastic flow of the most stressed section. Thus, recalling (7.311), (7.312), as well as k_6, and indicating with σ_{sh} the yield strength under hot temperature we can write

$$k_6 k_9 T l = \frac{1}{4} s^2 \sigma_{sh}; \tag{7.323}$$

then

$$F_h \le 0.393\,(D + d)\,\frac{\sigma_{sh} s^2}{k_6 k_9 l}; \tag{7.324}$$

or, since $\sigma_{sh} = 1.5 f_h$,

$$F_h \le 0.589\,(D + d)\,\frac{f_h s^2}{k_6 k_9 l}. \tag{7.325}$$

Let us also consider the following force:

$$F_{h0} = 0.393\,(D + d)\,\frac{f_h s^2}{k_6 k_9 l}, \tag{7.326}$$

that corresponds to the limit condition in terms of elastic behavior of the compensator.

If the value for F_h obtained from (7.322) is smaller than or equal to F_{h0} (because the value of Δ is smaller than the maximum allowable), when Δ goes to zero the reaction of the compensator onto the piping goes to zero as well.

On the other hand, if $F_h > F_{h0}$ when the expansion goes to zero, the compensator exerts a reaction F_c of opposite sign to F_h on the piping equal to

$$F_c = F_{h0} - F_h. \tag{7.327}$$

At this point one must set the criteria for the calculation of the thickness of the compensator. It must be such to ensure that the stresses caused by internal pressure be compatible with safety.

To that extent, we recall that (7.310) allows us to obtain the angle ϑ_0 where its positive bending moment assumes the maximum value due to the expansion Δ.

Based on Fig. 7.34, we see that it may be considered crucial for the sizing until about

$$\beta = \frac{r}{l} \le 0.3.$$

The distance between the two points of the semiwave when the moment reaches this maximum value is equal to

$$l_1 = l - 2r\,(1 - \cos\vartheta_0)\,. \tag{7.328}$$

The term in parenthesis varies between 0.87 and 0.625 for $\beta = 0.1 \div 0.3$.

In favor of resistance and factoring in the fact that for small values of β the second term on the right side of the equal sign in (7.328) has a limited impact on l_1, we adopt the following equation for l_1:

$$l_1 = l - 1.25r. \tag{7.329}$$

Thus

$$l_1 = \frac{D - d - 2.5r}{2}. \tag{7.330}$$

Because of expansion Δ, if $\beta \leq 0.3$ the compensator deforms plastically in correspondence of the points with maximum moment. As far as the rotations that the ideal beam of length l_1 (simply supported and subject to a uniformly distributed load) would show at the ends, these plastic rotations have the same direction on one end, and the opposite one at the other end. This indicates a behavior of the beam that may be equal to a beam that is perfectly clamped at one end, and simply supported at the opposite end. In fact, at the end where the plastic rotation has the same direction than the rotation of the beam that is simply supported, we are clearly faced with a plastic hinge. Viceversa, at the other end there are stresses of the opposite sign to those generated by a moment of clamped edge of the evenly loaded beam. Therefore, the beam is prestressed, and one can reasonably assume a perfect clamped edge and ignore the verification in this position.

In the case of a beam with one clamped edge and one simply supported edge, the maximum positive moment under an evenly distributed load p is equal to

$$M = \frac{9}{128} p l_1^2 \cong 0.07 p l_1^2. \tag{7.331}$$

We simply considered a strip of unitary length with $W = s^2/6$. Moreover, since these are primary bending stresses we may adopt a value of the allowable stress equal to $1.5 f_c$, corresponding to the yield strength.

So we have:

$$0.07 p l_1^2 = \frac{s^2}{6} 1.5 f_h = 0.25 s^2 f_h; \tag{7.332}$$

then

$$s = 0.53 l_1 \sqrt{\frac{p}{f_h}}; \tag{7.333}$$

and finally, recalling (7.331)

$$s = 0.265 l_1\,(D - d - 2.5r)\,\sqrt{\frac{p}{f_h}}. \tag{7.334}$$

Equation (7.335) is valid, as we said, for $\beta \leq 0.3$. On the other hand, if $\beta > 0.3$, considering that the minimum moment as a result of Δ is in absolute terms greater than the maximum one and becomes manifest at the top of the wave, the thickness s must be obtained from the following equation:

$$s = 0.265 l_1 (D - d) \sqrt{\frac{p}{f_h}}. \tag{7.335}$$

Based on (7.334), (7.335), and Fig. 7.35, we determine that it is convenient to adopt a large value for the radius r, as long as we do not have $\beta > 0.3$. In fact, as long as $\beta \leq 0.3$, if r increases a smaller thickness may be used, and the reduction of parameter k_{10} notwithstanding, the value for $\Delta_{\max}$ goes up. Viceversa, if $\beta > 0.3$, we must use (7.335) and the thickness will increase. Moreover, for $\beta > 0.35$, the value for k_{10} considerably decreases, and consequently the value of $\Delta_{\max}$ is also considerably reduced.

In summary, we would like to examine an example. Assuming that

$d = 500\,\text{mm}$
$D = 700\,\text{mm}$
$r = 25\,\text{mm}$
$p = 0.5\,\text{N}\,\text{mm}^{-2}$
$E = 210000\,\text{N}\,\text{mm}^{-2}$
$\sigma_{sc} = 210\,\text{N}\,\text{mm}^{-2}$
$\sigma_{sh} = 145\,\text{N}\,\text{mm}^{-2}$
$f_c = \frac{210}{1.5} = 140\,\text{N}\,\text{mm}^{-2}$
$f_h = \frac{145}{1.5} = 96.7\,\text{N}\,\text{mm}^{-2}$

We have

$l = \dfrac{700 - 500}{2} = 100\,\text{mm}$
$\alpha = \dfrac{D}{d} = 1.4$
$\beta = \dfrac{r}{l} = 0.25$
$k_6 = 1.133$ (according to (7.286))
$k_8 = 0.0228$ (according to Fig. 7.33)
$k_9 = 0.29$ (according to Fig. 7.34)
$k_{10} = 0.862$ (according to Fig. 7.35)

Applying (7.334) we obtain

$$s = 0.265\,(700 - 500 - 62.5)\sqrt{\frac{0.5}{96.7}} = 2.6\,\text{mm}.$$

Let us assume $s = 3\,\text{mm} > 2.6$. Based on (7.319)

$$\Delta_{\max} = \frac{0.862}{1.133}\,(140 + 96.7)\,\frac{100^2}{210000 \times 3} = 2.86\,\text{mm}.$$

We assume that an expansion of 10 mm has to be absorbed with a four-wave compensator.

We then have

$$\Delta = \frac{10}{4} = 2.5\,\text{mm} < 2.86\,\text{mm}.$$

Based on (7.322) we have

$$F_h = \frac{0.036}{0.0228} 210000\,(700 + 500) \left(\frac{3}{100}\right)^2 2.5 = 26858\,\text{N}.$$

From (7.324) we have

$$F_h \leq 0.393\,(700 + 500)\,\frac{145 \times 3^2}{1.133 \times 0.29 \times 100} = 18730\,\text{N}.$$

Applying (7.326) we obtain

$$F_{h0} = 0.393\,(700 + 500)\,\frac{96.7 \times 3^2}{1.133 \times 0.29 \times 100} = 12487\,\text{N}.$$

Since $F_{h0} < 26858$, we finally have

$$F_h = 18730\,\text{N}$$

$$F_c = 12487 - 26858 = -14371\,\text{N}$$

8

The Influence of Holes

8.1 Hole Lines on Cylinders, Spheres, and Cones

As we have seen in Sect. 2.3, the presence of a hole causes a clustering of the force lines as one gets closer to the edge of the hole itself. Thus, there is a stress peak in this area, and it is necessary to establish whether and how it should be factored in during the sizing of the vessel (see Fig. 8.1). Before examining this issue, or rather the effect of the so-called isolated hole, i.e., one not influenced by surrounding holes that will be addressed in Sect. 8.3, we will look at hole lines.

If different holes are located along a random hole line and are sufficiently close (we will discuss the conditions for this to be true shortly), the analysis turns out to be simpler than that of the isolated hole, contrary to what one may think. This is possible because of plastic collaboration of all fibers between holes.

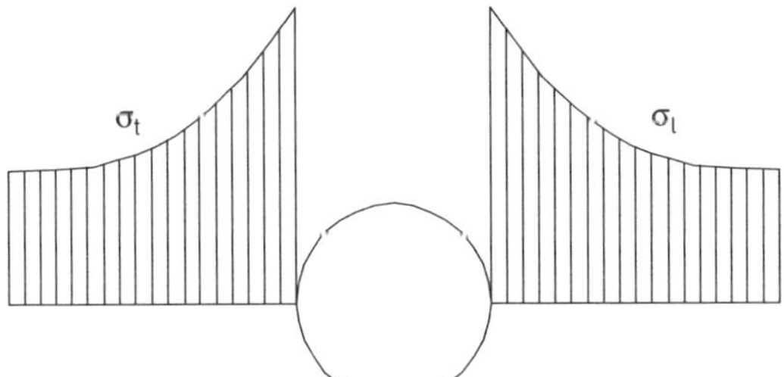

Fig. 8.1

In other words, based on the criterion discussed in Sect. 1.4, the peaks around the edges of the holes are ignored, and calculations are based on the average values of the stresses between holes. If there is only one stress present, all is reduced to a simple comparison of sections where the force lines run through, in the case of a wall with or without holes, as is the case, for instance, when the hole line is located on the generatrix of a cylinder (see Fig. 8.2).

In that case, if we consider a unitary thickness of the cylinder the circumferential force corresponding to the pitch l is given by

$$F = \sigma_{tm} l, \tag{8.1}$$

with σ_{tm} being the average hoop stress of the cylinder without holes.

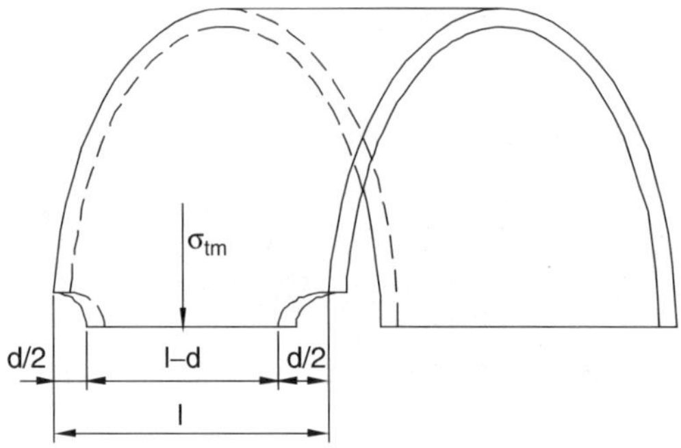

Fig. 8.2

If A is the interested cross-sectional area, the average stress σ'_{tm} between holes is therefore equal to

$$\sigma'_{tm} = \frac{F}{A} = \frac{F}{l-d} = \sigma_{tm}\frac{l}{l-d} = \frac{\sigma_{tm}}{z}. \tag{8.2}$$

The dimensionless quantity z is the efficiency of ligaments. It identifies the increase in value of the average value of the hoop membrane stress caused by the holes. If the hole is generically oriented with respect to the generatrixes, the problem is more complex, yet follows the same criteria that led us to (8.2).

Let us now look at Fig. 8.3 that shows two holes of diameter d at distance l between centers and a joining line that forms the angle α with the generatrixes. Finally, σ_{tm} and σ_a are the average hoop stress and the longitudinal stress in the area of the cylinder without holes, respectively.

The force F_1 exerted in circumferential direction on one section without holes with unitary thickness and width equal to the projection of the pitch l on the generatrix is given by

$$F_1 = \sigma_{tm} l \cos\alpha. \tag{8.3}$$

Assuming that

$$\sigma_a = m\sigma_{tm}, \tag{8.4}$$

force F_2 exerted in the longitudinal direction on one section without holes with a width equal to the projection of l on the perpendicular to the generatrix is given by

$$F_2 = m\sigma_{tm} l \sin\alpha. \tag{8.5}$$

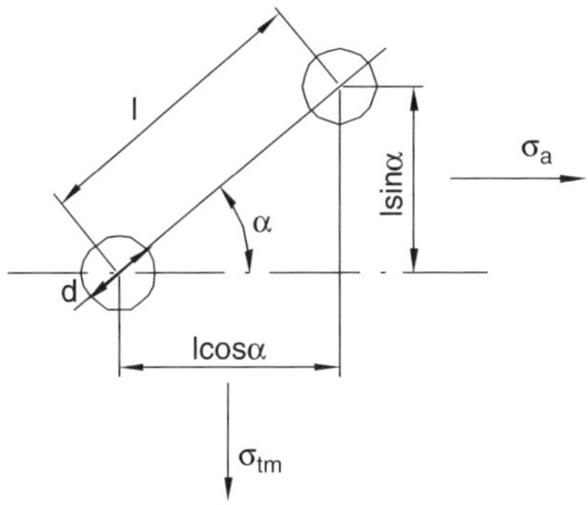

Fig. 8.3

Decomposing F_1 and F_2, as shown in Fig. 8.4, i.e., in the parallel and orthogonal components to the line joining the centers of both holes, we obtain

$$\begin{aligned} N_1 &= F_1 \cos\alpha = \sigma_{tm} l \cos^2\alpha \\ T_1 &= F_1 \sin\alpha = \sigma_{tm} l \sin\alpha\cos\alpha \\ N_2 &= F_2 \sin\alpha = m\sigma_{tm} l \sin^2\alpha \\ T_2 &= F_2 \cos\alpha = m\sigma_{tm} l \sin\alpha\cos\alpha \end{aligned} \tag{8.6}$$

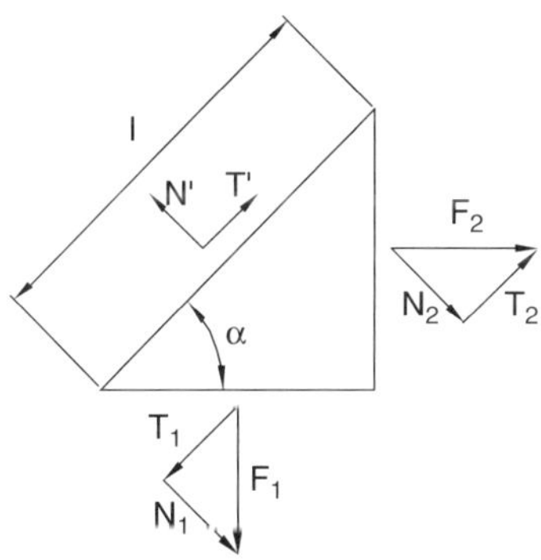

Fig. 8.4

The orthogonal force on the section going through the center of the holes is therefore given by

$$N' = N_1 + N_2 = \sigma_{tm} l \left[m + (1-m)\cos^2\alpha\right]. \tag{8.7}$$

The shear force is given by

$$T' = T_1 - T_2 = \sigma_{tm} l\,(1-m)\sin\alpha\cos\alpha. \tag{8.8}$$

As usual, let us consider the average values of the stresses. Observing that the resisting section is equal to $(l - d)$, we will see that the normal and shear stresses corresponding to N' and T' are given by

$$\sigma' = \frac{N'}{l-d} = \sigma_{tm} \frac{l}{l-d} \left[m + (1-m)\cos^2\alpha\right] \tag{8.9}$$

and

$$\tau' = \frac{T'}{l-d} = \sigma_{tm} \frac{l}{l-d} (1-m)\sin\alpha\cos\alpha. \tag{8.10}$$

Let us now consider the normal force exerted on a section without holes of unitary thickness and width orthogonal to the joining line of the two holes. From Fig. 8.5 we have

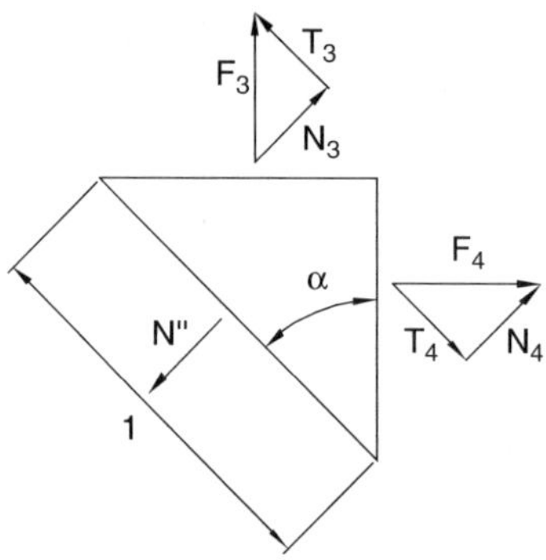

Fig. 8.5

$$F_3 = \sigma_{tm} \sin\alpha; \tag{8.11}$$
$$F_4 = m\sigma_{tm} \cos\alpha. \tag{8.12}$$

The components F_3 and F_4 parallel to the joining line between the two holes are given by

$$\begin{aligned} N_3 &= \sigma_{tm} \sin^2\alpha \\ N_4 &= m\sigma_{tm} \cos^2\alpha \end{aligned} \tag{8.13}$$

At the same time, the normal force N'' on the considered section, (i.e., the normal stress on that section) is given by

$$N'' = N_3 + N_4 = \sigma_{tm} \left[m + (1-m)\sin^2\alpha\right]. \tag{8.14}$$

In the area between the two holes the value of this stress is influenced by the reciprocal position of the holes. If the holes are very far apart from one another, i.e., if $l = \infty$, the value of the stress is as in (8.14). Viceversa, if $l = d$

the value of this stress is zero. If the holes are very close, the value of the stress is very small because the force lines go around the pair of holes without affecting the area between them in any significant way.

Siebel and Schweigerer, the authors of the analysis above, proposed the introduction of a corrective factor $(l-d)/l$ that accounts somewhat for the phenomenon. It corresponds to the above limit conditions since it is equal to 1 for $l=\infty$ and equal to 0 for $l=d$. We believe, though, that it does not reflect the phenomenon in a sufficiently approximated way. Note that the behavior of stress $\sigma_{\rho y}$ in Fig. 2.11 shows similar characteristics.

For example, if we had a second hole with the center at distance $4r$ from the center of the drawn hole and with equal radius, we would have $l=4r$ and $(l-d)/l=0.5$. Intuitively, the average value of the stress $\sigma_{\rho y}$ in the area between the two holes would be, in fact, considerably smaller than 0.5σ.

We believe to be closer to reality by introducing a corrective factor to substantially reduce the value of the stress when the holes are very close. Therefore, based on (8.14) and with reference to the area between both holes we write:

$$\sigma''=N''\left(\frac{l-d}{l}\right)^2=\sigma_{tm}\left(\frac{l-d}{l}\right)^2\left[m+(1-m)\sin^2\alpha\right], \qquad (8.15)$$

and stress τ'' must be assumed to be equal to τ'.

We recall that based on Sect. 3.1 the average value of the radial stress may be conventionally assumed to be $-p/2$. Thus, we may build the three characteristic circles shown in Fig. 8.6 on Mohr's plane. The circle based on σ', σ'', τ' and τ'' has a radius equal to

$$R=\sqrt{\left(\frac{\sigma'-\sigma''}{2}\right)^2+\tau'^2}. \qquad (8.16)$$

The three principle average stresses σ_I, σ_{II} and σ_{III} are consequently identified, and their values increase going from σ_I to σ_{III}, according to the usual convention (see Sect. 1.3).

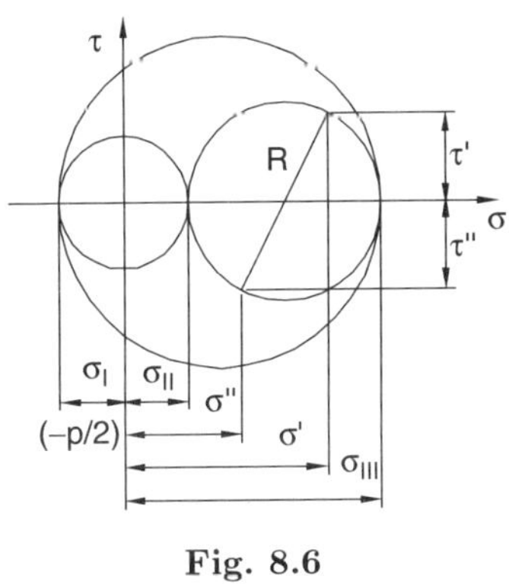

Fig. 8.6

Specifically, σ_{III}, that is of greatest interest to us, has the following equation:

$$\sigma_{III} = \frac{\sigma' + \sigma''}{2} + R = \frac{1}{2}\left[\sigma' + \sigma'' + \sqrt{(\sigma' - \sigma'')^2 + 4\tau'^2}\right]; \tag{8.17}$$

where z represents the ratio between the hoop stress σ_{tm} and the maximum principal average stress σ_{III}.

This ratio, called efficiency of ligaments, identifies the impact of the holes on the behavior of the stresses. One should not forget that the maximum principal stress σ_{III} coincides with σ_{tm} in the absence of holes. Therefore, we have

$$z = \frac{\sigma_{tm}}{\sigma_{III}}; \tag{8.18}$$

and

$$\sigma_{III} = \frac{\sigma_{tm}}{z}. \tag{8.19}$$

As usual, following Guest's failure theory, the average ideal membrane stress is given by

$$\sigma_{id(m)} = \sigma_{III} - \sigma_I = \frac{\sigma_{tm}}{z} + \frac{p}{2}. \tag{8.20}$$

Recalling the meaning of σ_{tm} at this point, as far as cylinders we obtain the following from (8.20):

$$\sigma_{id(m)} = \frac{p}{z}\frac{D_e - 2s}{2s} + \frac{p}{2}. \tag{8.21}$$

Assuming that the ideal stress is equal to the basic allowable stress f, from (8.21) we obtain

$$p\,(D_e - 2s + zs) = 2fzs \tag{8.22}$$

and

$$s\,[2fz + p\,(2 - z)] = pD_e. \tag{8.23}$$

Finally

$$s = \frac{pD_e}{2fz + p\,(2 - z)}. \tag{8.24}$$

The value z is smaller than or at the most equal to one. To further favor resistance (8.24) may be reduced to

$$s = \frac{pD_e}{2fz + p}. \tag{8.25}$$

If we compare (8.25) with (3.88), we establish the similarity of the equations that differ because of the introduction of the efficiency of ligaments z to factor in the presence of holes.

Based on the analysis above, we now obtain the equation for z valid for a hole line, regardless of its orientation. First of all, based on (8.17) and (8.18) we have

$$z = \frac{\sigma_{tm}}{\sigma_{III}} = \frac{2}{\dfrac{\sigma'}{\sigma_{tm}} + \dfrac{\sigma''}{\sigma_{tm}} + \sqrt{\left(\dfrac{\sigma'}{\sigma_{tm}} - \dfrac{\sigma''}{\sigma_{tm}}\right)^2 + 4\left(\dfrac{\tau'}{\sigma_{tm}}\right)^2}}. \tag{8.26}$$

We now introduce the quantities A, B, and C. Recalling (8.9) we write

$$A = \frac{\sigma'}{\sigma_{tm}} = \frac{m + (1-m)\cos^2\alpha}{\dfrac{l-d}{l}}, \tag{8.27}$$

and recalling (8.15) we write

$$B = \frac{\sigma''}{\sigma_{tm}} = \left(\frac{l-d}{l}\right)^2 \left[m + (1-m)\sin^2\alpha\right]. \tag{8.28}$$

Finally, from (8.10)

$$C = \frac{\tau'}{\sigma_{tm}} = \frac{(1-m)\sin\alpha\cos\alpha}{\dfrac{l-d}{l}}. \tag{8.29}$$

Equation (8.26) then becomes

$$z = \frac{2}{A + B + \sqrt{(A-B)^2 + 4C^2}} \tag{8.30}$$

Note that if the stress σ_a is caused by pressure only, or if the increase in value of σ_a caused by pressure from external forces (weight of the vessel, thrusts coming from pipes, etc.) is negligible, the value of m may be obtained from the analysis in Sect. 3.1.
In fact,

$$\sigma_{tm} = \frac{p(D-2s)}{2s} = \frac{p}{a-1} \tag{8.31}$$

and

$$\sigma_a = \frac{p}{a^2-1}. \tag{8.32}$$

Thus

$$m = \frac{1}{a+1}. \tag{8.33}$$

Recalling the meaning of a, we see that the value of m is always smaller than 1/2. Assuming the latter value for m, the process favors resistance.

In this case, the equations for A, B and C become, respectively

$$A = \frac{1}{2}\frac{1+\cos^2\alpha}{\dfrac{l-d}{l}}, \tag{8.34}$$

$$B = \frac{1}{2}\left(\frac{l-d}{l}\right)^2 \left(1+\sin^2\alpha\right), \tag{8.35}$$

and

$$C = \frac{1}{2} \frac{\sin \alpha \cos \alpha}{\dfrac{l-d}{l}}. \tag{8.36}$$

Interestingly, if the hole is along a generatrix $\alpha = 0$ and $C = 0$. Therefore, from (8.30) we have

$$z = \frac{1}{A} = \frac{l-d}{l}. \tag{8.37}$$

Of course, we find the same equation for z we had obtained directly through (8.12). As an example, if the holes are placed along a circumference $\alpha = 90°$ and $C = 0$. Therefore, we have

$$z = \frac{1}{A} = 2\frac{l-d}{l}. \tag{8.38}$$

As one can see, if the value of the pitch and the diameter are equal, the value of z on the circumference is twice that of holes aligned along a generatrix. This result was easily predictable by observing that the holes along a generatrix cause an increase in hoop stress (between holes), while those along a circumference cause an increase in longitudinal stress, conventionally assumed to be half of the previous one.

Basically, the holes can be located along the wall of the vessel in such a way to form hole lines both along the generatrix and the circumference or along an oblique line. In this case, one must compute the efficiency of ligaments along all hole lines and assume the smallest of the resulting values, as long as it is not greater than one, of course (this may occur, e.g., with a single hole line along a circumference if the pitch is greater than twice the diameter).

Now we introduce the following dimensionless quantities:

$$\beta = \frac{l-d}{l}; \tag{8.39}$$

$$k_1 = \frac{4}{1 + \cos^2\alpha + \beta^3\left(1 + \sin^2\alpha\right) + \sqrt{\left[1 + \cos^2\alpha - \beta^3\left(1 + \sin^2\alpha\right)\right]^2 + 4\sin^2\alpha\cos^2\alpha}} \tag{8.40}$$

Based on (8.30) and (8.34)–(8.36) we obtain

$$z = k_1\beta. \tag{8.41}$$

k_1 is shown in Fig. 8.7. Of course, for $\alpha = 0°$ we have $k_1 = 1$, whereas for $\alpha = 90°$ we have $k_1 = 2$. When $k_1 > 1/\beta$ the curves have no more meaning since $z > 1$. Therefore, they have been interrupted.

It may be interesting to note that the set of curves in Fig. 8.7 may be substituted by the one curve in Fig. 8.8. If k_2 is the approximate value of k_1 obtained from this curve, we can see that it differs from the exact value by plus or minus 4% at most, for $\beta = 0.3 \div 1$ and $z \leq 1$. Assuming that the error does not exceed ±6%, it is also possible to use the following equation:

$$k_2 = \frac{2}{1 + \cos^2\alpha\left(1 + \sin^2\alpha\right)}, \tag{8.42}$$

the latter being much simpler than (8.40).

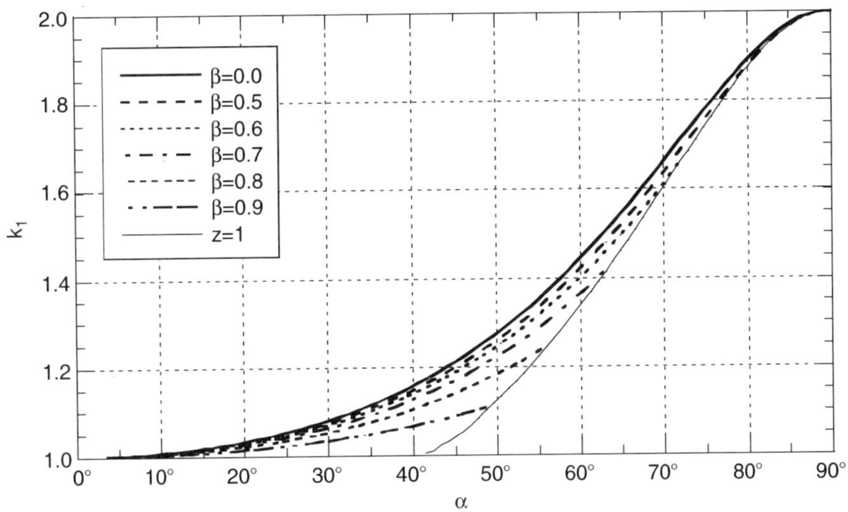

Fig. 8.7

Hence, if we assume to compute the efficiency z with a relatively small margin of error, it is possible to write

$$z = k_2\beta = k_2\frac{l-d}{l}, \tag{8.43}$$

where k_2 is only a function of angle α.

For thick cylinders, the average ideal stress between holes in correspondence of the internal fiber may exceed the yield strength or the rupture stress per creep if one adopts (8.25), i.e., if one only assumes that the ideal average membrane stress is equal to f.

In agreement with Sects. 1.5 and 3.2, if we also assume that the average ideal stress between holes in correspondence of the internal fiber does not exceed $1.5f$ instead, one shall proceed as follows.

First of all, let us remember that the radial stress in correspondence of the internal fiber is $-p$. Indicating the hoop stress in the cylinder without holes and in the same position with σ_{ti}, by similarity with (8.19) we write

$$\sigma_{III} = \frac{\sigma_{ti}}{z}. \tag{8.44}$$

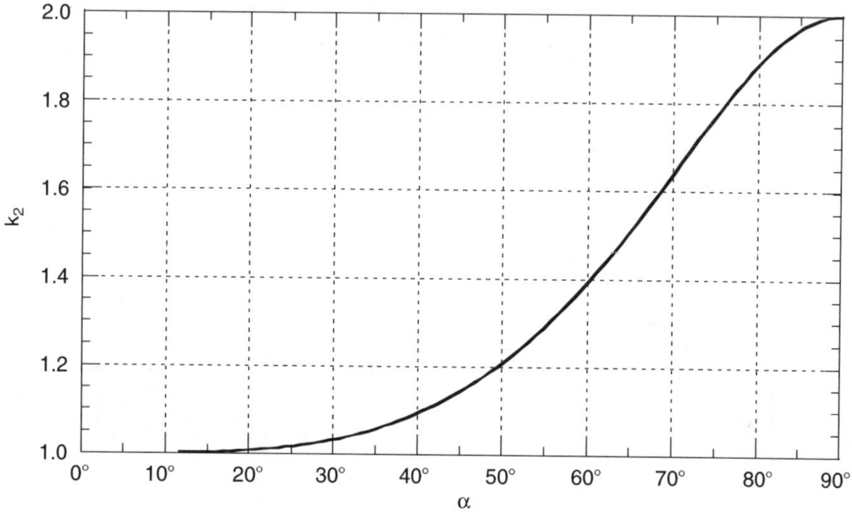

Fig. 8.8

This way the variation of σ_t through the thickness is factored in, but the stress peaks at the edges of the holes are still ignored.

The corresponding ideal stress is equal to

$$\sigma_{id} = \sigma_{III} - \sigma_I = \frac{\sigma_{ti}}{z} + p. \tag{8.45}$$

On the other hand, based on Sect. 3.1 we know that σ_{ti} in the cylinder without holes is equal to

$$\sigma_{ti} = p\frac{a^2+1}{a^2-1}; \tag{8.46}$$

hence, from (8.45)

$$\sigma_{id} = \frac{p}{z}\frac{a^2+1}{a^2-1} + p. \tag{8.47}$$

Assuming that $\sigma_{id} = 1.5f$, (that corresponds to the yield strength) we have

$$\frac{p}{z}\frac{a^2+1}{a^2-1} + p = 1.5f. \tag{8.48}$$

From (8.48) through steps that are not shown we obtain

$$a = \sqrt{\frac{1.5fz + p\,(1-z)}{1.5fz - p\,(1+z)}}. \tag{8.49}$$

On the other hand, recalling the meaning of a,

$$s = \left(1 - \frac{1}{a}\right)\frac{D_e}{2}, \tag{8.50}$$

and from (8.49) we obtain

$$s = \left(1 - \sqrt{\frac{1.5fz - p\,(1+z)}{1.5fz + p\,(1-z)}}\right)\frac{D_e}{2}, \tag{8.51}$$

or

$$s = \left(1 - \sqrt{\frac{fz - 0.666p\,(1+z)}{fz + 0.666p\,(1-z)}}\right)\frac{D_e}{2}. \tag{8.52}$$

By comparing (8.52) with (3.94), the two equations coincide for $z = 1$.

Of course, the former substitutes (8.25) only when the former results in greater thickness. Fig. 8.9 highlights the range of validity of both equations. Since these are very thick cylinders, note that if (8.34)–(8.36) may always be

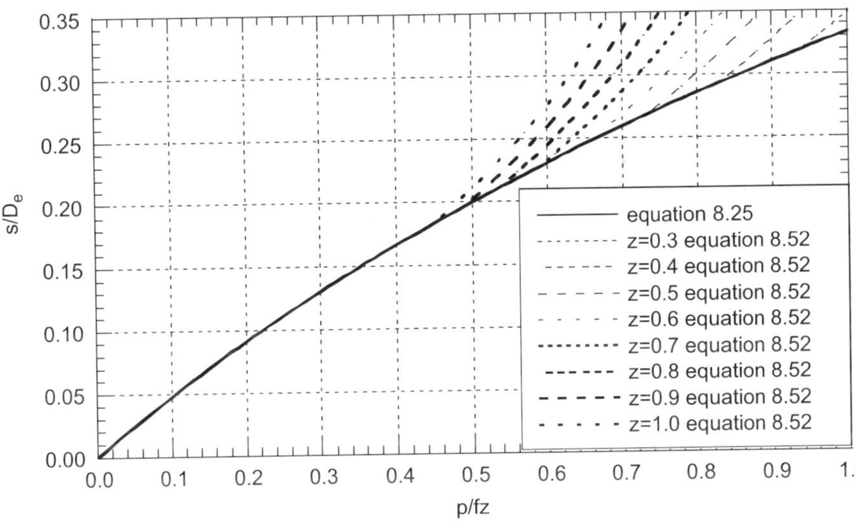

Fig. 8.9

used to compute z, the value of coefficient m may be considerably smaller than 1/2. Based on (8.25) the value of m is given by (8.33). Based on (8.52) m is equal to the ratio σ_a/σ_{ti}; thus, based on (8.32) and (8.46)

$$m = \frac{1}{a^2+1}. \tag{8.53}$$

Figure 8.10 shows m, in both instances, as a function of a. As we can see, it can be considerably smaller than the approximate value 1/2 that is central to (8.34)–(8.36). Therefore, it may be best to refer to (8.27), (8.28), and (8.29) present in the equation for z to compute the coefficients A, B and C and adopt the actual value of m.

This is important only in the case of oblique hole lines or those along a circumference, whereas it has no impact in the case of hole lines along a generatrix. So far, we focused on cylinders with holes. The considerations that were made can be extended to spheres and cones, as well. When we consider spheres, a rigorous analysis according to Sect. 5.2 produces a complex equation. For the sake of simplicity and to favor resistance (without increasing the thickness necessarily) we may proceed as follows.

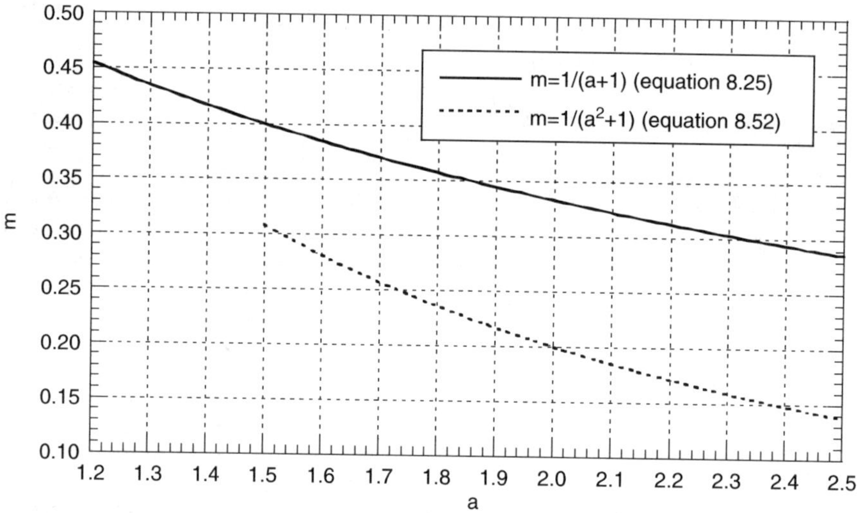

Fig. 8.10

The adoption of (5.63) for the sphere without holes corresponds to considering the following as ideal membrane stress:

$$\sigma_{id(m)} = \frac{p}{4}\frac{D_e - s}{s}. \tag{8.54}$$

This was already pointed out in Sect. 5.2. We may consider the following as average ideal membrane stress between holes, where z has the known meaning

$$\sigma_{id(m)} = \frac{p}{4z}\frac{D_e - s}{s}. \tag{8.55}$$

Through the coefficient z we do not simply factor in the increase in circumferential stress between holes, but we also have an increase in radial stress that is not influenced by the holes. In other words, the ideal stress is assumed to be equal to

$$\sigma_{id(m)} = \sigma_{III} - \frac{\sigma_I}{z}. \tag{8.56}$$

This is in contrast to (8.20) relative to cylinders and takes a more conservative approach. From (8.55) assuming $\sigma_{id(m)} = f$ we obtain

$$s = \frac{pD_e}{4fz + p}. \tag{8.57}$$

By comparing (8.57) with (5.63) we see that, as expected, the two equations coincide for $z = 1$.

Note that (8.25) obtained from (8.24) also matches the simplified and cautionary concept to increase the radial stress σ_r through z. As far as the value z, given the equality of the principal circumferential and meridian stresses, the coefficient m in (8.27) and the ensuing ones is equal to one.

We establish that $C = 0$ and that based on (8.30) in any case $z = 1/A$. On the other hand,

$$A = \frac{l}{l - d}; \tag{8.58}$$

hence, in any direction of the hole line

$$z = \frac{l - d}{l}. \tag{8.59}$$

This is according to expectations and could have been obtained following the same computation process that lead to (8.2), as well. Similar considerations to those relative to cylinders can be made for thick spheres.

Indicating the circumferential stress in the sphere without holes in correspondence of the internal fiber with σ_{ti}, we know from Sect. 5.1 that

$$\sigma_{ti} = p\frac{\frac{a^3}{2} + 1}{a^3 - 1}. \tag{8.60}$$

Hence, corresponding to the hole line we have

$$\sigma_{id} = \frac{p}{z}\frac{\frac{a^3}{2} + 1}{a^3 - 1} + p; \tag{8.61}$$

assuming that $\sigma_{id} - 1.5f$, we obtain

$$\frac{p}{z}\frac{\frac{a^3}{2} + 1}{a^3 - 1} + p = 1.5f. \tag{8.62}$$

From (8.62) through steps that are not shown here we obtain

$$a = \sqrt[3]{\frac{1.5fz + p(1 - z)}{1.5fz - p(0.5 + z)}}. \tag{8.63}$$

On the other hand, recalling the meaning of a

$$s = \left(1 - \frac{1}{a}\right)\frac{D_e}{2}; \tag{8.64}$$

then, from (8.63)

$$s = \left(1 - \sqrt[3]{\frac{1.5fz - p(0.5 + z)}{1.5fz + p(1 - z)}}\right)\frac{D_e}{2} \tag{8.65}$$

or

$$s = \left(1 - \sqrt[3]{\frac{fz - 0.666p(0.5 + z)}{fz + 0.666p(1 - z)}}\right)\frac{D_e}{2}. \tag{8.66}$$

Comparing (8.66) with 5.75, we see that they coincide when $z = 1$.

Of course, (8.66) substitutes (8.57) only when it generates a greater thickness. Figure 8.11 shows the relative range of applicability.

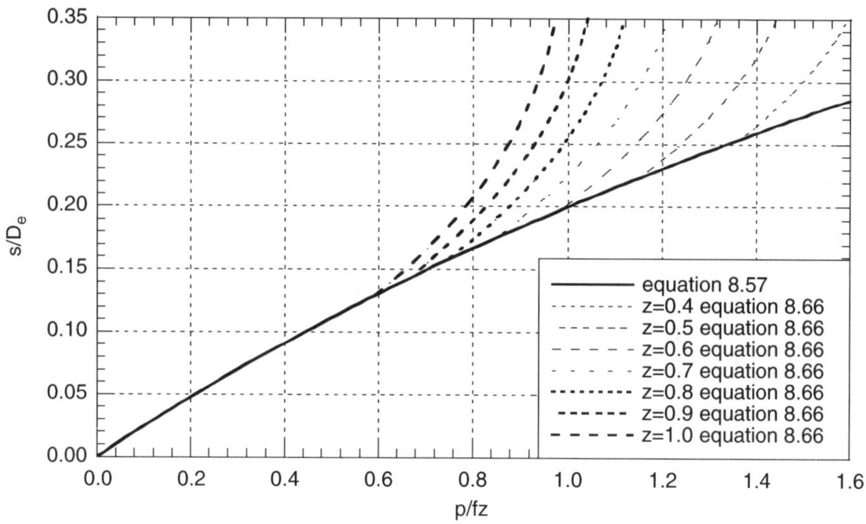

Fig. 8.11

As far as cones, from (2.16) we know that considering the cone as a membrane the meridian stress is equal to half the circumferential stress, similarly to the cylinder where the longitudinal stress is half the hoop one. The same considerations already made for cylinders relative to the efficiency of ligaments z of the hole lines are true for cones, as well.

As far as the fundamental equation of cones with holes, based on the analysis so far as well as (6.67), we can write

$$s = \frac{pD_i}{2fz - p}\frac{1}{\cos\alpha}, \tag{8.67}$$

remembering that the angle α in (8.67) is the semiangle at the tip of the cone.

The need to size a thick cone is unlikely, and for this reason we shall not discuss it here. At this point, we must set the conditions that determine when we are faced with hole lines as opposed to a series of isolated holes (adopting the calculation criteria introduced here).

As we have seen, the analysis of hole lines is based on the approach that considers all fibers between holes as collaborating. For this to happen without generating intolerable stress peaks at the edge of the holes, the distance between the holes must not exceed certain values. A rigorous study of the phenomenon is possible on a case by case basis only, or with finite element analysis programs, or through experimental research. If we make some hypothesis as to the behavior of the phenomenon in a sufficiently realistic way, it is nonetheless possible to put together a simple theoretical calculation, capable of delivering acceptable values for the characteristic distance among holes.

First of all, we observe that the clustering of the force lines at the edge of the hole causes greater deformations in this location, and they produce a more significant increase in radius compared to that in areas far away from the hole. This phenomenon is reminiscent of the analysis in Sect. 2.2 about edge effects, and that will help us develop the calculation criteria to follow. The fact that the size of the area of a collaborating wall around a hole may be correlated to the quantity $\sqrt{D_m s}$, typical of edge effects, is after all confirmed by experimental research, and the codes of many countries refer to this principle as well.

Furthermore, we observe that until now we analyzed vessels with holes disregarding the fact that a nozzle is welded in correspondence of a hole. The presence of the nozzle, besides having potential influence on the efficiency of ligaments z, as we shall see in Sect. 8.2, represents a stiffening of the wall of the vessel, in the sense that it goes against a possible rotation of the wall itself in correspondence of the edge of the hole.

Finally, we should not forget that, in contrast to what happens for edge effects that involve the entire circumference, the phenomenon under study is limited to the area along the hole line. In view of these considerations, we can realistically say that there is no rotation at the edge of the hole. Based on (2.45) we establish that if $y' = 0$ for $x = 0$, it is reduced to the following:

$$y = C_1 e^{-\beta x} (\sin \beta x + \cos \beta x) \, . \tag{8.68}$$

Indicating the radial deflection with y_0 for $x = 0$, we have

$$y = y_0 e^{-\beta x} (\sin \beta x + \cos \beta x) \, . \tag{8.69}$$

We recall that the hoop stress ${\sigma''}_t$ generated by the deflection y is proportional to the deflection. Hence, indicating the value of the stress mentioned above with ${\sigma''}_{to}$ for $x = 0$, from (8.69) we may write

$${\sigma''}_t = {\sigma''}_{t0} e^{-\beta x} (\sin \beta x + \cos \beta x) \, . \tag{8.70}$$

The average value of ${\sigma''}_t$ in the area ranging from the edge of the hole ($x = 0$) and the fiber at distance L from the edge is given by

$$\sigma''_{tm} = \frac{\sigma''_{t0}}{L} \int_0^L e^{-\beta x} (\sin \beta x + \cos \beta x)\, dx; \tag{8.71}$$

resolving the integral we obtain

$$\sigma''_{tm} = \sigma''_{t0} \frac{1 - e^{-\beta L} \cos \beta L}{\beta L}. \tag{8.72}$$

We define the ideal membrane stress in the areas without holes of the vessel as $\sigma'_{id(m)}$.

If z is the efficiency of ligaments of the hole line under examination, critical in terms of the sizing of the vessel, based on the previous analysis

$$\sigma'_{id(m)} = zf. \tag{8.73}$$

Recalling (8.69), in correspondence of the holes the average ideal membrane stress is given by

$$\sigma_{id(m)} = \sigma'_{id(m)} + \sigma''_{tm} = zf + \sigma''_{t0} \frac{1 - e^{-\beta L} \cos \beta L}{\beta L}. \tag{8.74}$$

If we consider all fibers between the edge of the hole and the fiber at distance L from it as collaborating, $\sigma_{id(m)}$ must be equal to the basic allowable stress f. Therefore, from (8.74) we obtain

$$\frac{\sigma''_{t0}}{f} \frac{1 - e^{-\beta L} \cos \beta L}{\beta L} = 1 - z. \tag{8.75}$$

In correspondence of the edge of the hole the ideal membrane stress is equal to

$$\sigma_{id} = zf + \sigma''_{t0}; \tag{8.76}$$

in agreement with the analysis in Sect. 1.5, we assume that it is equal to $1.5f$ (that corresponds to the yield strength) and we obtain

$$\frac{\sigma''_{t0}}{f} = 1.5 - z. \tag{8.77}$$

Finally, from (8.75) and (8.77) we obtain

$$\frac{1 - e^{-\beta L} \cos \beta L}{\beta L} = \frac{1 - z}{1.5 - z}. \tag{8.78}$$

Equation (8.78) allows us to identify the values of βL that correspond to different values of z.

We recall the meaning of β (2.41) where

$$\beta L = 1.817 \frac{L}{\sqrt{D_m s}}, \tag{8.79}$$

and D_m is the average diameter.

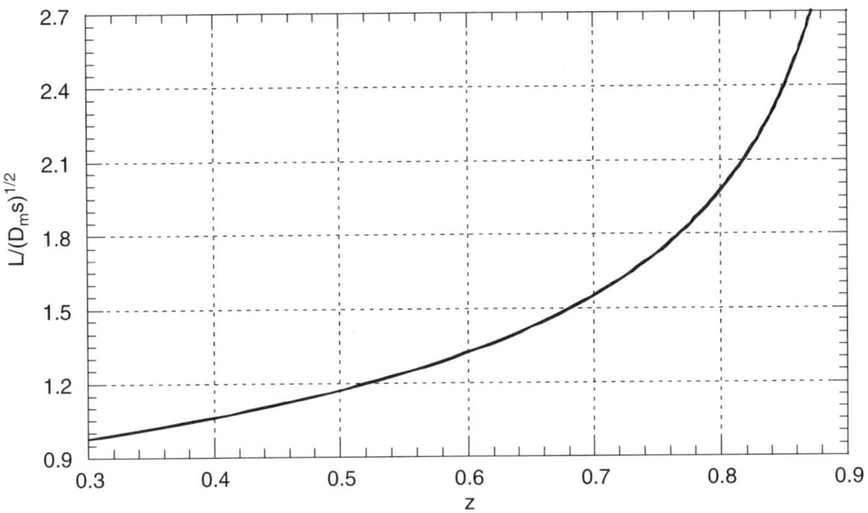

Fig. 8.12

It is possible therefore to build the curve in Fig. 8.12 that plots the value $L/\sqrt{D_m s}$ as a function of z. The length L represents half the maximum length between holes for them to be considered part of a hole line. The physical meaning of the above analysis is in fact as follows.

If the distance between holes is equal to $2L$ and if the thickness of the wall is such that the average ideal membrane stress is f, the ideal membrane stress at the edge of the hole is $1.5f$. Interestingly, the value of $L/\sqrt{D_m s}$ increases with z. We draw the following conclusions: in the case of equal ratio between pitch and diameter of the holes, the distance L is the same. For this reason, small holes that are close to each other more easily make up a hole line than large and distant ones. In the case of equal diameter, it is possible that holes that are even relatively distant from one another may be considered to belong to a hole line because an increase of z is matched by an increase of the characteristic length L.

Note that the maximum allowable distance between holes for $z = 0.4 \div 0.8$ (i.e., the most likely range for z) is equal to

$$2L = (2.1 \div 4)\sqrt{D_m s}. \tag{8.80}$$

Some codes mention only one value of $2L$, regardless of the value of z, and equal to $3\sqrt{D_m s}$. As we can see, it is in between the extreme values in (8.80), but it is also too conservative for large values of z and too large for small values of the efficiency of ligaments of the hole line. It is better to correlate the value of $2L$ to the value of z, as is shown in Fig. 8.12.

At this point it is important to verify if the maximum stress due to the bending moment at the edge of the hole is compatible with the criteria discussed in Sect. 1.5. From (2.48) and recalling that $y_0 = C_1$, we have, for $x = 0$

$$\left(\frac{d^2y}{dx^2}\right)_{x=0} = -2\beta^2 C_1 = -2\beta^2 y_0. \tag{8.81}$$

On the other hand

$$\sigma''_{t0} = \frac{2Ey_0}{D_m}, \tag{8.82}$$

and

$$\left(\frac{d^2y}{dx^2}\right)_{x=0} = -\frac{\beta^2 \sigma''_{t0} D_m}{E}. \tag{8.83}$$

Then, from (2.37), indicating the moment at the edge of the hole with M_0

$$\left(\frac{d^2y}{dx^2}\right)_{x=0} = -\frac{M_0\left(1-\mu^2\right)}{EI} = -\frac{12M_0\left(1-\mu^2\right)}{Es^3}. \tag{8.84}$$

Furthermore, based on (8.83)

$$M_0 = \frac{\beta^2 \sigma''_{t0} D_m s^3}{12\left(1-\mu^2\right)}. \tag{8.85}$$

Given that the maximum stress σ_a caused by M_0 is equal to $6M_0/s^2$, from (8.85) we obtain

$$\sigma_a = \frac{\beta^2 \sigma''_{t0} D_m s}{2\left(1-\mu^2\right)}. \tag{8.86}$$

Recalling the meaning of β, with $\mu = 0.3$ we have

$$\sigma_a = 1.816\sigma''_{t0}. \tag{8.87}$$

Finally, from (8.77) we have

$$\sigma_a = (2.727 - 1.816z)\, f \tag{8.88}$$

Note that in any case $\sigma_a < 3f$, as it should be based on the analysis in Sect. 1.5.

8.2 High Thickness Nozzles and Equivalent Diameter

If the nozzle welded to the wall of a cylinder, a sphere or a cone in correspondence of a hole has a thickness greater than the required minimum value, it acts as a reinforcement of the wall itself, and is able to partially or totally compensate for the portion of wall taken away by the hole. We indicate γ as the ratio between the minimum thickness required s_{to} of the nozzle and the actual thickness s_t of the same, that is,

$$\gamma = \frac{s_{t0}}{s_t}. \tag{8.89}$$

Moreover, if $\sigma'_{id(m)}$ is the ideal membrane stress in the nozzle that is assumed to be isolated from the vessel, it is easy to see that (8.73) is valid even in this case by substituting z with γ.

Assuming as in Sect. 8.1 that the rotation at the end of the nozzle welded to the vessel is zero, (8.78) and Fig. 8.12 are valid even in this case, provided z is substituted with γ and the average diameter of the nozzle d_m and the thickness s_t are considered instead of D_m and s. Therefore, the length L would be the length of the nozzle to consider as far as the compensation for the hole.

In fact, one cannot ignore that the ideal membrane stress σ_{id} in correspondence of the end of the nozzle that is limited to $1.5f$ through the computation process presented in Sect. 8.1 (refer to (8.76) and the ensuing considerations) has the characteristics of a local membrane stress. Based on the criteria in Sect. 1.5, if it is correct to limit its value to $1.5f$ (that corresponds to the yield strength), it is also required that at a distance $0.5\sqrt{r_m s_t}$ from the end of the nozzle the membrane stress does not exceed $1.1f$. As we shall see later on, this condition is more restrictive than the previous one, and must be taken into consideration to determine the value of the useful length of the reinforcement.

We indicate this length with l; at a distance l_1 it is required that $\sigma_{id} = 1.1f$. We have

$$\beta l_1 = \frac{1.817}{\sqrt{d_m s_t}} 0.5\sqrt{\frac{d_m}{2} s_t} = 0.645. \tag{8.90}$$

Substituting x with l_1 in (8.70), based on (8.90) one obtains

$$\sigma''_t = 0.7345\sigma''_{t0}. \tag{8.91}$$

Recalling (8.73) and the considerations at the beginning of the section, one can write

$$\sigma_{id} = \gamma f + 0.7345\sigma''_{t0} = 1.1f \tag{8.92}$$

and

$$\frac{\sigma''_{t0}}{f} = 1.5 - 1.36\gamma; \tag{8.93}$$

(8.75) is written as follows:

$$\frac{\sigma''_{t0}}{f} \frac{1 - e^{-\beta l} \cos \beta l}{\beta l} = 1 - \gamma. \tag{8.94}$$

From (8.93) and (8.94) we obtain the following:

$$\frac{1 - e^{-\beta l} \cos \beta l}{\beta l} = \frac{1 - \gamma}{1.5 - 1.36\gamma}. \tag{8.95}$$

Recalling that

$$\beta l = 1.817 \frac{l}{\sqrt{d_m s_t}}, \tag{8.96}$$

(8.95) allows to plot the curve in Fig. 8.13. From it we obtain the useful length l of the nozzle as a function of γ to compensate for the hole.

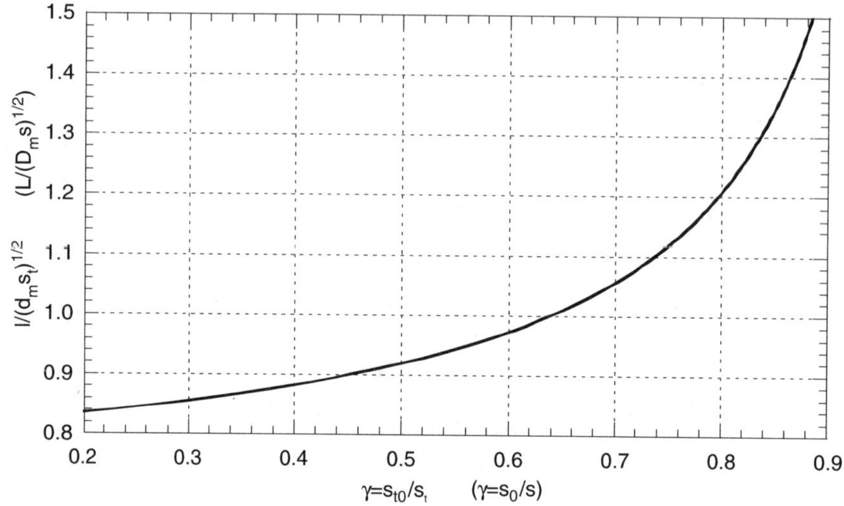

Fig. 8.13

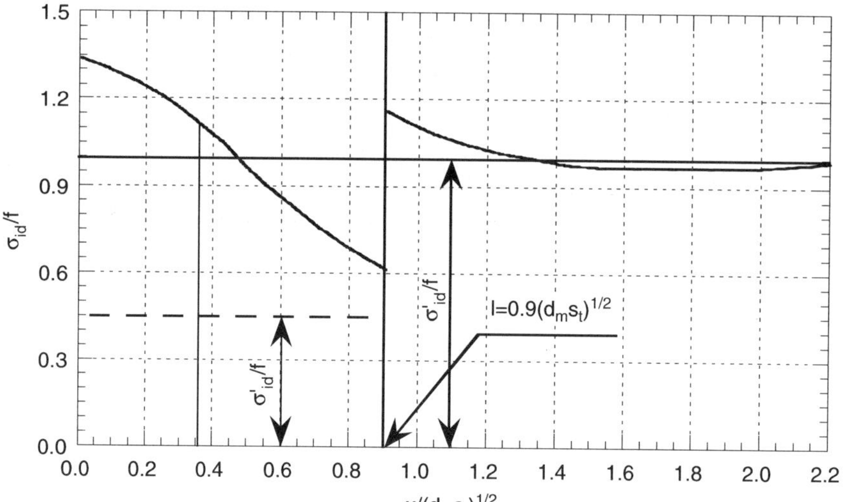

Fig. 8.14

By comparing Fig. 8.13 with Fig. 8.11 we see that $l < L$. Then, adopting l as the useful length, $\sigma_{id} < 1.5f$ at the end of the nozzle. In addition, $\sigma_{id} = 1.1f$ in the characteristic position intended for local membrane stresses. As an example, in Fig. 8.14 we show the behavior of stress σ_{id} in a nozzle with the following characteristics:

$$\frac{s_t}{s_{t0}} = 2.2 \qquad follows \qquad \gamma = 0.4545$$

From Fig. 8.13 we obtain $l = 0.9\sqrt{d_m s_t}$. In addition, let us assume that the nozzle has thickness s_t for $x < l$, whereas for $x > l$ the thickness is equal to s_{to}. Thus, for $x \leq l$ we have $\sigma'_{id} = 0.4545f$, whereas for $x > l$ $\sigma'_{id} = f$. We determine at the end of the nozzle that $\sigma_{id} \approx 1.34f < 1.5f$. Moreover, for $x = 0.353\sqrt{d_m s_t}$ ($x = 0.5\sqrt{r_m s_t}; \beta x = 0.645$) we have $\sigma_{id} = 1.1f$.

Finally, the maximum value for σ_{id} in the area of the nozzle with thickness equal to s_{to} is approximately $1.16f$.

Of course, the average value of σ_{id} for $x = 0 \div l$ is f, as it should be, based on the analysis in Sect. 8.1 that leads to (8.94). Based on the curve in Fig. 8.13, we see that the value of l increases as γ increases. This means that the smaller the contribution of the nozzle to the compensation of the hole, the greater the length l that may be considered useful for the reinforcement.

We also establish that in any case $l > 0.8\sqrt{d_m s_t}$. Many codes use the coefficient 0.8 as a fixed value to determine l. This value, that also represents the limit of $l/\sqrt{d_m s_t}$ for $\gamma = 0$, may therefore be taken into consideration as a simple and conservative criterion. More adequately, one may correlate the value l to γ, as shown in Fig. 8.13.

This approach will help us to determine the sizing criterion of isolated nozzles, as we shall see in Sect. 8.3. They also allow us to establish the calculation criterion for hole lines with nozzles with increased thickness.

In fact, in view of our analysis, the reinforcement area is represented by overthickness $(s_t - s_{to})$ multiplied by $2l$, if we consider nozzles that do not stick out protrude inside the vessel. If that is the case, one shall add the area given by $2cs_t$, where c is the height of the protruding part provided $c \leq l$.

Figure 8.15 visually explains the concept. It also is intended to show that the wall of the vessel can be reinforced with a plate. In that case even the length L can be obtained from Fig. 8.13 by substituting l with L, d_m with D_m, s_t with s and s_{to} with s_0. Area A of the reinforcement partially or totally compensates for the hole. In case (a) the area A' that is not compensated for is given by

$$A' = d_i s_1 - A. \tag{8.97}$$

Therefore, it is possible to identify an equivalent (or ideal) diameter given by

$$d = \frac{A'}{s_1} = d_i - \frac{A}{s_1}. \tag{8.98}$$

In case (b) we have

$$A' = d_e s_1 - A; \tag{8.99}$$

thus

$$d = \frac{A'}{s_1} = d_e - \frac{A}{s_1}. \tag{8.100}$$

Once the equivalent diameter of the holes d is computed, it is possible to calculate the efficiency of ligaments z by introducing the value of this diameter according to the criteria presented in Sect. 8.1. Of course, the conditions that

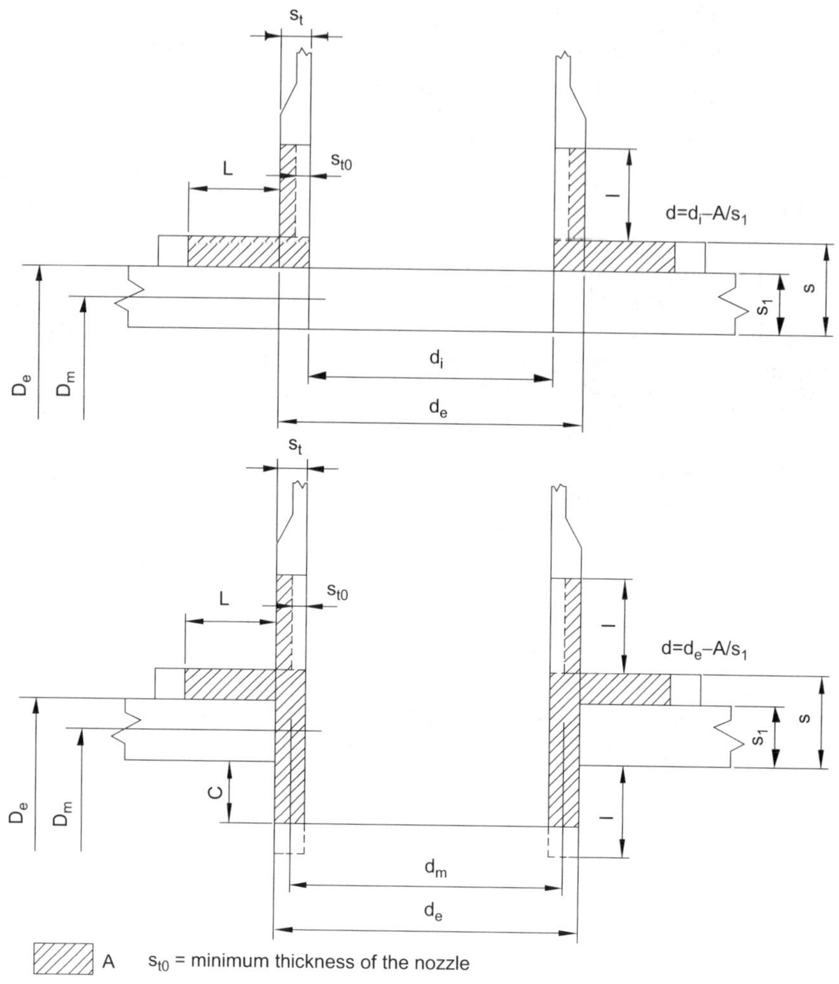

Fig. 8.15

determine an actual hole line must be verified. In reference to Fig. 8.15, note that the thickness to consider in Fig. 8.11 is the one relative to the wall of the vessel, i.e., s_1.

The introduction of the equivalent diameter is recommended for hole lines, especially when they lie along a circumference or an oblique line, such as cylinders and cones. That way, one implicitly and correctly factors in the impact of the different values that the two principal stresses have. Based on (8.98) and (8.100) $d \leq 0$ may result. This means that area A completely compensates for the hole, and therefore the thickness of the vessel can be the minimum for a wall without holes (of course, only if there are no other hole lines, so that $z < 1$ and if the weld joint efficiency is equal to one).

8.3 Isolated Holes on Cylinders, Spheres, and Cones: Y and T Branches

If an isolated hole is present on a cylinder, a sphere (or a dished head) and a cone, or if the conditions in Sect. 8.1 are not satisfied, even though there is a hole line, one must apply calculation criteria similar to the ones introduced in Sect. 8.2.

Basically, one must apply the so-called rule of the rectangle (see Fig. 8.16). This simple criterion consists of considering useful to compensate for the hole the extra thickness compared to the material required for the wall of the vessel without holes and the nozzle. This extra thickness is shown in the rectangle in Fig. 8.16, the boundaries of which are the length L and l, typical of the vessel and the nozzle, respectively.

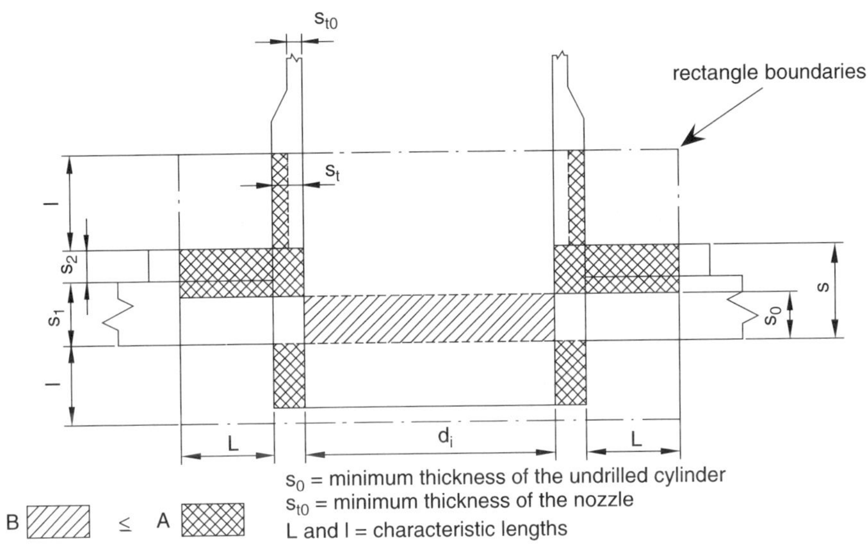

Fig. 8.16

If we indicate the minimum required thickness of the vessel without holes with s_0 and the internal diameter of the nozzle with d_i, the cross-sectional area B is removed as a result of the drilling:

$$B = d_i s_0. \tag{8.101}$$

If the thickness of the wall of the vessel s_1 is greater than s_0, at the sides of the hole there is an extra thickness that may totally or partially compensate for the missing B.

Moreover, if there is a reinforcement plate of thickness s_2, this contributes to compensate for B, at least along length L. Finally, if the thickness of the nozzle s_t is greater than the minimum required thickness s_{to}, the area corresponding to the extra thickness $(s_t - s_{to})$ contributes to compensate for B. If the nozzle projects inside the vessel, as shown in Fig. 8.16, the entire projection area contributes to compensate, as well.

If A is the sum of these reinforcement areas, the hole is compensated, and the resistance of the vessel in the area around the isolated hole is ensured as long as

$$A \geq B. \tag{8.102}$$

As far as both length L and l, we refer back to Sect. 8.2. Specifically, length l relative to the nozzle can be obtained from Fig. 8.13 as a function of the ratio s_{to}/s_t and also based on the average diameter of the nozzle d_m as well as its thickness. To be conservative, one can adopt a value for l that is independent from γ and is given by

$$l = 0.8\sqrt{d_m s_t}. \tag{8.103}$$

Even the length L may be obtained from Fig. 8.13 by substituting l with L, d_m with D_m, s_t with s, and s_{to} with s_o.

Note that the thickness s is the sum of the thickness of the wall s_1 and the thickness of the additional reinforcement plate s_2 (if present), as long as its width is at least equal to L. If this is not the case one must proceed as follows. By assuming an arbitrary value for L and indicating the width of the plate with c, one obtains a virtual value s given by

$$s' = \frac{cs_2 + Ls_1}{L}. \tag{8.104}$$

This thickness, that was introduced to replace s_t in the parameters of Fig. 8.13, results in value for L that does not match with the one considered for the calculation of s' through (8.104).

The correct value for L is identified through successive approximations. Even for L it is possible to adopt a conservative value independently from the ratio s_0/s, that is

$$L = 0.8\sqrt{D_m s} \tag{8.105}$$

or

$$L = \sqrt{D_m s}, \tag{8.106}$$

if $s/s_0 < 1.5$.

On the other hand, if one factors in the behavior of the curve in Fig. 8.13 a number of considerations can be made. The more the compensation for the hole depends on the extra thickness of the wall of the vessel or on the reinforcement plate, the more the value of L decreases and l increases. The opposite occurs if the compensation for the hole depends on the extra thickness of the nozzle.

Figure 8.13 may be used to determine L, but caution is required. Let us consider the case where the nozzle has a large diameter with respect to the vessel. Specifically, let us assume that the compensation mostly depends on the extra thickness of the wall of the vessel so that the ratio between thickness of the nozzle and the wall is small. In these cases the assumption that there will be no rotation at the edge of the hole does not correspond to reality. Thus, Fig. 8.13 is no longer valid.

Let us examine the situation where the bending moment at the edge of the hole is zero and, consequently, the rotation is at a maximum. Of course, actual situations will be somewhere in between. The following considerations involve an extreme situation.

If the bending moment is equal to zero in (2.45), then $C_1 = 0$. According to the same viewpoint presented in Sect. 8.1, we establish that

$$\sigma''_t = \sigma''_{t0} e^{-\beta x} \cos \beta x. \tag{8.107}$$

The average value of σ''_t for $x = 0 \div L$ is given by

$$\sigma''_{tm} = \sigma''_{t0} \frac{1 + e^{-\beta L} (\sin \beta L - \cos \beta L)}{2\beta L}. \tag{8.108}$$

The ideal membrane stress in the areas without holes of the vessel is equal to

$$\sigma'_{id(m)} = \gamma f. \tag{8.109}$$

In correspondence of the nozzle we have

$$\sigma_{id(m)} = \gamma f + \sigma''_{t0} \frac{1 + e^{-\beta L} (\sin \beta L - \cos \beta L)}{2\beta L}. \tag{8.110}$$

Assuming that $\sigma_{id(m)} = f$ we have

$$\frac{\sigma''_{t0}}{f} \frac{1 + e^{-\beta L} (\sin \beta L - \cos \beta L)}{2\beta L} = 1 - \gamma. \tag{8.111}$$

Then, recalling (8.77) and substituting z with γ we have

$$\frac{\sigma''_{t0}}{f} = 1.5 - \gamma. \tag{8.112}$$

From (8.111) and (8.112) we obtain

$$\frac{1 + e^{-\beta L} (\sin \beta L - \cos \beta L)}{2\beta L} = \frac{1 - \gamma}{1.5 - \gamma}. \tag{8.113}$$

Recalling (8.79) one can build curve A in Fig. 8.17. Based on (8.107) and (8.90) and assuming $x = l_1$, we obtain

$$\sigma''_t = 0.419 \sigma''_{t0}. \tag{8.114}$$

Similarly to (8.91), one writes

$$\sigma_{id} = \gamma f + 0.419 \sigma''_{t0} = 1.1 f; \tag{8.115}$$

then

$$\frac{\sigma''_{t0}}{f} = 2.62 - 2.38\gamma. \tag{8.116}$$

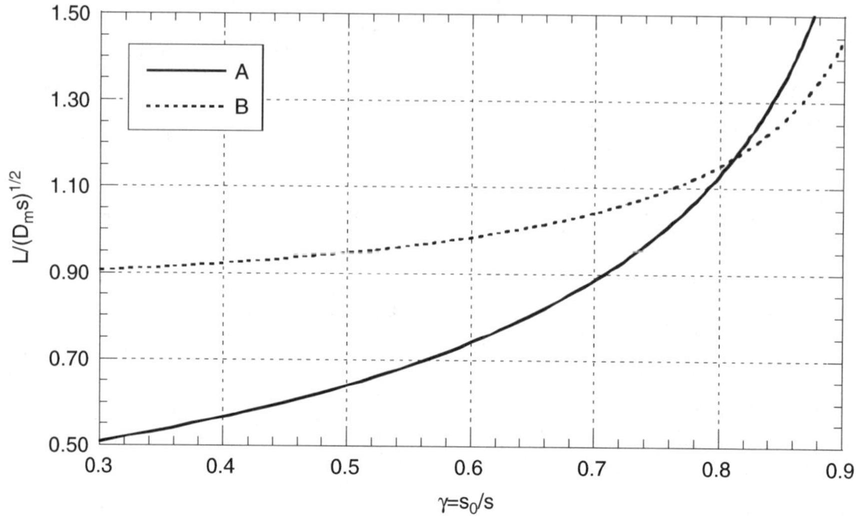

Fig. 8.17

Recalling (8.111), one obtains the following:

$$\frac{1+e^{-\beta L}\left(\sin\beta L-\cos\beta L\right)}{2\beta L}=\frac{1-\gamma}{2.62-2.38\gamma}. \tag{8.117}$$

From (8.117) it is possible to plot B in Fig. 8.17.

In summary, curve A plots the values of L for which the ideal membrane stress at the edge of the hole is equal to $1.5f$. Curve B plots the values for L so that, at distance $0.5\sqrt{r_m s}$ from the edge of the hole, the ideal membrane stress is equal to $1.1f$. The first condition is almost always crucial (except for high values of γ) in contrast to when the rotation at the edge is zero.

The values of L are considerably smaller than those obtained from Fig. 8.13, as well. The values of L are particularly small when the values of γ are small. On the other hand, if the nozzle is large, γ is likely to be considerably smaller than one, unless the reinforcement is used predominantly through the extra thickness of the nozzle.

Still aware that Fig. 8.17 refers to an extreme condition, it must be considered for large nozzles. It is, of course, impossible to provide more detailed instructions because only a direct investigation of a specific case or the analysis of a vast number of case studies may offer reliable quantitative information. Note that some codes assume that

$$L=0.75\sqrt{D_m s} \tag{8.118}$$

for $d_i/D_i > 1/3$, where d_i and D_i are the internal diameter of the nozzle and the vessel, respectively.

It is interesting to consider the possibility of a nozzle of thickness equal to the required minimum be integrated into the wall without having to increase the thickness of the vessel compared to the minimum required thickness for a wall without holes. To that extent, observe that based on (8.93) if $\gamma = 1$ (i.e., $s = s_0$) we have:

$$\sigma''_{t0} = 0.14f. \tag{8.119}$$

On the other hand, recalling (8.70) and integrating σ''_t for $x = 0 \div \infty$ (see the analysis in Sect. 4.4) one obtains

$$\int_0^\infty \sigma''_t dx = \frac{\sigma''_{t0}}{\beta} = 0.14f\frac{\sqrt{D_m s}}{1.817} = 0.077f\sqrt{D_m s}. \tag{8.120}$$

This force must balance the half hole (with reference to unitary thickness). Indicating the inside diameter of the nozzle with d_i, one writes

$$\frac{d_i}{2}f = 0.077f\sqrt{D_m s}; \tag{8.121}$$

then

$$d_i = 0.154\sqrt{D_m s}. \tag{8.122}$$

If the inside diameter of the nozzle is smaller or equal to the value obtained from (8.122), it is not necessary to increase the thickness neither of the nozzle nor of the wall of the vessel.

This possibility is quite interesting in terms of all nozzles with a small diameter to be found in vessels for manometers, thermometers, and so on. The previous calculation criteria allow us to create diagrams of considerable practical usefulness.

In fact, based on the rule of the rectangle (see Fig. 8.16) and ignoring the potential greater thickness of the wall with respect to s_0 in correspondence of the nozzle in favor of resistance, the compensation for non protruding nozzles inside the vessel is obtained with reference to the half hole if

$$\frac{d_i}{2}s_0 = L(s - s_0) + l(s_t - s_{t0}). \tag{8.123}$$

By indicating the values visible on the ordinate in Fig. 8.13 with α and α_t for the wall of the vessel and the nozzle, respectively, (8.123) may be written as follows:

$$\frac{d_i}{2}s_0 = \alpha\sqrt{D_m s}(s - s_0) + \alpha_t\sqrt{d_m s_t}(s_t - s_{t0}), \tag{8.124}$$

or

$$\frac{d_i}{2}s_0 = \sqrt{D_m s_0}\left[\alpha\sqrt{\frac{s}{s_0}}(s - s_0) + \alpha_t\sqrt{\frac{d_m s_t}{D_m s_0}}(s_t - s_{t0})\right]. \tag{8.125}$$

Conversely,

$$\frac{s_{t0}}{s_0} = \frac{d_m}{D_m}. \tag{8.126}$$

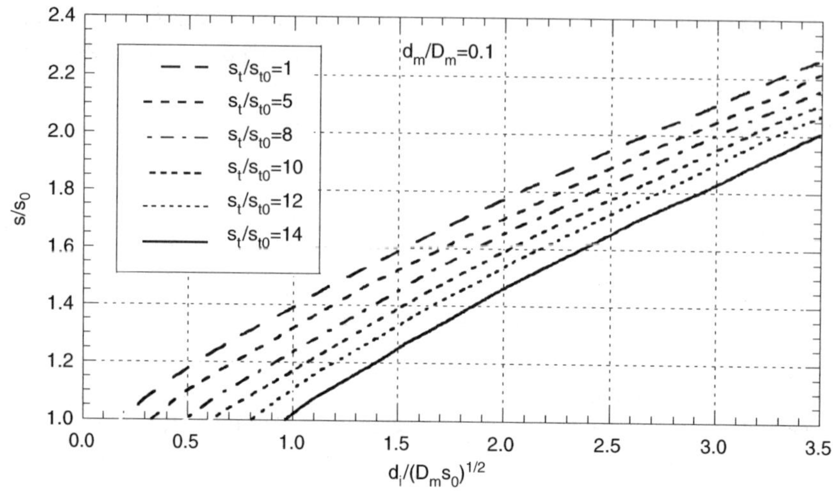

Fig. 8.18

Through a series of steps from (8.125) we obtain the following:

$$\frac{d_i}{\sqrt{D_m s_0}} = 2\left[\alpha\sqrt{\frac{s}{s_0}}\left(\frac{s}{s_0}-1\right)+\alpha_t\left(\frac{d_m}{D_m}\right)^2\sqrt{\frac{s_t}{s_{t0}}}\left(\frac{s_t}{s_{t0}}-1\right)\right]. \quad (8.127)$$

We recall that based on Sect. 8.2, α_t is a function of s_t/s_{to} (Fig. 8.13). Similarly to what was said earlier in this section, α may be considered a function of s/s_0 through the same diagram as long as the ratio between diameter of the nozzle and of the vessel is not very high.

Then, based on (8.127) we see that the ratio s/s_0 between the necessary thickness of the vessel and the minimum required one for a wall without holes is a function of d_m/D_m, s_t/s_{to} and $d_i/\sqrt{D_m s_o}$. Based on Fig. 8.13 it was possible to plot the curves in Figs. 8.18–8.20 relative to $d_m/D_m = 0.1$, 0.2 and 0.3. They help to rapidly obtain the thickness s of the vessel through a few attempts.

In fact, once the inside or outside diameter of the vessel is known, one adopts a thickness s after a first approximation. Then it is possible to compute D_m and through the known equation, based on the value of the pressure p and the basic allowable stress f, the thickness s_0. Based on the size of the nozzle d_i and s_t are known and d_m as well as s_{to} can be computed.

This way all parameters showing in the diagrams are known, and it is possible to obtain s that will generally not coincide with the one assumed. A second attempt makes it possible to correctly compute the required thickness. The plots in these figures lead to interesting considerations. First of all, the curve corresponding to $s_t/s_{to} = 1$ does not go through the origin, and this agrees with (8.122).

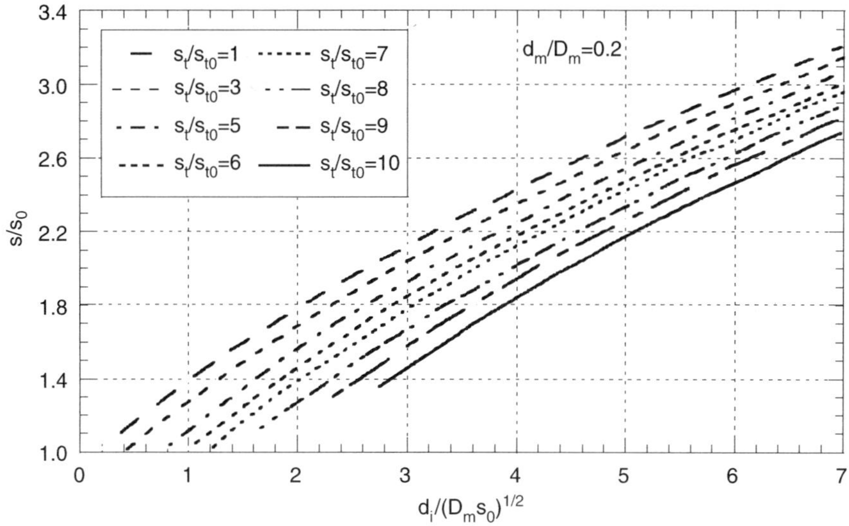

Fig. 8.19

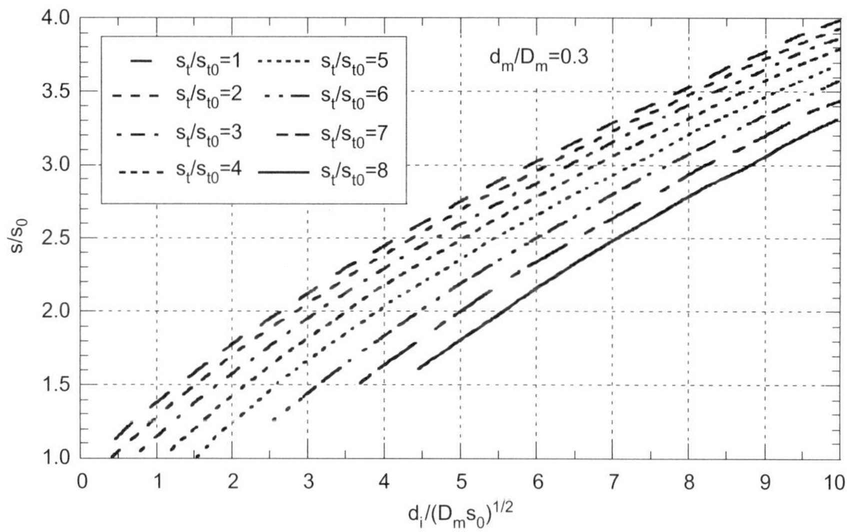

Fig. 8.20

Moreover, the condition according to which the compensation can be obtained exclusively with the extra thickness of the nozzle is at the intersection of the curves with the axis of the abscissa. Finally, note that some curves have been truncated to impose the limitation $s_t/s \leq 1.5$. Up to now we focused on openings with nozzles having an orthogonal axis to the wall of the vessel. However, nozzles that do not satisfy this condition are quite frequent. It is therefore necessary to provide a few suggestions as to how to proceed in this case.

A rigorous sizing criterion would require a case by case analysis, or if we want to elaborate a more general rule, it would be necessary to have access to experimental data about a sufficiently large number of case studies. But we can elaborate an empirical calculation method that fulfills reliable criteria reasonably well, and yet is simple and conservative. If we examine Fig. 8.21 about a nozzle with an orthogonal axis to the wall of the cylinder, we see that the application of the rule of the rectangle corresponds to the following principles. Considering only hoop stresses for the sake of simplicity, we see that areas A_1 and A_4 corresponding to the minimum required thickness of the cylinder without holes, are correlated to areas B_1 and B_4 by the following relationship:

$$A_1 = A_4 = \frac{p}{f} B_1 = \frac{p}{f} B_4. \tag{8.128}$$

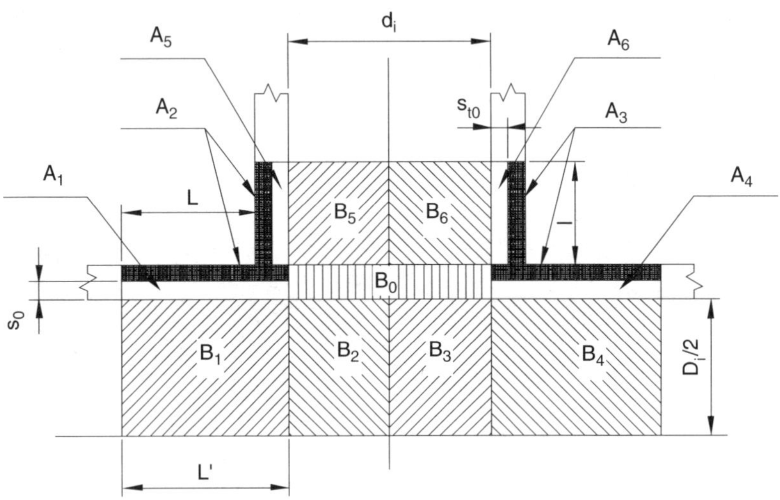

Fig. 8.21

In fact, (8.128) corresponds to the following:

$$s_0 L' = \frac{p}{f}\frac{D_i}{2} L'; \tag{8.129}$$

then

$$s_0 = \frac{pD_i}{2f}, \tag{8.130}$$

also known as Mariotte's equation.

Analogously,

$$A_5 = A_6 = \frac{p}{f} B_5 = \frac{p}{f} B_6 \tag{8.131}$$

As far as area B_0 included in the thickness of the vessel, it may be excluded from the correlation between B areas and A areas because the corresponding thrusts are applied to the wall of the hole, thus along the axis of the vessel in the section under examination.

Given the great rigidity of its wall, the ensuing circumferential stresses are irrelevant. The areas A_2 and A_3 represented by the extra thickness of the wall of the vessel and the nozzle that constitute the reinforcement of the opening are intended to compensate for the opening itself, and have the following relation:

$$A_2 = A_3 = \frac{p}{f} B_2 = \frac{p}{f} B_3. \tag{8.132}$$

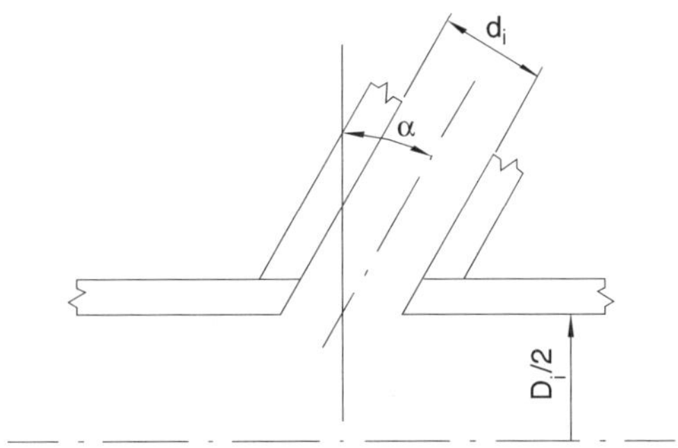

Fig. 8.22

In fact, we have

$$A = A_2 + A_3 = \frac{p}{f} (B_2 + B_3) = \frac{pD_i}{2f} d_i; \tag{8.133}$$

and recalling (8.130)

$$A = d_i s_0. \tag{8.134}$$

Equation (8.134) corresponds to (8.101) and (8.102) that identify the rule of the rectangle. We now look at Fig. 8.22 about a tilted nozzle where α is the angle that the axis of the nozzle forms with the orthogonal line to the wall of the vessel. We now consider an ideally isolated nozzle (see Fig. 8.23). We see that it may be treated as a cylinder at which end there is half a hole.

Three areas, B_5, B_6 and B_7, are identified. The thrust caused by pressure p corresponding to area B_5 is balanced by the thrusts corresponding to the areas B_6 and B_7, the sum of which is equal to B_5. It is quite evident that while the area B_5 is correlated to area A_5 and area B_6 is correlated to area A_6, the area B_7 requires to be compensated by a reinforcement area indicated as A_7 in the figure. Considering the nozzle connected to the vessel, this area becomes part of the entire reinforcement area, thus increasing the thickness of the nozzle or the wall of the vessel or both.

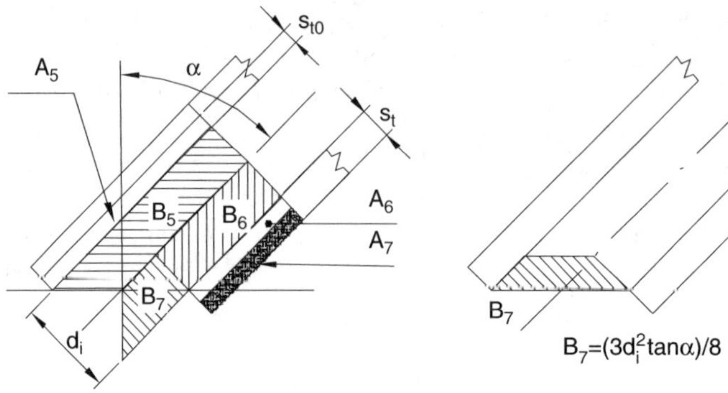

Fig. 8.23

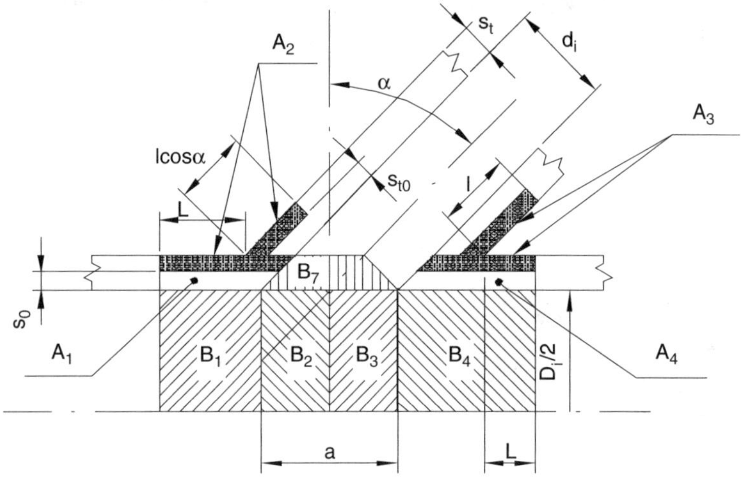

Fig. 8.24

We establish that area B_7 in Fig. 8.23 corresponds to area B_7 in Fig. 8.24. The area of reinforcement A_3 must be correlated to area B_3 (as in the case of the nozzle orthogonal to the wall) and area B_7. In view of the considerations above we have

$$A_3 = \left(\frac{D_i}{2}\frac{a}{2} + \frac{3d_i a \sin\alpha}{8}\right)\frac{p}{f}. \tag{8.135}$$

Then, noting that $a = d_i/\cos\alpha$

$$A_3 = \frac{D_i d_i}{4\cos\alpha}\left(1 + \frac{3}{2}\frac{d_i}{D_i}\sin\alpha\right)\frac{p}{f}. \tag{8.136}$$

Recalling (8.130),

$$A_3 = \left(1 + \frac{3}{2}\frac{d_i}{D_i}\sin\alpha\right)\frac{d_i s_0}{2\cos\alpha}, \tag{8.137}$$

and assuming that

$$k_1 = 1 + \frac{3}{2}\frac{d_i}{D_i}\sin\alpha, \tag{8.138}$$

we have

$$A_3 = k_1 \frac{d_i s_0}{2\cos\alpha}. \tag{8.139}$$

Note that with $k_1 = 1$, area A_3 would be correlated to area B_3 only, given that $a = d_1/\cos\alpha$. Therefore, k_1 highlights the influence of the tilted nozzle, or to put it differently, the influence of area B_7. k_1 is shown in Fig. 8.25. As far as the reinforcement area A_2, it must be correlated to area B_2 like the nozzle orthogonal to the wall of the vessel.

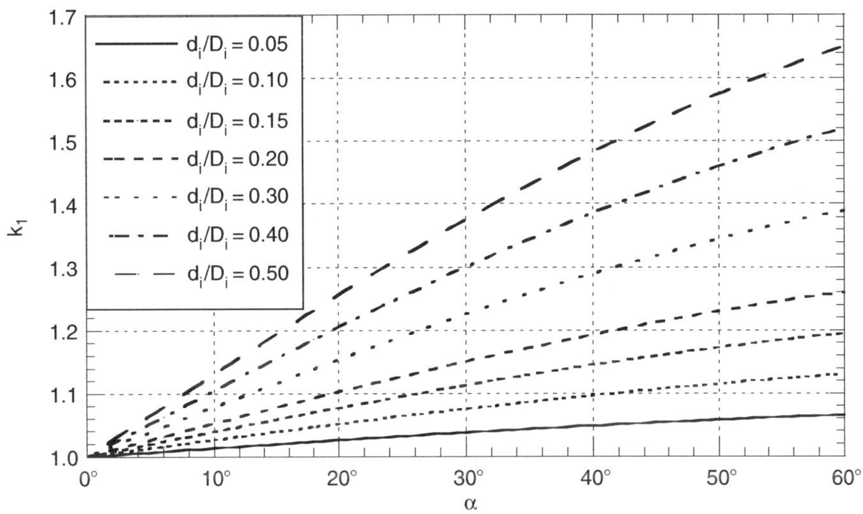

Fig. 8.25

However, note that the wall of the nozzle forms an obtuse angle with the wall of the vessel, and we believe that one should take a reduced effectiveness of the reinforcement area, consisting of the extra thickness of the nozzle, into account. Therefore, it is intended for the useful length of this reinforcement to be limited to $l\cos\alpha$ instead of l. Hence, we will have

$$A_2 = \frac{d_i s_0}{2\cos\alpha}, \tag{8.140}$$

with the limitation mentioned earlier regarding the useful reinforcement area of the nozzle.

We now consider the similar case of a nozzle tilted on a sphere or a dished head (in the spherical portion) (see Fig. 8.26). Proceeding as before, area A_3

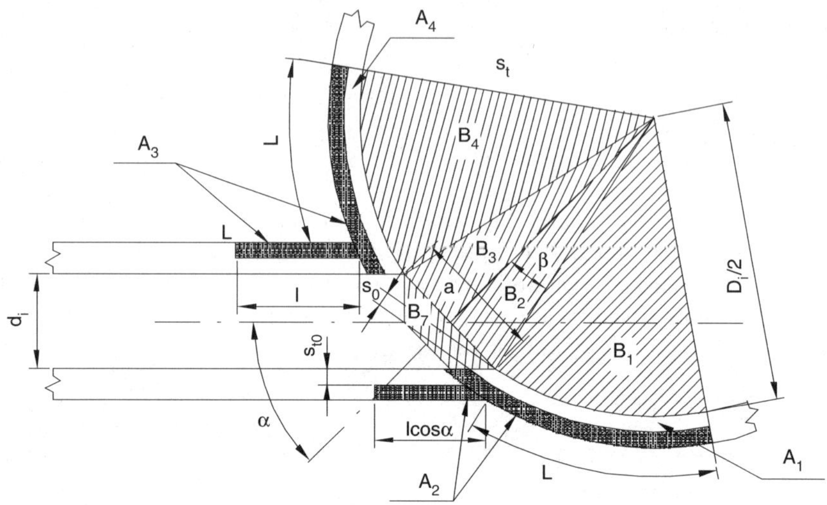

Fig. 8.26

will have to be correlated to the areas B_3 and B_7. By indicating the inside diameter of the sphere or twice the inside radius of the dished head with D_1 we have

$$B_3 = \frac{D_i a}{8} \cos \beta. \tag{8.141}$$

On the other hand

$$\cos \beta = \sqrt{1 - \left(\frac{a}{D_i}\right)^2} = \sqrt{1 - \left(\frac{d_i}{D_i \cos \alpha}\right)^2}; \tag{8.142}$$

then

$$B_3 = \frac{D_i d_i}{8 \cos \alpha} \sqrt{1 - \left(\frac{d_i}{D_i \cos \alpha}\right)^2}. \tag{8.143}$$

In addition

$$B_7 = \frac{3 d_i a \sin \alpha}{8} = \frac{3 d_i^2 \tan \alpha}{8}. \tag{8.144}$$

It is easy to see that in the case of the nozzle with an axis orthogonal to the wall, the area A_3 is correlated to area B_3 only.

Therefore, with reference to (8.139) we can write

$$A_3 = k_2 \frac{d_i s_0}{2 \cos \alpha}, \tag{8.145}$$

given that

$$k_2 = \frac{B_3 + B_7}{B_3}. \tag{8.146}$$

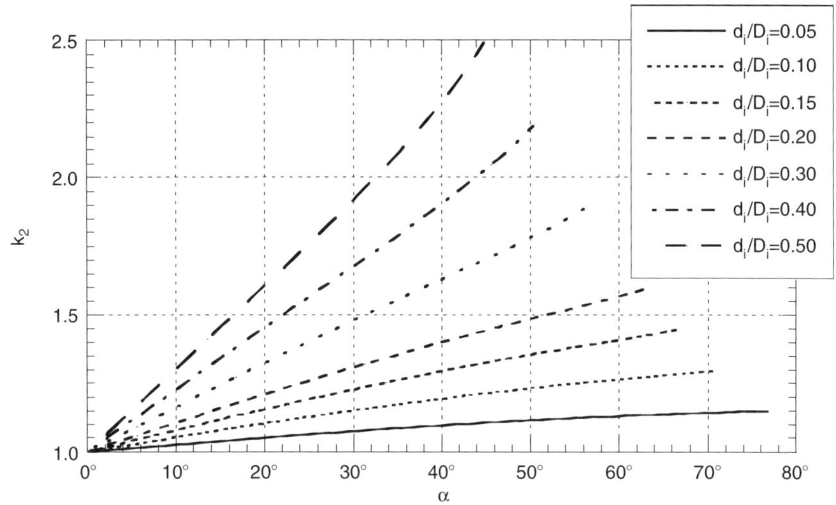

Fig. 8.27

Recalling (8.143) and (8.144) we obtain

$$k_2 = 1 + 3\frac{d_i}{D_i}\frac{\sin\alpha}{\sqrt{1-\left(\dfrac{d_i}{D_i\cos\alpha}\right)^2}}. \tag{8.147}$$

k_2 is shown in Fig. 8.27. Equation (8.140) is true in this case, as well, with the indications as far as area A_2 is concerned.

Note (see Fig. 8.28) that the maximum value of the ratio d_i/D_i as a function of the angle α is equal to

$$\frac{d_i}{D_i} = \cos^2\alpha. \tag{8.148}$$

By substituting this value in (8.147) we obtain

$$k_{2(\max)} = 1 + 3\frac{d_i}{D_i}. \tag{8.149}$$

Figure 8.26 can also represent a nozzle inserted on a cylinder and tilted in the orthogonal plane to its axis. The previous considerations made about the nozzle tilted on the sphere are still valid, but the thickness s_0 shall be half the minimum required thickness of the cylinder without holes. This is true both for (8.145) and to estimate the useful extra thickness of the wall of the vessel required for reinforcement.

In fact, for the section under study one must consider the longitudinal stresses in the cylinder that, as we know, have a value equal to about half

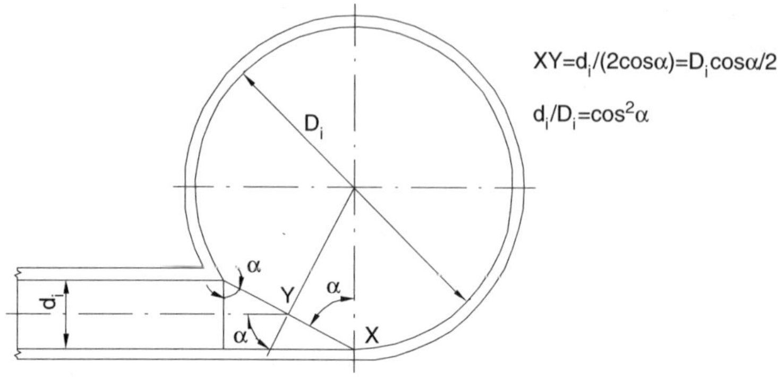

Fig. 8.28

the hoop stresses that determine the value of the minimum thickness s_0 of the cylinder. In this case one must consider a thickness s_0 equal to half the computed one.

Of course, it is also necessary to carry out a verification along the circumferential direction based on the computational thickness s_0 relative to the criteria presented for the non tilted nozzle. In that case we have $a = d_1$ and $\alpha = 0$. By adopting criteria that are similar to those outlined for tilted nozzles, it is possible to elaborate a sizing criterion for Y and T branches.

First of all, it is necessary to distinguish between Y branches with a characteristic angle $\alpha \leq 60°$ (see Fig. 8.29) and $\alpha > 60°$ (see Fig. 8.30). In the first instance, we establish that the areas B_1 and B_2 are equal to

$$B_1 = B_2 = \frac{d_i^2}{2} \tan \frac{\alpha}{2}; \tag{8.150}$$

area B_3 is equal to

$$B_3 = \frac{d_i^2}{8} \left(\frac{2}{\tan \alpha} - \tan \frac{\alpha}{2} \right). \tag{8.151}$$

The sum of the areas B_1 and B_2 must correspond to the reinforcement area A_1, and as far as what we saw before, it must be

$$A_1 = (B_1 + B_2) \frac{p}{f} = \frac{d_i^2}{4} \frac{p}{f} \tan \frac{\alpha}{2}. \tag{8.152}$$

Finally, recalling (8.130) we have

$$A_1 = k_3 \frac{d_1 s_0}{2}, \tag{8.153}$$

given that

$$k_3 = \tan \frac{\alpha}{2}. \tag{8.154}$$

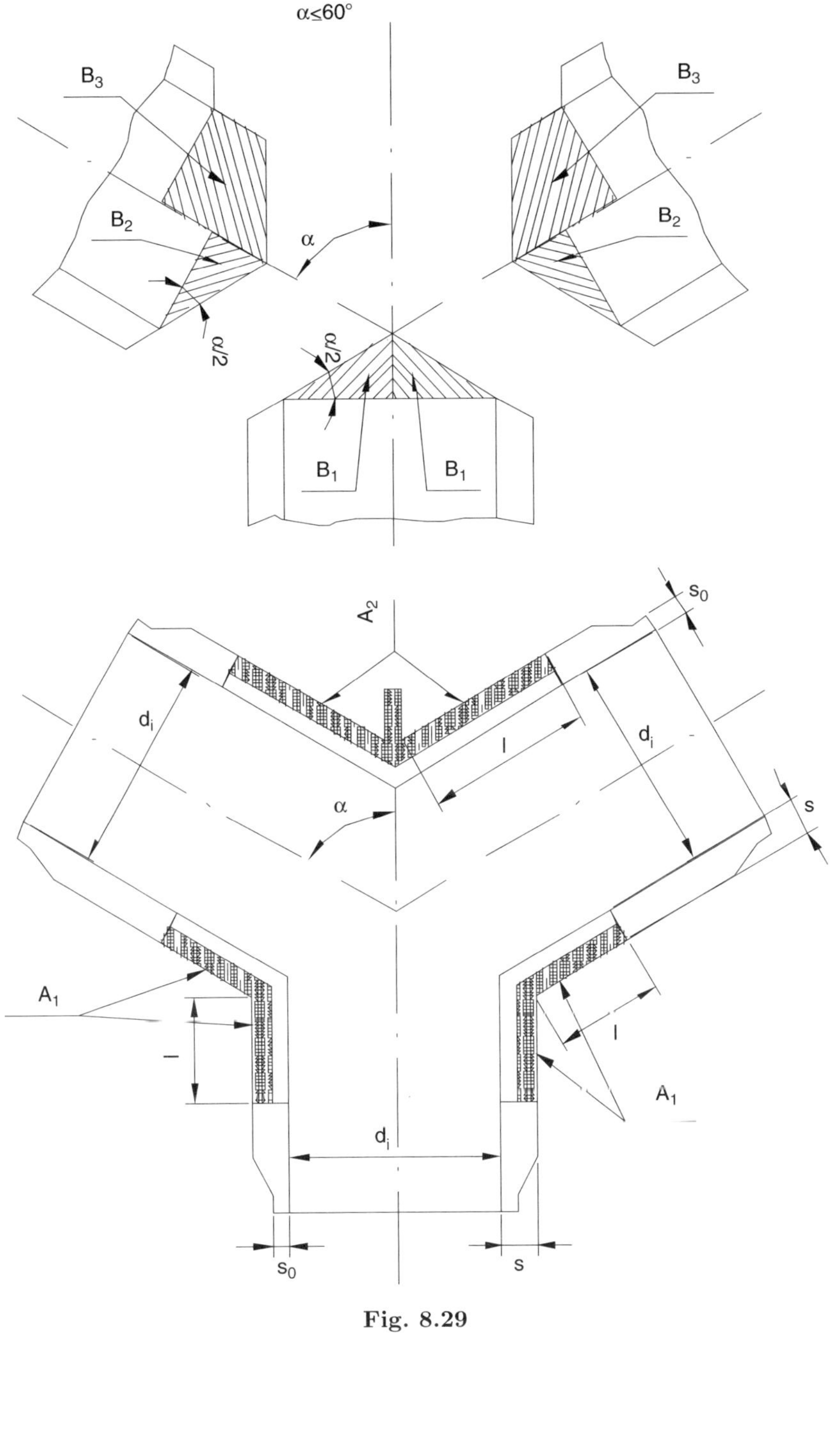

Fig. 8.29

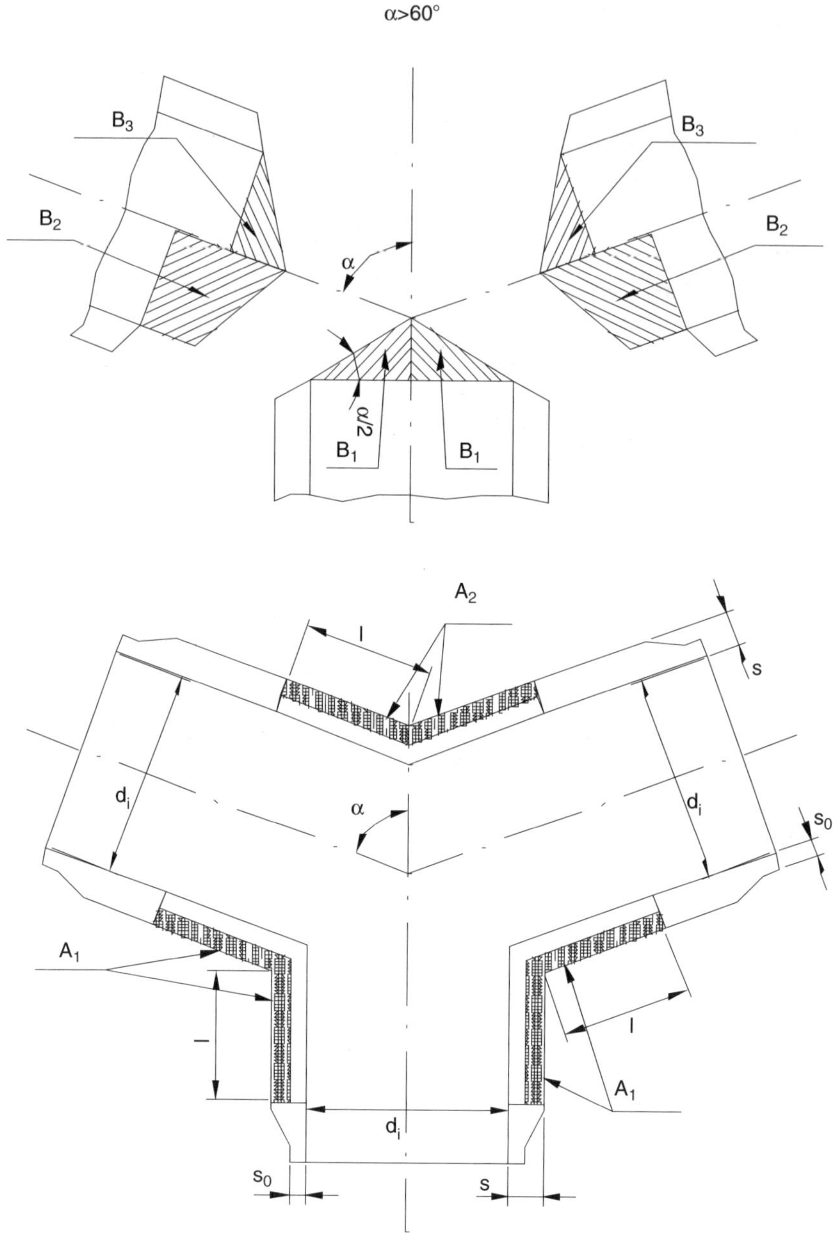

Fig. 8.30

The reinforcement area A_2 that may also include a semiannular reinforcement, as shown in the figure, must correspond to twice the area B_3 instead, so that

$$A_2 = 2B_3\frac{p}{f} = \frac{d_i^2}{4}\left(\frac{2}{\tan\alpha} - \tan\frac{\alpha}{2}\right)\frac{p}{f}. \qquad (8.155)$$

And finally, recalling (8.130)

$$A_2 = k_4\frac{d_1 s_0}{2}; \qquad (8.156)$$

given that

$$k_4 = \frac{2}{\tan\alpha} - \tan\frac{\alpha}{2}. \qquad (8.157)$$

If the angle α is greater than 60°, with reference to Fig. 8.29 similarly we have

$$B_1 = \frac{d_i^2}{8}\tan\frac{\alpha}{2}, \qquad (8.158)$$

$$B_2 = \frac{d_i^2}{8}\left(2\tan\frac{\alpha}{2} - \frac{1}{\tan\alpha}\right), \qquad (8.159)$$

and

$$B_3 = \frac{d_i^2}{8\tan\alpha}. \qquad (8.160)$$

Consequently

$$A_1 = (B_1 + B_2)\frac{p}{f} = \frac{d_i^2}{8}\left(3\tan\frac{\alpha}{2} - \frac{1}{\tan\alpha}\right)\frac{p}{f}, \qquad (8.161)$$

and, based on (8.130)

$$A_1 = k_3\frac{d_1 s_0}{2} \qquad (8.162)$$

with

$$k_3 = \frac{3}{2}\tan\frac{\alpha}{2} - \frac{1}{2\tan\alpha}. \qquad (8.163)$$

In addition

$$A_2 = 2B_3\frac{p}{f} = \frac{d_i^2}{4\tan\alpha}\frac{p}{f}. \qquad (8.164)$$

Finally,

$$A_2 = k_4\frac{d_1 s_0}{2}, \qquad (8.165)$$

with

$$k_4 = \frac{1}{\tan\alpha}. \qquad (8.166)$$

In summary, the minimum value of the reinforcement area A_1 is given by

$$A_1 = k_3\frac{d_1 s_0}{2}; \qquad (8.167)$$

given that

$$k_3 = \tan\frac{\alpha}{2} \quad (\alpha \le 60°) \tag{8.154 bis}$$

and

$$k_3 = \frac{3}{2}\tan\frac{\alpha}{2} - \frac{1}{2\tan\alpha} \quad (\alpha > 60°)\,. \tag{8.163 bis}$$

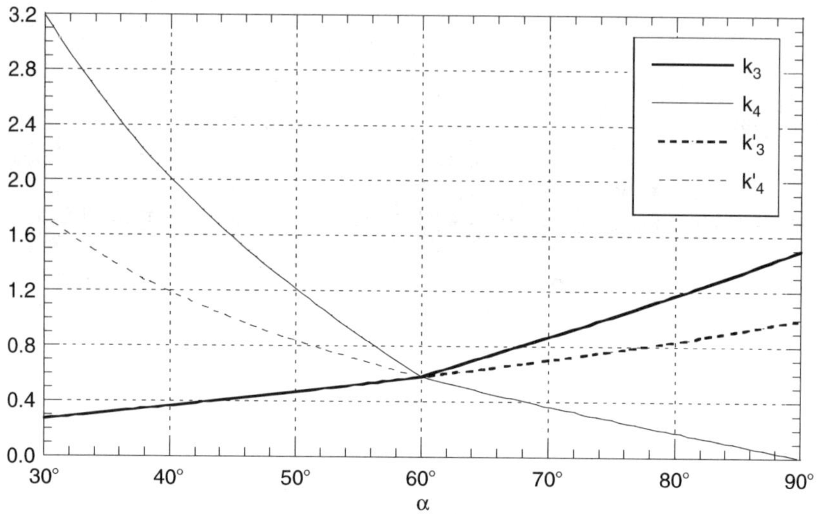

Fig. 8.31

The minimum value of the reinforcement area A_2 is given by

$$A_2 = k_4\frac{d_1 s_0}{2}, \tag{8.168}$$

given that

$$k_4 = \frac{2}{\tan\alpha} - \tan\frac{\alpha}{2} \quad (\alpha \le 60°)\,; \tag{8.157 bis}$$

$$k_4 = \frac{1}{\tan\alpha} \quad (\alpha > 60°)\,, \tag{8.166 bis}$$

where, as usual, s_0 is the minimum required thickness of the tube with internal diameter d_1. k_3 and k_4 are shown in Fig. 8.31. For $\alpha = 60°$ the values of k_3 and k_4 obviously coincide and we have

$$k_3 = k_4 = \frac{1}{\sqrt{3}}. \tag{8.169}$$

The minimum value of the entire reinforcement area corresponds to that angle and is given by

$$A_{tot} = 2A_1 + A_2 = 3k_3\frac{d_1 s_0}{2} = \frac{\sqrt{3}}{2}d_1 s_0. \tag{8.170}$$

For $\alpha = 90°$, the branch takes on the shape of a T.

Note that as far as the reinforcement area A_2, we have $k_4 = 0$ and $A_2 = 0$. In fact, in that case we have a straight pipe where a nozzle of equal diameter has been inserted. k_4 and its corresponding area A_2 refer to the wall of the tube opposite to the nozzle, that obviously does not require reinforcement.

As far as the area A_1, we have $k_3 = 1.5$; therefore

$$A_1 = \frac{3}{4} d_1 s_0. \tag{8.171}$$

This may seem to be in contrast with (8.134) based on which it should be that $k_3 = 1$ and

$$A_1 = \frac{d_1 s_0}{2}. \tag{8.172}$$

In fact, one should consider that in this case the diameter of the nozzle is equal to the one of the pipe in which it is inserted.

In our view, this is a branch that requires caution in terms of sizing to justify an increase of 50% of the reinforcement area, with respect to the one required by the pure and simple application of the rule of the rectangle. In summary, note that if $\alpha \leq 60°$ (i.e., for elements with a Y shape and a small α angle) one abandons the value of k_4 given by (8.166) (valid for $\alpha > 60°$) and adopts the greater one given by (8.157) instead, in order to operate conservatively because of the high concentration of stresses in the corner. By analogy, if $\alpha > 60°$ (i.e., for T-shaped or similar shaped elements), k_3 given by (8.154) (valid for $\alpha \leq 60°$) is abandoned in favor of the greater one given by (8.163) to operate conservatively because of the equality of the diameters of both the tube and the inserted nozzle.

8.4 Flat Head with Central Hole

In Sect. 6.4 we realized that the analysis of a flat head is nothing but simple given the mutual influence between head and cylinder and the need to ensure, albeit through different verification criteria, both the resistance of the head and that at the edge of the cylinder.

Intuitively, it is clear that a rigorous discussion of the stress in a flat head with a central hole is quite difficult, especially if one aims at achieving a generally valid calculation method. Therefore, we will just focus on a relatively simple yet conservative calculation method.

First of all, we observe that the presence of a central hole in a flat head is quite frequent. It may be intended for inspection of the vessel, and in that case the cap is placed on the inside and pressed against the head by the same pressure in the vessel. Possibly, the head may be drilled to engage a piping or a nozzle with a flange and blind counterflange.

In any case, the thrust which would be transmitted to the area occupied by the hole, if the head were not drilled, is not zeroed: it is simply transferred

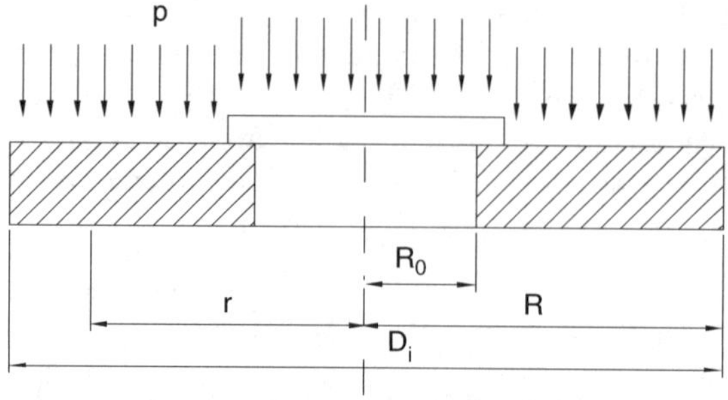

Fig. 8.32

to the edge of the hole or through the gasket of the cap, or through its corresponding axial thrust on the piping or on the nozzle mentioned earlier (see Fig. 8.32). Even if the head is drilled, we still use (6.110) and thus (6.115). If R_0 is the radius of the hole and we assume the head to be simply supported, for $r = R_0$ and $r = R$ it must be $M_r = 0$.

On the other hand, based on the first equation in (6.99) and (6.115)

$$M_r = K\left[-\frac{3pr^2}{16K} + \frac{C_1}{2} - \frac{C_2}{r^2} + \mu\left(-\frac{pr^2}{16K} + \frac{C_1}{2} + \frac{C_2}{r^2}\right)\right], \tag{8.173}$$

that can also be written as

$$M_r = K\left[-\frac{pr^2}{16K}(3+\mu) + \frac{C_1}{2}(1+\mu) - \frac{C_2}{r^2}(1-\mu)\right]; \tag{8.174}$$

hence

$$\begin{aligned} -\frac{pR_0^2}{16K}(3+\mu) + \frac{C_1}{2}(1+\mu) - \frac{C_2}{R_0^2}(1-\mu) = 0 \\ -\frac{pR^2}{16K}(3+\mu) + \frac{C_1}{2}(1+\mu) - \frac{C_2}{R^2}(1-\mu) = 0 \end{aligned} . \tag{8.175}$$

Let us introduce now the following dimensionless quantity:

$$\alpha_0 = \frac{R_0}{R}. \tag{8.176}$$

From (8.175) we obtain

$$-\frac{pR^2}{16K}(3+\mu)\left(\alpha_0^2 - 1\right) = \frac{C_2}{R^2}(1-\mu)\left(\frac{1}{\alpha_0^2} - 1\right), \tag{8.177}$$

then

$$C_2 = \frac{pR^4}{16K}\frac{3+\mu}{1-\mu}\alpha_0^2. \tag{8.178}$$

From the second equation in (8.175) we also obtain

$$C_1 = \frac{pR^2}{16K}\frac{3+\mu}{1+\mu}\left(1+\alpha_0^2\right). \tag{8.179}$$

Then, from (8.174),

$$M_r = -\frac{pr^2}{16}(3+\mu) + \frac{pR^2}{16}(3+\mu)\left(1+\alpha_0^2\right) - \frac{pR^4}{16r^2}(3+\mu)\,\alpha_0^2 \tag{8.180}$$

We introduce the dimensionless quantity α given by the ratio between the generic radius and the radius of the head

$$\alpha = \frac{r}{R}; \tag{8.181}$$

(8.180) may also be written as

$$M_r = \frac{3+\mu}{16}pR^2\left(1+\alpha_0^2 - \frac{\alpha_0^2}{\alpha^2} - \alpha^2\right). \tag{8.182}$$

By analogy, the moment M_c is given by

$$M_c = K\left[-\frac{pr^2}{16K}(1+3\mu) + \frac{C_1}{2}(1+\mu) + \frac{C_2}{r^2}(1-\mu)\right]. \tag{8.183}$$

Recalling (8.178) and (8.179)

$$M_c = -\frac{pr^2}{16}(1+3\mu) + \frac{pR^2}{16}(3+\mu)\left(1+\alpha_0^2\right) - \frac{pR^4}{16r^2}(3+\mu)\,\alpha_0^2, \tag{8.184}$$

and finally

$$M_c = \frac{3+\mu}{16}pR^2\left(1+\alpha_0^2 + \frac{\alpha_0^2}{\alpha^2} - \frac{1+3\mu}{3+\mu}\alpha^2\right). \tag{8.185}$$

Let us recall that $M_r = 0$ at the edge of the hole and at the edge of the head, i.e., for $\alpha = \alpha_0$ and $\alpha = 1$. The moment is positive between these two values of α and reaches a maximum that we want to identify in terms of position and value. Deriving (8.174) and assuming the derivative is zero, we obtain

$$\frac{dM_r}{d\alpha} = \frac{3+\mu}{16}pR^2\left(2\frac{\alpha_0^2}{\alpha^3} - 2\alpha\right); \tag{8.186}$$

then

$$\alpha = \sqrt{\alpha_0}. \tag{8.187}$$

The corresponding value of M_r is equal to

$$M_{r(\max)} = \frac{3+\mu}{16}pR^2\left(1+\alpha_0\right)^2. \tag{8.188}$$

Let us compare (8.188) with (6.121) that computes the maximum value of M_r at the center of the head without holes.

As one can see, the maximum value of M_r is lower in the drilled plate and corresponds to the value in a head without holes for $\alpha_0 = 0$. We now calculate the derivative of M_c with respect to α and set it to zero. We obtain

$$\alpha^4 = -\frac{3+\mu}{2(1+\mu)}\alpha_0^2. \tag{8.189}$$

Equation (8.189) does not have real solutions. Hence, the moment M_c has no minimum or maximum values. The greatest value of M_c occurs at the edge of the hole, i.e., for $\alpha = \alpha_0$. From (8.185) we obtain

$$M_{c(\max)} = \frac{3+\mu}{8}pR^2\left(1 + \frac{1-\mu}{3+\mu}\alpha_0^2\right). \tag{8.190}$$

Let us now compare (8.190) with (6.123) that is about the maximum value of M_c at the center of the head without holes. In the drilled head the value of the moment is twice the one in a head without holes. For $\alpha_0 = 0$, i.e., in the absence of holes through (8.190) one obtains a value twice the latter. This is absurd from a physical point of view, and it shows how (8.190), contrary to (8.188), may not be used in extreme situations.

Let us now examine the behavior of the plate with a very small hole. To be clear, we choose an actual case assuming that $\alpha_0 = 0.01$.

Calculations of M_r and M_c through (8.182) and (8.185) lead to Fig. 8.33 where one can also see the values for these moments in a head without holes. As one can verify, M_r and M_c practically coincide with the values of the head without holes, except for a small portion of the head around the hole. In this area the value of M_r drops abruptly until it becomes zero at the edge of the hole, while M_c becomes about twice the one with a not drilled head.

This way the local influence of the hole is clearly highlighted. Based on the considerations made about plastic collaboration, the peak can and must be ignored. In other words, the presence of a tiny hole cannot and must not provoke an increase in thickness of the head. One must also consider that if the thickness of the head does not increase, the value of the peak, equal to twice the ideal stress based on which the plate is sized, is equal to $2 \times 1.5f = 3f$, according to Sect. 1.5.

An increase in radius of the hole corresponds to an increase in the maximum value of M_c, too. Fig. 8.34 shows the curves relative to M_r and M_c for $\alpha_0 = 0.4$ and for $\alpha_0 = 0.8$. As one can see, by increasing the radius of the hole the values of M_r decrease, while those of M_c increase. The entity of the peak, with respect to the average value of M_c, decreases, too. Undoubtedly, one must consider the moment M_c, but the issue is how to factor it in without stressing local effects around the hole but yet limiting the area with plastic flow. In our view, for the sizing it is acceptable to consider the value of M_c that corresponds to the maximum value of M_r. It is called M_{rif} in Figs. 8.33 and 8.34.

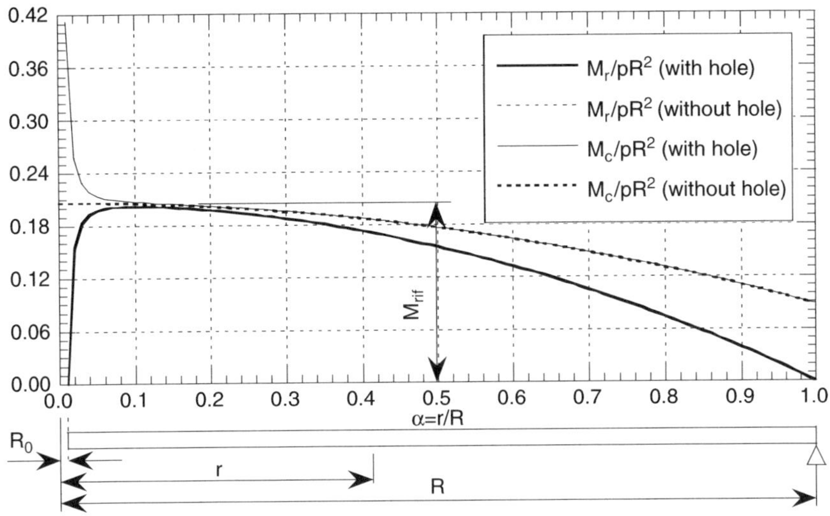

Fig. 8.33

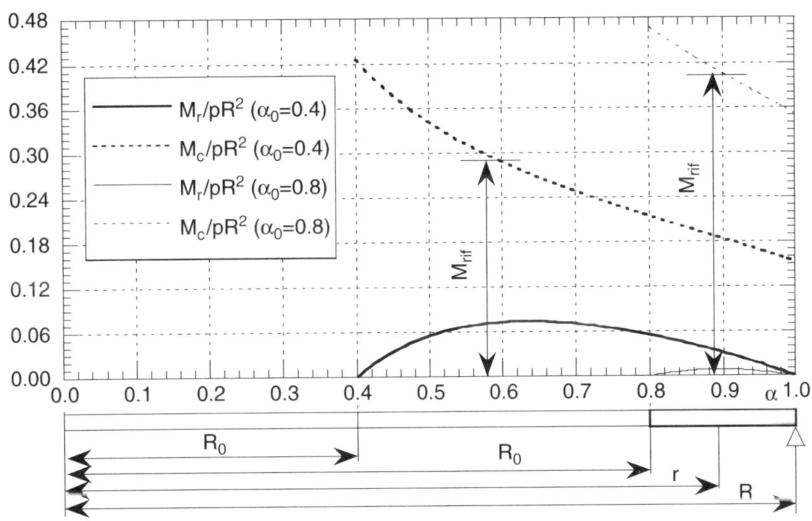

Fig. 8.34

Assuming that the corresponding stress is equal to $1.5f$, the fibers around the hole show signs of yielding, but in any case the stress is less than $3f$. This means that increasing the radius of the hole is equivalent to extending plastic flow to a greater area of the plate. Note that the entity of the peak (referred to M_{rif}) decreases and the values of the radial stress considerably decrease, too. Thus, we believe that the suggested criterion may be taken into consideration.

If we introduce the maximum value from (8.187) (max M_r) into (8.185) we obtain

$$M_c = \frac{3+\mu}{16} pR^2 \left(1 + \alpha_0^2 + 2\frac{1-\mu}{3+\mu}\alpha_0\right). \tag{8.191}$$

The corresponding stress coincides with σ_{id}, as stated previously about heads without holes, and recalling that $W = s^2/6$ we have

$$\sigma_{id} = \frac{3\,(3+\mu)}{8} \frac{pR^2}{s^2} \left(1 + \alpha_0^2 + 2\frac{1-\mu}{3+\mu}\alpha_0\right). \tag{8.192}$$

As usual, assuming that $\sigma_{id} = 1.5f$ (that corresponds to the yield strength) and introducing the diameter D_i of the head, we have

$$s = \frac{\sqrt{3+\mu}}{4} \sqrt{1 + \alpha_0^2 + 2\frac{1-\mu}{3+\mu}\alpha_0} D_i \sqrt{\frac{p}{f}}. \tag{8.193}$$

Recalling (6.84) and (6.136), we may write

$$s = CyD_i\sqrt{\frac{p}{f}}; \tag{8.194}$$

given that $\mu = 0.3$ and that $\alpha_0 = d/D_i$

$$y = \sqrt{1 + \left(\frac{d}{D_i}\right)^2 + 0.424\frac{d}{D_i}}. \tag{8.195}$$

The values of y are shown in Fig. 8.35 (curve A).

Under equal conditions, y represents the ratio between the required thickness for the drilled plate and the one required for a plate without holes.

At this point, some perplexity may arise about the possibility to size according to the same criteria, regardless of whether we have a head with a small hole or a head with a hole so large that one should actually talk about a circular crown. In fact, examining both Fig. 8.33 and 8.34 it is evident that in the case of a small hole, the effect is clearly localized with values of M_r and M_c that vary only in proximity of the hole compared to the rest of the head. If the hole is instead quite large, M_r ceases to dominate, and the moments M_c become predominant. Their behavior is rather smooth through the crown, and even though the values increase towards the hole, one cannot talk of a localized effect. For this reason it is best for the sake of global resistance of the crown to consider the average value M_{cm} of M_c. Through (8.185) one obtains the following equation of M_{cm}:

$$M_{cm} = \frac{pR^2}{6} \frac{1-\alpha_0^3}{1-\alpha_0}, \tag{8.196}$$

where the average value of σ_c (as well as σ_{id}) in the fiber most distant from the neutral axis is given by

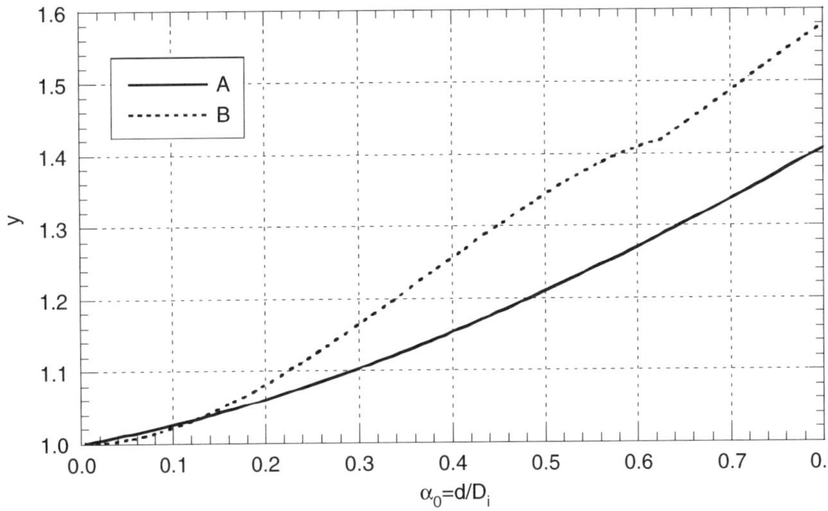

Fig. 8.35

$$\sigma_{id(m)} = \sigma_{cm} = \frac{pR^2}{s^3}\frac{1-\alpha_0^3}{1-\alpha_0}. \tag{8.197}$$

As usual, if we assume that $\sigma_{id(m)} = 1.5f$ we have

$$s = \frac{\sqrt{3+\mu}}{4}\frac{2}{\sqrt{1.5\,(3+\mu)}}\sqrt{\frac{1-\alpha_0^3}{1-\alpha_0}}D_i\sqrt{\frac{p}{f}}. \tag{8.198}$$

Recalling (6.136) we have

$$y = \frac{2}{\sqrt{1.5\,(3+\mu)}}\sqrt{\frac{1-\alpha_0^3}{1-\alpha_0}} = 0.899\sqrt{\frac{1-\left(\dfrac{d}{D_i}\right)^3}{1-\dfrac{d}{D_i}}}. \tag{8.199}$$

We see that the values of y obtained equation from (8.199) are always lower than those given by (8.195). The latter equation and curve A in Fig. 8.35 guarantee that the average value among the maximum values of σ_{id} does not exceed $1.5f$, too.

This way the global resistance of the crown is assured. The analysis above assumes that the head is simply supported. Of course, this is an extreme situation that does not correspond to reality with a head welded to a cylinder, but in certain instances this is still not far from the truth. This may occur when the ratio p/f between pressure and basic allowable stress is small, especially if the thickness of the cylinder is equal to the minimum required one or slightly greater. As far as Fig. 6.28, under these conditions the value of C is rather large (we know that this is due to the fact that the verification of resistance at the edge of the cylinder is crucial).

To proceed with the investigation, it is important to verify how the central hole influences the rotation at the edge of the head. Still assuming that the head is simply supported based on (6.116), (8.178), and (8.179), we obtain the following for $r = R$ and $\mu = 0.3$:

$$\varphi_R = \frac{pR^3}{16K}\left(1.538 + 7.252\alpha_0^2\right). \tag{8.200}$$

Remembering the meaning of K, we finally obtain

$$\varphi_R = \frac{p}{E}\left(\frac{R}{s}\right)^3\left(1.05 + 4.95\alpha_0^2\right) \tag{8.201}$$

By indicating the ratio between the rotation at the edge of the drilled plate and the plate without holes with k_φ, with equal thickness, we have:

$$k_\varphi = 1 + 4.71\alpha_0^2. \tag{8.202}$$

If the thickness is to be increased through y instead, based on (8.194), (8.195), and (8.201) we obtain

$$k_\varphi = \frac{1 + 4.71\alpha_0^2}{\left(1 + \alpha_0^2 + 0.424\alpha_0\right)^{3/2}}. \tag{8.203}$$

Equation (8.203) is shown in Fig. 8.36. We see that except for the values of d/D_i that are smaller than 0.21, in spite of the increase in thickness, the rotation at the edge of the drilled simply supported head is greater than that of a head without holes. Given that the head is connected to the cylinder, this greater rotation generates a greater moment at its edge than that occurring with a head without holes. Note that the head behaves almost like a simply supported one, precisely in those cases when a verification at the edge of the cylinder is crucial.

This means that the sizing based on y in (8.195) or curve A in Fig. 8.35 may not be conservative enough to guarantee the resistance of the cylinder. Let us then examine the problem from a different viewpoint, i.e., assuming that the thickness of the drilled head is such that it will have the same rotation φ_R of a not drilled head for an equal moment M_{rR} at the edge. Remember that according to the first equation in (8.175) the moment M_r must be zero for $r = R_0$, as it should be.

Hence

$$C_2 = -\frac{pR_0^4}{16K}\frac{3+\mu}{1-\mu} + \frac{C_1R_0^2}{2}\frac{1+\mu}{1-\mu}; \tag{8.204}$$

from (8.174) and (8.204) and for $r = R$, we have

$$M_{rR} = -\frac{pR^2}{16}\left(3+\mu\right)\left(1-\alpha_0^4\right) + \frac{C_1K}{2}\left(1+\mu\right)\left(1-\alpha_0^2\right) \tag{8.205}$$

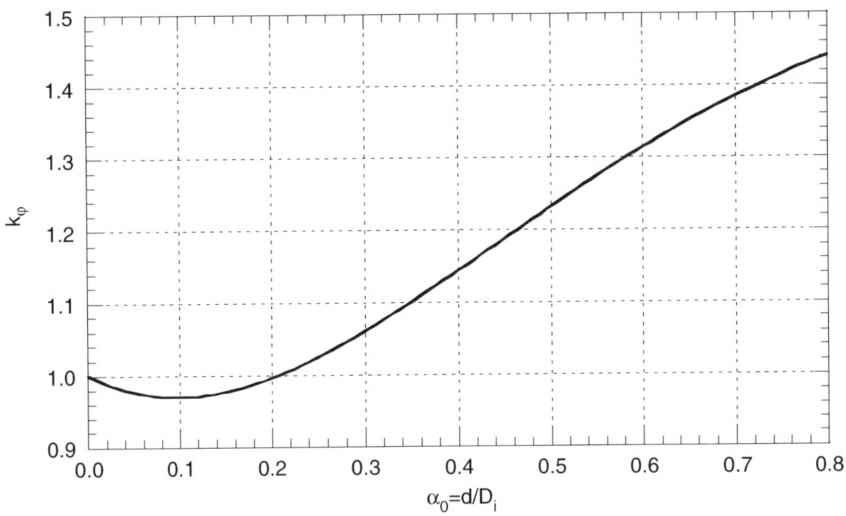

Fig. 8.36

from (6.115) and (8.204) and $r = R$, we obtain

$$\varphi_R = -\frac{pR^3}{16K}\left(1 + \frac{3+\mu}{1-\mu}\alpha_0^4\right) + \frac{C_1 R}{2}\left(1 + \frac{1+\mu}{1-\mu}\alpha_0^2\right); \tag{8.206}$$

from (8.205) and (8.206) through a series of steps we have

$$\frac{M_{rR}}{pR^2} = \frac{K\varphi_R}{pR^3}\frac{\left(1-\mu^2\right)\left(1-\alpha_0^2\right)}{1-\mu+(1+\mu)\,\alpha_0^2} - \frac{1}{8}\frac{\left[1-\mu+(3+\mu)\,\alpha_0^2\right]\left(1-\alpha_0^2\right)}{1-\mu+(1+\mu)\,\alpha_0^2}. \tag{8.207}$$

With $\mu = 0.3$ from (8.207) we have

$$\frac{M_{rR}}{pR^2} = \frac{K\varphi_R}{pR^3}\frac{0.91\left(1-\alpha_0^2\right)}{0.7+1.3\alpha_0^2} - \frac{1}{8}\frac{\left(0.7+3.3\alpha_0^2\right)\left(1-\alpha_0^2\right)}{0.7+1.3\alpha_0^2}, \tag{8.208}$$

and, recalling the meaning of K, we finally have

$$\frac{M_{rR}}{pR^2} = \frac{E}{p}\varphi_R\left(\frac{s}{R}\right)^3\frac{\left(1-\alpha_0^2\right)}{12\left(0.7+1.3\alpha_0^2\right)} - \frac{1}{8}\frac{\left(0.7+3.3\alpha_0^2\right)\left(1-\alpha_0^2\right)}{0.7+1.3\alpha_0^2}. \tag{8.209}$$

First of all, we establish that for $M_{rR} = 0$ (8.209) corresponds to (8.201). In addition, for $\varphi_R = 0$ we obtain

$$M_{rR} = -\frac{1}{8}\frac{\left(0.7+3.3\alpha_0^2\right)\left(1-\alpha_0^2\right)}{0.7+1.3\alpha_0^2}pR^2, \tag{8.210}$$

that leads to the value of M_{rR} in the case of a perfectly clamped edge. Specifically, for $\alpha_0 = 0$, i.e., a nondrilled head, we rediscover (6.129).

In the case of a nondrilled head, (8.209) is reduced to

$$\frac{M_{rR}}{pR^2} = 0.119\frac{E}{p}\varphi_R\left(\frac{s}{R}\right)^3 - 0.125. \tag{8.211}$$

Assuming that

$$\beta = 0.119\frac{E}{p}\varphi_R\left(\frac{s}{R}\right)^3, \tag{8.212}$$

we write

$$\frac{M_{rR}}{pR^2} = \beta - 0.125. \tag{8.213}$$

If s is the thickness of the undrilled head, the thickness of the drilled head will be equal to ys, where y is the usual quantity the value of which we want to identify. Based on (8.209), where we substitute y with ys, and recalling (8.212) we write

$$\frac{M_{rR}}{pR^2} = \beta y^3 \frac{1-\alpha_0^2}{1.428\left(0.7+1.3\alpha_0^2\right)} - \frac{1}{8}\frac{\left(0.7+3.3\alpha_0^2\right)\left(1-\alpha_0^2\right)}{0.7+1.3\alpha_0^2}. \tag{8.214}$$

Assuming that

$$\gamma = \frac{1-\alpha_0^2}{1.428\left(0.7+1.3\alpha_0^2\right)}, \tag{8.215}$$

and

$$\delta = \frac{1}{8}\frac{\left(0.7+3.3\alpha_0^2\right)\left(1-\alpha_0^2\right)}{0.7+1.3\alpha_0^2} \tag{8.216}$$

from (8.214) we obtain

$$\frac{M_{rR}}{pR^2} = \beta y^3\gamma - \delta \tag{8.217}$$

Assuming that M_{rR} is equal in both instances, we have:

$$\beta - 0.125 = \beta y^3\gamma - \delta; \tag{8.218}$$

then

$$y = \sqrt[3]{\frac{\beta + \delta - 0.125}{\beta\gamma}}. \tag{8.219}$$

Quantities γ and δ are a function of α_0 while the value of β depends on the characteristics of the constraint between head and cylinder. In view of the analysis in Sect. 6.4, common practice shows that the value of β varies between 0.06 and 0.12 (the latter almost corresponds to simply supported edge). This made it possible to plot the curves in Fig. 8.37.

As one can see, β modestly influences the value of y as long as $\alpha_0 < 0.65$. For greater values of α_0, the progression of the curves changes completely as β changes. Specifically, for $\beta = 0.06$, the necessary thickness of a drilled

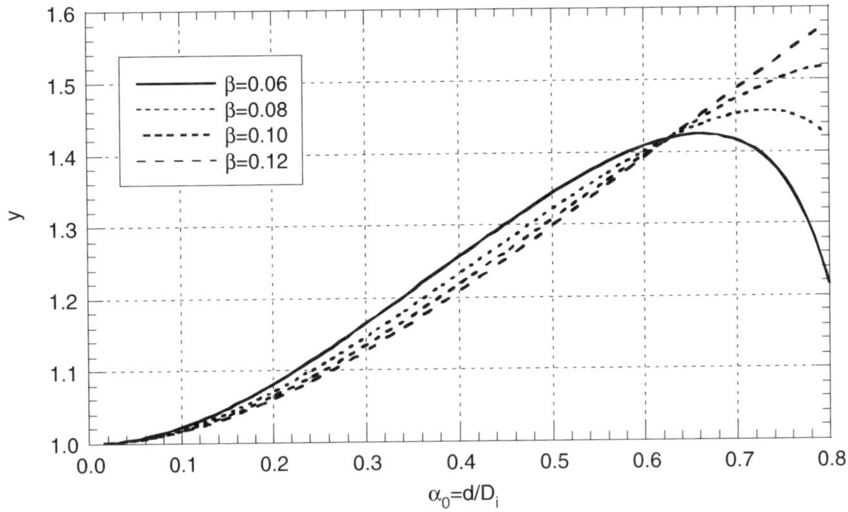

Fig. 8.37

head is considerably reduced as α_0 increases. On the other hand, in order to adopt a simple and general calculation criterion one must analyze worst case scenarios.

In summary, the set of curves in Fig. 8.37 may be substituted by the one curve in Fig. 8.35 where the values of y are greater than those to be obtained from curve A. The adoption of curve B therefore guarantees the resistance of the head even under extreme conditions characterized by simple supporting, as well as the resistance of the cylinder at the connection with the head. In summary, the sizing of the plate must be done based on (8.194) that we rewrite as

$$s = CyD_i\sqrt{\frac{p}{f}}. \tag{8.194 bis}$$

C can be obtained from Fig. 6.28, while y is derived from curve B in Fig. 8.35.

8.5 Drilled Plates

As in the case of the head with a central hole in Sect. 8.4, a rigorous analysis of the topic is extremely complicated, and we can only suggest a very simple yet conservative calculation criterion. Figure 8.38 shows a circular plate with a series of holes. First of all, we assume the presence of pressure only on the plate. This occurs, either directly in the not drilled areas of the plate, or indirectly through the thrusts caused by the pressure transmitted to the plate by the tubes or the nozzles connected to the plate itself.

In other words, there are no thrusts of another nature, such as, e.g., those caused by prevented dilations of the tubes that connect opposite plates.

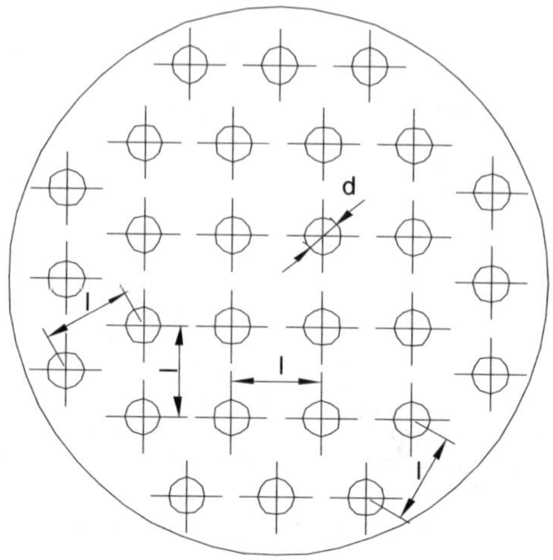

Fig. 8.38

Essentially, there are two problems. First of all, when examining the behavior of the plate and its connected cylinder, one must ensure that no stresses unacceptable for resistance are generated at the edge of the latter. To prevent this from happening, the rigidity of the drilled plate must at least be equal to that of the undrilled plate that guarantees the resistance of the cylinder, according to the criteria discussed in Sect. 6.4.

Let l be the pitch between holes in Fig. 8.38 and s be the thickness of the drilled plate. With reference to the length l, the moment of inertia I of the plate, given the presence of the holes with a diameter d, is equal to

$$I = \frac{(l-d)\,s^3}{12}. \tag{8.220}$$

With reference to the same length l, the moment of inertia I_0 of a undrilled plate of thickness s_0 would be equal to

$$I = \frac{l s_0^3}{12}. \tag{8.221}$$

Assuming that $I = I_0$, we have

$$\left(\frac{s}{s_0}\right)^3 = \frac{l}{l-d}. \tag{8.222}$$

If y_1 (see Sect. 8.4) is the ratio between the thickness of the drilled plate and of the undrilled plate, we obtain

$$y_1 = \sqrt[3]{\frac{l}{l-d}}. \tag{8.223}$$

Therefore, recalling (6.84) the sizing of the plate can be obtained through the following equation:

$$s = Cy_1 D_i \sqrt{\frac{p}{f}} \tag{8.224}$$

where C is taken from Fig. 6.28.

If the stress at the edge of the cylinder is crucial for the sizing of the plate without holes, the sizing done according to our analysis guarantees resistance, since the increase in thickness establishes again the original rigidity of the plate. It may happen, though, that the stress of the plate itself is crucial for the undrilled plate. In that case the increase in thickness obtained through y_1 is insufficient. In fact, the section modulus of the area between the holes is smaller than the one corresponding to the undrilled plate of thickness s_0.

Thus, it is necessary to further increase the thickness by assuming this time the equality of the section modulus.

After a similar process we have

$$W = \frac{(l - d)\, s^2}{6} \tag{8.225}$$

for the drilled plate, and

$$W_0 = \frac{l s_0^2}{6} \tag{8.226}$$

for the undrilled plate. Assuming that $W = W_0$, we obtain the following equation of y_2 that represents the ratio s/s_0:

$$y_2 = \sqrt{\frac{l}{l - d}}. \tag{8.227}$$

It would seem natural to write

$$s = Cy_2 D_i \sqrt{\frac{p}{f}}, \tag{8.228}$$

but this equation is illogical for a number of reasons. Always referring to undrilled plates, if the stress at the edge of the cylinder is critical (with large values of C), the increase in thickness caused by the introduction of y_2 is an unnecessary burden.

On the other hand, if the stress in the plate (small values of C that are orientatively below 0.43) is critical, the thickness is increased through y_2 beyond what is required to reestablish the rigidity of the plate. This works in favor of the resistance of the cylinder but decreases the moment of clamping of the plate, and as a result the moments at the center go up.

The value of C obtained from Fig. 6.28 may not be the best choice anymore to guarantee the resistance of the plate. Therefore, (8.228) may not be used if C represents the value of the form factor obtained through the figure mentioned above.

A safe and conservative calculation criterion requires to consider the extreme condition of the simply supported plate with a form factor equal to 0.454 (Sect. 6.4). In other words, we recommend the following approach.

The thickness of the plate must be the largest among those obtained from

$$s = Cy_1 D_i \sqrt{\frac{p}{f}} \tag{8.224 bis}$$

and

$$s = 0.454 y_2 D_i \sqrt{\frac{p}{f}}, \tag{8.229}$$

where C is the form factor obtained from Fig. 6.28, while y_1 and y_2 are taken from (8.223) and (8.227). A sizing criterion of the plate integrated in the concepts above that also takes C into account for values below 0.454, is as follows.

The three values of the thickness below must be computed

$$s_1 = Cy_1 D_i \sqrt{\frac{p}{f}}; \tag{8.230}$$

$$s_2 = 0.454 y_2 D_i \sqrt{\frac{p}{f}}; \tag{8.231}$$

$$s_3 = \frac{2C + 0.454}{3} y_2 D_i \sqrt{\frac{p}{f}}. \tag{8.232}$$

As you can see, the first two equations have already been considered. The sizing criterion is as follows.

Assuming a basic thickness s_2, the thickness s may be assumed to be equal to s_1 if $s_1 > s_2$. Moreover, if $s_3 < s_2$ the thickness s may be assumed to be equal to s_3. This possibility differentiates the latter criterion from the previous one, and induces a less burdensome sizing of the plate when the stress in the plate is crucial in terms of resistance. Given the behavior of C and the influence of y_1 and y_2, the thickness s of the plate may correspond to either s_1 or s_3, as we shall point out further on during the analysis of Fig. 8.41.

This calculation criterion (where d is the diameter of the hole) can always be used, provided, of course, that the conditions mentioned at the beginning of the section are met. Specifically, the criterion must be used if there are expanded pipes inserted in the holes, or if there are welded pipes or nozzles with a thickness equal to the minimum required value.

If the thickness exceeds the minimum, it is possible to factor in the reinforcement represented by the extra thickness by introducing the concept of the equivalent diameter, already discussed in Sect. 8.2.

With reference to Fig. 8.39, if A is the dashed line in the figure, it is possible to identify an equivalent diameter given by

$$d = d_i - \frac{A}{s} \ (case\ a)$$

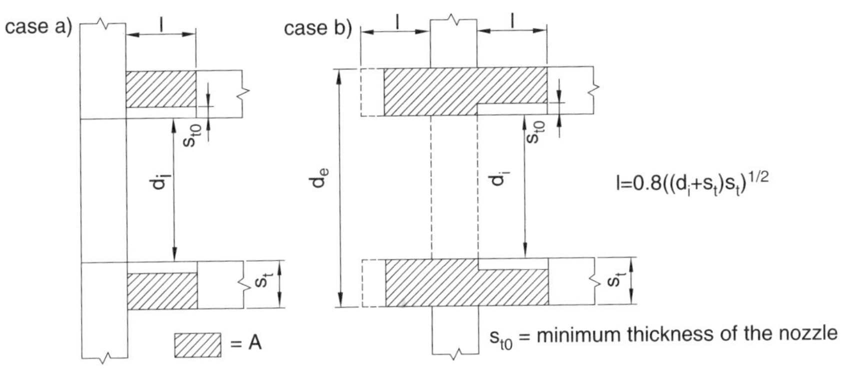

Fig. 8.39

and by

$$d = d_e - \frac{A}{s} \text{ (case b).}$$

This diameter may be plugged into (8.223) and (8.227) to compute y_1 and y_2. This procedure is both defensible and conservative. One should not forget, in fact, that through y_1 and y_2 it is possible to provide the plate with the same moment of inertia and the same section modulus of the plate without holes. Evidently, the reinforcement area represented by extra thickness of the nozzle is more effective to that extent than the extra thickness of the plate. Finally, some reasonable doubt may stem from the fact that once again we have completely ignored the existence of stress peaks at the edge of the holes.

In Sect. 8.4, we were able to establish that at the center of the plate (where we have the maximum positive values for the moments M_r and M_c) the presence of a hole causes a peak, the value of which never exceeds twice the stress occurring in the absence of holes. In any case, it is never larger than $3f$ in agreement with Sect. 1.5 about peaks caused by bending moments.

On the other hand, M_r and M_c have the same value. Observe Fig. 2.13 about two orthogonal stresses of equal value and where the peak is twice the basic stress. In certain cases one may also consider another more complex and scientifically more significant calculation criterion that may be useful for tubesheets under high pressure and nuclear reactors.

Let us consider the tubesheet in Fig. 8.40 that shows holes at regular distance according to the classic triangular display. The tubes are expanded, possibly including a slight welding for a better tightness, while l and d_i are the pitch and the inside diameter of the tubes, respectively. It was experimentally established that if the thickness of the plate is at least twice the pitch, in terms of deformations it can be treated as a plate without holes, if the elasticity modulus E is substituted with a virtual modulus.

It is given by

$$E' = E\frac{l - d_i}{l}. \tag{8.233}$$

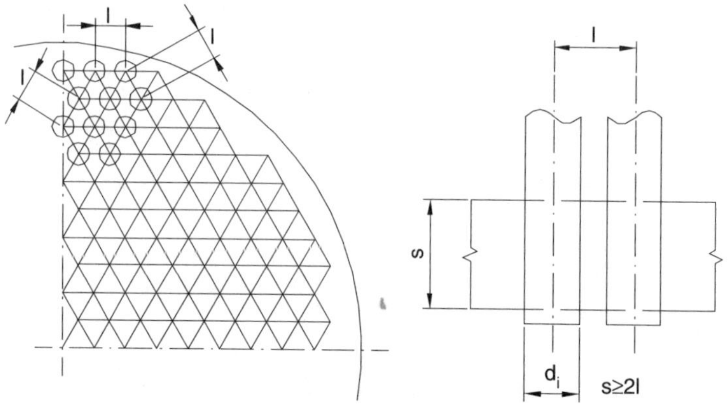

Fig. 8.40

As far as the known correlations between deformations and stresses where Poisson's modulus comes into play, they are essentially valid even for the drilled plate (except for the substitution of E with E'), so that the value of μ does not require corrections.

Finally, once the bending moments in the plate are known, the average stresses between holes are obtained as follows: if σ is the stress corresponding to a plate without holes, the average stress σ' between holes is given by

$$\sigma' = \sigma \frac{l}{l - d_i}. \tag{8.234}$$

Of course, corresponding to the edges of the holes there are stress peaks characteristic of secondary stresses that can be ignored.

Using the efficiency of ligaments z (with reference in this specific case to the internal diameter of the tube and not of the hole), we may write:

$$z = \frac{l - d_i}{l}. \tag{8.235}$$

Then, we have

$$E' = zE, \tag{8.236}$$

and

$$\sigma' = \frac{\sigma}{z}. \tag{8.237}$$

This way, except for required adaptations, it is possible to do the sizing of the drilled plate by adopting the same calculation criteria discussed in Sect. 6.4.

We establish that in order to factor in (8.236), it suffices to adopt the following values in (6.176) corrected by the quantities k_1 and k_3.

$$\begin{aligned} k_1' &= \frac{k_1}{z} \\ k_3' &= \frac{k_3}{z} \end{aligned} \tag{8.238}$$

After this modification and through (6.178) and (6.179), one obtains the values Z and S valid for a drilled plate. As far as (6.185) and (6.186), based on (8.237) it suffices to consider the following values instead of the resulting values of σ_r and σ_c:

$$\begin{aligned} \sigma'_r &= \frac{\sigma_r}{z} \\ \sigma'_c &= \frac{\sigma_c}{z} \end{aligned} . \tag{8.239}$$

No change is required for σ_a that is equal to $-p$ even in the case of a drilled plate. Once σ'_c, σ'_r and σ_a at the center of the plate are known, one computes the value of the ideal stress σ_{id} according to criteria already discussed several times, to verify if it is smaller than $1.5f$.

As far as the edge of the plate, given the presence of holes it is best if the ideal stress does not exceed $1.5f$. As far as the cylinder, there is nothing to add to the analysis in Sect. 6.4 (especially (6.92) and (6.93)).

The verification criteria are identical, i.e., an ideal stress equal to $3f$ is allowable at the edge of the cylinder. However, if it is greater than $1.5f$, it is necessary to reduce the value of the bending moment at the edge accordingly with regard to bending moments at the center of the plate, in agreement with the criterion discussed in the section in question. Therefore, similarly to the

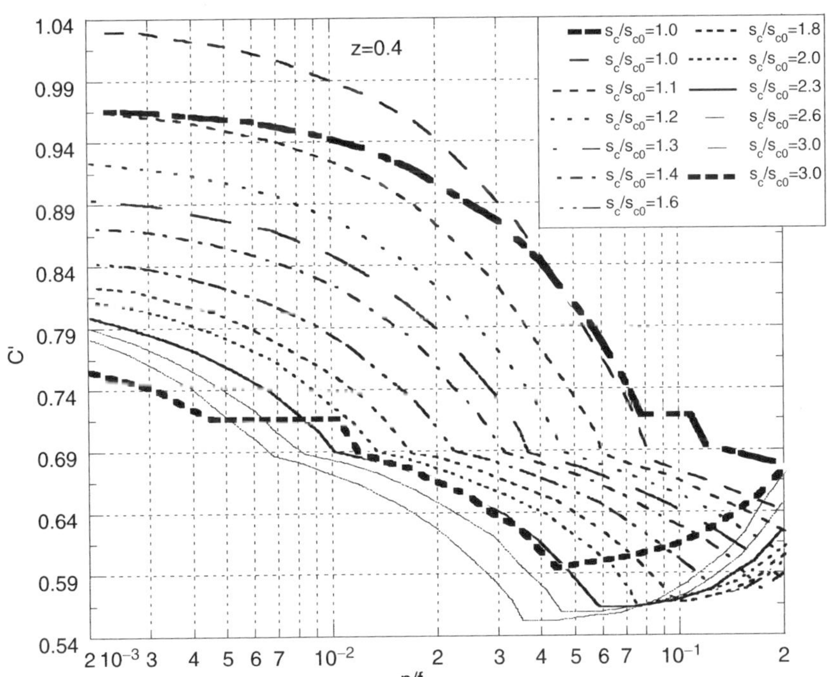

Fig. 8.41

analysis in Sect. 6.4, for different values of z it is possible to compute a form factor C' for a drilled plate so that the thickness of the plate can be assumed to be equal to

$$s = C'D_i\sqrt{\frac{p}{f}}. \tag{8.240}$$

Figure 8.41 shows an example, including the calculation of the values of C' for $z = 0.4$. We also included two dashed curves that show the extreme values of s_c/s_{co}, i.e., s_c/s_{co} equal to 1 and 3, respectively. They correspond to the approximated sizing criterion shown previously, based on s_1, s_2 and s_3 ((8.230)–(8.232)), having assumed that $y_1 = 1/\sqrt[3]{z}$ and $y_2 = 1/\sqrt{z}$.

Specifically, the left portion of the curves corresponds to $s = s_1$, the portion parallel to axis of the abscissa corresponds to $s = s_2$, and finally the remaining portion of the curves corresponds to $s = s_3$.

The values of C' that correspond to the approximated sizing criterion are generally greater than the exact ones, so that this criterion can be considered conservative. In fact, for $s = s_1$ this condition is not true: this means that it is not sufficient to reconstitute the rigidity of the plate through an increase in thickness to factor in the presence of holes with regard to the limitation to $3f$ of the value of the ideal stress at the edge of the cylinder.

Note that the differences between the exact values of C' and the ones corresponding to $s = s_1$ are rather modest and result in differences in thickness equal to a few percentage points.

We conclude that this approximate criterion is indeed acceptable, the issues shown in Fig. 8.41 notwithstanding. In addition, note that the value of z to compute the values of C' in that figure is quite small. For greater values of z, the differences mentioned earlier are obviously less sensitive.

Finally, note that for expanded tubes, whilst to compute z (8.235) one considers the internal diameter of the tube, to compute y_1 and y_2, to favor resistance, one considers the diameter of the hole, thus $y_1 > 1/\sqrt[3]{z}$ and $y_2 > 1/\sqrt{z}$.

8.6 Holes in Quadrangular Vessels

First of all, we consider a hole line with an axis that is parallel to the vessel axis (Fig. 8.42).

Since we are looking at a flat wall, it is useful to refer back to Sect. 2.3. Specifically, we are interested in the first equation in (2.76), given that the axial stress can be ignored if compared to the stresses orthogonal to the axis that are generated by the bending moment present along the hole line.

Let us keep in mind that this equation refers to an isolated hole. If σ is the constant stress that would be exerted in a section parallel to the axis of a vessel without holes, and σ' the stress in different positions in the case of an isolated hole (this case is about values of the stresses generated by a bending

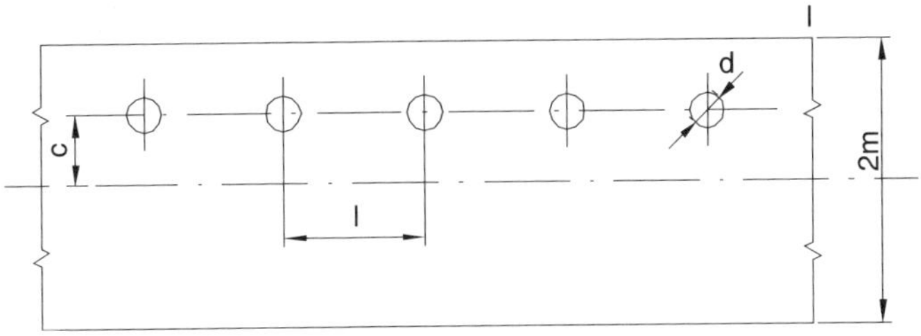

Fig. 8.42

moment in the fibers that are farthest from the neutral axis), and finally x the abscissa measured along the parallel to the axis, starting from the center of the hole, based on the above equation we can write

$$\sigma' = \sigma\left[1 + \frac{1}{2}\left(\frac{r}{x}\right)^2 + \frac{3}{2}\left(\frac{r}{x}\right)^4\right]. \tag{8.241}$$

Now, if we consider two holes and a distance l between the two centers (Fig. 8.42) with sufficient approximation, we may write

$$\sigma' = k\sigma\left[1 + \frac{1}{2}\left(\frac{r}{x}\right)^2 + \frac{3}{2}\left(\frac{r}{x}\right)^4 + \frac{1}{2}\left(\frac{r}{l-x}\right)^2 + \frac{3}{2}\left(\frac{r}{l-x}\right)^4\right], \tag{8.242}$$

where the value k will be specified later on.

We integrate σ' between r and $(l - r)$, i.e., for the entire section between the holes. By indicating the integral with I and based on (8.242) we have

$$I = \int_r^{l-r} \sigma' dx = \frac{k\sigma}{2}\left[2x - \frac{r^2}{x} - \frac{r^4}{x^3} + \frac{r^2}{l-x} + \frac{r^4}{(l-x)^3}\right]_r^{l-r}. \tag{8.243}$$

Finally, we obtain

$$I = ko\left[l - \frac{r^2}{l-r} - \frac{r^4}{(l-r)^3}\right]. \tag{8.244}$$

On the other hand, for the sake of equilibrium the value of the integral must correspond to the product of the stress σ by the pitch l, given that it is reduced to that in the absence of holes.

Thus, we have

$$k\sigma\left[l - \frac{r^2}{l-r} - \frac{r^4}{(l-r)^3}\right] = \sigma l; \tag{8.245}$$

then

$$k = \frac{1}{1 - \dfrac{\left(\frac{r}{l}\right)^2}{1 - \frac{r}{l}} - \dfrac{\left(\frac{r}{l}\right)^4}{\left(1 - \frac{r}{l}\right)^3}}. \tag{8.246}$$

Similarly to what was seen about hole lines in cylinders, spheres and cones, we introduce the efficiency of ligaments given by

$$z = \frac{l-d}{l} = \frac{l-2r}{l}. \tag{8.247}$$

Equation (8.246) can also be written as

$$k = \frac{(1+z)^3}{(1+z)^3 - \frac{1}{2}(1-z^2)^2 - \frac{1}{2}(1-z)^4}. \tag{8.248}$$

k is shown in Fig. 8.43.

One obtains the maximum value of σ' for $x = r$ and $x = l - r$. Based on (8.242) we have

$$\sigma'_{\max} = k\sigma \left[3 + \frac{1}{2}\left(\frac{\frac{r}{l}}{1-\frac{r}{l}}\right)^2 + \frac{3}{2}\left(\frac{\frac{r}{l}}{1-\frac{r}{l}}\right)^4 \right]; \tag{8.249}$$

hence

$$k_1 = \frac{\sigma'_{\max}}{\sigma} = k \left[3 + \frac{1}{2}\left(\frac{\frac{r}{l}}{1-\frac{r}{l}}\right)^2 + \frac{3}{2}\left(\frac{\frac{r}{l}}{1-\frac{r}{l}}\right)^4 \right]. \tag{8.250}$$

The ratio k_1 is shown in Fig. 8.44. It is equal to 3 for $r/l = 0$, i.e., $l = \infty$. In that case we refer back to the isolated hole in Sect. 2.3, and we find the value in the first equation in (2.78) and in Fig. 2.11.

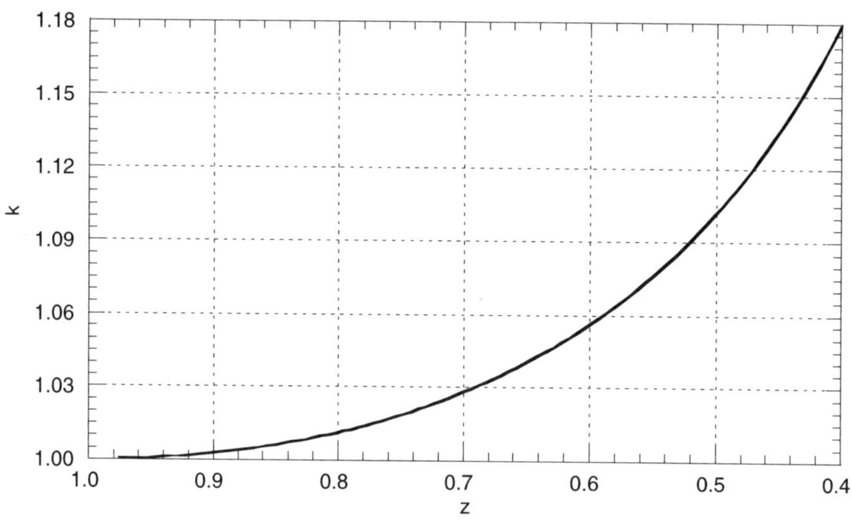

Fig. 8.43

If $r/l > 0$ the value of k_1 is greater than 3 given the influence of the second hole. Our interest is actually not focused on this ratio but rather on the ratio between maximum stress at the edge of the holes and the average stress between holes. In fact, compared to the vessel without holes, the thickness is necessarily increased to limit the average value of the stress between holes to the allowable level, according to the criteria discussed in Sect. 1.5. Returning to the considerations made about (8.245) and indicating the average value of σ' with σ'_m, we have

$$\sigma'_m (l - d) = \sigma l; \tag{8.251}$$

then

$$\sigma'_m = \sigma \frac{l}{l-d} = \sigma \frac{l}{l-2r} = \frac{\sigma}{z}. \tag{8.252}$$

Therefore, based on (8.249) we have

$$k_2 = \frac{\sigma'_{\max}}{\sigma'_m} = k \left[3 + \frac{1}{2} \left(\frac{\frac{r}{l}}{1 - \frac{r}{l}} \right)^2 + \frac{3}{2} \left(\frac{\frac{r}{l}}{1 - \frac{r}{l}} \right)^4 \right] \left(1 - \frac{2r}{l} \right); \tag{8.253}$$

or

$$k_2 = kz \left[3 + \frac{1}{2} \left(\frac{1-z}{1+z} \right)^2 + \frac{3}{2} \left(\frac{1-z}{1+z} \right)^4 \right]. \tag{8.254}$$

The ratio k_2 is shown in Fig. 8.44, as well. It highlights how $\sigma'_{\max}$ varies as a function of the average value of the stress relative to the space between holes and as a function of the ratio r/l or the efficiency of ligaments z. It is

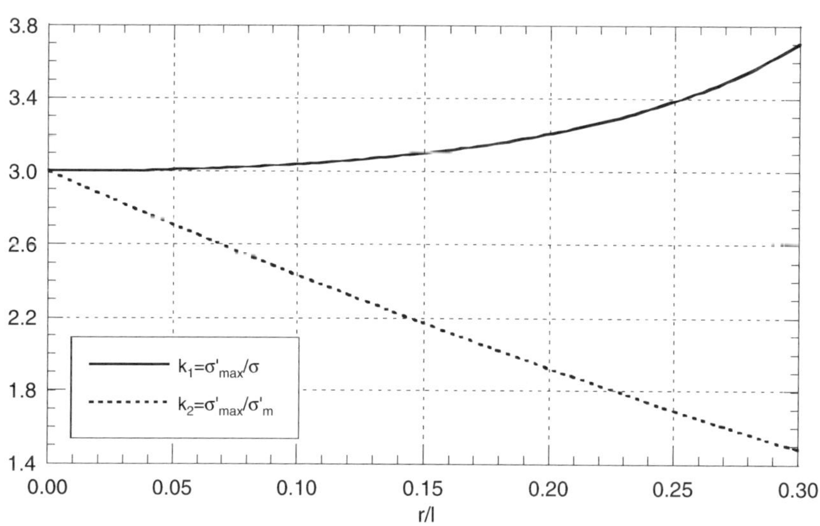

Fig. 8.44

equal to 3 for $l = \infty$, i.e., if there is an isolated hole. In fact, in that case the average value of the relative stress coincides with σ. For $r/l > 0$, the value of k_2 decreases as the holes get closer and the value of z consequently becomes smaller.

As we already pointed out in Sect. 7.3, the bending moment present in the section is crucial for the sizing of quadrangular vessels, so that the allowable stress is assumed to be equal to $1.5f$ (that corresponds to the yield strength). In the case of a drilled vessel with a hole line like the one we are considering, it will have to be at maximum level, based on the criteria in Sect. 1.5. Then

$$\sigma'_m = 1.5f; \tag{8.255}$$

still based on the same criteria it will have to be

$$\sigma'_{\max} \leq 3f; \tag{8.256}$$

thus

$$k_2 = \frac{\sigma'_{\max}}{\sigma'_m} \leq 2. \tag{8.257}$$

Based on Fig. 8.44 we establish that this condition is not always true. More appropriately, it is true only if

$$z \leq 0.63.$$

In other words, if we use the efficiency of ligaments in agreement with (8.255), in terms of the average value of the stress between holes, it is not at all guaranteed that the limit value of the maximum stress equal to $3f$ is not exceeded at the edge of the hole.

To eliminate this problem, one can introduce a virtual efficiency z' equal to z for $z \leq 0.63$, given by the following equation for $z > 0.63$ instead

$$z' = \frac{2}{k_1}. \tag{8.258}$$

Recalling the meaning of k_1 (8.250), in that case based on (8.252) (with z' instead of z) as it should be, we have

$$\frac{\sigma'_{\max}}{\sigma'_m} = \frac{\sigma'_{\max}}{\sigma} z' = 2. \tag{8.259}$$

Equation (8.258) is shown in Fig. 8.45. We see that, if the holes are rather distant from one another and z is high, it is incorrect to base the sizing of the vessel on z, and it is necessary to reduce its value that should never exceed 2/3. To exemplify this and to better explain the phenomenon Fig. 8.46 shows the value of the different stresses for $z = 0.6$. Recalling our discussion in Sect. 7.3, let us proceed with the sizing of the vessel when a hole line, like the one described above, is present.

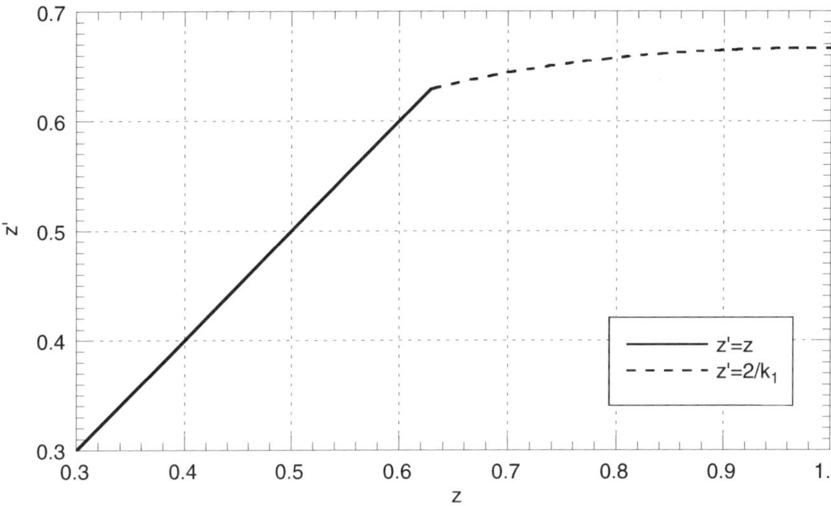

Fig. 8.45

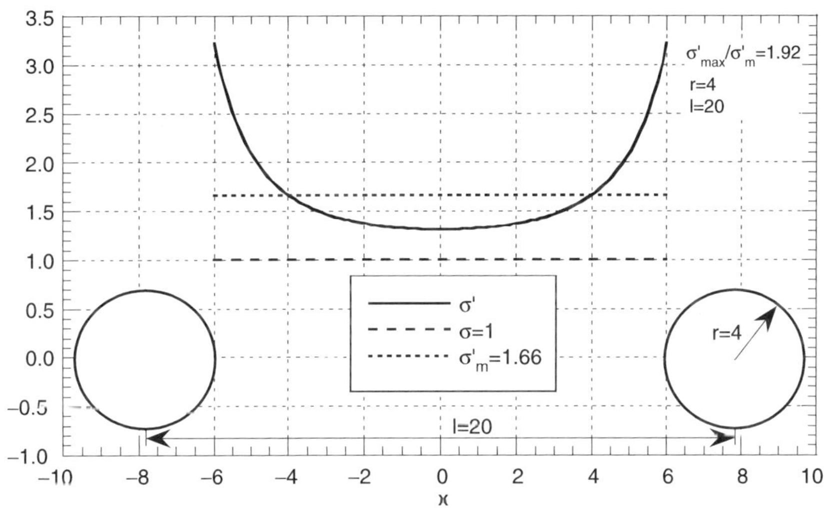

Fig. 8.46

If the hole line is at distance c from the center line of the wall (Fig. 8.47), the moment corresponding to the pitch l is equal to

$$M = M_c l, \tag{8.260}$$

where M_c is obtained from (7.53). Similarly, the traction force N referred to the pitch l is given by

$$N = pnl, \tag{8.261}$$

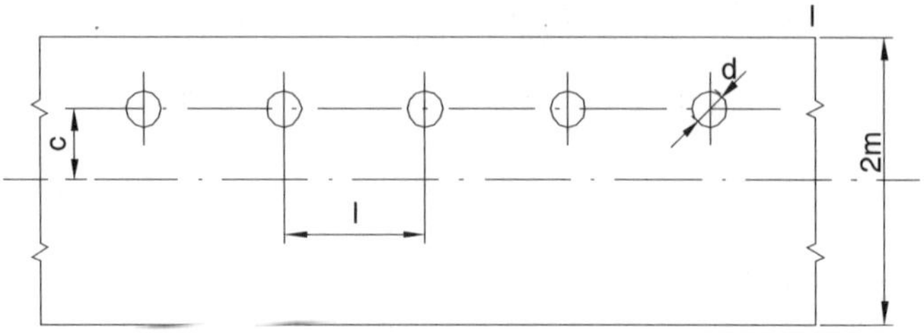

Fig. 8.47

under the assumption that the hole line will be on the wall of the vessel with a 2 m width (Fig. 7.7).

The section modulus for the area between the two holes is equal to

$$W = \frac{1}{6}(l-d)\,s^2, \tag{8.262}$$

and s is the thickness of the wall.

The cross-sectional area between the holes is equal to:

$$A = (l-d)\,s. \tag{8.263}$$

The ideal average stress $\sigma_{id(m)}$ between the holes in correspondence of the most stressed fibers (i.e., fibers at maximum distance from the neutral axis) is given by

$$\sigma_{id(m)} = \frac{pnl}{(l-d)\,s} + \frac{6\,|M_c|\,l}{(l-d)\,s^2}. \tag{8.264}$$

Based on the criteria in Sect. 1.5, one must assume that $\sigma_{id(m)} = 1.5f$. Introducing the efficiency of ligaments z equal to

$$z = \frac{l-d}{l} \tag{8.265}$$

one obtains

$$\frac{pn}{zs} + \frac{6\,|M_c|}{zs^2} = 1.5f; \tag{8.266}$$

then

$$s = \frac{\dfrac{pn}{3z} + \sqrt{\left(\dfrac{pn}{3z}\right)^2 + 4\dfrac{|M_c|}{z}f}}{f}. \tag{8.267}$$

As usual, ignoring the first term under the root because it is very small compared to the second, in the end we have

$$s = \frac{pn}{3fz} + \sqrt{\frac{4|M_c|}{fz}}. \tag{8.268}$$

As a result of the considerations made about the behavior of stresses, the efficiency z must be substituted with the virtual efficiency z' that can be obtained from Fig. 8.45 and that coincides with z only if $z \leq 0.63$. Moreover, note that based on Sect. 7.3, as a result of the knuckle, the moments in the central area of the wall are in fact slightly greater than the ones computed through (7.73). The phenomenon can be ignored, as it is generally of relatively modest entity. If the vessel has equal sides and if the drilling is positioned in correspondence of the center line, one can take it into account by introducing k_2 obtained from Fig. 7.10.

In that case we have

$$s = \frac{pm}{3fz'} + \sqrt{\frac{4k_2\,|M_c|}{fz'}}. \tag{8.269}$$

Alternatively, recalling the equation of M_c in a vessel with equal sides and $c = 0$:

$$s = \frac{pm}{3fz'} + m\sqrt{\frac{2}{3}\frac{pk_2}{fz'}}. \tag{8.270}$$

Finally, note that to factor in the influence of holes with a great diameter on moments in the area of interest, some codes require the following expressions of z with regard to the term under the root:

$$\begin{aligned} z &= \frac{l-d}{l} \qquad d \leq 0.6\,m \\ z &= \frac{l-0.6m}{l} \qquad d > 0.6\,m \end{aligned} \tag{8.271}$$

We are then left with the case of offset drillings across the center line of the wall (Fig. 8.48).

An in-depth theoretical analysis of the behavior of stresses is quite difficult.

Therefore, we simply suggest to follow the indications included in a few of the existing codes on this topic. Not discounting the considerations made about the expressions of z (in this instance l is the distance between the center of the holes as a diagonal), the moment M_c in (8.268) is

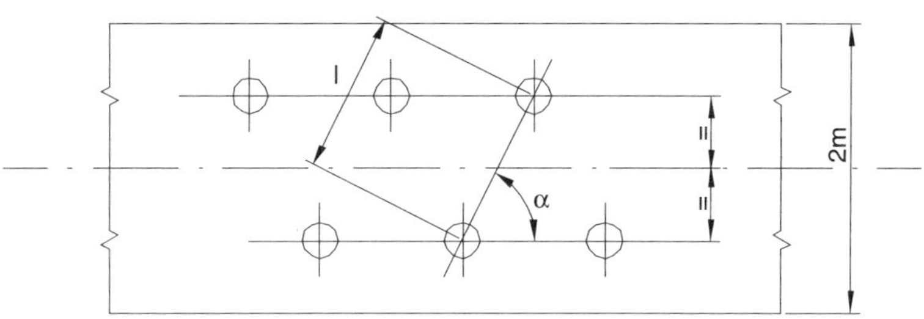

Fig. 8.48

$$M_c = \left(A - \frac{m^2}{2}\right) p \cos\alpha, \tag{8.272}$$

and α is the angle that the connecting line of the center of holes forms with the axis of the vessel (Fig. 8.48). Even in this case we refer you back to the considerations made about the substitution of z with z' if $z > 0.63$.

In summary, note that one may consider an equivalent diameter that factors in the reinforcement represented by the potential presence of nozzles with a thickness greater than the minimum required one. This had been already done for the flat drilled plates (Sect. 8.5). The calculation to identify the value of the equivalent diameter is identical to that shown in Sect. 8.5.

9

The Influence of Supports

9.1 Cylindrical Vessels on Saddle Supports

The most comprehensive and qualified study about the behavior of a horizontal cylindrical vessel on two symmetrical saddle supports, was done by Zick (1951). With reference to Fig. 9.1, F is the total load on the support, L the length of the cylinder measured between the tangency lines of the heads, D_e the outside diameter of the cylinder, H the height of the head, A the distance between the head's tangency line and the support's center line, b the width of the support, s_c the thickness of the cylinder in the section to be verified, s_r the thickness of the potential reinforcement plate, s_f the thickness of the head, z_c the weld joint efficiency of the cylinder, α the angle corresponding to the arch of the support.

The verifications to be done are as follows.

(1) Verification of the cylinder with combined compressive and bending stress in correspondence of the upper edges of the support without pressure. The following inequality shall be satisfied:

$$\frac{0.25F}{s_c\left(b + 1.1\sqrt{D_e s_c}\right)} + \frac{\beta u k_2 F}{s_c^2} \leq 1.25f, \tag{9.1}$$

where f is the basic allowable stress of the material to compute based on Sect. 1.2. In addition, the coefficient β is equal to 0.25 for $L \geq 4D_e$; otherwise, it will be equal to D_e/L. The coefficient u is equal to

$$\begin{array}{ccl} u = 1.5 & \text{for} & A \leq 0.25D_e \\ u = 18A/D_e - 3 & & 0.25D_e < a < 0.5D_e \\ u = 6 & & A \geq 0.5D_e \end{array}$$

Finally, note that the thickness s_c of the cylinder must have the same value beyond the edges of the supports for a length equal to at least $0.55\sqrt{D_e s_c}$. k_2 may be obtained from Fig. 9.2. It is a very complex function of the angle α (see Zick's paper).

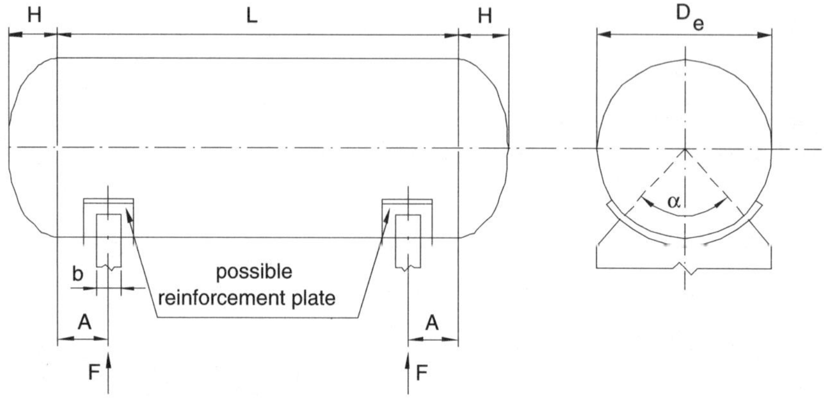

Fig. 9.1

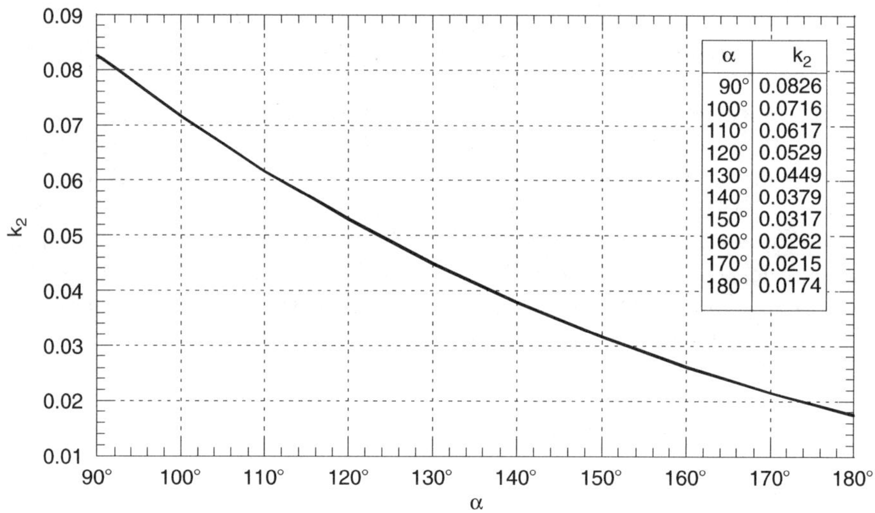

Fig. 9.2

If a reinforcement plate of thickness s_r is to be used, and if it extends beyond the edges of the support for a width equal to at least $0.55\sqrt{D_e s_c}$, and if $A \leq 0.25D_e$, one can replace the thickness s_c with the sum $s_c + s_r$. The plate must extend itself circumferentially beyond the support for a length of at least $0.05D_e$; it must also be welded to the cylinder with continuous seams and its thickness shall not be greater than $1.5s_c$.

(2) Verification of the cylinder to the compressive stress in correspondence of the support. The following relation must be satisfied:

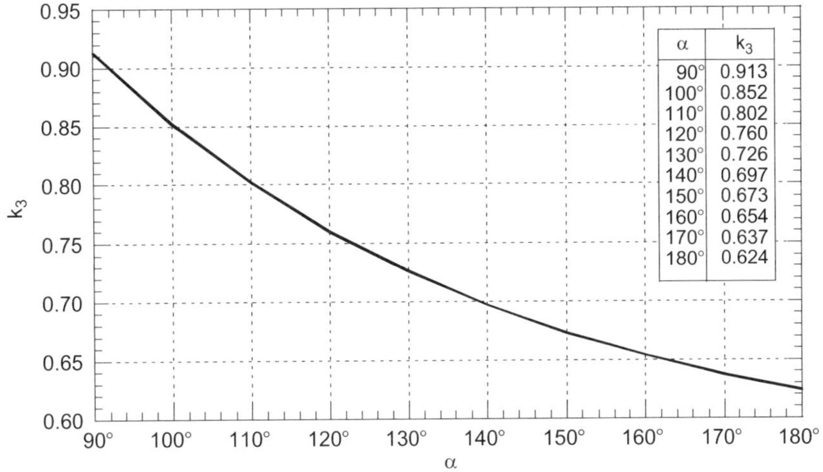

Fig. 9.3

$$\frac{k_3 F}{S_c\left(b+1.1\sqrt{D_e S_c}\right)} \leq 0.75 f \tag{9.2}$$

As far as the extension of the constant thickness s_c of the cylinder, we refer to the verification above. The same is true as far as the possibility to substitute the thickness of the cylinder in (9.2) with the sum of the thickness of the cylinder and of the reinforcement plate. During this verification one can consider s_r even though $A > 0.25D_e$. k_3 is derived from Fig. 9.3. As far as its analytical equation, we refer back to the original publication.

(3) Verification of the cylinder under longitudinal stress in correspondence of the support. The bending moment in this location has the following equation:

$$M = -\frac{2F}{L+\frac{4}{3}H}\left(\frac{A^2}{2}+\frac{3}{2}HA-\frac{D_e^2-4H^2}{16}\right) \tag{9.3}$$

The section modulus is equal to

$$W = \frac{\pi}{4} D_e^2 s_c . \tag{9.4}$$

This value of W may be adopted only if the possibility of ovalization of the section can be ruled out. This is guaranteed if there is a rigidity ring in correspondence of the center line of the support, or two rings at the sides of the support (see Fig. 9.4). Moreover, if the saddle support is very close to the head ($A < 0.25D_e$) the head itself prevents the ovalization of the section from happening.

Otherwise, a portion of the upper part of the section (see Fig. 9.5) must be considered ineffective as far as the bending moment, and one must assume a reduced value of the section modulus given by

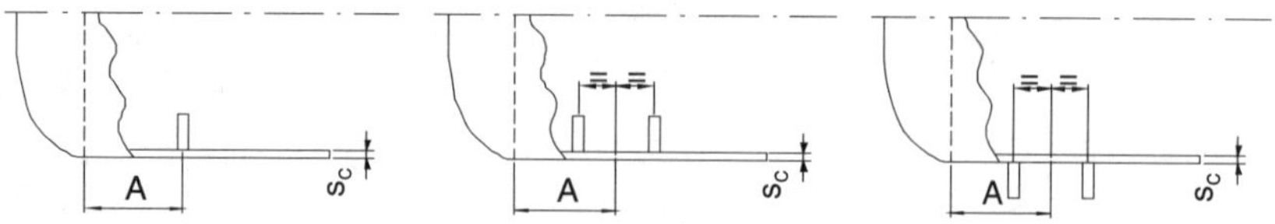

Fig. 9.4

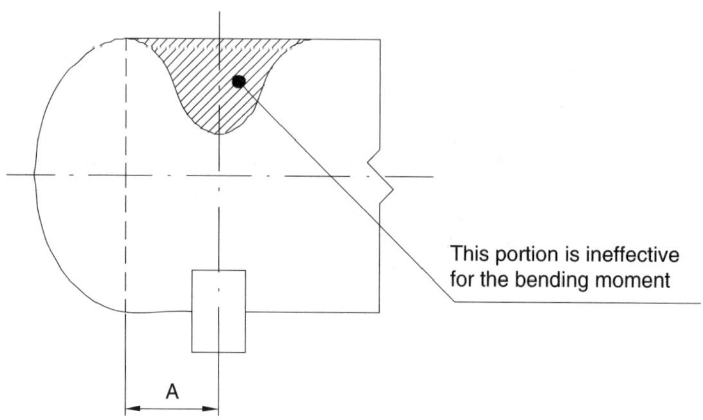

Fig. 9.5

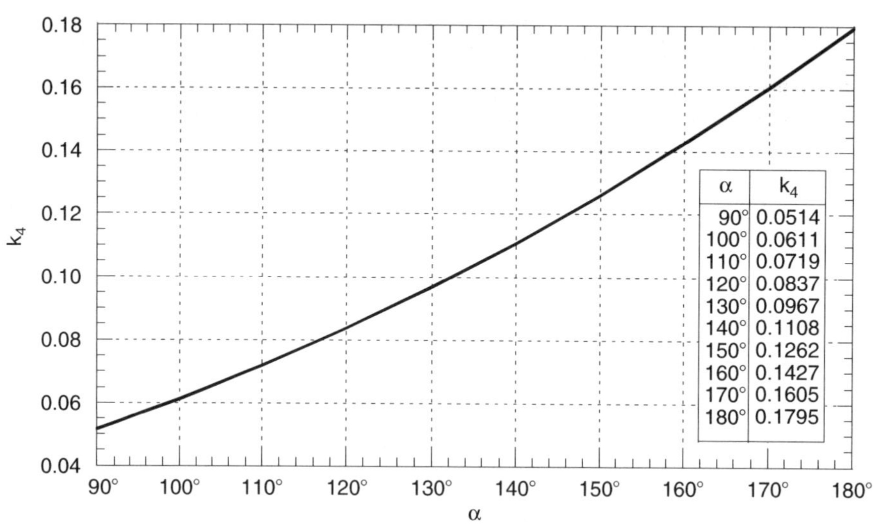

Fig. 9.6

$$W = k_4 D_e^2 s_c, \tag{9.5}$$

where k_4 is shown in Fig. 9.6 The following relation must be satisfied for the most stressed fiber (in the upper area of the section):

$$\frac{|M|}{W} + \frac{pD_e}{4s_c} \leq z_c f. \tag{9.6}$$

The second term on the left of the inequality corresponds to the longitudinal stress caused by the thrust of the pressure on the heads.

For verification in correspondence of the most stressed compressed fiber in the lower area of the section, the more dangerous condition occurs when there is no pressure in the vessel. Therefore, the following inequality shall hold:

$$\frac{|M|}{W} \leq f_c. \tag{9.7}$$

The allowable stress f_c must factor in the danger of buckling given that this is a compressive stress. Thus, the value of f_c must be correlated to the modulus of elasticity and the ratio s_c/D_e. It is possible to proceed as follows. Assuming that

$$\gamma = \frac{Es_c}{14.5D_e}, \tag{9.8}$$

if

$$\gamma \geq 0.75f \tag{9.9}$$

we assume that

$$f_c = 0.75f. \tag{9.10}$$

By contrast,

$$f_c = 2\gamma - \frac{4}{3}\frac{\gamma^2}{f}. \tag{9.11}$$

For instance, for $\gamma = 0.3f$ it follows that $f_c = 0.48f$.

(4) Verification of the cylinder under longitudinal stresses in correspondence of the center line. The bending moment in this position has the following equation:

$$M = -\frac{2F}{L + \frac{4}{3}H}\left[\frac{(L-2A)^2}{8} - \frac{A^2}{2} - \frac{3}{2}HA + \frac{D_e^2 - 4H^2}{16}\right]. \tag{9.12}$$

Equation (9.4) is valid as far as the section modulus. The necessary verifications are the same as the ones pointed out above, i.e., the inequalities in (9.6) and (9.7) must be satisfied.

(5) Verification of the cylinder under shear stresses in correspondence of the lateral edges of the support. One must distinguish between a cylinder reinforced from the head or not, depending on its distance from the support. Therefore, if $A > 0.25D_e$ the following inequality shall hold:

$$\frac{2k_5F(L-2A-H)}{D_es_c(L+H)} \leq 0.8f, \tag{9.13}$$

where $k_5 = 1/\pi$ if there is a rigidity ring in correspondence of the center line of the support; otherwise k_5 can be obtained from Fig. 9.7. If $A \leq 0.25D_e$ the

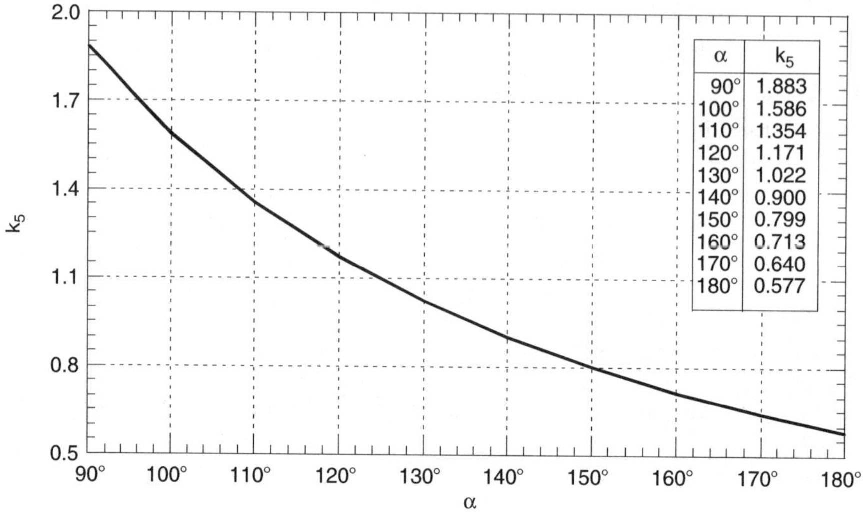

Fig. 9.7

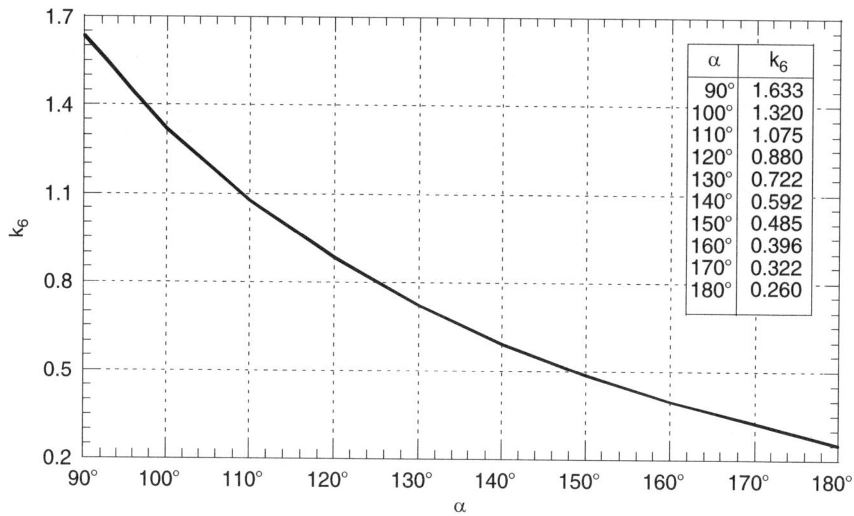

Fig. 9.8

following relation must be satisfied:

$$\frac{2k_6 F}{D_e s_c} \leq 0.8f, \tag{9.14}$$

where k_6 can be derived from Fig. 9.8.

(6) Verification of the head under shear and tensile stresses. This verification must be performed only if $A \leq 0.25D_e$. Thus, the head contributes

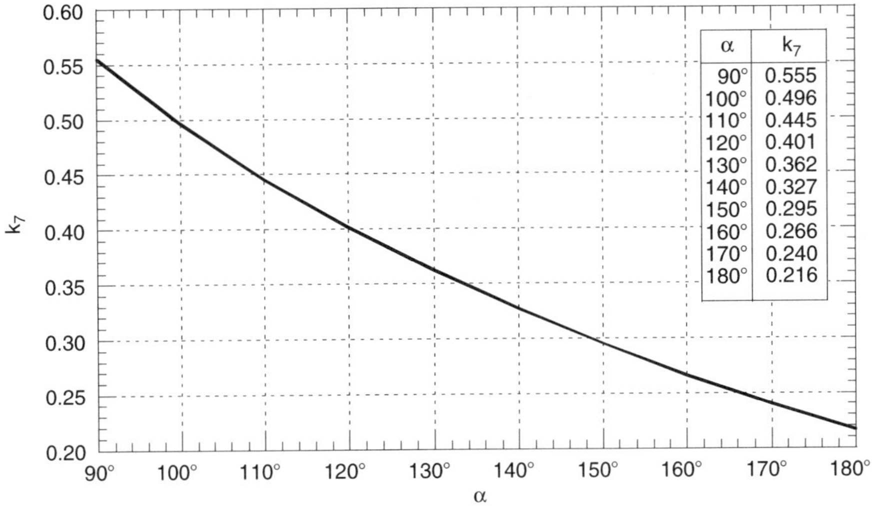

Fig. 9.9

to the stiffening of the cylinder in correspondence of the support. As far as the verification in terms of pure shear we refer to the same equation already discussed for the cylinder except, of course, for the thickness s_f of the head instead of s_c.

In other words, the following relation must be verified:

$$\frac{2k_6F}{D_e s_f} \leq 0.8f. \tag{9.15}$$

In addition, the following relation must be verified, as well

$$\frac{2k_7F}{D_e s_f} + \frac{pD_e}{2s_f}C \leq 1.5\frac{f}{x}, \tag{9.16}$$

where C is the form factor (see Sect. 6.2), k_7 is derived from Fig. 9.9, while we recommend to adopt $x = 1.2$ during verification of the working conditions and 1.1 during verification of the conditions in the hydraulic test. (7) Verification of the stiffening rings.

The stiffening ring placed in correspondence of the center line of the support must be verified considering both the bending moment and the normal force given by the following equations:

$$M = \frac{k_2 F D_e}{2} \tag{9.17}$$

and

$$N = k_8 F, \tag{9.18}$$

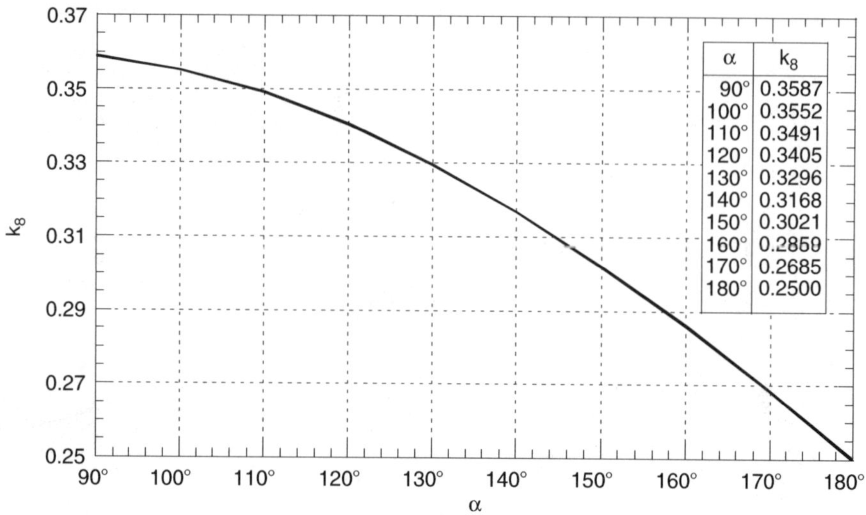

Fig. 9.10

where the value of k_2 is obtained from Fig. 9.2, whereas k_8 is shown in Fig. 9.10. If there are two rings at the sides of the support instead (at a distance from the support or the potential reinforcement plate not greater than $0.55\sqrt{D_e s_c}$, yet at a distance from the center line of the support in any case not greater than $0.25D_e$), a bending moment as well as a normal stress are applied on each one given by

$$M = \frac{k_9 F D_e}{4}; \tag{9.19}$$

$$N = \frac{1}{2} k_{10} F. \tag{9.20}$$

k_9 and k_{10} are shown in Figs. 9.11 and 9.12. If W_a and A_a are the minimum section modulus and the cross-sectional area of the ring, respectively, the following relationship must be verified:

$$\frac{M}{W_a} + \frac{N}{A_a} \leq f_r, \tag{9.21}$$

where f_r is the allowable stress of the ring.

During calculation of W_a and A_a it is possible to consider a portion of shell collaborating with the ring. The maximum extension of the collaborating shell at both sides of the ring is equal to $0.55\sqrt{D_e s_c}$. In that case if $f_r > f$, one must substitute f_r with f in (9.21).

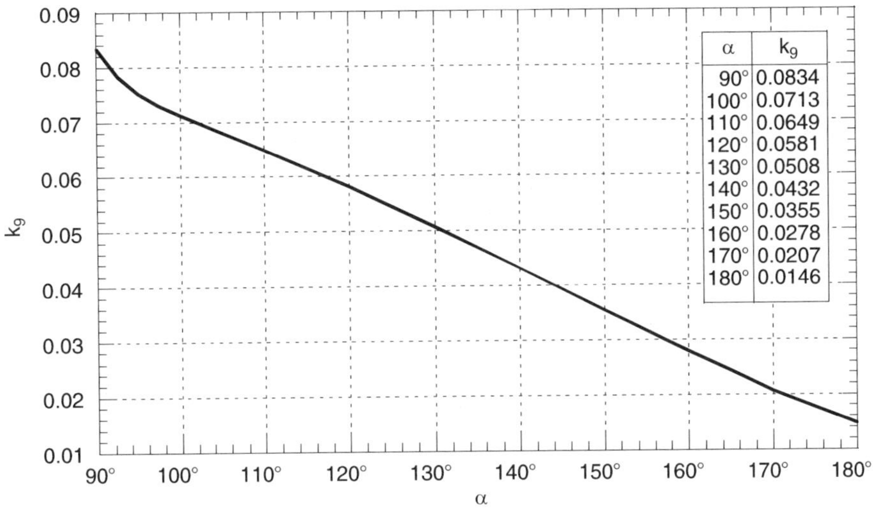

Fig. 9.11

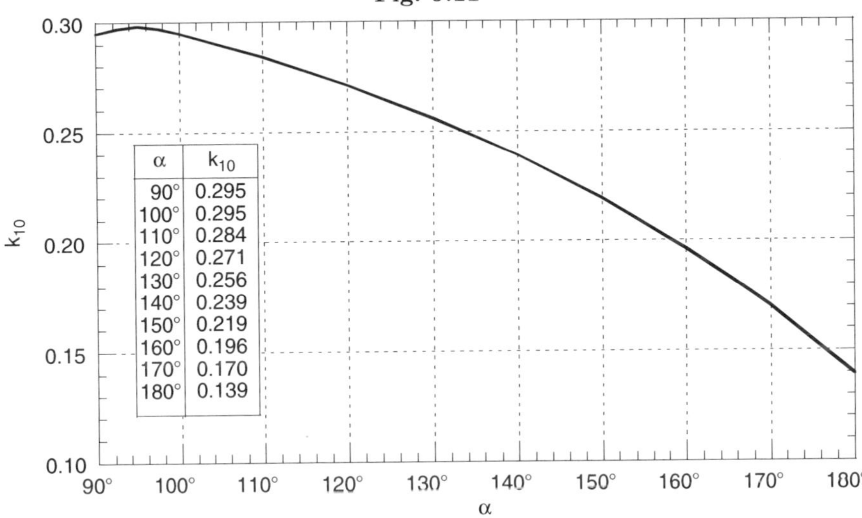

Fig. 9.12

9.2 Spherical Vessels Resting on a Parallel

We examine the spherical vessel shown in Fig. 9.13 containing a liquid of specific weight γ and resting on a series of columns through a ring placed along the parallel $n-n$.

Assuming that the vessel is full, the pressure caused by the hydrostatic load of the liquid is equal to

$$p = R\,(1-\cos\varphi)\,, \tag{9.22}$$

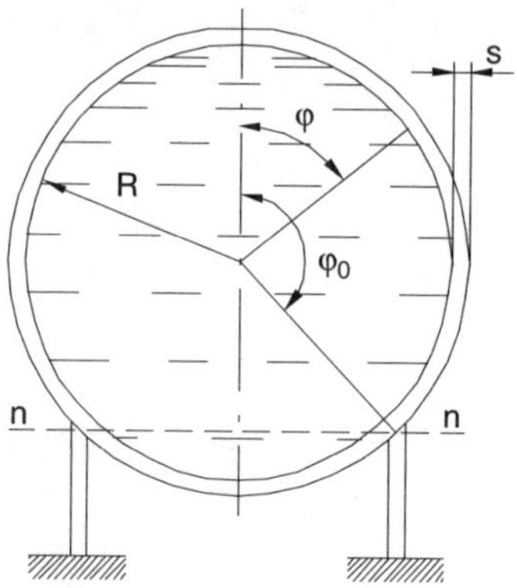

Fig. 9.13

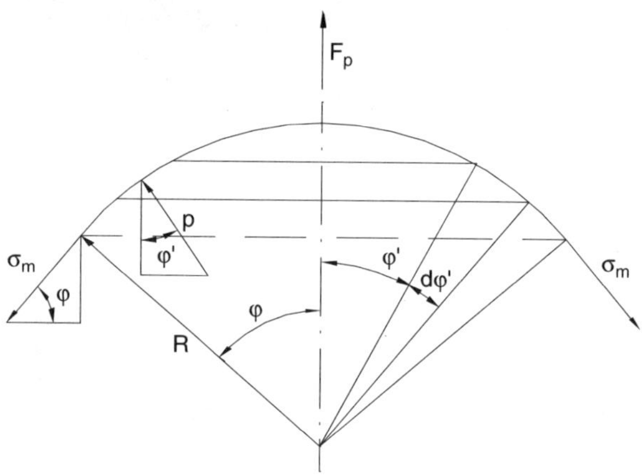

Fig. 9.14

where φ is the angle created by the orthogonal line to the surface and the vertical axis of the vessel.

Let us consider the portion of the vessel above the parallel $n - n$. With reference to the generic top portion identified by the angle φ, the resultant F_p of the pressures is obtained as follows.

For the elementary crown identified by the angle φ' (see Fig. 9.14) the resultant is equal to

$$dF_p = p\cos\varphi' 2\pi R\sin\varphi' Rd\varphi' = 2\pi pR^2 \sin\varphi' \cos\varphi' d\varphi'; \tag{9.23}$$

and recalling (9.22)

$$dF_p = 2\pi\gamma R^3 \left(1 - \cos\varphi'\right) \sin\varphi' \cos\varphi' d\varphi'. \tag{9.24}$$

The resultant F_p is therefore given by

$$F_p = 2\pi\gamma R^3 \int_0^{\varphi} (1 - \cos\varphi') \sin\varphi' \cos\varphi' d\varphi'. \tag{9.25}$$

Solving the integral we obtain

$$F_p = 2\pi\gamma R^3 \left(\frac{\sin^2\varphi}{2} + \frac{\cos^3\varphi}{3} - \frac{1}{3} \right). \tag{9.26}$$

By indicating the meridian stresses with σ_m and the thickness of the vessel with s, the resultant F_σ of the σ_m acting at the edge of the top portion is equal to

$$F_\sigma = 2\pi R \sin\varphi s \sigma_m \sin\varphi = 2\pi R \sin^2\varphi s \sigma_m. \tag{9.27}$$

For the equilibrium $F_\sigma = F_p$; therefore, based on (9.26) and (9.27) we obtain

$$\sigma_m = \frac{\gamma R^2}{6s} \frac{3\sin^2\varphi + 2\cos^3\varphi - 2}{\sin^2\varphi}, \tag{9.28}$$

that may also be written as

$$\sigma_m = \frac{\gamma R^2}{6s} \frac{1 - \cos\varphi}{1 + \cos\varphi} (1 + 2\cos\varphi). \tag{9.29}$$

At this point we refer back to Sect. 2.1, and specifically to (2.10) that we rewrite as

$$\frac{\sigma_m}{R_1} + \frac{\sigma_t}{R_2} = \frac{p}{s}. \tag{9.30}$$

In the specific case where $R_1 = R_2 = R$,

$$\sigma_t = \frac{pR}{s} - \sigma_m. \tag{9.31}$$

Based on (9.22) and (9.29) we obtain

$$\sigma_t = \frac{\gamma R^2}{6s} \frac{1 - \cos\varphi}{1 + \cos\varphi} (5 + 4\cos\varphi). \tag{9.32}$$

Considering a bottom portion below the parallel $n - n$, (9.24) is still valid if one considers the resultant positive when it is directed upward, as done in the previous instance. It is therefore given by

$$F_p = 2\pi\gamma R^3 \int_{\varphi}^{\pi} (1 - \cos\varphi') \sin\varphi' \cos\varphi' d\varphi'. \tag{9.33}$$

Resolving the integral, we obtain

$$F_p = -2\pi\gamma R^3 \left(\frac{1}{3} + \frac{\sin^2\varphi}{2} + \frac{\cos^3\varphi}{3} \right), \tag{9.34}$$

and recalling (9.27) and assuming that $F_p = -F_\sigma$, given the convention of signs, we obtain

$$\sigma_m = \frac{\gamma R^2}{6s} \frac{2 + 3\sin^2\varphi + 2\cos^3\varphi}{\sin^2\varphi}, \tag{9.35}$$

that may also be written as

$$\sigma_m = \frac{\gamma R^2}{6s} \frac{5 - 5\cos\varphi + 2\cos^2\varphi}{1 - \cos\varphi}. \tag{9.36}$$

Based on (9.31) and recalling (9.22), we finally obtain

$$\sigma_t = \frac{\gamma R^2}{6s} \frac{1 - 7\cos\varphi + 4\cos^2\varphi}{1 - \cos\varphi}. \tag{9.37}$$

In conclusion we may write:

For the area above the parallel $n - n$

$$\begin{aligned} \sigma_m &= k_1 \frac{\gamma R^2}{s} \\ \sigma_t &= k_2 \frac{\gamma R^2}{s} \end{aligned} \tag{9.38}$$

For the area below the parallel $n - n$

$$\begin{aligned} \sigma_m &= k_3 \frac{\gamma R^2}{s} \\ \sigma_t &= k_4 \frac{\gamma R^2}{s} \end{aligned} \tag{9.39}$$

given that

$$\begin{aligned} k_1 &= \frac{1}{6} \frac{1 - \cos\varphi}{1 + \cos\varphi} (1 + 2\cos\varphi) \\ k_2 &= \frac{1}{6} \frac{1 - \cos\varphi}{1 + \cos\varphi} (5 + 4\cos\varphi) \\ k_3 &= \frac{1}{6} \frac{5 - 5\cos\varphi + 2\cos^2\varphi}{1 - \cos\varphi} \\ k_4 &= \frac{1}{6} \frac{1 - 7\cos\varphi + 2\cos^2\varphi}{1 - \cos\varphi} \end{aligned} \tag{9.40}$$

where k_1, k_2, k_3, and k_4 are shown in Figs. 9.15 and 9.16.

The values of k_3 and k_4 have been considered for the angles between 90° and 180°, given that we rule out that the parallel $n - n$ may correspond to an angle φ below 90°.

We considered the values of k_1 and k_2 for the angles between 0° and 120°. Even though in principle one may also consider a position of the parallel $n - n$

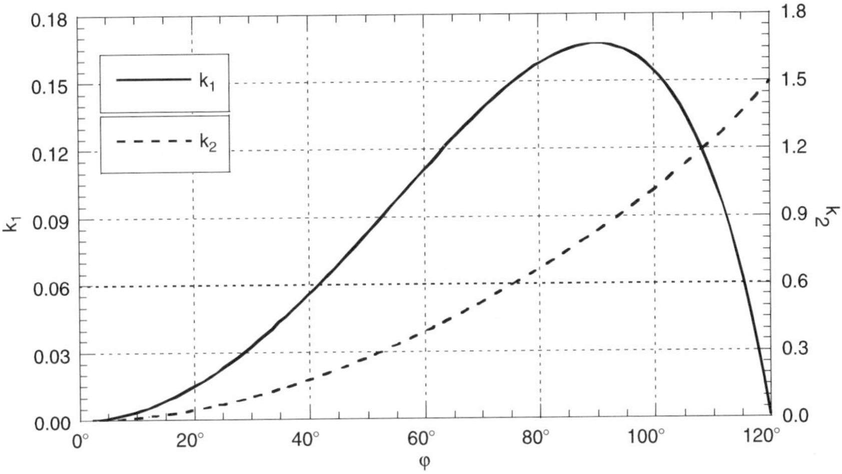

Fig. 9.15

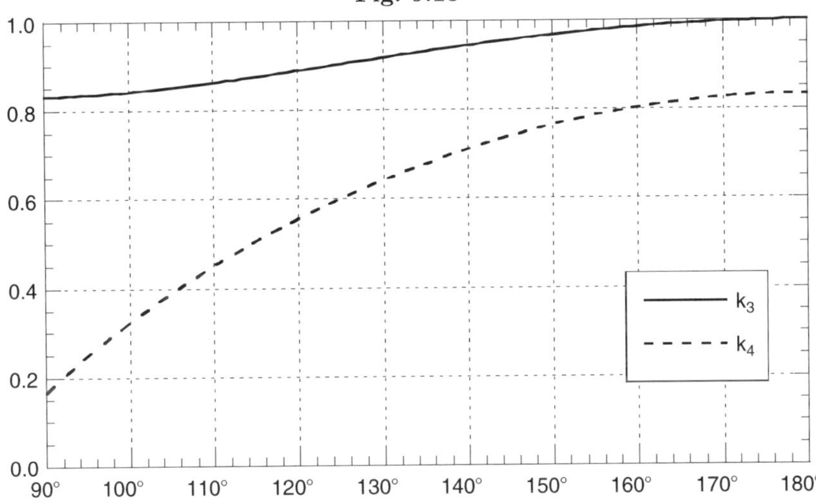

Fig. 9.16

corresponding to an angle greater than 120°, it is, however, not recommended, given that in this case k_1 becomes negative. This means that the meridian stresses become negative as well, and this is quite dangerous as far as the risk of buckling of the vessel characterized by reduced thickness. We establish that for the same φ the value of k_1 does not coincide with the one of k_3. Similarly, the value of k_2 does not coincide with that of k_4.

Both σ_m and σ_t show a discontinuity in correspondence of the parallel $n-n$. The discontinuity of σ_m depends on the reactions of the support ring.

Recalling that the weight of the liquid P is equal to

$$P = \frac{4}{3}\pi\gamma R^3. \tag{9.41}$$

Indicating the angle φ that corresponds to the parallel $n-n$ with φ_0, a vertical force q_v is exerted on the ring per length unit equal to

$$q_v = \frac{P}{2\pi R \sin\varphi_0} = \frac{2}{3}\gamma\frac{R^2}{\sin\varphi_0}. \tag{9.42}$$

In addition, the following radial force q_0 directed from the outside towards the inside is exerted on the ring per length unit:

$$q_0 = -\frac{q_v}{\tan\varphi_0} = -\frac{2}{3}\gamma\frac{R^2\cos\varphi_0}{\sin^2\varphi_0}. \tag{9.43}$$

As far as the discontinuity of the stresses σ_t, it is a consequence of the calculation method used here where the sphere was considered as a membrane.

In order to establish the correct behavior of the stresses in correspondence of the parallel $n-n$, it is necessary to recall the considerations in Sect. 2.2 about edge effects. With reference to that section we observe that the difference between stresses σ_t in correspondence of the angle φ_0 is given by

$$\Delta\sigma_t = (k_2 - k_4)\frac{\gamma R^2}{s}, \tag{9.44}$$

with k_2 and k_4 computed for $\varphi = \varphi_0$. Assuming that

$$k_5 = k_2 - k_4, \tag{9.45}$$

we have

$$\Delta\sigma_t = k_5\frac{\gamma R^2}{s}; \tag{9.46}$$

k_5 is shown in Fig. 9.17.

The difference in radial deflection Δy between the two portions of the sphere is equal to

$$\Delta y = \frac{\Delta\sigma_t}{E}R = k_5\frac{\gamma R^3}{Es}. \tag{9.47}$$

Recalling the presentation on edge effects, a radial force F_0 is generated per unit of length at the edge of both portions to reestablish congruence. The force is centripetal in the upper portion and centrifugal in the lower one. By contrast, no moment is generated, given that the two connected portions share the same dimensional characteristics (radius and thickness). Therefore, force F_0 must be able to generate a radial deflection equal to $\Delta y/2$. Recalling (2.66) we have

$$y_0 = \frac{\Delta y}{2} = 2.57\frac{F_0}{E}\left(\frac{R}{s}\right)^{3/2}. \tag{9.48}$$

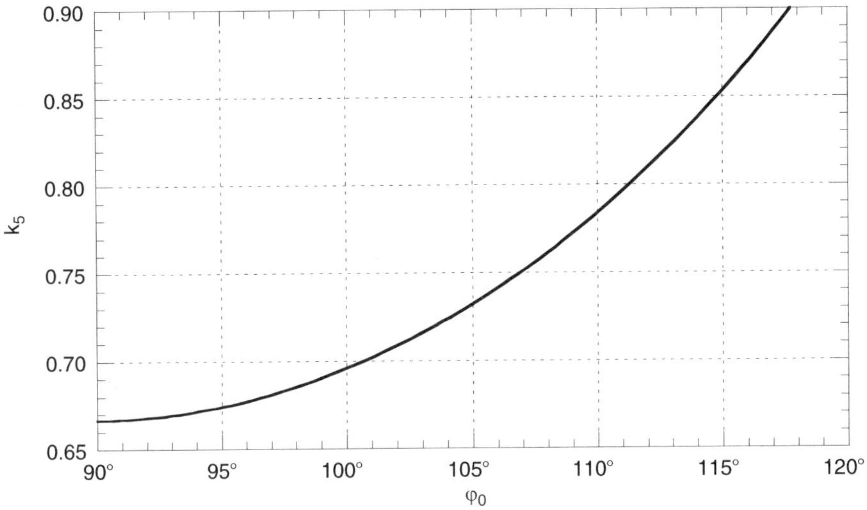

Fig. 9.17

Based on (9.47) we obtain

$$F_0 = 0.1945 k_5 \gamma \sqrt{sR^3}. \tag{9.49}$$

Finally, based on (2.53) we have

$$M_x = \pm \frac{0.1945}{\beta} k_5 \gamma \sqrt{sR^3} e^{-\beta x} \sin \beta x. \tag{9.50}$$

The plus or minus sign refers to the portion of the sphere above or below the parallel $n-n$, respectively, by conventionally considering positive the moments that stretch the external fibers of the vessel.

Recalling (2.41), from (9.50) we finally have:

$$M_x = \pm 0.1514 k_5 \gamma \sqrt{s} R^2 e^{-\beta x} \sin \beta x. \tag{9.51}$$

The meridian stress corresponding to M_x, given $W = s^2/6$, is therefore given by

$$\sigma''_m = \pm 0.908 k_5 \frac{\gamma R^2}{s} e^{-\beta x} \sin \beta x. \tag{9.52}$$

In (9.52) the plus sign refers to the external fibers of the upper portion and to the internal fibers of the lower portion.

Based on Sect. 2.2, the maximum value of σ''_m is reached for $\beta x = \pi/4$. Thus, from (9.52) we have

$$|\sigma''_m|_{(\max)} = 0.293 k_5 \frac{\gamma R^2}{s}. \tag{9.53}$$

Moreover, we observe that

$$\beta x = \frac{1.285}{\sqrt{Rs}} R\Delta\varphi = 1.285\sqrt{\frac{R}{s}}\Delta\varphi, \tag{9.54}$$

where $\Delta\varphi$ is the (always positive) difference in radiants between the angle φ, characteristic of the point of the sphere in question, and the angle φ_0 corresponding to the parallel $n - n$. The stress σ''_m is added to the membrane stress σ_m obtained from both (9.38) and (9.39).

Based on the values of k_3 and k_5, for $\varphi_0 = 90° \div 120°$ and based on (9.62), we observe that the impact of σ''_m is relatively modest, if we consider that this is a secondary stress (it is in fact the result of the respect of congruence) for which high values are allowed with respect to the allowable stress of the material, as shown in Sect. 1.5.

Given the respect of congruence, the onset of stresses σ''_m is matched by circumferential stresses σ''_t equal to $\mu\sigma''_m$. Thus, based on (9.52) and assuming that $\mu = 0.3$, they have the following equation:

$$\sigma''_t = \pm 0.272 k_5 \frac{\gamma R^2}{s} e^{-\beta x} \sin\beta x. \tag{9.55}$$

The sign convention is the one already described for σ''_m, and again these are secondary stresses. Similarly to σ''_m, the maximum value of σ''_t is equal to

$$|\sigma''{}_t|_{(\max)} = 0.0878 k_5 \frac{\gamma R^2}{s}. \tag{9.56}$$

Based on the values of k_2 and k_4, for $\varphi_0 = 90° \div 120°$ and based on (9.56), we observe that the impact of σ''_t is quite modest and not critical for the resistance of the sphere. As far as the circumferential membrane stresses, we recall Sect. 2.2 about the behavior of deflection y, and more specifically (2.45), that we rewrite as

$$y = e^{-\beta x}\left(C_1 \sin\beta x + C_1 \cos\beta x\right). \tag{9.57}$$

If there are no bending moments at the edge, as in this case, $C_1 = 0$ as already pointed out in Sect. 2.2. Then, if y_0 is the deflection corresponding to $x = 0$, we can write:

$$y = y_0 e^{-\beta x} \cos\beta x. \tag{9.58}$$

Therefore, recalling (9.47) and (9.48) we have

$$y = \pm \frac{k_5}{2} \frac{\gamma R^3}{Es} e^{-\beta x} \cos\beta x, \tag{9.59}$$

with the plus or minus sign for the lower or upper portion, respectively, and having considered the deflections directed from the inside to the outside of the sphere as positive. The stress σ'_t corresponding to the deflection y is equal to

$$\sigma'_t = \pm \frac{Ey}{R} = \pm \frac{k_5}{2} \frac{\gamma R^2}{s} e^{-\beta x} \cos\beta x. \tag{9.60}$$

To obtain the actual values of the circumferential membrane stress, the stress σ'_t must be added to the stress σ_t obtained through (9.38) and (9.39).

Note that for $\beta x = \pi/2$ the value of σ'_t is zero. Based on (9.54), it is matched by the following value of $\Delta\varphi$:

$$\Delta\varphi_0 = 1.222\sqrt{\frac{s}{R}}\,(rad) = 70\sqrt{\frac{s}{R}}\,(^\circ)\,. \tag{9.61}$$

Basically, the maximum value of the circumferential membrane stress is present in the upper portion of the sphere in correspondence of the parallel where σ'_t is equal to zero. Therefore, it may be obtained through the second equation in (9.38) by computing k_2 for the angle $(\varphi_0 - \Delta\varphi_0)$.

From (9.61) and the equation above, we obtain the maximum value of σ_t as a function of the ratio s/R for every angle φ_0 that we may write as

$$\sigma_{t(\max)} = k_6\frac{\gamma R^2}{s}\,. \tag{9.62}$$

k_6 is shown in Fig. 9.18.

Note that for the same φ_0 the value of k_6 is greater for small values of s/R. The verification of the sphere must be done considering both $\sigma_{t(\max)}$ obtained from (9.62) and σ_t or σ_m (that coincide) for $\varphi = 180°$. In fact, in this position one has a second relative maximum value for both the circumferential and meridian stresses.

From the point of view of best usage of the resistance of the sphere, the most effective solution is to make sure that $k_6 = 1$, given that k_3 and k_4 are equal to one for $\varphi = 180°$. Of course, the resulting angle φ_0 may turn

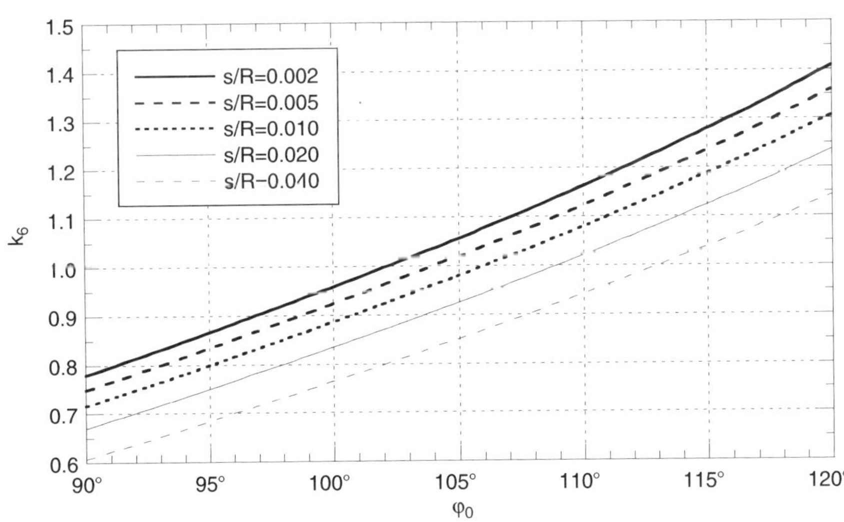

Fig. 9.18

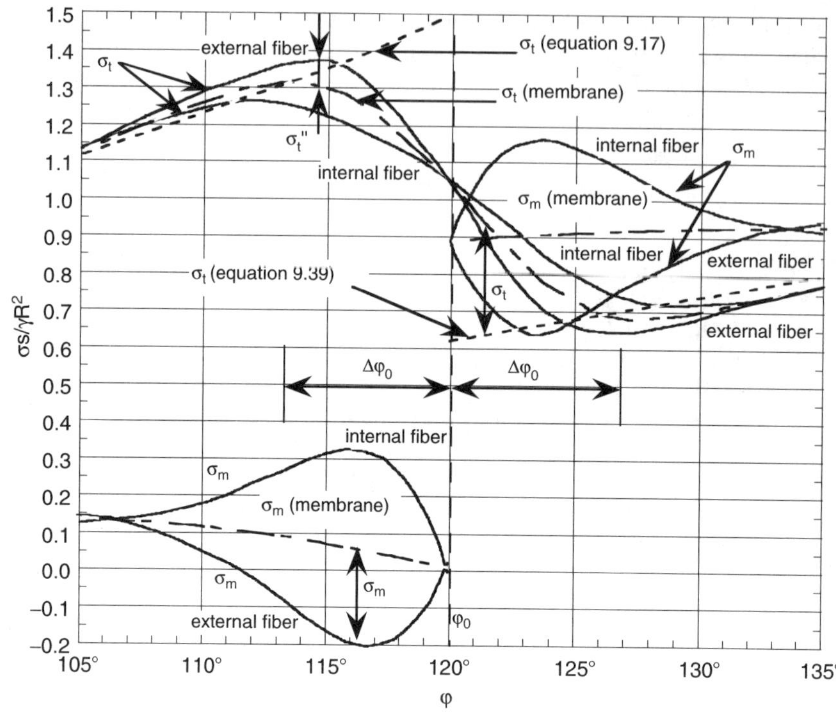

Fig. 9.19

out to be unacceptable in terms of dimensional characteristics and cost of the structure supporting the sphere. Figure 9.19 shows the behavior of the stresses if $\varphi_0 = 120°$ and $s/R = 0.01$ to better illustrate the phenomenon.

9.3 Local Effects of Forces and Moments on Cylinders

Locally, the cylinder may be under bending moments or forces that generate membrane or bending stresses in addition to those caused by internal pressure or external forces, such as weight, wind, seismic effects, and so on. This is due to the presence of isolated supports that sustain cylinders with a vertical or horizontal axis (the case of the two saddle supports was already discussed in Sect. 9.1). It is therefore necessary to estimate the impact of these stresses and to determine the criteria required to decide whether they are dangerous for the resistance of the cylinder.

The issue has been studied in detail by Bijlaard (1954, 1955) at Cornell University. First of all, let us consider the case of a radial force F (Fig. 9.20) active in correspondence of the center line of the cylinder. This force uniformly impacts a rectangular area of sides a and b along the generatrix of the cylinder and the circumference, respectively. L, D_e, and s be the length, the

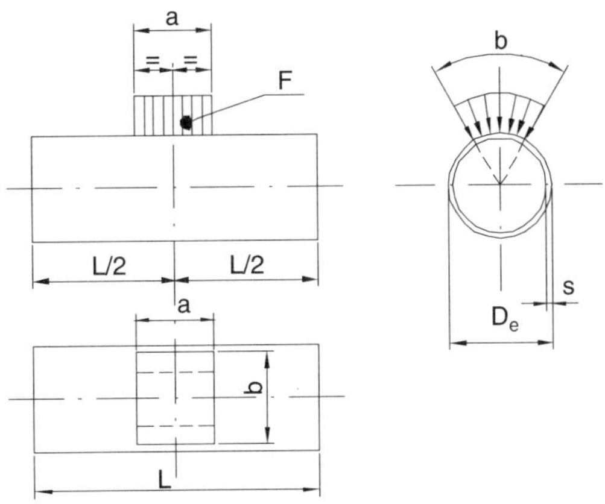

Fig. 9.20

outside diameter and the thickness of the cylinder. We introduce the following dimensionless quantities:

$$\alpha = 32\frac{a^2}{D_e s} \tag{9.63}$$

$$\beta = \frac{a}{L} \tag{9.64}$$

$$\gamma = \frac{b}{a} \tag{9.65}$$

In addition, N_t and M_t are the normal force and the bending moment along the circumference, while N_a and M_a are the normal force and the bending moment along the axis. Both forces and moments are in reference to the unity of length. We obtain N_t at the edge of the area loaded as a function of α, β and γ through Fig. 9.21.

M_t can be derived in the same position through Fig. 9.22 instead.

As far as N_a and M_a, they can be obtained through both Fig. 9.23 and 9.24. If force F is not applied to the center line of the cylinder (Fig. 9.25), one must consider an equivalent length L_e given by

$$L_e = L - \frac{4c^2}{L}, \tag{9.66}$$

that replaces L in (9.64). Moreover, if force F has an impact on a circular area of diameter d, one may conventionally assume that:

$$a = b = 0.85d, \tag{9.67}$$

and proceed as shown.

Note that the sign of M_t and M_a is positive if the moments generate positive stresses on the internal fiber of the cylinder. The sign of N_t and N_a is positive if there are tensile forces.

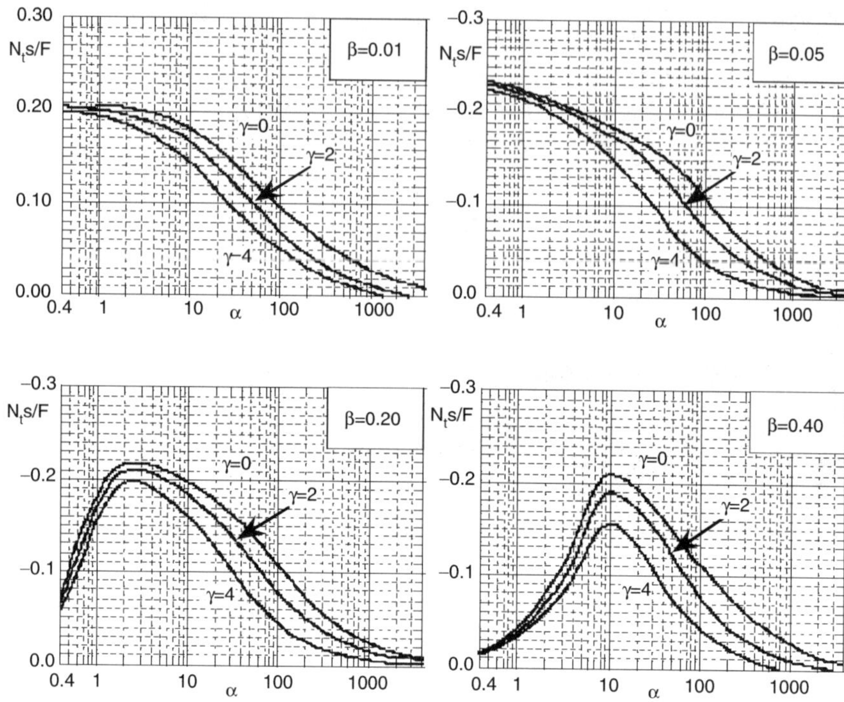

Fig. 9.21

If σ'_{tm} is the hoop membrane stress in the cylinder not generated locally by force F (generally caused by internal pressure only), the hoop membrane stress σ_{tm} at the edge of the area impacted by force F is given by

$$\sigma_{tm} = \sigma'_{tm} + \frac{N_t}{s}. \tag{9.68}$$

By analogy, if σ'_{am} is the longitudinal membrane stress in the cylinder not generated locally by F (generally caused by pressure and a potential bending moment concerning the entire section due to its own weight, wind, and so on), the longitudinal membrane stress σ_{am} at the edge of the area impacted by F is given by

$$\sigma_{am} = \sigma'_{am} + \frac{N_a}{s}. \tag{9.69}$$

Both σ_{tm} and σ_{am} do not belong to the category of local membrane stresses according to Sect. 1.5, their local impact notwithstanding. The local membrane stresses originate from the respect of congruence or discontinuities that highlight general membrane stresses locally (e.g., nozzles). By contrast, these stresses are a function of N_t and N_a and together with moments M_t and M_a shall balance force F instead. Therefore, considering the nature and behavior of the components σ_{tm} and σ_{am} caused by N_t and N_a, a value of the stress

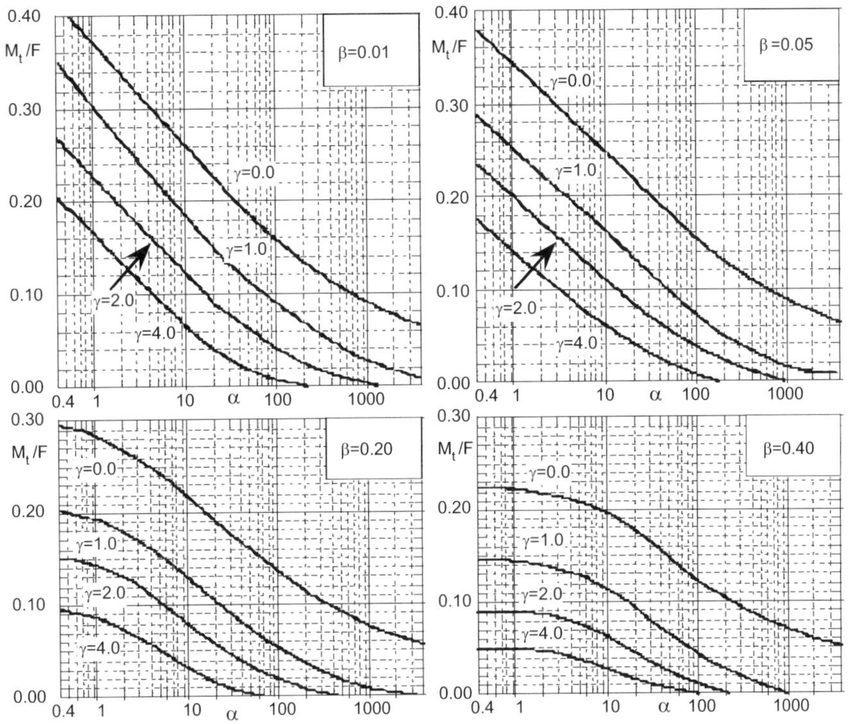

Fig. 9.22

greater than f is allowable, even though we believe one should never approach $1.5f$ (that corresponds to the yield strength). We suggest that if $b \leq \frac{\pi}{3} D_e$ the allowable stress is equal to $1.5f/1.2 = 1.25f$.

By introducing the weld joint efficiency z, if the welded seams are impacted by the stresses in question, we have

$$\begin{aligned} \sigma_{tm} &\leq 1.25zf \\ \sigma_{am} &\leq 1.25zf \end{aligned} \tag{9.70}$$

Of course, $z = 1$ if the verification does not concern the weldings.

Moreover, if σ_{tm} and σ_{am} carry the opposite sign, in the corner of the area impacted by force F, the ideal membrane stress according to Guest–Tresca is greater than σ_{tm} and σ_{am}. It is important then to also verify that:

$$\sigma_{id(m)} = |\sigma_{tm} - \sigma_{am}| \leq 1.25zf. \tag{9.71}$$

The verifications must always be done considering the most unfavorable condition (presence or absence of pressure, wind, seismic effects, and so on).

If σ'_{tf} is the potential hoop bending stress on the internal fiber of the cylinder not generated locally by F, the total bending stress is given by

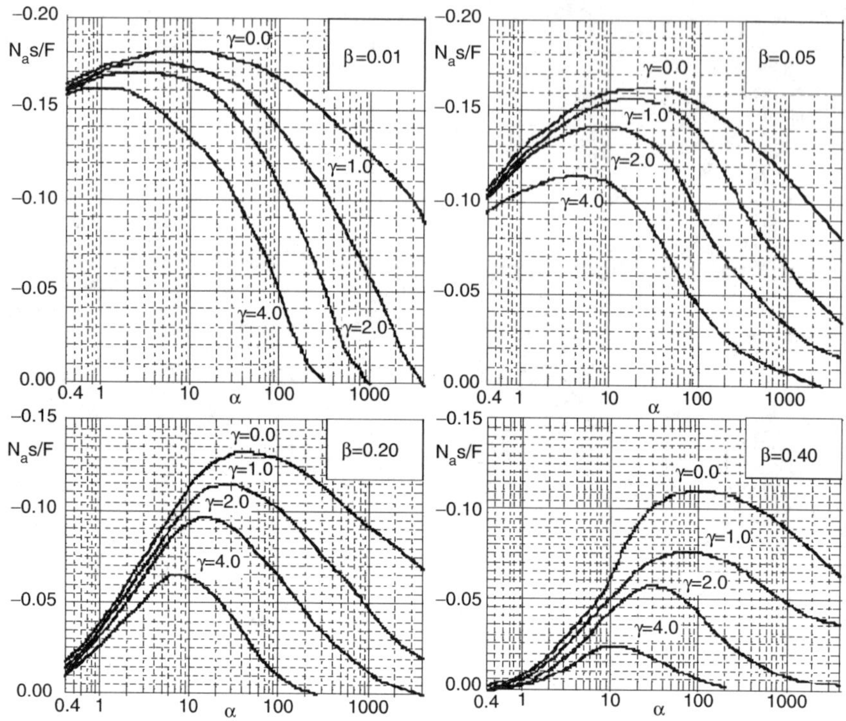

Fig. 9.23

$$\sigma_{tf} = \pm\left({\sigma'}_{tf} + \frac{6M_t}{s^2}\right), \tag{9.72}$$

with a plus or minus sign, depending on whether this is referred to the internal or the external fiber.

Similarly, in the longitudinal direction if σ'_{af} is the potential bending stress on the internal fiber not generated by F, the total bending stress is given by

$$\sigma_{af} = \pm\left({\sigma'}_{af} + \frac{6M_a}{s^2}\right). \tag{9.73}$$

At this point we consider a rectangular section of unitary width under tensile force and bending moment with total plastic flow (see Fig. 9.26). The corresponding normal force is given by

$$N_t = (s - 2a)\,\sigma_s, \tag{9.74}$$

whereas the bending moment is given by

$$M_t = a\,(s - a)\,\sigma_s. \tag{9.75}$$

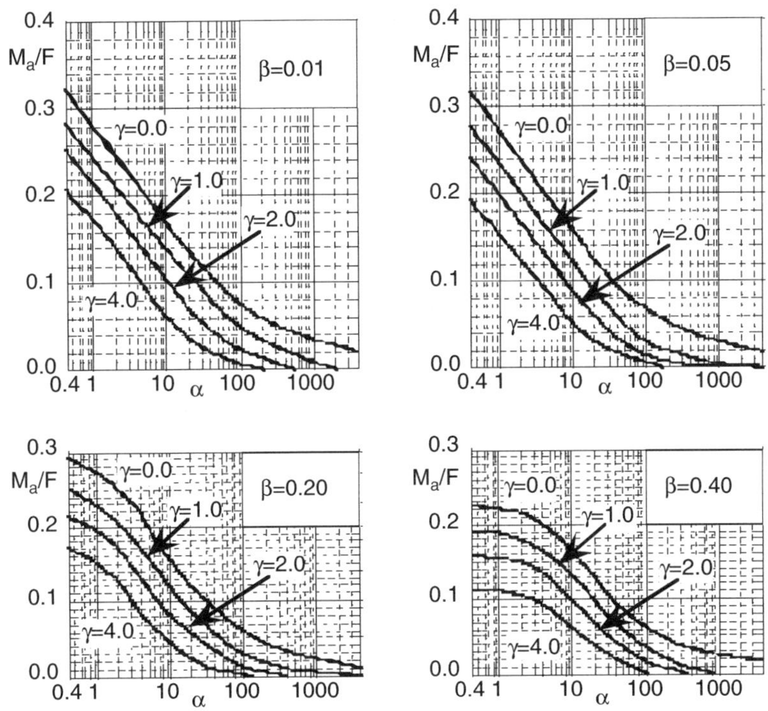

Fig. 9.24

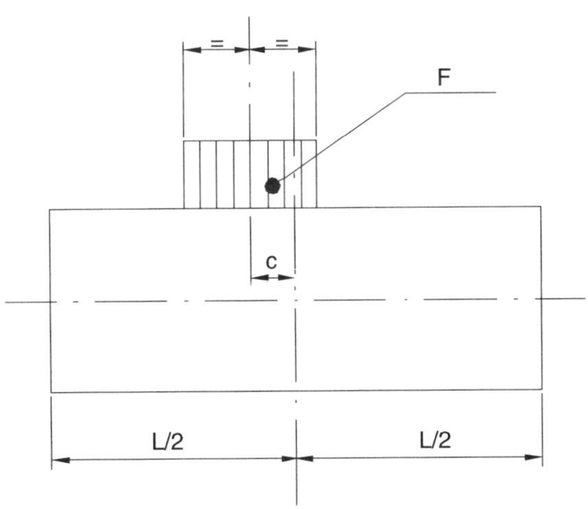

Fig. 9.25

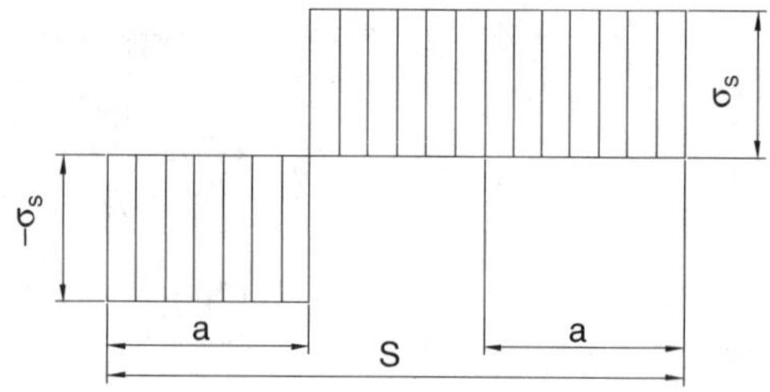

Fig. 9.26

The stresses corresponding to N_t and M_t computed according to the laws of elasticity are given by

$$\sigma_{tm} = \frac{N_t}{s} = \left(1 - 2\frac{a}{s}\right)\sigma_s$$
$$\sigma_{tf} = \frac{6M_t}{s^2} = 6\frac{a}{s}\left(1 - \frac{a}{s}\right)\sigma_s \tag{9.76}$$

We obtain

$$\frac{a}{s} = \frac{1 - \dfrac{\sigma_{tm}}{\sigma_{tf}}}{2}, \tag{9.77}$$

and, finally,

$$\frac{\sigma_{tf}}{\sigma_s} = \frac{3}{2}\left[1 - \left(\frac{\sigma_{tm}}{\sigma_s}\right)^2\right]. \tag{9.78}$$

Considering that based on Sect. 1.2 in general $\sigma_s = 1.5f$, from (9.78) we obtain

$$\sigma_{tf} = \left[2.25 - \left(\frac{\sigma_{tm}}{f}\right)^2\right] f. \tag{9.79}$$

Performing the calculation in the elastic field, the moment M_t corresponding to the value of σ_{tf} obtained through (9.79) causes total plastic flow of the section as if σ_{tm} were equal to $1.5f$ by itself. Following a similar procedure to that described for membrane stresses, one introduces the safety factor 1.2, obtaining the following condition that factors in potential welded seams:

$$\sigma_{tf} \leq \left[1.875 - 0.833\left(\frac{\sigma_{tm}}{f}\right)^2\right] zf. \tag{9.80}$$

Similarly, one may write

$$\sigma_{af} \leq \left[1.875 - 0.833\left(\frac{\sigma_{am}}{f}\right)^2\right] zf. \tag{9.81}$$

Conventionally extending the criterion above to the case of biaxial stresses, one may proceed as follows. Assuming that

$$\sigma_{id} = |\sigma_{tf} - \sigma_{af}|, \tag{9.82}$$

and also requiring that

$$\sigma_{id(f)} \leq \left[1.875 - 0.833\left(\frac{\sigma_{id(m)}}{f}\right)^2\right] zf. \tag{9.83}$$

Note, e.g., in reference to (9.80), that if σ_{tm} is equal to the maximum allowable value ($\sigma_{tm} = 1.25zf$), we have

$$\sigma_{tf} \leq 0.573zf; \tag{9.84}$$

hence, at the most

$$\sigma_t = \sigma_{tm} + \sigma_{tf} = 1.823zf \leq 1.5zf \tag{9.85}$$

that provokes modest plastic flow limited to the fibers at farthest distance from the neutral axis for which σ_{tm} and σ_{tf} have the same sign.

If the loaded area is such that $b \geq \frac{\pi}{2}D_e$, it is best to further reduce the allowable stresses. We recommend to increase the previously introduced safety factor from 1.2 to 1.5, and to verify the following conditions. As far as membrane stresses:

$$\sigma_{tm}, \sigma_{am}, \sigma_{id(m)} \leq zf, \tag{9.86}$$

and as far as bending stresses:

$$\begin{aligned} \sigma_{tf} &\leq \left[1.5 - 0.667\left(\frac{\sigma_{tm}}{f}\right)^2\right] zf \\ \sigma_{af} &\leq \left[1.5 - 0.667\left(\frac{\sigma_{am}}{f}\right)^2\right] zf. \\ \sigma_{id(f)} &\leq \left[1.5 - 0.667\left(\frac{\sigma_{id(m)}}{f}\right)^2\right] zf \end{aligned} \tag{9.87}$$

For values of b between $\frac{\pi}{3}D_e$ and $\frac{\pi}{2}D_e$ it is possible for the allowable stresses to adopt values that are between those discussed earlier and those obtained from (9.86) and (9.87).

Up to this point we implicitly assumed that the stresses σ_t and σ_a are principal stresses, as usually is the case. Otherwise, due to the presence of shear stresses τ in the section considered, the ideal stresses must, of course, be computed according to the well-known criteria and adopting the failure theory of Guest–Tresca. (As we shall see later on about bending moments applied locally, a circumferential moment generates a twisting moment on the entire section of the cylinder).

The maximum values of local stresses due to force F occur at the edge of the area in question, thus it is not necessary to extend the analysis outside of this area. However, this requires the absence of reciprocal interferences among local stresses, as it may be the case if various forces are applied in close succession along a circumference or along a generatrix of the cylinder.

In order to rule out these interferences, the distance between the centers of the areas under load must be equal to or greater than k_1b in the circumferential direction, and equal to or greater than k_2a in the longitudinal direction. The values of k_1 and k_2 are shown in Table 9.1. If these conditions are not satisfied, it is necessary to examine the behavior of forces and moments along the circumference or along the generatrix.

As far as the first investigation, we want to compute the state of stress at point A (see Fig. 9.27) identified by angle φ (in radiants). The values for N_ts/F, M_t/F, N_as/F and M_a/F are computed using the diagrams shown above.

Based on these values, through Figs. 9.28–9.31 (depending on the parameters in question) one determines on the abscissa a value of Z that we call Z_0

Table 9.1

α	β	k_1	k_2
0.4	0.01	8.0	8.0
	0.05	6.0	8.0
	0.20	3.0	4.0
	0.40	1.5	2.0
10	0.01	3.0	8.0
	0.05	2.5	8.0
	0.20	1.5	3.0
	0.40	1.5	2.0
200	0.01	≈ 0	5.0
	0.05		5.0
	0.20		2.5
	0.40		1.75
3200	any	≈ 0	2.5

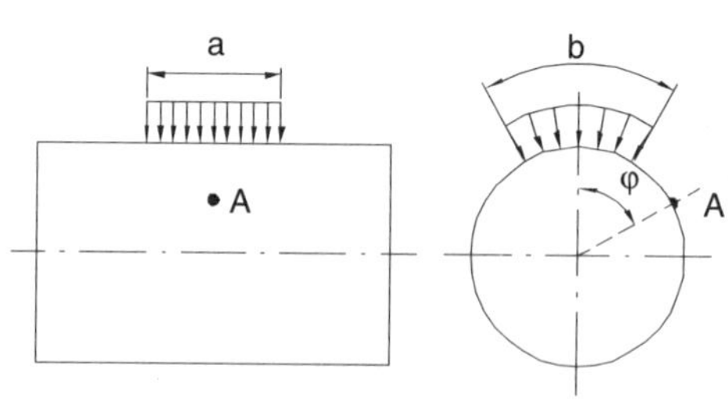

Fig. 9.27

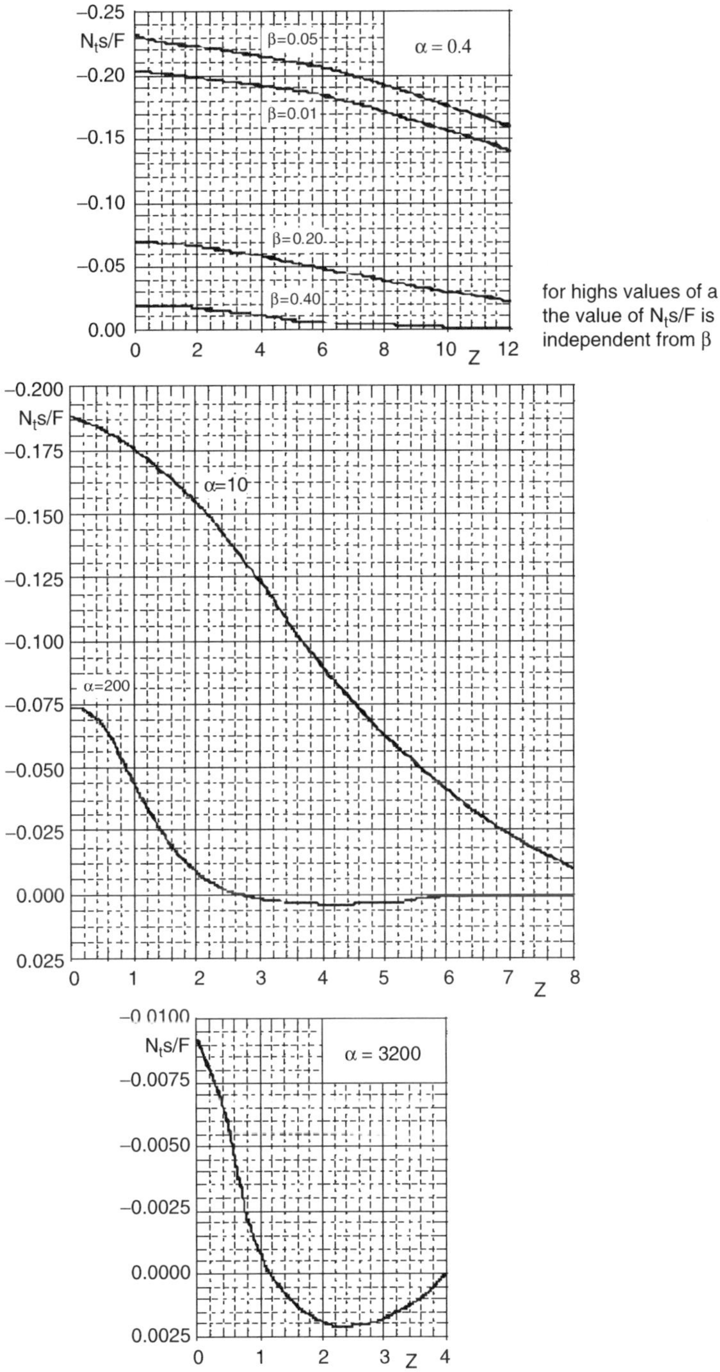
−0.25
Nts/F
−0.20
−0.15
−0.10
−0.05
0.00
β=0.05
β=0.01
β=0.20
β=0.40
α = 0.4
0
2
4
6
8
10
12
Z
for highs values of a the value of Nts/F is independent from β
−0.200
Nts/F
−0.175
−0.150
−0.125
−0.100
−0.075
−0.050
−0.025
0.000
0.025
α=10
α=200
0
1
2
3
4
5
6
7
8
Z
Nts/F
−0.0075
−0.0050
−0.0025
0.0000
0.0025
α = 3200
0
1
2
3
4
Z

Fig. 9.28

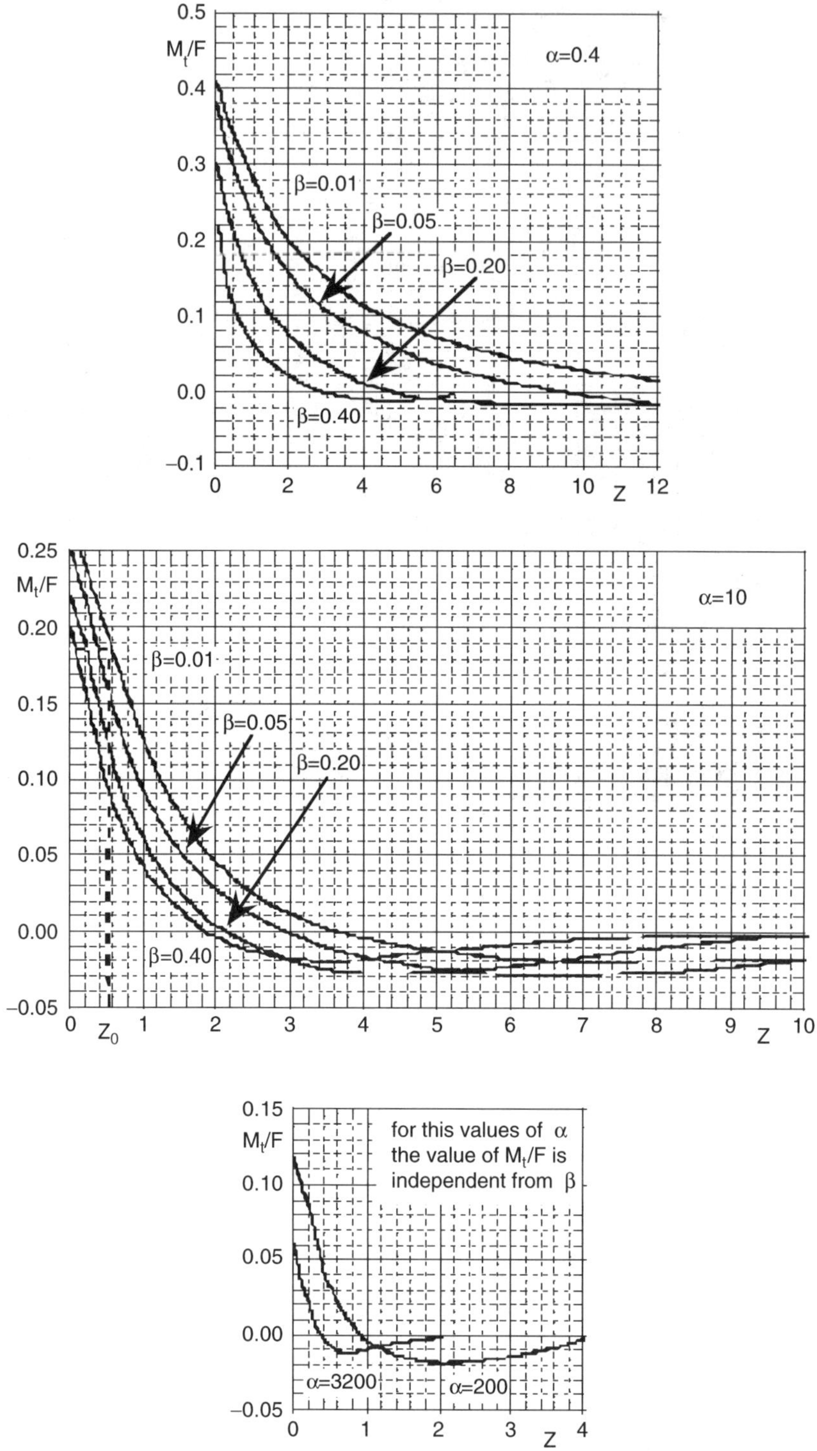
0.5
M_t/F
0.4
0.3
0.2
0.1
0.0
−0.1
α=0.4
β=0.01
β=0.05
β=0.20
β=0.40
0
2
4
6
8
10
Z
12
0.25
M_t/F
0.20
0.15
0.10
0.05
0.00
−0.05
α=10
β=0.01
β=0.05
β=0.20
β=0.40
0
Z_0
1
2
3
4
5
6
7
8
9
Z
10
0.15
M_t/F
0.10
0.05
0.00
−0.05
for this values of α
the value of M_t/F is
independent from β
α=3200
α=200
0
1
2
3
Z
4

Fig. 9.29

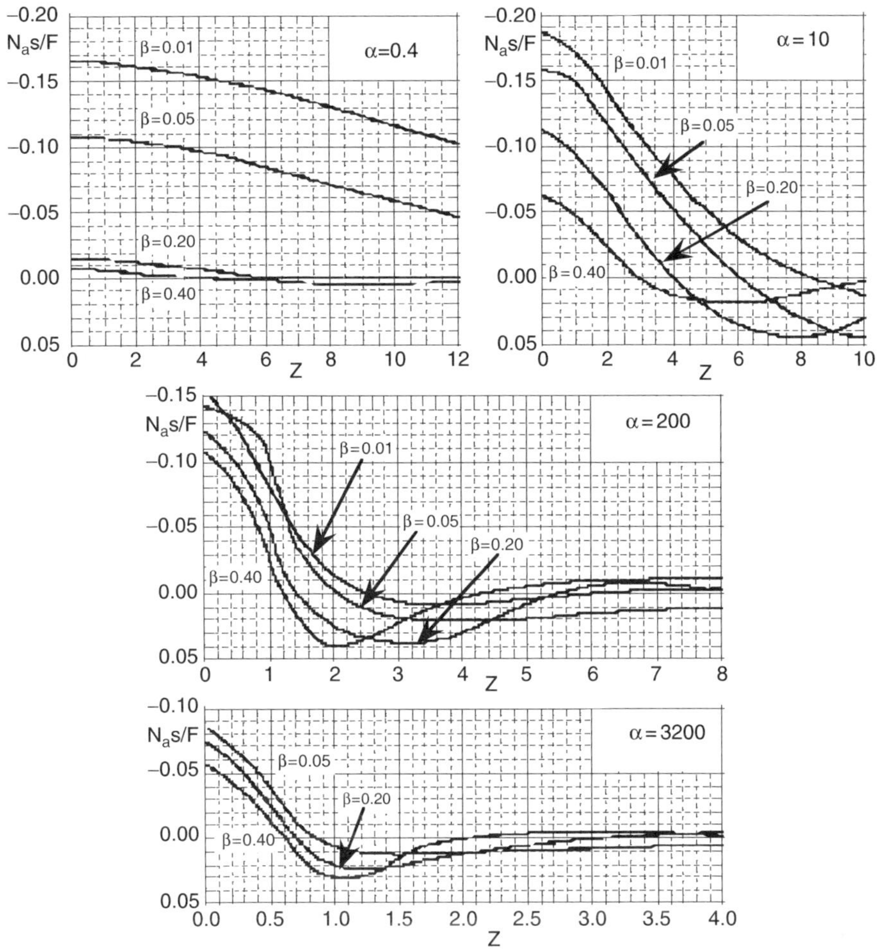

Fig. 9.30

(e.g., the dotted line in Fig. 9.29). With reference to the position of point A one computes

$$Z = \frac{D_e \varphi}{a}. \tag{9.88}$$

Finally, one computes Z_1 given by

$$Z_1 = Z + Z_0 - \frac{b}{a}, \tag{9.89}$$

and based on the values of Z_1 on the abscissa in the diagrams in question, one can read in the ordinate the values for $N_t s/F$, etc. corresponding to point A.

Figures 9.28 and 9.31 are referred to the condition for which $b = 0$. As far as the evaluation of the forces and moments in a point A along the generatrix

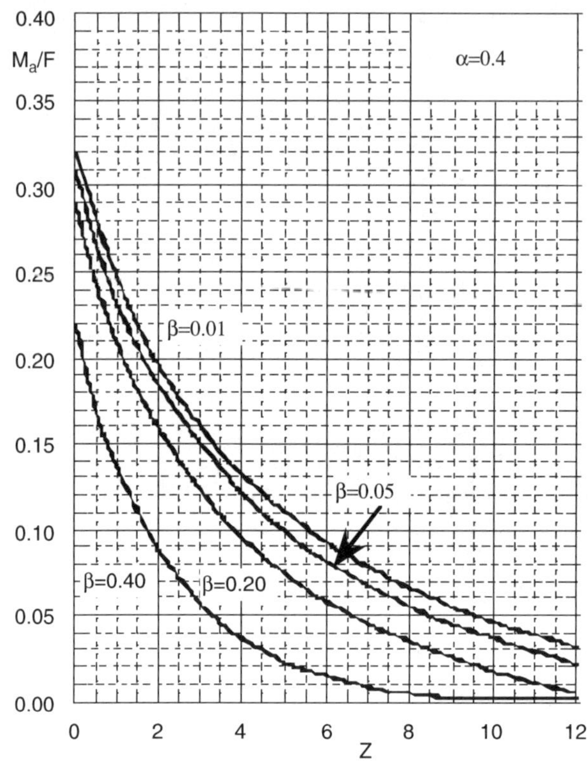

for this values of the values of M_a/F are independent from β

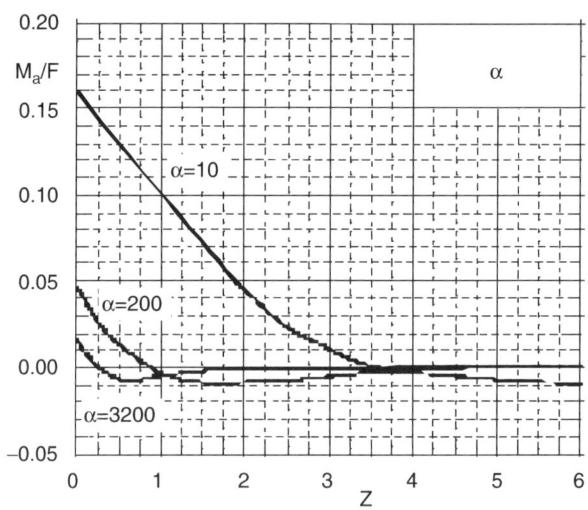

Fig. 9.31

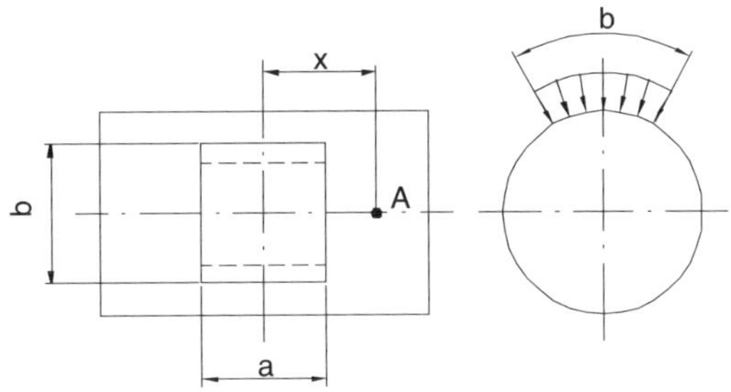

Fig. 9.32

(Fig. 9.32) characterized by the distance x, one should proceed in the following way. By using the fundamental diagrams based on the values of α, β and γ, one can read the values of the parameters correlated to N_t, M_t, N_a and M_a on the ordinate. P_1 is generally anyone of these parameters. The values of the parameters P may also be obtained from the same diagrams but assuming $\gamma = 0$ instead.

Let P_0 be the generic parameter identified this way. The values of the parameters P that we indicate with P_2 may be identified through Fig. 9.33–9.36 as a function of $2x/a$.

The value of the generic parameter P with reference to point A is given by

$$P = P_2 \frac{P_1}{P_0}. \tag{9.90}$$

Figures 9.33 and 9.36 refer to the condition for which $b = 0$. If a longitudinal moment M (Fig. 9.37) covering an area with sides g and b along the generatrix and the circumference, respectively, is active on the cylinder, one can proceed as follows. Consider two identical areas with sides a and b, with $a = g/3$, and subject to two forces F of equal value but opposite sign equal to

$$F = \pm \frac{1.5M}{g}. \tag{9.91}$$

The center lines of both areas are at a distance from each other of $2g/3$.

If the moment is exerted on a circular area of diameter d, the circle is considered as a square with the sides:

$$b = g = 0.85d. \tag{9.92}$$

The forces and maximum moments occur on the external edges of the areas mentioned above and are given by

$$\begin{aligned} N_t &= N_{t1} - N_{t2} \\ M_t &= M_{t1} - M_{t2} \\ N_a &= N_{a1} - N_{a2} \\ M_a &= M_{a1} - M_{a2} \end{aligned}. \tag{9.93}$$

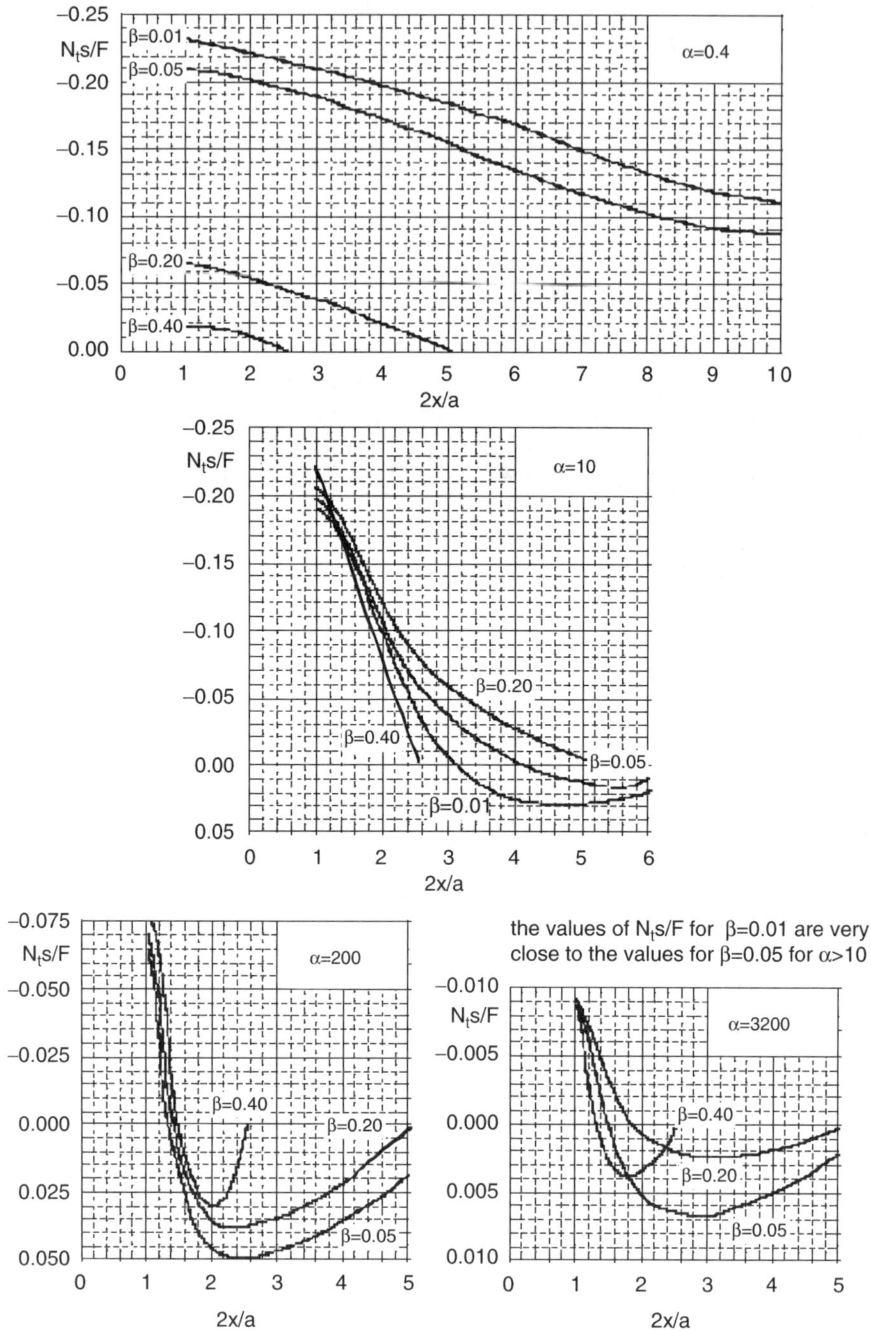

α=0.4
β=0.01
β=0.05
β=0.20
β=0.40
N_t s/F
−0.25
−0.20
−0.15
−0.10
−0.05
0.00
0 1 2 3 4 5 6 7 8 9 10
2x/a
α=10
β=0.20
β=0.40
β=0.05
β=0.01
N_t s/F
−0.25
−0.20
−0.15
−0.10
−0.05
0.00
0.05
0 1 2 3 4 5 6
2x/a
α=200
β=0.40
β=0.20
β=0.05
N_t s/F
−0.075
−0.050
−0.025
0.000
0.025
0.050
0 1 2 3 4 5
2x/a
the values of N_t s/F for β=0.01 are very close to the values for β=0.05 for α>10
α=3200
β=0.40
β=0.20
β=0.05
N_t s/F
−0.010
−0.005
0.000
0.005
0.010
0 1 2 3 4 5
2x/a

Fig. 9.33

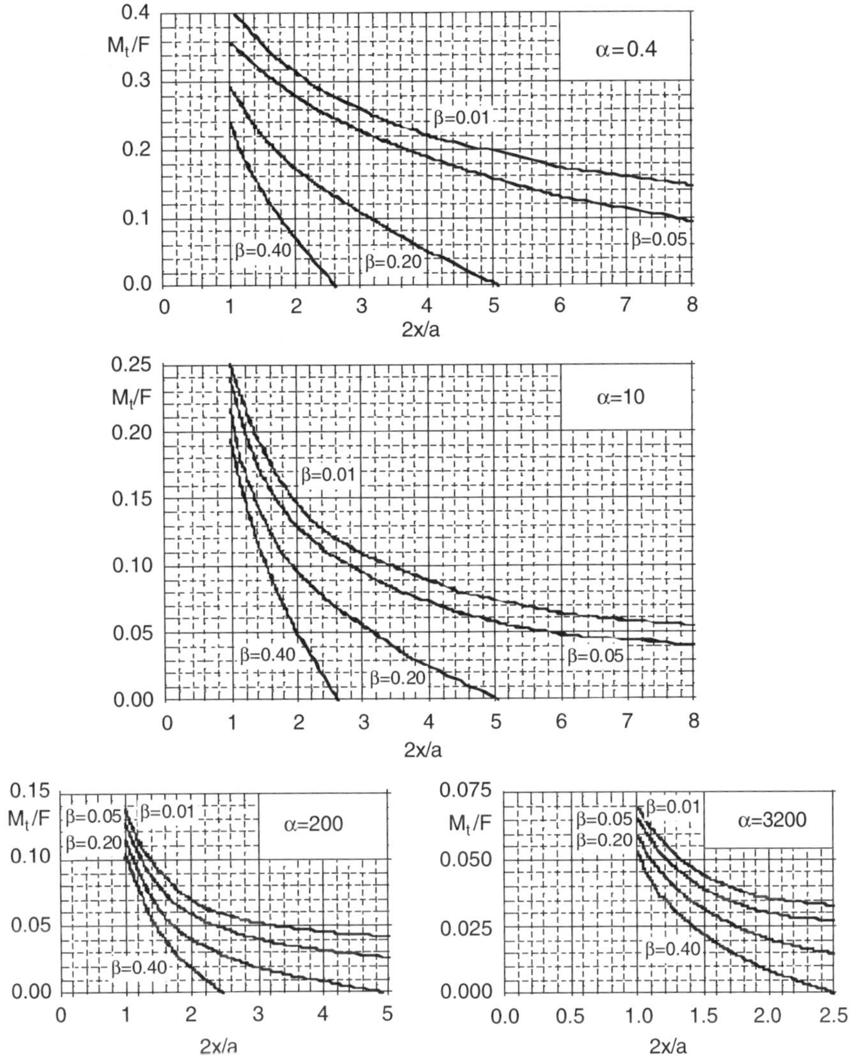

Fig. 9.34

The forces and moments with subscript 1 are derived from Figs. 9.21–9.24 based on the values of F, a and b above. The forces and moments with subscript 2 may be neglected if the value for k_2 in Table 9.1 is smaller than 5. Otherwise, they must be taken into consideration and correspond to the values of the normal forces and the bending moments that take place at a distance $x = 2.5a$ from the center of the area subject to force F.

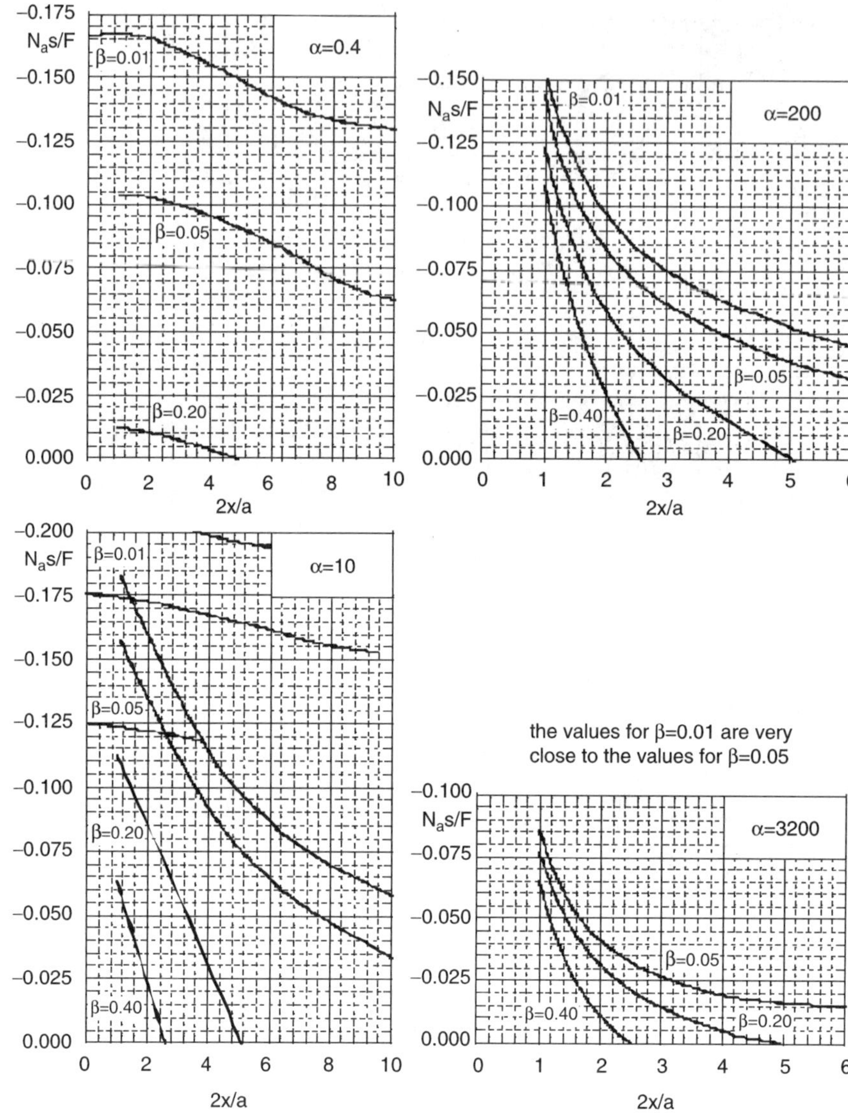

Fig. 9.35

Recalling what was said about (9.90) based on Figs. 9.33 and 9.36, it is possible to determine the value of P_2 that corresponds to $2x/a = 5$. For P_0 and P_1 we refer back to what was said earlier, and through (9.90) it is possible to compute the parameters P through which we obtain the forces and

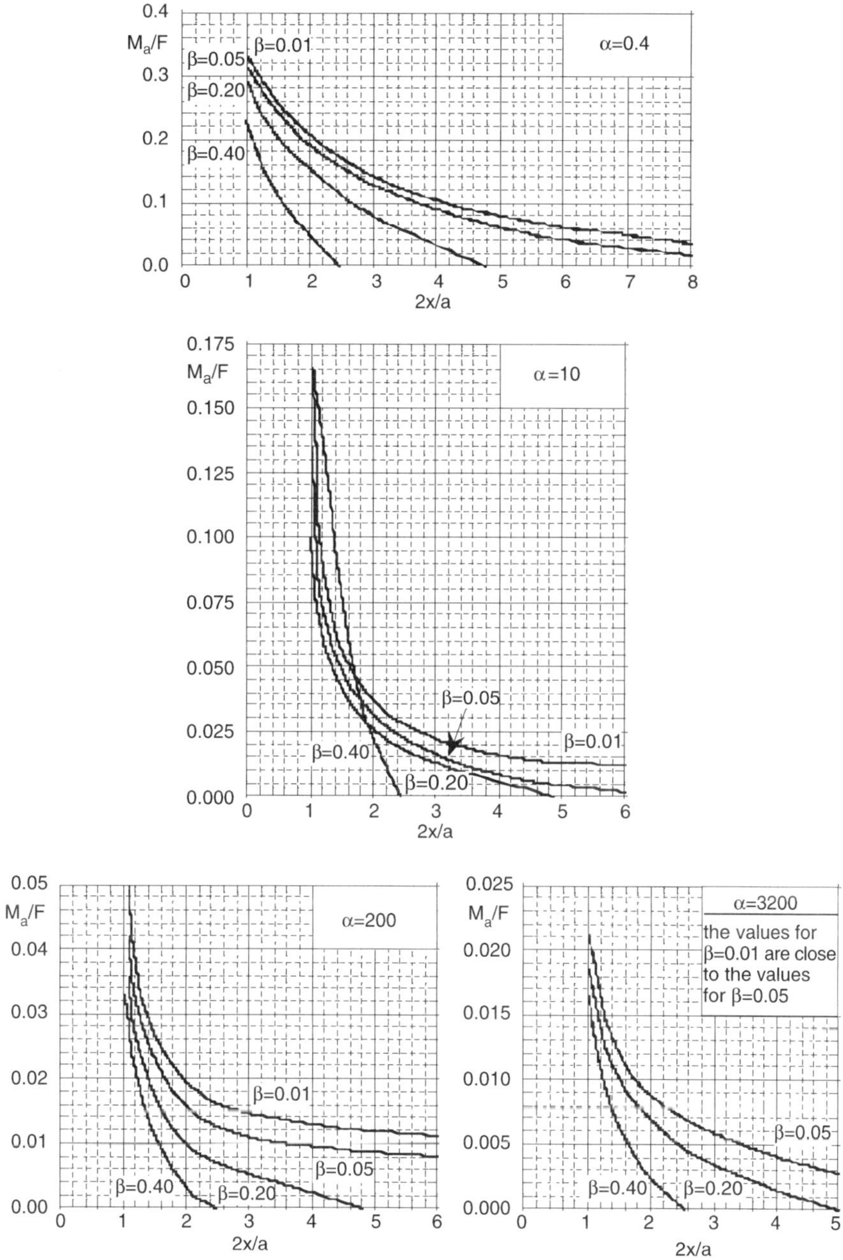
M_a/F
α=0.4
β=0.01
β=0.05
β=0.20
β=0.40
0.4
0.3
0.2
0.1
0.0
0 1 2 3 4 5 6 7 8
2x/a
M_a/F
α=10
β=0.05
β=0.01
β=0.40
β=0.20
0.175
0.150
0.125
0.100
0.075
0.050
0.025
0.000
0 1 2 3 4 5 6
2x/a
M_a/F
α=200
β=0.01
β=0.05
β=0.40
β=0.20
0.05
0.04
0.03
0.02
0.01
0.00
0 1 2 3 4 5 6
2x/a
M_a/F
α=3200
the values for
β=0.01 are close
to the values
for β=0.05
β=0.05
β=0.40
β=0.20
0.025
0.020
0.015
0.010
0.005
0.000
0 1 2 3 4 5
2x/a

Fig. 9.36

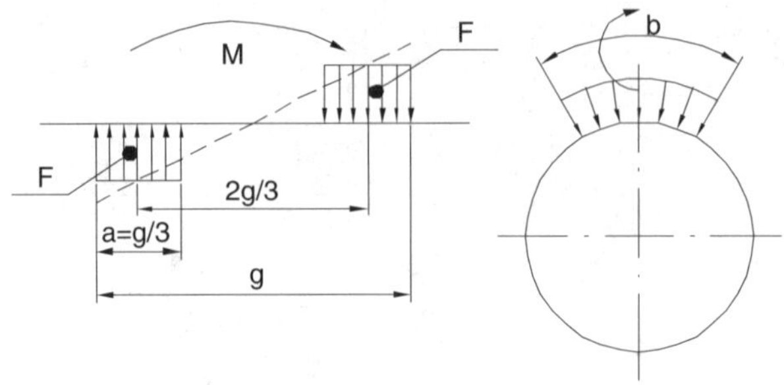

Fig. 9.37

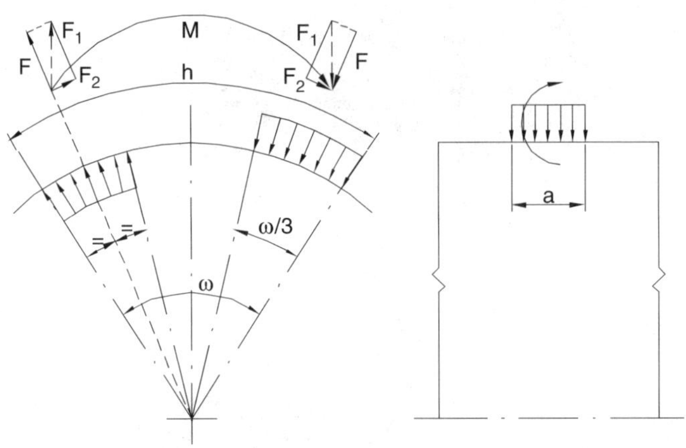

Fig. 9.38

moments with subscript 2. Note that during calculations one must introduce the equivalent length L_e given by (9.66) that may be different for both areas subject to forces F.

If the cylinder is under a circumferential moment (Fig. 9.38) that covers an area with sides a and h along the generatrix and the circumference, respectively, one considers two identical areas with sides a and b, given that $b = h/3$ subject to forces F of opposite sign obtained as follows. With reference to (9.82) we establish that

$$F_1 = \frac{F}{\cos\dfrac{\omega}{3}} \tag{9.94}$$

hence

$$M = 2F_1 r \sin\frac{\omega}{3} = 2Fr\tan\frac{\omega}{3}. \tag{9.95}$$

Given that

$$h = \omega r, \tag{9.96}$$

we finally obtain

$$F = \pm \frac{1.5M}{h} \frac{\dfrac{\omega}{3}}{\tan \dfrac{\omega}{3}}, \tag{9.97}$$

with ω expressed in radiants.

Furthermore, we establish that

$$F_2 = F \tan \frac{\omega}{3} = \frac{1}{2} \frac{M}{r}. \tag{9.98}$$

With respect to the center of the vessel the two forces exerted in the same direction F_2 generate a moment M_c equal to

$$M_c = 2F_2 r = M, \tag{9.99}$$

as it should be, since the forces F are radial. From a global point of view, the moment M_c is a twisting moment that impacts the entire section of the cylinder.

If the angle ω is smaller or equal to $\pi/4$, it is possible to simplify (9.97) (to favor resistance and making an error smaller than 3%) assuming that

$$F = \pm \frac{1.5M}{h}. \tag{9.100}$$

If the area is circular one may proceed as above, that is,

$$a = h = 0.85d. \tag{9.101}$$

The equations in (9.93) are still valid for normal forces and bending moments.

The parameters with subscript 1 are obtained, as usual, based on the values of F, a and b. The parameters with subscript 2 may be ignored if the value of k_1 in Table 9.1 is smaller than 5. Otherwise, they must be factored in and correspond to the values for normal forces and bending moments occurring in a point identified from angle φ, that the radius in question forms with the one going through the center of the area at hand, given that $\varphi = 5b/D_e$.

Recalling what was said about (9.89), first of all we determine the value of Z_0. The value of Z is equal to

$$Z = \frac{5b}{D_e} \frac{D_e}{a} = 5\frac{b}{a}. \tag{9.102}$$

Therefore, based on (9.89) we have

$$Z_1 = 4\frac{b}{a} + Z_0. \tag{9.103}$$

Through the process described earlier we obtain the values of the forces and moments with subscript 2.

The allowable stresses are those pointed out before, keeping in mind, however, that the value of h (in lieu of b) is crucial to determine which equations shall be used ($h > \pi/3D_e$ or $h < \pi/3D_e$). If the calculation of N_t, M_t, N_a and M_a highlights the fact that the stresses on the cylinder are not compatible with its resistance, since the conditions described earlier about the allowable values of the stresses are not fulfilled, it is necessary to reinforce the cylinder locally with a plate (Fig. 9.39). If l and m are the sides of the plate along the generatrix and the circumference, respectively, one may assume with a sufficient degree of approximation for practical cases that the stress at the edges of the plate corresponds to the one taking place if the entire plate were loaded. In other words, it is sufficient to replace a with l (or g with l in the case of an axial bending moment), and b with m (or h with m in the case of a circumferential bending moment) and to perform the calculation as described.

As far as the stresses at the edges of the loaded area (Fig. 9.39), by indicating the thickness of the cylinder and the plate, with s_c and s_p, respectively, one proceeds in a similar way based on the dimensions of the area and assuming that $s = s_c + s_p$. If N is the generic normal force computed as discussed, and N_c and N_p are the forces relative to the cylinder and the plate, respectively, we have:

$$N_c = N\frac{s_c}{s_c + s_p} \qquad N_p = N\frac{s_p}{s_c + s_p} \tag{9.104}$$

By analogy, if M is the generic bending moment, by indicating the moments relative to the cylinder and the plate with M_c and M_p, the following conservative values may be adopted:

$$M_c = M\frac{\sqrt{s_c}}{\sqrt{s_c} + \sqrt{s_p}} \qquad M_p = M\frac{\sqrt{s_p}}{\sqrt{s_c} + \sqrt{s_p}} \,. \tag{9.105}$$

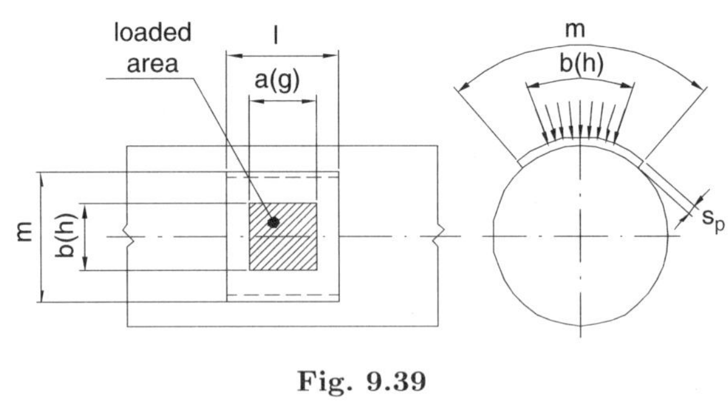

Fig. 9.39

9.4 Local Effects of Forces and Moments on Spheres

Prof. Bijlaard at Cornell University also studied local effects of forces and bending moments on spheres in depth (Bijlaard (1957a,b)). First of all, it is important to point out that the local stresses in question rapidly decrease moving away from the loaded area, as is the case with edge effects (see Sect. 2.2). The calculation criteria relative to spheres are transferable to hemispherical heads and the central spherical portion of paraelliptical heads, as well. The same is true for elliptical heads, as long as one adopts the radius corresponding to the center of the loaded area as the radius of reference.

To be able to apply the calculation criteria below, one must rule out the possibility that the local stresses present in correspondence of the head-cylinder junction influence the stresses in question. To that extent assuming that $\Delta = 2.5\sqrt{rs}$, and given that s is the thickness of the head, the subsequent calculation criteria may be adopted only if the distance between the edge of the loaded area and the junction head-cylinder for the hemispherical head, or the distance between the edge of the loaded area and the junction between the spherical portion and the torus knuckle for the paraelliptical head, is greater than or equal to Δ.

We advise to consider the elliptical head as a paraelliptical head and to do the checks usually done for that type of head. In that case, the radius r for the calculation of Δ is that of the ideal central spherical portion. If the loaded area is a square with a side equal to a, it is considered to be like a circular area with diameter d equal to a. On the other hand, if the area is rectangular with sides a and b it is considered like a circular area with $d = \sqrt{ab}$. Due to the effect of a radial force F applied to the sphere (see Fig. 9.40), the following normal forces and bending moments are generated in reference to the length

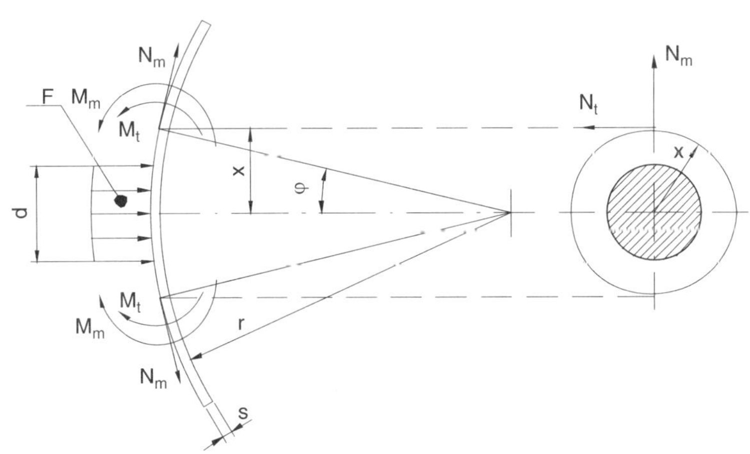

Fig. 9.40

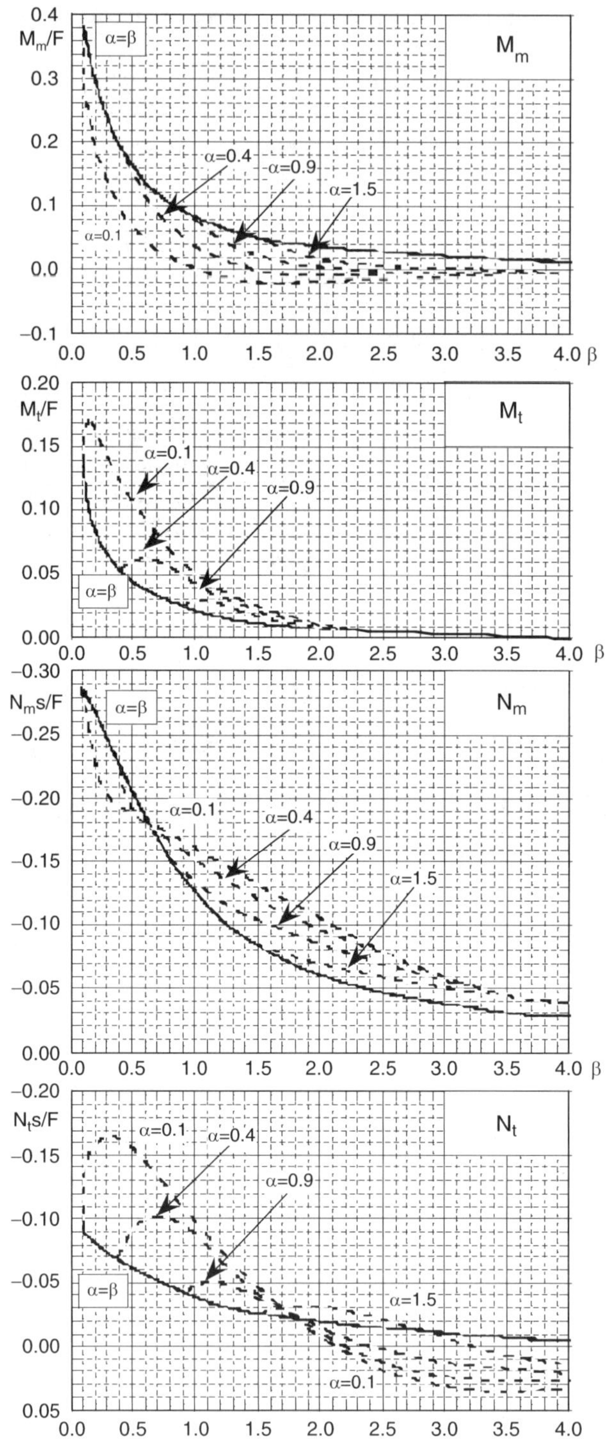

M_m/F
α=β
M_m
α=0.4
α=0.9
α=1.5
α=0.1
0.4
0.3
0.2
0.1
0.0
−0.1
0.0 0.5 1.0 1.5 2.0 2.5 3.0 3.5 4.0 β
M_t/F
M_t
α=0.1
α=0.4
α=0.9
α=β
0.20
0.15
0.10
0.05
0.00
0.0 0.5 1.0 1.5 2.0 2.5 3.0 3.5 4.0 β
N_m s/F
α=β
N_m
α=0.1
α=0.4
α=0.9
α=1.5
−0.30
−0.25
−0.20
−0.15
−0.10
−0.05
0.00
0.0 0.5 1.0 1.5 2.0 2.5 3.0 3.5 4.0 β
N_t s/F
N_t
α=0.1
α=0.4
α=0.9
α=β
α=1.5
α=0.1
−0.20
−0.15
−0.10
−0.05
0.00
0.05
0.0 0.5 1.0 1.5 2.0 2.5 3.0 3.5 4.0

Fig. 9.41

unit: a normal force N_m and a bending moment M_m along the meridian, and a normal force N_t as well as a bending moment M_t along the circumference. The diameter of the loaded area is correlated to a dimensionless quantity that we indicate with α and that is given by

$$\alpha = 0.91 \frac{d}{\sqrt{rs}}. \tag{9.106}$$

Similarly, if x is the distance of the point to be analyzed in terms of stress (see Fig. 9.40) from the center of the loaded area, consider the dimensionless quantity β correlated to x, given by

$$\beta = 1.82 \frac{x}{\sqrt{rs}}. \tag{9.107}$$

Through Fig. 9.41 it is possible to obtain the dimensionless quantities $N_m s/F$, M_m/F, $N_t s/F$ and $M_t s/F$ that lead to the unknown normal forces and bending moments as a function of α and β. The solid curves correspond to $\alpha = \beta$ and identify the values of the forces and moments in question at the edge of the loaded area. Generally, this is where there is the most unfavorable combination of normal forces and bending moments.

If the point to be checked is outside the area under load, then $\alpha < \beta$. This condition is reflected in the portion of dotted curves shown in Fig. 9.40. If bending moment M is applied on the sphere (Fig. 9.42), the value of normal forces and bending moments is a function of M, α, β, as well as angle ϑ. These values may be obtained through Fig. 9.43. Even in this case the solid curves refer to $\alpha = \beta$, i.e., to the stresses at the edge of the loaded area.

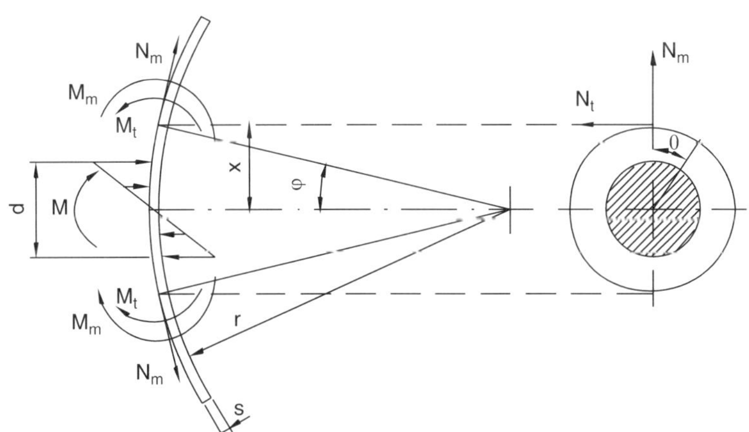

Fig. 9.42

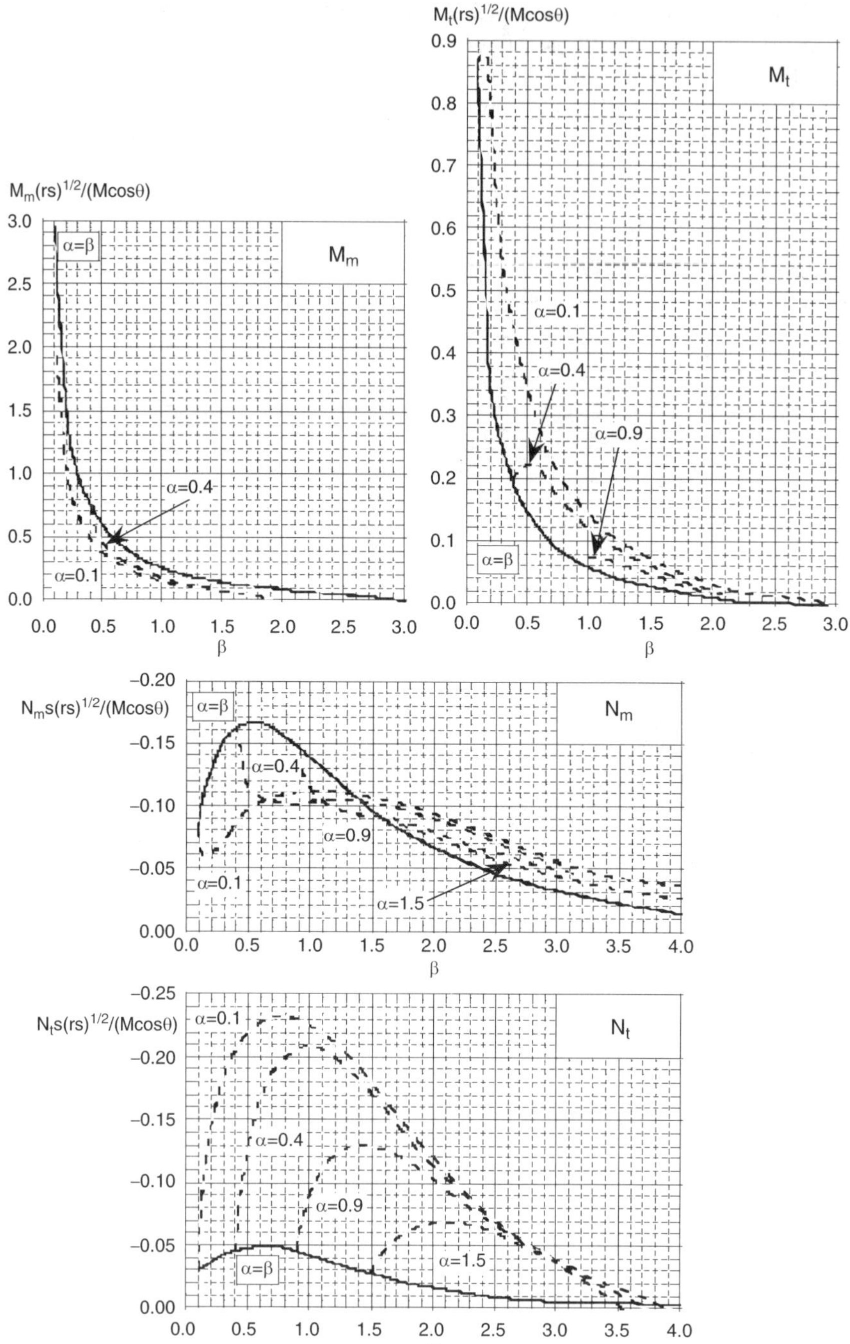

M_t(rs)^1/2/(Mcosθ)
M_t
α=0.1
α=0.4
α=0.9
α=β
β
M_m(rs)^1/2/(Mcosθ)
M_m
α=β
α=0.4
α=0.1
β
N_ms(rs)^1/2/(Mcosθ)
N_m
α=β
α=0.4
α=0.9
α=0.1
α=1.5
β
N_ts(rs)^1/2/(Mcosθ)
N_t
α=0.1
α=0.4
α=0.9
α=1.5
α=β
β

Fig. 9.43

As far as resistance verification, one may refer to Sect. 9.3 by substituting longitudinal stresses with meridian stresses. Note that even in this case it is always necessary to consider the bi-axial state of stresses. Specifically, if a moment M is applied to the sphere, one must identify the most unfavorable value of ϑ in that regard. As far as the need of a reinforcement plate and the criteria to be followed to examine the stresses at the edge of the plate, as well as the edge of the loaded area, considerations similar to those made about cylinders will apply.

10

Fatigue Analysis

10.1 General Approach

Fatigue phenomena will not be discussed in detail here. We will simply focus our attention on a few fundamental concepts, specific considerations relative to pressure vessels, and the calculation criteria to be used for fatigue analysis.

Fatigue is the degradation in the resistance of the material, as a result of repeated events over time. In contrast to creep, where, besides temperature, time is crucial, in the case of fatigue the crucial factor is the variation of stresses or deformations over time. Time plays a role only when it is matched by an increased number of cycles. In other words, the two fundamental parameters that may cause fracture because of fatigue are the wide range of variation among stresses and the number of these variations exerted on the material. The correlation between these two parameters is shown in Sect. 10.2.

The standards of the ASTM describe the fatigue phenomenon as "the behavior of materials subject to repeated stress or deformation cycles that cause a deterioration of the material that manifests itself through a progressive fracture." The progressive character of the damage is often visible because of the type of fracture. In fact, the ruptured surface usually shows two distinct areas. The first one is smooth and opaque and shows a series of waves, while the second one is wrinkled and irregular.

In proximity of the fracture, the material does not have plastic deformations, and the edges seem clear-cut without any sign of area reduction. The waves in the smooth area start from an initiation point. From that point, it progressively extends itself to the remaining section. When the area of the resisting section becomes unable to withstand the acting force or moment, the rupture occurs by disjunction. Thus, it has the characteristics of fragile fracture. If the rupture point corresponds to a point with high stress peaks, it is understandable how the progressive fracture may originate from there. Plastic flow develops, and as a result of alternate stresses, under certain conditions the material undergoes plastic flow cycles that lead to alternate, increasing deformations up to a crack.

At this point, the presence of a crack provokes further concentration of force lines, so that the fracture extends around the original point more and more, until it causes rupture due to the reduction in resisting area. In the absence of structural discontinuities, it is more difficult to identify the cause of fatigue fracture.

Generally, the phenomenon is explained as follows: a polycrystalline material always has a few crystals with preferred planes pointed in such a way to become subject to plastic flow cycles, even though the stress in the examined point is smaller than the yield strength. To this extent, different theories and schemes have been suggested, but we shall not dwell on them. In any case, experience tells us that if structural discontinuities and deriving stress peaks are a highly unfavorable element as far as resistance to fatigue, the phenomenon occurs at certain stress levels (or rather variations thereof) and number of cycles, even though the discontinuities are absent.

The fracture mechanism due to fatigue may manifest itself even in a perfectly smooth specimen. The variations in stress may be of different amplitude as well as different type. The literature considers two special conditions, but this does not rule out a series of intermediate situations. We are referring to pulsatory and alternate fatigue. The former requires stresses that vary in cycles from a value zero to a maximum value, either positive or negative. The latter shows stresses that vary in cycles from a negative to a positive value but equal in absolute value.

Pulsatory fatigue is characterized by higher values of the rupture stress compared to the alternate one. Considering pressure vessels where pressure is in fact the crucial parameter in terms of stress levels, it would seem at first that, at the most, they may be subject to pulsatory fatigue. Leaving aside special cases with reverse pressure (e.g., a lined vessel alternatively under pressure from the inside and the outside), during starts and stops the pressure varies from 0 to p or viceversa, and the corresponding stresses vary between 0 and the value corresponding to p, which makes this a pulsatory fatigue phenomenon.

Even assuming pressure fluctuations during working operations, the phenomenon still causes pressure fluctuations that keep the same sign. Thus, this condition is even less dangerous than pulsatory fatigue. In actuality, the working conditions of the vessel are generally quite different.

For instance, it suffices to consider the forces and moments on the vessel by connected piping run by warm fluid. If a cold spring is done during installation of the piping, the stresses will change sign when warm fluid runs through it, but it is not necessary to make the cold spring responsible for the change in sign, recalling Sect. 7.5.

If the expansion of the piping is one of the causes of potential alternate stresses, however, they are not the only or the most important reason for the inversion of sign of the stresses. In previous chapters we systematically referred to plastic collaboration of the material every time the exceeding of the deformation corresponding to the yield strength in a spot or a limited area of the vessel does not constitute danger.

Specifically, we allowed plastic flow of the material in the presence of secondary stresses. In fact, this is equivalent to assuming the allowable stress relative to the sum of primary and secondary stresses to be equal to $3f$, thus generally equal to $2\sigma_s$, where σ_s is the yield strength. As we pointed out in Sect. 1.4 and through Fig. 1.11, under these conditions the stress varies between a minimum equal to $-\sigma_s$ and a maximum equal to σ_s or viceversa after the first cycle (or the first cycles).

If the vessel is subject to cyclical phenomena in the point where this stress is present, we are faced with alternate fatigue. On the other hand, a vessel without secondary stresses is unthinkable. It suffices to think of the junction between cylinder and head, the different nozzles, not to mention potential hole lines and thermal stresses (particularly during transient phases).

Contrary to what a superficial study may lead one to believe, alternate fatigue is the rule and not the exception in the case of pressure vessels undergoing cyclical phenomena. Therefore, fatigue analysis is conducted, as we shall see in the following sections, in reference to allowable stress with alternate fatigue, as shown in Figs. 10.3 and 10.4. Fatigue analysis is not necessarily imperative for all vessels. In fact, if the number of cycles is modest, it will not be necessary.

Section 10.3 enumerates the necessary conditions to avoid fatigue analysis in detail.

10.2 Fatigue Curves

As known, the fatigue rupture stress, regardless of whether it is pulsatory or alternate, is a function of the number of cycles. From a historical point of view, remember that the issue of fatigue resistance mainly surfaced after the development of the railways around 1850.

Studies were done by Hodgkinson, Rankine, and Fairbain, but the first to perform a systematic investigation was Wöhler, an engineer of the Bavarian Railway, to whom we owe the first diagrams about alternate fatigue rupture stress per bending. For instance, Fig. 10.1 shows the qualitative behavior of the curve that correlates this rupture stress with the number of cycles. If such curves are drawn in logarithmic coordinates, the behavior is almost linear for a number of cycles below the critical value N_c beyond which the rupture stress remains constant (see Fig. 10.2). This means that there is a value of the stress below which fatigue rupture will never occur also for a large number of cycles (possibly even an infinite one). Generally, the value of N_c is within the range 10^6 and $2 \cdot 10^6$. Conventionally, the assumption is that $N_c = 10^6$.

While for power machinery the value of the rupture stress under fatigue corresponding to N_c is the most significant one, given the extremely high number of cycles, for pressure vessels the behavior of the rupture stress is

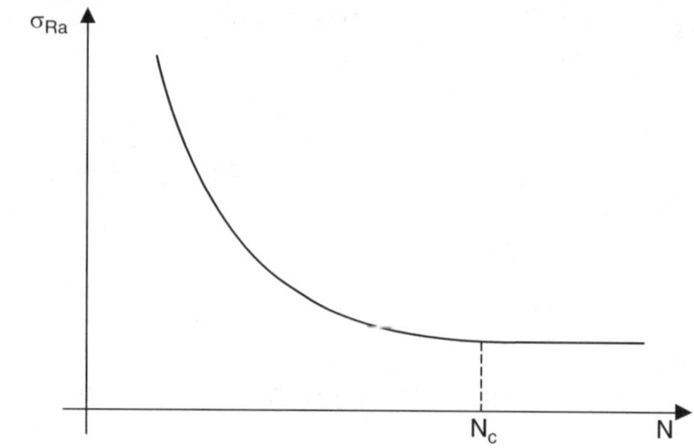

Fig. 10.1

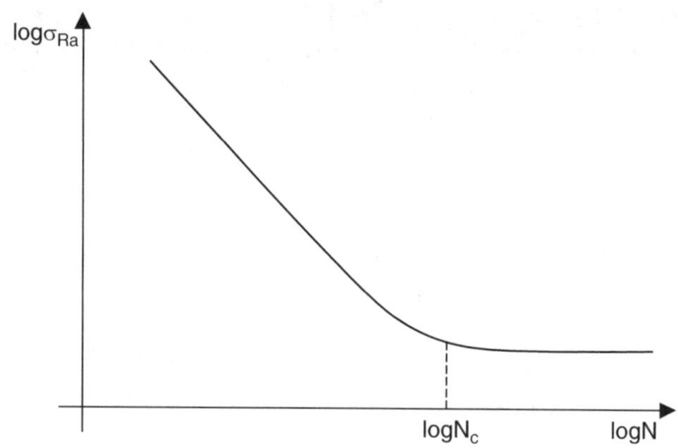

Fig. 10.2

of greatest interest for a number of cycles smaller than N_c instead. In fact, the number of cycles is rarely high, at least for the most significant stress fluctuations.

Thus, we comprehend the importance of fatigue curves mentioned earlier. They are shown in Figs. 10.3 and 10.4. Note that the rupture stress is not shown on the ordinate, but the allowable stress f_a under alternate fatigue instead, obtained through a safety factor equal to about 2. These curves may be used immediately comparing the maximum alternate stress σ_a present in a point of the vessel (see Sect. 10.4) with the allowable stress f_a.

Considering the issue from another point of view, once the value of σ_a is identified, and requiring that $f_a = \sigma_a$, they help establish the allowable number of cycles to be compared with the actual number of cycles. A few specific considerations are due.

The values of f_a are obtained through the safety factor, based on the alternate deformations that cause the fatigue rupture and multiplying these

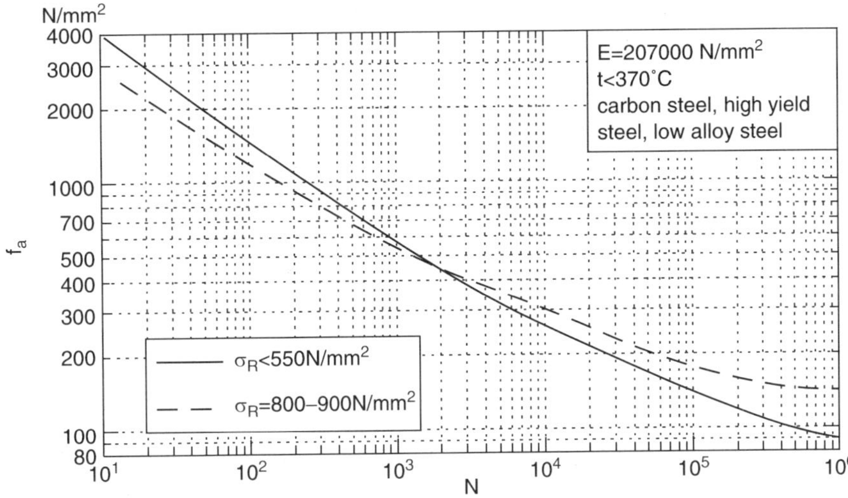

Fig. 10.3

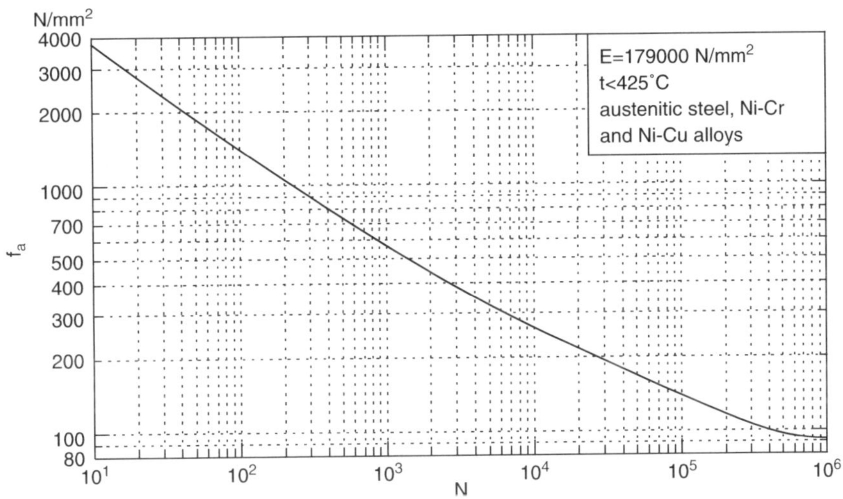

Fig. 10.4

deformations by the modulus of elasticity at room temperature. The figures indicate the reference value of E. If the value of E is smaller than the one indicated because of the temperature, it is necessary to correct the value of f_a through the following equation:

$$f'_a = \frac{E_c}{E} f_a, \tag{10.1}$$

where f'_a is the value to assume during calculations and E_c the modulus of elasticity for the actual temperature.

This process is allowed as long as the temperature is within the limits shown in the figures. If the calculation methodology (Sect. 10.4) requires to subtract the number of allowable cycles from σ_a instead, and the value of the modulus of elasticity is smaller than the one indicated in the figures, one shall correct the value of σ_a through

$$\sigma'_a = \frac{E}{E_c}\sigma_a, \tag{10.2}$$

and, setting $f_a = \sigma'_a$, subtract the requested allowable number of cycles from the curves. Finally, note that Fig. 10.3 shows two curves valid for rupture stress σ_R of the material less or equal to 550 N mm^{-2}, and for rupture stress between 800 and 900 N mm^{-2}, respectively.

If the rupture stress is between 550 and 800 N mm^{-2}, one shall proceed by interpolation. Finally, if the number of cycles is greater than 10^6, for f_a one assumes the value corresponding to $N = 10^6$. Similarly, if the calculation methodology (Sect. 10.4) requires to subtract the corresponding number of cycles from the value of the maximum alternate stress σ_a, then if σ_a (or σ'_a if it is necessary to adopt (10.2)) is smaller than the value of f_a that corresponds to 10^6, one assumes $N = 10^6$.

10.3 Vessels not Requiring Fatigue Analysis

A pressure vessel must not necessarily be verified for fatigue, even though it may undergo cyclical stresses. This type of verification is not required whenever the following conditions are satisfied. Any vessel will be put under pressure a number of times during its lifetime. Therefore, no vessel will escape cyclical stresses. Of course, the number of cycles is crucial to establish whether the phenomenon should be considered or not in terms of potential rupture caused by fatigue. Note that based on Sect. 1.5 and following sections about secondary stresses, if f is the basic allowable stress of the material, the corresponding allowable stress relative to the sum of primary and secondary stresses is equal to $3f$ (that corresponds to twice the yield strength). If the vessel was sized according to the criteria described throughout previous chapters, the maximum ideal stress computed according to the laws of elasticity that may be present in the vessel is, in fact, equal to $3f$. As we know, this stress varies from a minimum value equal to $-1.5f$ and a maximum value equal to $1.5f$ after the first cycle.

However, if the vessel was sized exploiting the resistance of the material to the fullest, in some points there may be an ideal alternate stress equal to $1.5f$. It would seem acceptable to have such a number of cycles so that the corresponding allowable stress under alternate fatigue would at least be equal to $1.5f$. However, it is best to adopt a more conservative criterion in view of the potential presence of other cyclical phenomena that we will address later on, leading the minimum required value of the allowable stress under fatigue to $3f$. If f is known, one assumes f_a to be equal to $3f$, and based on the fatigue curve relative to the type of steel used, from this value f_a it is possible to establish the allowable number of cycles.

If the actual number of cycles is smaller or equal to this value, the fatigue analysis will not be necessary. A similar process is put in place if the pressure fluctuates during work. Δp be the amplitude of the fluctuation, p the design pressure and f_a the allowable stress under fatigue corresponding to the number of predicted fluctuations. In contrast to the previous instance where the pressure varies from 0 to p with a range of fluctuation equal to p, in this case it is equal to Δp.

If we assume the value of f to be the stress of reference, while the range of fluctuation of such stress is indeed equal to f (or to the general membrane stress) in the previous instance, in this case it is equal to

$$\Delta\sigma_m = \frac{\Delta p}{p} f. \tag{10.3}$$

Similarly to the previous case, the allowable stress under fatigue must be at least three times the range of fluctuation of the stress of reference.

Hence, it shall be

$$f_a = 3\frac{\Delta p}{p} f; \tag{10.4}$$

from the latter one obtains the maximum allowable value of the range in fluctuation of the pressure given by

$$\Delta p = \frac{p}{3}\frac{f_a}{f}. \tag{10.5}$$

In other words, once the predicted number of fluctuations and the corresponding value of f_a have been established, the allowable value of Δp is given by (10.5).

In the event of starts and stops of the vessel that is alternatively put under pressure, sudden differences in temperature between points may occur due to the behavior of the warm fluid inside. It is necessary to identify the allowable value of Δt. One must examine the difference in temperature between any couple of adjacent points, defined as such when their distance is less than $2\sqrt{rs}$, r is the average radius and s the thickness of the vessel, the nozzle, or the area of the vessel where the two points are located.

Considering that this is a transient phase, we must assume that the highest temperature Δt in one of the two points versus the other will be located in the point itself, and stresses will be generated around the point that will completely nullify the dilations caused by Δt. Considering the two principal stresses σ_I and σ_{II} present on the plane normal to the radius, given that $\sigma_{III} = 0$ we must write

$$\begin{aligned} \varepsilon_I &= \frac{1}{E}(\sigma_I - \mu\sigma_{II}) + \alpha\Delta t = 0 \\ \varepsilon_{II} &= \frac{1}{E}(\sigma_{II} - \mu\sigma_I) + \alpha\Delta t = 0 \end{aligned} \tag{10.6}$$

First of all, from (10.6) we obtain

$$\sigma_I = \sigma_{II}; \tag{10.7}$$

then

$$\frac{\sigma_I}{E}(1-\mu) + \alpha\Delta t = 0; \tag{10.8}$$

and

$$\sigma_I = \sigma_{II} = -\frac{\alpha E\Delta t}{1-\mu}. \tag{10.9}$$

For a useful comparison we refer back to both (3.179) and (5.121). In the most unfavorable case in which the material is under plastic flow, $\mu = 0.5$ and

$$|\sigma_I| = |\sigma_{II}| = 2\alpha E\Delta t. \tag{10.10}$$

Now, if we assume that these stresses are at most equal to the allowable stress under alternate fatigue (given that $\sigma_{III} = 0$ they are also ideal stresses according to Guest), even though this is in fact a phenomenon of pulsatory fatigue, we obtain the maximum allowable value of Δt given by

$$\Delta t = \frac{f_a}{2\alpha E}. \tag{10.11}$$

On the other hand, if we assume that the stresses in question considered to be independent may lead to plastic flow of the material, as implicitly hypothesized with $\mu = 0.5$, to our knowledge their value varies between $-\alpha E\Delta t$ and $\alpha E\Delta t$. Therefore, based on (10.11) we see that alternate stresses equal to $f_a/2$ are generated.

As far as the influence of pressure during starts and stops, we established that the maximum possible value of the ideal alternate stress is equal to $1.5f$, whereas the allowable stress under fatigue has been assumed to be $3f$. Even in this case the alternate stress is equal to $f_a/2$. Actually, during starts and stops the simultaneous influence of pressure and Δt between adjacent points leads to a total alternate stress at the most equal to f_a, as logic expects it to be.

In fact, the considerations above that were made in terms of stresses out of practicality, are absolutely incorrect, given that we are referring to material under plastic flow. They are correct if we refer to alternate deformations as the sum of the deformations caused by pressure and Δt instead. In fact, one should not forget that the values of f_a were obtained taking alternate deformations that cause rupture by fatigue and multiplying them by the modulus of elasticity.

Returning to (10.11), one must establish the predictable number of cycles in correspondence of the differences in temperature Δt. Based on that, it is possible to determine the value of f_a of the fatigue curve. The maximum allowable value of Δt is given by (10.11). The procedure is similar for temperature fluctuations. Therefore, one considers the fluctuations of Δt relative to any (a random) couple of adjacent points under working conditions. Even in this case, if Δt is the fluctuation of the difference in temperature between two points, it must not exceed the value given by (10.11).

Finally, as far as mechanical loads, such as e.g., the forces and moments transmitted by the piping connected to the vessel, the total variation of the corresponding ideal stresses shall not exceed the value of f_a, obtained from the fatigue curve by the predicted number of cycles.

In summary, for fluctuations in pressure, temperature and the variations of mechanical loads, if the value of these fluctuations is smaller than the value of f_a relative to 10^6 cycles, the fluctuations are not significant. In fact, in that case they do not cause fatigue rupture, regardless of the number of cycles.

10.4 Basic Criteria for Fatigue Analysis

If the conditions in Sect. 10.3 are not satisfied, it is necessary to perform a fatigue analysis by examining its most significant points in terms of the intensity of cyclical variations of the stresses. These positions may coincide with those where the values of the stresses in static conditions are at a maximum, but this is not always necessarily true, as the influence of thermal transient conditions in different points, e.g., may be crucial.

Moreover, one should remember that the fatigue analysis does not exhaust the verifications to be done on the vessel. Of course, in any case the limits of the stresses described earlier in relation to static stresses (Sect. 1.5) must be respected. During the fatigue analysis it is important to distinguish between two instances.

In the first case, the direction of the primary stresses does not vary during the cycle. As usual, σ_I, σ_{II} and σ_{III} shall be the three principal stresses matched by three ideal stresses, according to Guest–Tresca, given the differences among the three principal stresses.

If σ_i and σ_k are the two principal stresses used to compute the generic ideal stress called σ_{ik}, we have

$$\sigma_{ik} = \sigma_i - \sigma_k. \tag{10.12}$$

Let us now consider the extreme values of σ_{ik} during the cycle that we shall call σ'_{ik} and σ''_{ik}, respectively. The variation of the generic ideal stress is therefore equal to

$$\Delta\sigma_{ik} = \left|\sigma'_{ik} - \sigma''_{ik}\right|. \tag{10.13}$$

The value of the generic alternate stress $\sigma_{a(ik)}$ is equal to half the variation above, so we have

$$\sigma_{a(ik)} = \frac{\Delta\sigma_{ik}}{2} = \frac{|\sigma'_{ik} - \sigma''_{ik}|}{2}. \tag{10.14}$$

Amongst the three stresses $\sigma_{a(ik)}$ one should consider the greatest and call it σ_a. It represents the maximum alternate stress present in the analyzed point.

If this is the only cyclical phenomenon the vessel must endure, the value of σ_a must be at the most equal to the value of f_a which is obtained from the fatigue curve. In other words, entering the fatigue curve with $f_a = \sigma_a$, one establishes the maximum number of allowable cycles (recall (10.2)).

If the actual number of cycles is smaller than that value, the verification is positive. During the cycle it is possible to register a change of direction of the principal stresses, besides a variation of their value. In that case, after adopting a certain orientation of the axes in the desired point, one must examine the six stresses, three normal stresses and three shear stresses that we shall indicate as σ_1, σ_2 and σ_3, τ_{12}, τ_{13} and τ_{23}, respectively.

During the cycle the value of these stresses varies and indicating the extreme values of the stress σ_1 as σ'_1 and σ''_1, respectively, the corresponding variation is given by

$$\Delta\sigma_1 = \sigma'_1 - \sigma''_1. \tag{10.15}$$

The variations of the other five stresses in question are obtained in a similar way.

Based on $\Delta\sigma_1$, $\Delta\sigma_2$, $\Delta\sigma_3$ and so on, it is possible to compute the variations of the corresponding principal stresses that we shall indicate with $\Delta\sigma_I$, $\Delta\sigma_{II}$ and $\Delta\sigma_{III}$. If $\Delta\sigma_i$ and $\Delta\sigma_k$ are two of such variations, the generic alternate stress $\sigma_{a(ik)}$ is therefore given by

$$\sigma_{a(ik)} = \frac{|\Delta\sigma_i - \Delta\sigma_k|}{2}. \tag{10.16}$$

The maximum alternate stress σ_a present in the examined point, is therefore given by the greatest of the three stresses $\sigma_{a(ik)}$ computed this way.

The process continues as described. It is easy to establish that the previously examined case, where the direction of the principal stresses does not change, is part of the most general case now examined. In fact, based on (10.12) we have

$$\begin{aligned} \sigma'_{ik} &= \sigma'_i - \sigma'_k \\ \sigma''_{ik} &= \sigma''_i - \sigma''_k, \end{aligned} \tag{10.17}$$

and, based on (10.13)

$$\Delta\sigma_{ik} = |\sigma'_i - \sigma'_k - \sigma''_i + \sigma''_k|; \tag{10.18}$$

then, based on (10.15)

$$\Delta\sigma_{ik} = |\Delta\sigma_i - \Delta\sigma_k|. \tag{10.19}$$

Finally, based on (10.14)–(10.19) we get (10.16).

Until now we assumed the necessity to deal with only one cyclical phenomenon. Actually, in most likelihood the vessel is subject to a series of cyclical phenomena of different nature and characteristics.

Therefore, we assume that the vessel is subject to m cyclical phenomena matched by numbers of cycles, such as n_1, n_2, $n_3 \ldots n_m$. For the time being, we assume that all phenomena cause stresses σ_a of equal value in the considered point. Assuming that $f_a = \sigma_a$, from the fatigue curve we obtain a single value of N of the number of allowable cycles. Evidently, in order for the verification to be positive we must have

$$n_1 + n_2 + n_3 + \ldots + n_m \leq N, \tag{10.20}$$

or

$$\frac{n_1}{N} + \frac{n_2}{N} + \frac{n_3}{N} + \ldots + \frac{n_m}{N} \leq 1. \tag{10.21}$$

If we realistically assume instead that the value of σ_a is different for every phenomenon, as usual setting $f_a = \sigma_a$, from the fatigue curve we obtain different values of the number of allowable cycles that we indicate with N_1, N_2, N_3, $\ldots N_m$ (remember (10.2)). According to (10.21), it seems reasonable to consider the verification positive if

$$\frac{n_1}{N_1} + \frac{n_2}{N_2} + \frac{n_3}{N_3} + \ldots + \frac{n_m}{N_m} \leq 1. \tag{10.22}$$

This way a characteristic factor, called cycle ratio, is highlighted; it is generically given by

$$U_i = \frac{n_i}{N_i}. \tag{10.23}$$

For the examined cyclical phenomenon, it corresponds to the ratio between the actual number of cycles and the allowable number of cycles based on the value of σ_a. For the m cyclical phenomena the vessel is subject to, it is necessary to verify the following condition:

$$\sum_{i=1}^{m} U_i < 1. \tag{10.24}$$

Two explanations are due.

If the value of σ_a is smaller than the value of f_a that corresponds to $N = 10^6$ cycles for the computation of U, we assume that $N = 10^6$. If, for instance, a cyclical phenomenon with 1.000 cycles causes a variation in stress from 0 to 300 $N\,mm^{-2}$, and a second cycle that overlaps the first one with 10.000 cycles causes a variation in stress from 0 to $-200\,\mathrm{N\,mm^{-2}}$, the values of σ_a to consider (of course, the variations in stress examined are the maximum ones in the point in question) are

$$\sigma_{a1} = \frac{[300 - (-200)]}{2} = 250\,\mathrm{N\,mm^{-2}}$$

relative to 1.000 cycles, and

$$\sigma_{a2} = \frac{(300-0)}{2} = 150\,\mathrm{N\,mm^{-2}}$$

relative to 9.000 cycles.

We must also consider the possibility that incremental plastic deformations may occur due to thermal cycles. Through specific combinations of static and cyclical stresses, it is possible for the deformation of material to increase by a constant quantity at every cycle giving way to big deformations.

For instance, a situation like this one may occur with flanged connections. If the material is under plastic flow in the most stressed areas, the constraint conditions change. During the following operational cycle, regardless of whether there has been manual intervention on the tightness of the bolts, further deformations may occur.

From a general point of view, some criteria must be adopted to limit the range of variation of thermal stresses as a function of the membrane stress caused by pressure, as well as the yield strength. $\Delta\sigma_T$ shall be the maximum variation of the thermal stress during the cycle, σ_m the maximum static membrane stress caused by pressure, and σ_s the yield strength. If the behavior of the temperature through the wall of the vessel is linear, the following limits shall be adopted:

$$\sqrt{\sigma_m \Delta\sigma_T} \leq \sigma_s \tag{10.25}$$

for $0 < \sigma_m \leq 0.5\sigma_s$, and

$$\sigma_m + 0.25\Delta\sigma_T \leq \sigma_s \tag{10.26}$$

for $0.5\sigma_s < \sigma_m \leq \sigma_s$.

If the behavior of the temperature through the wall is parabolic instead, we must have

$$\left(2.07 - 1.285\frac{\sigma_m}{\sigma_s}\right)\sigma_m + 0.106\Delta\sigma_T \leq \sigma_s, \tag{10.27}$$

for $0.3\sigma_s < \sigma_m \leq 0.615\sigma_s$, and

$$\sigma_m + 0.192\Delta\sigma_T \leq \sigma_s \tag{10.28}$$

for $0.615\sigma_s < \sigma_m \leq \sigma_s$.

Note that if the membrane stress is equal to f with $f = \sigma_s/1.5$, from (10.26) one obtains $\Delta\sigma_T \leq 1.33\sigma_s = 2f_s$. Thus, the sum shall be $\sigma_m + \Delta\sigma_T = 3f$, which is a well-known value in reference to the allowable stress in static conditions for the sum of primary and secondary stresses.

Finally, note that if σ_s is greater than twice the allowable stress under fatigue for 10^6 cycles, one must assume $2f_a$ instead of σ_s in the previous equations, whenever a high number of cycles is predicted. Instead of using (10.25)–(10.28), one may refer to Fig. 10.5 that is representative of them. In

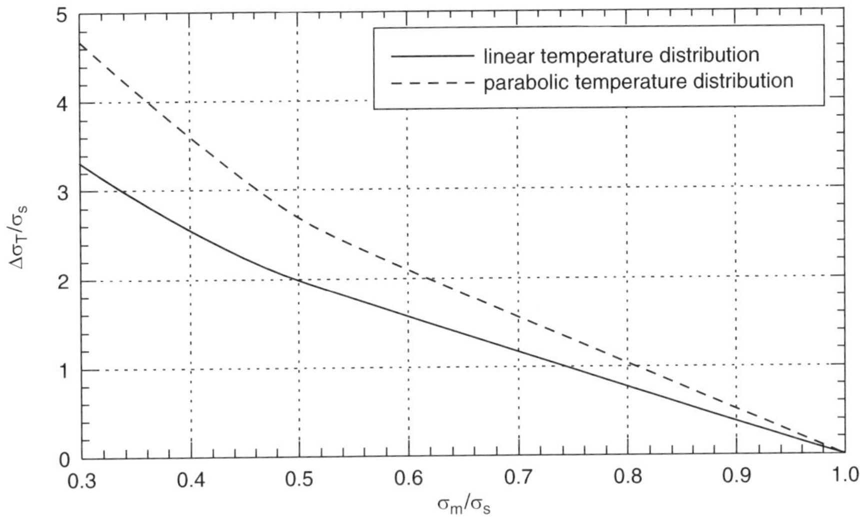

Fig. 10.5

these previous equations σ_m indicated the static membrane stress, i.e., the same symbol used in Sect. 1.5 to identify the general membrane stress. To avoid misunderstandings it is best to point out that the membrane stress in question may be either a general membrane stress or a local membrane stress.

As far as the latter, we recall that under static conditions the value of the allowable stress is equal to $1.5f$, as shown in Sect. 1.5, and therefore generally equal to the yield strength. Based on the equations above and Fig. 10.5, we establish that if $\sigma_m = \sigma_s$, then $\Delta\sigma_T = 0$. Therefore, if the local membrane stress reaches the maximum allowable value under static conditions, any cyclical phenomenon must be ruled out in terms of stresses of thermal origin. This is quite important because if this phenomenon occurs, it is absolutely necessary to intervene on the sizing of the vessel (e.g., as far as nozzles) in order to reduce the value of the local membrane stress.

Finally, it is worth noting that we are talking about potential fluctuations of thermal stresses during working conditions, i.e., during static pressure conditions. Any reference to starts and stops is out of question since there will be stress fluctuations in any case when there is a warm fluid inside. These should not be considered as incremental plastic deformations but simply within a general context of fatigue verifications by correlating them to the number of starts and stops.

Bibliography

Annaratone D., *Generatori di vapore*, CLUP editore

Baldacci R., Ceradini G., Giangreco E., *Plasticità*, vol. IIA, Italsider

Belluzzi O., *Scienza delle Costruzioni*, Zanichelli, Bologna

Bertolini I., *Costruzione di macchine*, Tamburini, Milano

Daddi I., *Materiali metallici*, Tamburini, Milano

Danusso A., *Scienza delle Costruzioni*, Tamburini, Milano

Harvey J.F., *Pressure Component Construction*, Van Nostrand, London

Roark R.J., *Formulas for stress and strain*, McGraw-Hill, New York

Schweigerer S., *Festigkeitsberechnung von Bauelementen des Dampfkessel-Behälter-und Rohrleitungsbaues*, Springer, Berlin Heidelberg New York

Timoshenko S., *Strength of materials*, Van Nostrand, London

Timoshenko S., *Theory of plates and Shells*, McGraw-Hill, New York

Wolf S., Dietrich, *Das Kalt-Biegen von Rohren*, Springer, Berlin Heidelberg New York

Zignoli V., *Costruzioni Metalliche*, UTET, Torino

AD Merkblatter B2,B3,B5,B9

ANCC, *Raccolta VSG*, ed. luglio 1973

ANCC, *Raccolta VSR*, ed. luglio 1978

ANCC, *Raccolta M*, ed. 1978

ANCC, *Raccolta S*, ed. 1978

ASME, *Boiler and Pressure Vessel Code, Sect. I, III, VIII Dampfkessel-Bestimmungen, Technische Vorschriften für Dampkesselanlagen, Teil 3 Berechnung* (German Code), Carl Heymanns, Koeln

BSI, *Unfired fusion welded pressure vessels* (English Code)

ISO TC/11, *Code de construction de chaudieres fixes*

ISO TC/11, *Pressure vessel Code Petroleum Refinery Piping*, ASA B31, ASME, New York

Albrecht W., *Zulässige Instationäre Wärmespannungen*, BWK, 1966

Annaratone D., *Calcolo di prima approssimazione delle tubazioni piane*, Rivista di Ingegneria, 1958

Annaratone D., *Tolleranze sui difetti costruttivi dei corpi cilindrici*, Il Calore, 1959

Annaratone D., *Metodo di calcolo per successive approssimazioni delle tubazioni piane*, La Termotecnica, 1961

Annaratone D., *Dimensionamento dei cilindri ad elevato spessore ed influenza del flusso termico sul dimensionamento*, Rivista di Ingegneria, 1964a

Annaratone D., *Un metodo di calcolo generale dei cilindri premuti dall'esterno*, Rivista di Ingegneria, 1964b

Annaratone D., *Metodo di calcolo semplificato dei cilindri premuti dall'esterno*, Rivista di Ingegneria, 1966

Annaratone D., *Criterio di calcolo approssimato del modulo di efficienza delle forature nei mantelli cilindrici*, Il Calore, 1967

Annaratone D., *Sul calcolo del modulo di efficienza delle forature oblique in conformità alle norme ANCC-VSG*, Il Calore, 1971

Annaratone D., *Considerazioni sulle norme vigenti per il dimensionamento dei cilindri sottoposti a pressione interna*, Il Calore, 1972

Annaratone D., *Considerazioni sul dimensionamento dei fondi piani*, Il Calore, 1975

Annaratone D., *Dimensionamento dei corpi sferici sottoposti a pressione interna*, Energia e Calore, 1981a

Annaratone D., *Fondi piani con gola di scarico*, Ingegneria, 1981b

Annaratone D., *Dimensionamento dei tronchetti dei recipienti in pressione sottoposti a sforzi esterni*, L'Industria Meccanica, 1982

Annaratone D., *Indagine sui fondi bombati sottoposti a pressione interna*, Rivista di Ingegneria, 1984

Annaratone D., *Cilindri cerchiati sottoposti a pressione*, Costruzioni Metalliche, 1988

Annaratone D., *Tensioni residue nei tubi curvati a freddo*, L'Industria Meccanica, 1991

Baileu R., Hichs R., *Behaviour of perforated plates under plane stress*, Journal Mechanical Engineering Sciences, 1960

Barthelemy J., *Calcul des fonds des réservoirs soumis à une pression intérieure-Résultats numériques et conclusions d'ordre general qu'on en tire*, Chaudronnerie Tolerie, 1961

Barthelemy J., *Calcul des fonds des réservoirs soumis à un pression intérieure*, Chaudronnerie Tolerie, 1962

Bijlaard P.P., *Stresses from radial loads in cylindrical pressure vessels*, Welding Journal, 1954

Bijlaard P.P., *Stresses from radial loads and external moments in cylindrical pressure vessels*, Welding Journal, 1955

Bijlaard P.P., *Local stresses in spherical shells from radial or moment loading*, Welding Journal, 1957

Bijlaard P.P., *On the stresses from local loads on spherical pressure vessels and pressure vessels heads*, Welding Research Bulletin, 1957

Bychawski Z., *Creep buckling of geometrically nonlinear circular plates*, Polska Akademia Nauk, 1966

Coates W.M., *The state of stress in full heads of pressure vessels*, Transactions of American Society of Mechanical Engineers, 1930

Crif, *Dimensionnement du fond plat avec rainure de décharge*, CRIF, Bruxelles, 1965

Decock J., *Etude du renforcement des ouvertures dans les réservoirs soumis à pression*, Chaudronnerie Tolerie, 1971

De Leiris H., Robert A., Venon M., *Calcul des contraintes dans les enveloppes de revolution chargées aximétriquement*, Chaudronnerie Tolerie, 1968

Denoor G., Tascher C., *Corps cylindriques métalliques sous pression realises par expansion limitée de leurs parois avec ou sans précontrainte*, Chaudronnerie Tolerie, 1970

Di Benedetto F., *Ripartizione delle tensioni in piastre circolari con foro rettangolare a spigoli arrotondati caricate esternamente con pressione uniforme*, Istituto di Costruzione di Macchine dell'Università di Palermo, 1957

Eisler B., *Berechnung der Wanddicken zylindrischer und kugelförmiger Kesselwandungen bei äußerem Überdruck*, Österreichische Ingegneur Zeitschrift, 1962

Galletly G.D., *Optimum design of thin circular plates on an elastic foundation*, The Institution of Mechanical Engineers, 1959

Gamberucci G., Guerrieri G., *Design of a tabulator reactor at a pressure of* 3200 kg/cm^2, SNAM Progetti

Garbato Q., *Contributo al calcolo della piastra a doppia curvatura*, Il Calore, 1960

Haenle S., *Beiträge zum Festigkeitsverhalten von Vorschweissflanschen und zur Ermittlung der Dichtkräfte für einige Flachdichtungen auf Asbestbasis*, Forschung auf dem Gebiete des Ingegnieurwesen, 1957

Hartemberg R.S., *The strength and stiffness on thin cylindrical shells on saddle support*, University of Wisconsin, 1941

Hoger R., *Berechnung der Wärmespannungen in Speichetrommeln bei nichtstationärer Temperaturverteilung*, BWK, 1963

Holmberg E.O., Axelson K., *Analysis of stresses in circular plates and rings*, Journal of Applied Mechanical Engineering, 1932

Hubert M., *Le calcul des corps cylindriques percés*, Chaudronnerie Tolerie, 1961

Jasper T.M.,Sullivan J.W., *The collapsing strength of steel tubes*, Transactions of American Society of Mechanical Engineers, 1931

Lemasson M.G., *Stabilité du réservoir cylindrique à axe horizontal*, Chaudronnerie Tolerie, 1971

Manzella G., *Risultati sperimentali sulla distribuzione delle tensioni in dischi o cilindri con fori eccentrici e con foro centrale soggetto a pressione interna*, Istituto di Costruzione di Macchine dell'Università di Palermo, 1951

Manzella G., *Tensioni dovute a pressione esterna in dischi, o cilindri, con foro centrale e fori eccentrici*, Istituto di Macchine dell'Università di Palermo, 1952

Maulbetsch J.L., Hetenyl M., *Stresses in pressure vessels*, Journal of Applied Mechanics, 1936

Monfort C., *Contribution à l'étude des contraintes termiques*, Revue Générale de termique, 1962

Odquist F.K.G., *Om barverkan vid tunna cylindrical skal ock karlvaggar*, Proceedings of Royal Swedish Institute For Engineers Research, 1942

Pich R., *Die Gestaltänderungsenergie Hypotese*, VGB, 1961

Reissner E., *Stresses and small displacements of shallow spherical shells*, Journal of Mathematics and Physics, 1947

Roark R.J., *The strength and stiffness of cylindrical shells under concentrated loading*, Journal Applied Mechanical, 1935

Roark R.J., *Stresses and deflections in thin shells and curved plates due to concentrated and variously distributed loading*, Natural Advances and Communication in Aeronautics, 1941

Roshal O., *Die Flossenwand, ein neues Bauelement für den modernen Hochleistung-Dampferzeuger*, VGB, 1963

Saunders H.E., Windemburg D.F., *Strength of thin cylindrical shell under external pressure*, Transactions of American Society Mechanical Engineers, 1931

Schorer H., *Design of large pipe lines*, Transactions American Society of Civil Engineers, 1933

Schweigerer S., *Importance des Formdehngrenzen pour le calcul de résistence des matériaux*, Caudronnerie Tolerie, 1961

Siebel E., Schweigerer S., *Berechnung von Kesseltrommeln mit versetzten Lochreihen*, Die Waerme, 1941

Soete W., Hebrant F., Dechaene R., Heirman J., *Le renforcement des ouvertures dans les réservoirs soumis à pression*, Chaudronnerie Tolerie, 1961

Soete W., Hebrant F., Dechaene R., Heirman J., *Pressure vessel research in Belgium*, Symposium on pressure vessel research towards better design, 1961

Sturm R.G., *A study of collapsing pressure of thin-walled cylinders*, Bulletin Series N. 329 – University of Illinois Engineering Experiments Station

Svensson N.L., *The bursting pressure of cylindrical and spherical vessel*, Journal of Applied Mechanics

Timoshenko S., Lessels, *Applied elasticity*, Westinghouse Technical Nigth School, 1925

Tondelli C., *Le condizioni di stabilità dei fasciami a sezione trasversale mistilinea simmetrica soggetti a pressione interna*, Il Calore, 1964

Ulrich E., *Résistance des coudes de section elliptique sous l'action d'une pression interne et d'un effort d'ouverture elastique*, Chaudronnerie Tolerie, 1961

Ulrich E., *Über die Grundlagen der Festigkeitsrechnung im Dampfkessel-Behälter und Rohrleitungsbau unter Berückrichtigung des Flissverfahrens der Werksoffe*, VGB, 1962

Ulrich E., *Über die Superpositionsprinzipe der Einzelspannungen und der Vergleichsspannungen und einige Anwendungen auf Spannungsprobleme des Kesselbaues*, Mitteilungen der Vereinigung der Grosskesselbesitzer, 1964

Winn L., *Eingeschweisste Stutzen in Zylindern*, Konstruktion, 1960

Winn L., *Calcul des réservoirs soumis à une pression interieure*, Chaudronnerie Tolerie, 1963

Zick L.P., *Stresses in large horizontal cylindrical pressure vessel on two saddle supports-welding*, Journal Research Supplement, 1951

Index

Printing: Krips bv, Meppel
Binding: Stürtz, Würzburg